Dietary Intake and Chronic Disease Prevention

Dietary Intake and Chronic Disease Prevention

Editors

Annalisa Noce
Annalisa Romani
Roberta Bernini

MDPI • Basel • Beijing • Wuhan • Barcelona • Belgrade • Manchester • Tokyo • Cluj • Tianjin

Editors
Annalisa Noce
Systems Medicine
University of Tor Vergata
Rome
Italy

Annalisa Romani
PHYTOLAB (Pharmaceutical, Cosmetic, Food Supplement, Technology and Analysis)
DiSIA
University of Florence
Florence
Italy

Roberta Bernini
Department of Agriculture and Forest Sciences (DAFNE)
University of Tuscia
Viterbo
Italy

Editorial Office
MDPI
St. Alban-Anlage 66
4052 Basel, Switzerland

This is a reprint of articles from the Special Issue published online in the open access journal *Nutrients* (ISSN 2072-6643) (available at: www.mdpi.com/journal/nutrients/special_issues/dietary_chronic).

For citation purposes, cite each article independently as indicated on the article page online and as indicated below:

LastName, A.A.; LastName, B.B.; LastName, C.C. Article Title. *Journal Name* **Year**, *Volume Number*, Page Range.

ISBN 978-3-0365-7309-0 (Hbk)
ISBN 978-3-0365-7308-3 (PDF)

Contents

About the Editors

Annalisa Noce

Annalisa Noce is an Associate Professor in Nephrology at the University Tor Vergata (Rome). Her principal scientific fields are nutrition in nephrology, oxidative stress in chronic kidney disease, endothelial dysfunction, and kidney ultrasound examination.

Annalisa Romani

Annalisa Romani was a Full Professor at the University of Florence (Italy). Her main areas of study were natural products, with special reference to polyphenols from extraction from plant matrices and agro-industrial by-products and characterizations by advanced analytical techniques to industrial applications. She has been a pioneer in the circular economy processes and in the relationship between University and Industry, fostering the technological transfer of the results obtained in university laboratories.

Roberta Bernini

Roberta Bernini is Associate Professor of Organic Chemistry and Chemistry of Natural Organic Products at the University of Tuscia (Viterbo, Italy). Her research topics include the design and synthesis of biologically active molecules using green chemistry methodologies; the preparation and characterization of natural extracts from plant matrices with antioxidant, antibacterial, antifungal, and anticancer properties; the valorization and reuse of bioactive compounds found in agro-industrial by-products for applications in the food, biomedical, and agronomic fields according to the principles of green chemistry and circular economy.

Preface to "Dietary Intake and Chronic Disease Prevention"

In this Special Issue, we emphasize the role of lifestyle changes and natural bioactive compounds in the prevention and clinical managment of non-communicable chronic degenerative diseases (such as chronic kidney disease, cancer, diabetes mellitus, metabolic syndrome, obesity, neurodegenerative diseases etc.).

We dedicate this Special Issue to our beloved and esteemed Full Professor Annalisa Romani, who recently passed away.

Annalisa Noce, Annalisa Romani, and Roberta Bernini
Editors

nutrients

MDPI

Editorial

Dietary Intake and Chronic Disease Prevention

Annalisa Noce [1,*], Annalisa Romani [2] and Roberta Bernini [3]

[1] UOC of Internal Medicine-Center of Hypertension and Nephrology Unit, Department of Systems Medicine, University of Rome Tor Vergata, Via Montpellier 1, 00 133 Rome, Italy
[2] PHYTOLAB (Pharmaceutical, Cosmetic, Food Supplement, Technology and Analysis), DiSIA, University of Florence Via Ugo Schiff 6, 50 019 Florence, Italy; annalisa.romani@unifi.it
[3] Department of Agriculture and Forest Sciences (DAFNE), University of Tuscia, Via San Camillo de Lellis, 01 100 Viterbo, Italy; roberta.bernini@unitus.it
* Correspondence: annalisa.noce@uniroma2.it; Tel.: +39-06-2090-2194; Fax: +39-06-2090-3362

Citation: Noce, A.; Romani, A.; Bernini, R. Dietary Intake and Chronic Disease Prevention. *Nutrients* **2021**, *13*, 1358. https://doi.org/10.3390/nu13041358

Received: 10 March 2021
Accepted: 7 April 2021
Published: 19 April 2021

Non-communicable diseases (NCDs) are non-infectious chronic pathologies. The most common are diabetes mellitus, obesity, metabolic syndrome, chronic kidney disease (CKD), cardiovascular (CV) diseases, cancer, and chronic respiratory diseases. Furthermore, their prevalence is likely to increase over time due to the aging population, urbanization, and lifestyle changes [1]. In industrialized and high-income countries, several studies have highlighted a direct correlation between socio-economic factors and health status; in particular, NCDs affect mainly the population with the lowest socio-economic level [2–5].

Before the COVID-19 pandemic, NCDs had spread all over the world, becoming an important public health problem even in developing countries. The "epidemiologic transition" observed from infectious diseases to NCDs in developing countries is related to a series of risk factors, mainly associated with economic development, such as the consumption of foods with high contents of saturated fats, salt, and sugars; low intake of fruit, vegetables, fiber, and ω-3 fatty acids; a sedentary lifestyle; smoking; and the unmoderated consumption of alcohol [6,7].

NCDs are responsible for high percentages of disability and mortality worldwide [8].

An unhealthy lifestyle, characterized by an unbalanced diet, together with insufficient sleep, physical inactivity, psychological stress, environmental pollution [9], smoking, or alcohol abuse contribute to cause metabolic alterations which can lead to the onset of NCDs.

In this context, a correct lifestyle and healthy dietary habits could exert protective effects, increasing the life expectancy. Then, nutrition plays an important role in NCDs prevention [10]. In particular, the Mediterranean diet, characterized by a high consumption of fruit, vegetables, extra virgin olive oil, cereals, legumes, and fish; a moderate intake of dairy products, eggs, and red wine; and a low intake of animal fats and red meat, represents a correct approach to prevent NCDs onset [11–16]. Moreover, pasta represents one of the basic foods of Mediterranean diet and, in this Special Issue, a preliminary study analyzes the antioxidant compounds present in three types of pasta and their biological activities on kidney cells, demonstrating that pasta's natural bioactive compounds play positive role in the protection of kidney cells from oxidative stress [17].

The beneficial effects are related to the presence of natural bioactive compounds, including antioxidants [18]. Epidemiological studies have demonstrated that an optimal daily intake of antioxidants such as polyphenols and vitamins is able to counteract the onset of NCDs and to slow their progression [18]. Polyphenols are a wide and complex group of compounds found in plant-derived foods, beverages, and agro-industrial byproducts; these bioactive molecules have important physiological effects on the prevention of several chronic diseases. For example, small phenols such as hydroxytyrosol found in extra-virgin olive oil and olive oil byproducts, catechins such as epigallocatechin found in green tea, and complex hydrolysable tannins such as punicalagin found in pomegranate peel and fruit, exhibit strong antioxidant, anti-inflammatory, antidiabetic, anti-obesity, anticancer, and antimicrobic activities [19–25]. In this context, an original article of this Special Issue demonstrated the gender-dependent positive antimicrobial action of *Castanea sativa* L. hydrolysable tannins in recurrent urinary infections in CKD patients [26].

The relationship between gut dysbiosis and the onset of NCDs has recently been highlighted [27,28]. Several studies have shown that polyphenols could influence the composition of the gut microbiota by promoting the growth of bacterial classes with positive effects and by inhibiting bacteria with negative effects on the microbiota composition [29,30]. Moreover, vitamins, and, in particular, vitamin C (ascorbic acid) and E (tocopherols), are natural compounds that play a pivotal role in preventing the NCDs onset, mainly for their antioxidant activity. Vitamin C is a water-soluble vitamin, able to protect from the cellular damage exerted by harmful oxidative compounds [31]. Vitamin E includes a group of lipid-soluble compounds with the highest antioxidant activity in vivo [32].

This Special Issue has contributed to better evidence of the role of the correct lifestyle and the natural bioactive compounds in preventing the NCDs onset and their treatment. In fact, some reviews and original articles have confirmed the cardioprotective role exerted by different dietary patterns and by natural bioactive compounds [33]. In particular, how a personalized Mediterranean diet in women can exert a positive action on the cardiovascular system [34], how ω-3 polyunsaturated fatty acids play a cardioprotective role in male obesity secondary hypogonadism (MOSH) patients [35], and how a caloric restriction diet can protect against organ damage induced by arterial hypertension, improving endothelial dysfunction [36]. Another study evaluated the possible relationship between dietary quality scores and cardiometabolic risk in a group of older Australian adults, demonstrating that a high intake of vegetables, grains, and non-processed red meat was associated with a better cardiometabolic risk profile [37]. An original review stressed a relationship between frailty, sarcopenia, and cardiovascular risk, underlining how the frail phenotype is associated with a poor outcome after cardiac surgery [38]. Healthy eating habits reduce also the risk of developing cancer [39–41] and other chronic NCDs (such as CKD and chronic respiratory diseases) [42–45], as evidenced by several papers of this Special Issue.

The Guest Editors would like to thank all the authors, reviewers who contributed to the success of this Special Issue, and the Nutrients team for their precious and constant support.

Author Contributions: Conceptualization, A.N., A.R. and R.B.; writing—original draft preparation, A.N., A.R. and R.B.; All authors have read and agreed to the published version of the manuscript.

Funding: This research received no external funding.

Institutional Review Board Statement: Not applicable.

Informed Consent Statement: Not applicable.

Conflicts of Interest: The authors declare no conflict of interest.

References

1. Yang, Y.; Sun, X.; Wang, J.; Yang, C.; Zhang, L. Incidence Rates of Four Major Non-Communicable Chronic Diseases in the Chinese Adult Population from 2007 to 2016: A Study Based on a National Commercial Claims Database. *Clin. Epidemiol.* **2020**, *12*, 215–222. [CrossRef]
2. Williams, J.; Allen, L.; Wickramasinghe, K.; Mikkelsen, B.; Roberts, N.; Townsend, N. A systematic review of associations between non-communicable diseases and socioeconomic status within low- and lower-middle-income countries. *J. Glob. Health* **2018**, *8*, 020409. [CrossRef]
3. Mackenbach, J.P.; Stirbu, I.; Roskam, A.J.; Schaap, M.M.; Menvielle, G.; Leinsalu, M.; Kunst, A.E. Socioeconomic inequalities in health in 22 European countries. *N. Engl. J. Med.* **2008**, *358*, 2468–2481. [CrossRef]
4. Marmot, M.G.; Kogevinas, M.; Elston, M.A. Social/economic status and disease. *Annu. Rev. Public Health* **1987**, *8*, 111–135. [CrossRef]
5. Zhu, Z.; Yang, X.; Fang, Y.; Zhang, J.; Yang, Z.; Wang, Z.; Liu, A.; He, L.; Sun, J.; Lian, Y.; et al. Trends and Disparities of Energy Intake and Macronutrient Composition in China: A Series of National Surveys, 1982–2012. *Nutrients* **2020**, *12*, 2168. [CrossRef] [PubMed]
6. Islam, S.M.; Purnat, T.D.; Phuong, N.T.; Mwingira, U.; Schacht, K.; Froschl, G. Non-communicable diseases (NCDs) in developing countries: A symposium report. *Glob. Health* **2014**, *10*, 81. [CrossRef]

7. Dessi, M.; Noce, A.; Bertucci, P.; Noce, G.; Rizza, S.; De Stefano, A.; Manca di Villahermosa, S.; Bernardini, S.; De Lorenzo, A.; Di Daniele, N. Plasma and erythrocyte membrane phospholipids and fatty acids in Italian general population and hemodialysis patients. *Lipids Health Dis.* **2014**, *13*, 54. [CrossRef]
8. Betlejewski, S. Social diseases, civilization diseases or lifestyle diseases? *Wiad. Lek.* **2007**, *60*, 489–492. [PubMed]
9. Bocedi, A.; Noce, A.; Marrone, G.; Noce, G.; Cattani, G.; Gambardella, G.; Di Lauro, M.; Di Daniele, N.; Ricci, G. Glutathione Transferase P1-1 an Enzyme Useful in Biomedicine and as Biomarker in Clinical Practice and in Environmental Pollution. *Nutrients* **2019**, *11*, 1741. [CrossRef] [PubMed]
10. Di Renzo, L.; Gualtieri, P.; Romano, L.; Marrone, G.; Noce, A.; Pujia, A.; Perrone, M.A.; Aiello, V.; Colica, C.; De Lorenzo, A. Role of Personalized Nutrition in Chronic-Degenerative Diseases. *Nutrients* **2019**, *11*, 1707. [CrossRef]
11. Di Daniele, N.; Noce, A.; Vidiri, M.F.; Moriconi, E.; Marrone, G.; Annicchiarico-Petruzzelli, M.; D'Urso, G.; Tesauro, M.; Rovella, V.; De Lorenzo, A. Impact of Mediterranean diet on metabolic syndrome, cancer and longevity. *Oncotarget* **2017**, *8*, 8947–8979. [CrossRef]
12. De Lorenzo, A.; Noce, A.; Bigioni, M.; Calabrese, V.; Della Rocca, D.G.; Di Daniele, N.; Tozzo, C.; Di Renzo, L. The effects of Italian Mediterranean organic diet (IMOD) on health status. *Curr. Pharm. Des.* **2010**, *16*, 814–824. [CrossRef]
13. Andreoli, A.; Lauro, S.; Di Daniele, N.; Sorge, R.; Celi, M.; Volpe, S.L. Effect of a moderately hypoenergetic Mediterranean diet and exercise program on body cell mass and cardiovascular risk factors in obese women. *Eur. J. Clin. Nutr.* **2008**, *62*, 892–897. [CrossRef]
14. Di Daniele, N.; Di Renzo, L.; Noce, A.; Iacopino, L.; Ferraro, P.M.; Rizzo, M.; Sarlo, F.; Domino, E.; De Lorenzo, A. Effects of Italian Mediterranean organic diet vs. low-protein diet in nephropathic patients according to MTHFR genotypes. *J. Nephrol.* **2014**, *27*, 529–536. [CrossRef]
15. Noce, A.; Marrone, G.; Urciuoli, S.; Di Daniele, F.; Di Lauro, M.; Pietroboni Zaitseva, A.; Di Daniele, N.; Romani, A. Usefulness of Extra Virgin Olive Oil Minor Polar Compounds in the Management of Chronic Kidney Disease Patients. *Nutrients* **2021**, *13*, 581. [CrossRef] [PubMed]
16. Noce, A.; Fabrini, R.; Bocedi, A.; Di Daniele, N. Erythrocyte glutathione transferase in uremic diabetic patients: Additional data. *Acta Diabetol.* **2015**, *52*, 813–815. [CrossRef]
17. Di Marco, F.; Trevisani, F.; Vignolini, P.; Urciuoli, S.; Salonia, A.; Montorsi, F.; Romani, A.; Vago, R.; Bettiga, A. Preliminary Study on Pasta Samples Characterized in Antioxidant Compounds and Their Biological Activity on Kidney Cells. *Nutrients* **2021**, *13*, 1131. [CrossRef]
18. Koch, W. Dietary Polyphenols-Important Non-Nutrients in the Prevention of Chronic Noncommunicable Diseases. A Systematic Review. *Nutrients* **2019**, *11*, 39. [CrossRef] [PubMed]
19. Pandey, K.B.; Rizvi, S.I. Plant polyphenols as dietary antioxidants in human health and disease. *Oxid. Med. Cell Longev.* **2009**, *2*, 270–278. [CrossRef]
20. Bernini, R.; Gilardini Montani, M.S.; Merendino, N.; Romani, A.; Velotti, F. Hydroxytyrosol-Derived Compounds: A Basis for the Creation of New Pharmacological Agents for Cancer Prevention and Therapy. *J. Med. Chem.* **2015**, *58*, 9089–9107. [CrossRef] [PubMed]
21. Bernini, R.; Carastro, I.; Palmini, G.; Tanini, A.; Zonefrati, R.; Pinelli, P.; Brandi, M.L.; Romani, A. Lipophilization of Hydroxytyrosol-Enriched Fractions from *Olea europaea* L. Byproducts and Evaluation of the in Vitro Effects on a Model of Colorectal Cancer Cells. *J. Agric. Food Chem.* **2017**, *65*, 6506–6512. [CrossRef] [PubMed]
22. Romani, A.; Ieri, F.; Urciuoli, S.; Noce, A.; Marrone, G.; Nediani, C.; Bernini, R. Health Effects of Phenolic Compounds Found in Extra-Virgin Olive Oil, By-Products, and Leaf of *Olea europaea* L. *Nutrients* **2019**, *11*, 1776. [CrossRef]
23. Mastrogiovanni, F.; Mukhopadhya, A.; Lacetera, N.; Ryan, M.T.; Romani, A.; Bernini, R.; Sweeney, T. Anti-Inflammatory Effects of Pomegranate Peel Extracts on In Vitro Human Intestinal Caco-2 Cells and Ex Vivo Porcine Colonic Tissue Explants. *Nutrients* **2019**, *11*, 548. [CrossRef] [PubMed]
24. Romani, A.; Bernini, R.; Noce, A.; Urciuoli, S.; Di Lauro, M.; Pietroboni Zaitseva, A.; Marrone, G.; Di Daniele, N. Potential Beneficial Effects of Extra Virgin Olive Oils Characterized by High Content in Minor Polar Compounds in Nephropathic Patients: A Pilot Study. *Molecules* **2020**, *25*, 4757. [CrossRef]
25. Romani, A.; Campo, M.; Urciuoli, S.; Marrone, G.; Noce, A.; Bernini, R. An Industrial and Sustainable Platform for the Production of Bioactive Micronized Powders and Extracts Enriched in Polyphenols From *Olea europaea* L. and *Vitis vinifera* L. Wastes. *Front. Nutr.* **2020**, *7*, 120. [CrossRef] [PubMed]
26. Noce, A.; Di Daniele, F.; Campo, M.; Di Lauro, M.; Pietroboni Zaitseva, A.; Di Daniele, N.; Marrone, G.; Romani, A. Effect of Hydrolysable Tannins and Anthocyanins on Recurrent Urinary Tract Infections in Nephropathic Patients: Preliminary Data. *Nutrients* **2021**, *13*, 591. [CrossRef]
27. Noce, A.; Marrone, G.; Di Daniele, F.; Ottaviani, E.; Wilson Jones, G.; Bernini, R.; Romani, A.; Rovella, V. Impact of Gut Microbiota Composition on Onset and Progression of Chronic Non-Communicable Diseases. *Nutrients* **2019**, *11*, 1073. [CrossRef]
28. Annalisa, N.; Alessio, T.; Claudette, T.D.; Erald, V.; de Antonino, L.; Nicola, D.D. Gut microbioma population: An indicator really sensible to any change in age, diet, metabolic syndrome, and life-style. *Med. Inflamm.* **2014**, *2014*, 901308. [CrossRef]
29. Parkar, S.G.; Stevenson, D.E.; Skinner, M.A. The potential influence of fruit polyphenols on colonic microflora and human gut health. *Int. J. Food Microbiol.* **2008**, *124*, 295–298. [CrossRef]
30. Duda-Chodak, A. The inhibitory effect of polyphenols on human gut microbiota. *J. Physiol. Pharmacol.* **2012**, *63*, 497–503.

31. Chen, Q.; Espey, M.G.; Krishna, M.C.; Mitchell, J.B.; Corpe, C.P.; Buettner, G.R.; Shacter, E.; Levine, M. Pharmacologic ascorbic acid concentrations selectively kill cancer cells: Action as a pro-drug to deliver hydrogen peroxide to tissues. *Proc. Natl. Acad. Sci. USA* **2005**, *102*, 13604–13609. [CrossRef] [PubMed]
32. Bruins, M.J.; Van Dael, P.; Eggersdorfer, M. The Role of Nutrients in Reducing the Risk for Noncommunicable Diseases during Aging. *Nutrients* **2019**, *11*, 85. [CrossRef] [PubMed]
33. Pickering, R.T.; Bradlee, M.L.; Singer, M.R.; Moore, L.L. Higher Intakes of Potassium and Magnesium, but Not Lower Sodium, Reduce Cardiovascular Risk in the Framingham Offspring Study. *Nutrients* **2021**, *13*, 269. [CrossRef]
34. Di Renzo, L.; Cinelli, G.; Dri, M.; Gualtieri, P.; Attina, A.; Leggeri, C.; Cenname, G.; Esposito, E.; Pujia, A.; Chiricolo, G.; et al. Mediterranean Personalized Diet Combined with Physical Activity Therapy for the Prevention of Cardiovascular Diseases in Italian Women. *Nutrients* **2020**, *12*, 3456. [CrossRef]
35. Noce, A.; Marrone, G.; Di Daniele, F.; Di Lauro, M.; Pietroboni Zaitseva, A.; Wilson Jones, G.; De Lorenzo, A.; Di Daniele, N. Potential Cardiovascular and Metabolic Beneficial Effects of omega-3 PUFA in Male Obesity Secondary Hypogonadism Syndrome. *Nutrients* **2020**, *12*, 2519. [CrossRef]
36. Di Daniele, N.; Marrone, G.; Di Lauro, M.; Di Daniele, F.; Palazzetti, D.; Guerriero, C.; Noce, A. Effects of Caloric Restriction Diet on Arterial Hypertension and Endothelial Dysfunction. *Nutrients* **2021**, *13*, 274. [CrossRef] [PubMed]
37. Owen, A.J.; Abramson, M.J.; Ikin, J.F.; McCaffrey, T.A.; Pomeroy, S.; Borg, B.M.; Gao, C.X.; Brown, D.; Liew, D. Recommended Intake of Key Food Groups and Cardiovascular Risk Factors in Australian Older, Rural-Dwelling Adults. *Nutrients* **2020**, *12*, 860. [CrossRef] [PubMed]
38. Pisano, C.; Polisano, D.; Balistreri, C.R.; Altieri, C.; Nardi, P.; Bertoldo, F.; Trombetti, D.; Asta, L.; Ferrante, M.S.; Buioni, D.; et al. Role of Cachexia and Fragility in the Patient Candidate for Cardiac Surgery. *Nutrients* **2021**, *13*, 517. [CrossRef]
39. Nakanishi, K.; Kobayashi, T.; Sugimoto, T.; Murase, T.; Itoh, T.; Kosaka, K. Does pan-pancreatic involvement occur in IDDM? *Diabetes Care* **1988**, *11*, 100–101. [CrossRef]
40. Skroza, N.; Proietti, I.; Marchesiello, A.; Volpe, S.; Balduzzi, V.; Bernardini, N.; Maddalena, P.; Mambrin, A.; Michelini, S.; Tolino, E.; et al. Do Diet and Lifestyles Play a Role in the Pathogenesis of NMSCs? *Nutrients* **2020**, *12*, 3459. [CrossRef] [PubMed]
41. Edefonti, V.; La Vecchia, C.; Di Maso, M.; Crispo, A.; Polesel, J.; Libra, M.; Parpinel, M.; Serraino, D.; Ferraroni, M.; Bravi, F. Association between Nutrient-Based Dietary Patterns and Bladder Cancer in Italy. *Nutrients* **2020**, *12*, 1584. [CrossRef] [PubMed]
42. Ma, E.; Ohira, T.; Yasumura, S.; Nakano, H.; Eguchi, E.; Miyazaki, M.; Hosoya, M.; Sakai, A.; Takahashi, A.; Ohira, H.; et al. Dietary Patterns and Progression of Impaired Kidney Function in Japanese Adults: A Longitudinal Analysis for the Fukushima Health Management Survey, 2011–2015. *Nutrients* **2021**, *13*, 168. [CrossRef] [PubMed]
43. Lopez-Olmedo, N.; Jonnalagadda, S.; Basto-Abreu, A.; Reyes-Garcia, A.; Alish, C.J.; Shamah-Levy, T.; Barrientos-Gutierrez, T. Adherence to Dietary Guidelines in Adults by Diabetes Status: Results From the 2012 Mexican National Health and Nutrition Survey. *Nutrients* **2020**, *12*, 3464. [CrossRef] [PubMed]
44. Lee, S.A.; Joshi, P.; Kim, Y.; Kang, D.; Kim, W.J. The Association of Dietary Macronutrients with Lung Function in Healthy Adults Using the Ansan-Ansung Cohort Study. *Nutrients* **2020**, *12*, 2688. [CrossRef]
45. Periz, M.; Perez-Cano, F.J.; Cambras, T.; Franch, A.; Best, I.; Pastor-Soplin, S.; Castell, M.; Massot-Cladera, M. Attenuating Effect of Peruvian Cocoa Populations on the Acute Asthmatic Response in Brown Norway Rats. *Nutrients* **2020**, *12*, 2301. [CrossRef]

nutrients MDPI

Article

Effect of Hydrolysable Tannins and Anthocyanins on Recurrent Urinary Tract Infections in Nephropathic Patients: Preliminary Data

Annalisa Noce [1,*], Francesca Di Daniele [1,2], Margherita Campo [3], Manuela Di Lauro [1], Anna Pietroboni Zaitseva [1], Nicola Di Daniele [1], Giulia Marrone [1,2] and Annalisa Romani [3,*]

1 UOC of Internal Medicine-Center of Hypertension and Nephrology Unit, Department of Systems Medicine, University of Rome Tor Vergata, Via Montpellier 1, 00133 Rome, Italy; francesca.didaniele@gmail.com (F.D.D.); dilauromanuela@gmail.com (M.D.L.); annapietroboni@icloud.com (A.P.Z.); didaniele@med.uniroma2.it (N.D.D.); giul.marr@gmail.com (G.M.)
2 PhD School of Applied Medical, Surgical Sciences, University of Rome Tor Vergata, Via Montpellier 1, 00133 Rome, Italy
3 PHYTOLAB (Pharmaceutical, Cosmetic, Food Supplement, Technology and Analysis), DiSIA, University of Florence, Via Ugo Schiff 6, 50019 Sesto Fiorentino, Florence, Italy; margherita.campo@unifi.it
* Correspondence: annalisa.noce@uniroma2.it (A.N.); annalisa.romani@unifi.it (A.R.); Tel.: +39-06-2090-2194 (A.N.); +39-055-457377 (A.R.)

Abstract: Urinary tract infections (UTIs) are caused by uropathogenic microorganism colonization. UTIs often require an antibiotic therapy that can cause the selection of antibiotic-resistant bacterial strains. A natural bioactive compound may represent a valid therapeutic adjuvant approach, in combination with drug therapy. In this paper, we present a pilot study, based on the administration of an oral food supplement (OFS), containing chestnut tannins and anthocyanins, to nephropathic patients suffering from recurrent UTIs (16 treated patients with 1 cp/day and 10 untreated patients). We performed laboratory tests and quality of life and body composition assessments, at T0 (baseline) and T1 (after 6 weeks OFS assumption). The analysis of OFS was performed by HPLC-DAD-MS for its content in polyphenols and by in vitro tests for its antioxidative and anti-free radical activities. In each capsule, polyphenol content was 6.21 mg (4.57 mg hydrolysable tannins, 0.94 mg anthocyanosides, 0.51 mg proanthocyanidins, 0.18 mg quercetin derivatives). A significant reduction of erythrocyte sedimentation rate was observed only in male patients. Urinalysis showed a significant reduction of leukocytes in both genders, whereas urinary bacterial flora at T1 significantly decreased only in male subjects. Tannins seem to exert an antimicrobial action according to gender, useful to counteract the recurrence of UTIs.

Keywords: hydrolysable tannins; anthocyanins; urinary tract infections; chronic kidney disease; quality of life; cranberry; Sweet Chestnut

Citation: Noce, A.; Di Daniele, F.; Campo, M.; Di Lauro, M.; Pietroboni Zaitseva, A.; Di Daniele, N.; Marrone, G.; Romani, A. Effect of Hydrolysable Tannins and Anthocyanins on Recurrent Urinary Tract Infections in Nephropathic Patients: Preliminary Data. *Nutrients* **2021**, *13*, 591. https://doi.org/10.3390/nu13020591

Academic Editor: Vassilios Liakopoulos
Received: 27 January 2021
Accepted: 7 February 2021
Published: 11 February 2021

1. Introduction

Urinary tract infections (UTIs) are a set of different clinical conditions due to the colonization of the urinary tract by uropathogenic microorganisms able to cause inflammatory and infective processes in the renal parenchyma or in excretory tract, as well. UTIs are the most frequent nephro-urological pathologies and represent the most common bacterial infections [1].

UTIs are a widespread global health problem and their prevalence is estimated at 0.7% worldwide [2]. UTIs occur more frequently in female subjects. In fact, a recent study observed that 40% of women present at least one episode of UTI during their lifetime and they have 30 times higher risk than men of developing UTIs [3].

The main UTI risk factors are age, female gender [4], sexual activity and the use of antibiotics [5]. Factors favoring the pathogenic invasion include anatomical abnormalities, medical devices such as urinary catheter, and favorable conditions related to the host (immunosuppression, diabetes mellitus, pregnancy). UTIs also represent a frequent condition in kidney transplant patients [6]. Furthermore, the onset of UTIs could have a genetic basis. Indeed, patients with a positive family history of UTI in first-degree relatives have an higher risk of developing them compared to the general population [7].

UTI can be caused by the invasion of various microorganisms, both Gram-negative and Gram-positive, and the most common pathogenic bacterium is *Escherichia coli* [8]. The diagnosis of UTI is made by combining symptoms with a positive urine culture [9]. In most patients, the threshold for bacteriuria is 1000 colony-forming units (CFU)/mL. However, in 20% of women with classic urinary symptoms, the urine culture can be negative, which mainly depends on the laboratory cut-off value [10].

Typical symptoms of UTIs can be systemic or local; the first include fever with chills and flank pain, and the second include dysuria, stranguria, pollakiuria, suprapubic pain and hematuria [11].

Recurrent UTIs are identified as two or more episodes of uncomplicated UTI of the lower urinary tract within the past 6 months, or 3 or more episodes over the past 12 months [12]. UTIs are a relatively frequent condition with a high impact on the quality of life and on healthcare costs, including visits, diagnostic tests and therapeutic prescriptions [13].

Recent studies have shown that recurrent UTIs are able to cause a worsening of the quality of life [14]. In fact, recurrent UTIs have a negative impact on daily habits, on sexual activity, on social and personal relationships, on the possibility of freely practicing sports, and on a decline in work productivity [15].

UTIs appear to be constantly growing in frequency, and this phenomenon is favored by the inappropriate use of antibiotics widely employed in both outpatient and hospital settings. This justifies the attention of research towards studies aimed to identifying new therapeutic strategies based on natural bioactive compounds, free from side effects such as nephrotoxicity or hepatotoxicity [16,17], and able to effectively counteract the recurrence of UTI.

Natural bioactive compounds exhibit well-known beneficial properties (such as antioxidants, anti-inflammatory and antimicrobial) which are mainly found in plant-based foods, such as fruit and vegetables [15,18–20]. Among these, the most studied are polyphenols, a wide group of substances that can be grouped into over 20 classes of organic compounds [21]. Recent studies suggest that long-term consumption of polyphenols, both in the form of fresh foods and oral food supplements, may have positive implications for human health. Specifically, several studies have demonstrated that polyphenols are able to reduce the incidence of chronic non-communicable diseases, such as cardiovascular diseases, obesity, diabetes mellitus, neurodegenerative diseases, chronic kidney disease (CKD) and some types of cancer [22–25].

Tannins belong to the class of secondary polyphenolic metabolites and they are found in a wide variety of foods, including cereals (such as sorghum, millet and barley) and legumes, but also in wine, green tea and coffee [26]. Thanks to their antioxidant and antimicrobial properties, tannins can be suitable for several innovative uses in various sectors, such as foods, cosmetics, phytotherapics, nutraceuticals and agronomics products [27].

Of clinical relevance are Chestnut tannins. In fact, several studies suggest that Chestnut tannins seem to have an important effect on human health, as they have known antioxidant, antitumor, antimicrobial, antifungal effects [28–30]. Tannins can also be involved in the reduction of triglycerides and total cholesterol levels and in the suppression of lipogenesis by insulin. Moreover, they present important astringent actions in the gastrointestinal tract [31]. In the literature, several studies are available concerning the antimicrobial activity of anthocyanins, proanthocyanidins and hydrolysable tannins from Cranberry and Sweet Chestnut. These studies report in vitro and in vivo actions towards bacteria and fungi such as *Escherichia coli*, *Klebsiella pneumoniae*, *Enterococcus faecalis*, *Pseudomonas aeruginosa*, *Candida* spp., which are among the main microorganisms responsible for UTIs. Moreover, both our and other research groups have previously tested these compounds in vitro as natural extracts, demonstrating that they can exert synergistic activities in combination with the traditional antibiotics or antifungals, or if administered as phytocomplexes [32–39]. For this reason, in the present study, an oral food supplement (OFS) containing extracts from Cranberry and Sweet Chestnut, was formulated and it was tested in vivo on patients with recurrent UTIs. The aim of our pilot study was to evaluate the anti-inflammatory, antimicrobial and antioxidant efficacy of hydrolysable Chestnut tannins and anthocyanins, administered as OFS, in a population of CKD patients affected by recurrent UTIs.

2. Materials and Methods

2.1. Oral Food Supplement

The studied OFS is referred to by the trade name "prosta-tan" and it is based on natural extracts rich in hydrolysable tannins obtained from Sweet Chestnut, furnished by Gruppo Mauro Saviola s.r.l. (Radicofani, Siena, Italy), Saviola Holding S.r.l. (Viadana, Mantova, Italy). Specifically, the OFS pharmaceutical form is a capsule containing a mixture of dry extracts from: *Castanea sativa* Mill. (22% p/p); *Serenoa repens* (W. Bartram) Small (20% p/p); *Vaccinium macrocarpon*, Ait. (11% p/p).

2.2. Chemicals

HPLC-grade solvents, formic acid (ACS reagent) and EGCG are from Sigma Aldrich Chemical Company Inc. (Milwaukee, WI, USA). Gallic acid and ellagic acid, cyanidin-3-*O*-glucoside chloride, quercetin and (±)-catechin hydrate, analytical grade, are from Sigma-Aldrich (St. Louis, MO, USA). HPLC-grade water was prepared via double-distillation and purification with a Labconco Water Pro PS polishing station (Labconco Corporation, Kansas City, MO, USA).

2.3. Extraction

Two extraction procedures at different pHs were optimized for anthocyanosidic and non-anthocyanosidic polyphenols. The powder present in one capsule was precisely weighed (416 ± 4 mg) and extracted in 4.0 m of a solution 70:30 EtOH:H_2O acidified by HCOOH (pH 3.2 for non-anthocyanosidic polyphenols, pH 1.8 for anthocyanosides). The mixtures were kept under stirring at room temperature, protected from light, for 1h, then centrifuged at 5000 rpm for 5 min to separate the solid matrices from the extracts.

2.4. HPLC-DAD-MS Analysis

The analysis was performed on the extracts without dilution. The extracts were analyzed with a HP-1260 liquid chromatograph equipped with a DAD detector and a HP MSD API-electrospray (Agilent Technologies, Santa Clara, CA, USA) in negative and positive ionization mode. The chromatographic separation was performed by using a column Luna, C18 250 × 4.60 mm, 5 μm (Phenomenex, Torrance, CA, USA), operating at 26 °C. The eluents were H_2O (pH 3.2 by HCOOH) and CH_3CN. A four-step linear solvent gradient from 100% H_2O up to 100% CH_3CN was applied with a flow rate of 0.8 mL/min over a 55 min period, as previously described [40,41]. Mass spectrometer operating conditions were: gas temperature 350 °C, flow rate of 10.0 L/min, nebulizer pressure 30 psi, quadrupole temperature 30 °C and capillary voltage 3500 V. Fragmentor 120 eV. The identification was performed according to chromatographic, spectrometric and spectrophotometric data, by comparison with the specific standards available. Five-point calibration curves ($r^2 \geq 0.999$) were used, built with the specific standards. The correction of molecular weights was performed by multiplying each calibration result by the ratio between the molecular weight of the quantified compound and the molecular weight of the standard. Gallic acid and its derivatives were calibrated at 280 nm with gallic acid; ellagic acid and its derivatives were calibrated at 254 nm with ellagic acid; proanthocyanidins were calibrated at 280 nm with (±)-catechin hydrate, anthocyanosides were calibrated at 520 nm with cyanidin 3-*O*-glucoside; quercetin and its derivatives were calibrated at 350 nm with quercetin.

2.5. In Vitro Assays

Folin-Ciocalteu in vitro antioxidant capacity: Total phenols and polyphenols content was evaluated by spectrophotometric Folin-Ciocalteu assay, by measuring the absorbance at 725 nm of a sample solution containing Folin-Ciocalteu reagent, and 20% Na_2CO_3 after 40 min incubation. The five-point calibration curve was performed in gallic acid. The phenols content of each sample is reported as GAEs and correlated with the in vitro antioxidant activity [42,43].

In vitro assay with stable radical DPPH• (1,1-diphenyl-2-picrylhydrazyl): The anti-radical activity was evaluated by stable radical DPPH• test, according to the previously reported procedure [44] with slight modifications. The extract was diluted and added 1:1 to an ethanolic solution of DPPH• (0.025 mg/mL).

The absorbance was measured at 517 nm with a DAD 8453 spectrophotometer (Agilent Technologies) at time 0 and every 2 min for 20 min. Antiradical activity % (AR%) was obtained through the relationship: [AR% = 100 × (A0 − A20)/A0]. A0 and A20 were the absorbance of DPPH• at 0 min and 20 min, respectively, after the addition of the diluted extract. The EC_{50} was the molar concentration in polyphenols of the solution that inhibits the DPPH• activity by 50%, determined by measuring the AR% for five different dilutions of the sample.

2.6. Patients

Twenty-six nephropathic patients affected by recurrent UTIs were recruited for the in vivo pilot study. Of these, sixteen patients (eight males and eight females) were treated with the OFS supplementation, as described below, and ten (five males and five females) represented the untreated subjects (control group). Each group were divided into two subgroups according to gender, homogeneous for age, body mass index (BMI) and CKD stage [45].

The inclusion criteria were age over 18 years, both sexes, signature and acceptance of informed consent, history of recurrent UTIs. The exclusion criteria were neoplastic subjects, patients with HIV positive infection, patients with liver disease and chronic viral hepatitis, patients with inflammatory and/or infectious pathologies in the acute phase, malnutrition (BMI < 18.5 kg/m^2); pregnancy and end stage renal disease.

At the time of enrollment, the selected patients with recurrent UTI history allhad a negative urine culture but increased microbial flora and leukocytes in the urine sediment examination, in the absence of urinary symptoms.

The patients of the OFS group were instructed to consume 1 capsule per day of OFS based on Chestnut tannins for six weeks. Blood and urinary parameters and the body composition assessments were monitored at two times during the study, at T0 (baseline) and at T1 (after six weeks), in both groups. Figure 1 shows the in vivo study flow-chart.

The study protocol complied with the declaration of Helsinki and was approved by the Ethical Committee of University Hospital Policlinico Tor Vergata (PTV) of Rome (project identification code 78/18 on 13 June 2018).

2.7. Laboratory Parameters

At baseline and after six weeks, we assessed the renal function through the evaluation of creatinine and estimated glomerular filtration rate (e-GFR). At the same time-points we evaluated the inflammatory status with C-reactive protein (CRP) and erythrocyte sedimentation rate (ESR). All patients underwent urinalysis to check UTIs signs. Furthermore, Free Oxygen Radical Test (FORT) and Free Oxygen Radical Defense (FORD) test were performed by CR4000, on capillary blood samples, to evaluate the oxidative stress [46] and the total antioxidant defense capacity [47], respectively.

A Dimension Vista 1500 (Siemens Healthcare Diagnostics, Milano, Italy) instrument was used to monitor all parameters. Standard enzymatic colorimetric techniques (Roche Modular P800, Roche Diagnostics, Indianapolis, IN, USA) were used to assess the lipid profile.

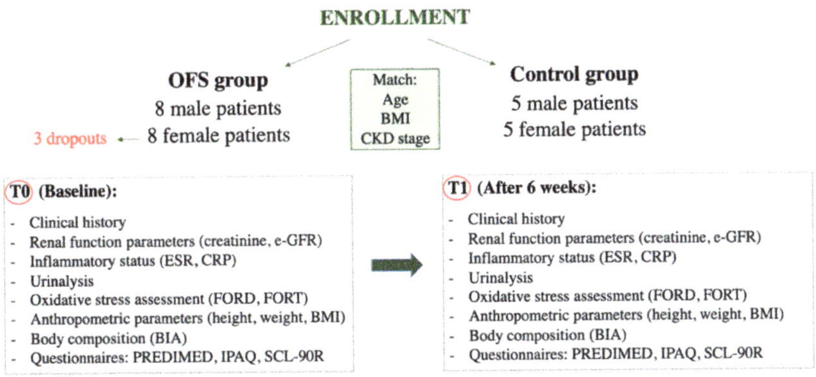

Figure 1. Flow-chart of in vivo pilot study. Abbreviations: BIA, bioelectrical impedance analysis; BMI, Body mass index; CKD, Chronic kidney disease; CRP, C reactive protein; e-GFR, Estimated-glomerula filtration rate; ESR, Erythrocyte sedimentation rate; FORD, Free oxygen radical defense; FORT, Free oxygen radical test; IPAQ, International physical activity questionnaire; OFS, Oral food supplement; PREDIMED, Prevención con Dieta Mediterránea; SCL-90R; Symptoms checklist-90 revised.

All other parameters were analyzed according to standard procedures of Clinical Biochemical Laboratories of University Hospital PTV of Rome.

2.8. Anthropometic and Body Composition Parameters

At the two time-points of the study (T0 and T1), an assessment of anthropometric parameters, such as height, weight and BMI, was performed. Body weight (kg) was measured to the nearest 0.01 kg with a balance scale (Seca 711, Hamburg, Germany), height (m) was measured with stadiometer to the nearest 0.1 cm (Seca 220, Hamburg, Germany). Standard methods were used to collect the anthropometric parameters [48]. BMI was calculated as body weight divided by height squared (kg/m^2). Moreover, all enrolled patients underwent the evaluations of body composition by bioelectrical impedance analysis (BIA) using a BIA 101S instruments, Akern/RIL System-Florence. Resistance, reactance, impedance and phase angle at 50 KHz frequency were measured at T0 and T1. For the monitoring of hydration status, we evaluated total body water (TBW), intracellular water (ICW) and extracellular water (ECW) [49].

2.9. Questionnaires

To assess the possible biases induced by lifestyle changes, at baseline and at T1, we administered two questionnaires, the Prevención con Dieta Mediterránea (PREDIMED) questionnaire for the evaluation of adherence to the Mediterranean diet [50] and the International Physical Activity Questionnaire (IPAQ) for the evaluation of weekly physical activity [51], to all enrolled patients.

In addition, we administered a questionnaire for the evaluation of quality of life: the Symptoms Checklist-90 Revised (SCL-90R) [52]. SCL-90R assesses the presence and severity of psychological distress symptoms. Specifically, the questionnaire consists of 90 items and allows to detect nine different symptoms spheres: somatization, obsessive-compulsive disorder, interpersonal sensitivity, depression, anxiety, hostility, phobic anxiety, paranoid ideation and psychoticism [53]. The analyzed spheres of this study were somatization, anxiety and depression.

2.10. Statistical Analysis

All parametric variables are reported as means ± standard deviation, while non-parametric variables are reported as median (range minimum-maximum). We checked the normality of data for all continuous variables using the Kolmogorov-Smirnov test.

The significance between T0 and T1 of parametric variables was tested with paired *t*-test, while the Wilcoxon test was used for the non-parametric variables. A *p*-value < 0.05 was considered statistically significant. The homogeneity of the subgroups was assessed using univariate ANOVA with a covariate for continuous parametric variables. Moreover, the short PREDIMED, IPAQ and SCL-90 data matrices were analyzed according to McNemar's test [54]. Statistical analysis was performed with the Statistical Package for the Social Sciences Windows, version 15.0 (SPSS, Chicago, IL, USA). The graphic result visualization was obtained using GraphPad Prism (La Jolla, CA, USA).

3. Results

3.1. Supplement Characterization and In Vitro Study

The 1 h extraction procedure (see Section 2) was optimized and validated by comparing the quali-quantitative compositions of extracts prepared in the same conditions, but kept under stirring for 24 h, both for anthocyanosides and for the other polyphenols. Specifically, the OFS powder was extracted at pH 1.9 and pH 3.2 for 1 h and for 24 h. The HPLC-DAD-MS analyses (not reported here) showed a similar composition for the extracts at pH 3.2, whereas anthocyanosidic compounds extracted at pH 1.9 underwent a partial degradation with the longer time of extraction. Figure 2 A, B shows the chromatographic profiles of the two OFS extracts. The first one, acquired at 520 nm, is the profile of anthocyanosidic compounds extracted at pH 1.9, where six compounds were detected, identified and quantified (Table 1), the most abundant of which was cyanidin 3-*O*-arabinoside (0.435 ± 0.005 mg/g powder). Cyanidin was also found as its 3-*O*-galactoside and 3-*O*-glucoside (compounds 1–3 in Figure 2). Additionally, peonidin 3-*O*-galactoside, peonidin 3-*O*-glucoside and peonidin 3-*O*-arabinoside were present (compounds 4–6); peonidin 3-*O*-galactoside in the same amount as cyanidin 3-*O*-arabinoside. Total anthocyanosides were 1.89 ± 0.03 mg/g powder. These results are consistent with those previously reported in the literature for cranberry [55,56].

Table 1. Polyphenol content in the tested OFS. Results in mg/g powder, with absolute errors.

Polyphenols	mg/g
Cyanidin 3-*O*-galactoside	0.347 ± 0.004
Cyanidin 3-*O*-glucoside	0.205 ± 0.003
Cyanidin 3-*O*-arabinoside	0.435 ± 0.005
Peonidin 3-*O*-galactoside	0.435 ± 0.006
Peonidin 3-*O*-glucoside	0.066 ± 0.002
Peonidin 3-*O*-arabinoside	0.397 ± 0.005
Vescalin	0.51 ± 0.01
Castalin	0.340 ± 0.009
Pedunculagin I	0.705 ± 0.008
Monogalloyl glucose I	0.198 ± 0.005
Gallic acid	1.34 ± 0.03
Monogalloyl glucose II	0.65 ± 0.02
Vescalagin	1.57 ± 0.02
Castalagin	1.15 ± 0.03
Gallic acid derivatives	2.68 ± 0.04
Proanthocyanidins	1.04 ± 0.03
Quercetin derivatives	0.364 ± 0.008
Total polyphenols	**12.4 ± 0.2**

The second chromatographic profile, acquired at 280 nm, shows the presence of a large variety of non-anthocyanosidic polyphenols and two peaks of proanthocyanosidic compounds ("P" peaks). Total polyphenols extracted at pH 3.2 are 10.5 ± 0.2 mg/g powder: 9.1 ± 0.2 mg/g hydrolyzable tannins, 1.04 ± 0.03 mg/g proanthocyandins, 0.364 ± 0.008 mg/g quercetin derivatives. Gallic acid is the most represented polyphenol according to the number of moles, but the high molecular weights of the main hydrolysable tannins of Sweet Chestnut, vescalagin and castalagin, make them very abundant in weight, with vescalagin being the highest one (1.57 ± 0.02 mg/g powder) [40,41]. Quercetin derivatives were present in relatively low amounts, and come from the leaf component in Sweet Chestnut extract. The other hydrolysable tannins, complex gallic and ellagic acid esthers with glucose molecules, are typical compounds of Sweet Chestnut wood extracts, once generally indicated as "tannic acid". Proanthocyanidins are typical of cranberry extracts, whereas fatty acids from Serenoa repens, its bioactive compounds, were not detectable and in any case cannot be efficiently extracted using the described procedures. Therefore, in one capsule containing 500 mg of powder, the total polyphenol content is 6.21 mg (4.57 mg hydrolysable tannins, 0.94 mg anthocyanosides, 0.51 mg proanthocyanidins, 0.18 mg quercetin derivatives, taking into account the absolute errors reported above).

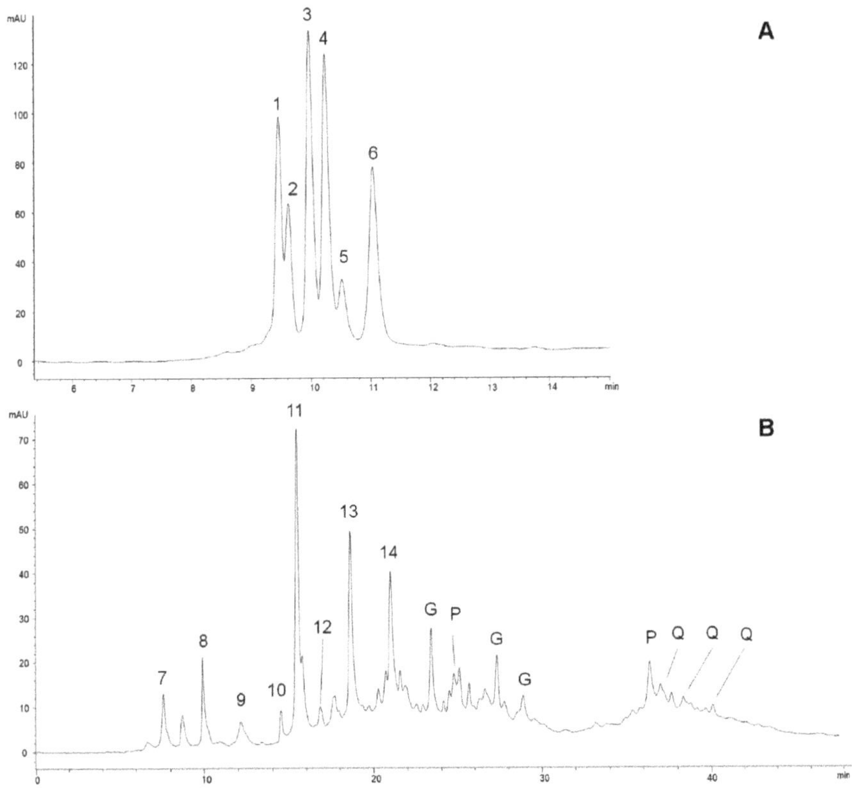

Figure 2. Chromatographic profiles of the OFS extracts (see Section 2). (**A**) pH 1.9, acquired at 520 nm: Anthocyanosidic compounds. 1. Cyanidin 3-*O*-galactoside; 2. Cyanidin 3-*O*-glucoside; 3. Cyanidin 3-*O*-arabinoside; 4. Peonidin 3-*O*-galactoside; 5. Peonidin 3-*O*-glucoside; 6. Peonidin 3-*O*-arabinoside. (**B**) pH 3.2, acquired at 280nm. Non-anthocyanosidic polyphenols. 7. Vescalin; 8. Castalin; 9. Pedunculagin I; 10. Monogalloyl glucose I; 11. Gallic acid; 12. Monogalloyl glucose II; 13. Vescalagin; 14. Castalagin; G. Gallic acid derivatives; P. Proanthocyanidins; Q. Quercetin derivatives.

Total antioxidant capacity and total polyphenols were evaluated by spectrophotometric assay with the Folin-Ciocalteu reagent, which allows for the determination of total phenols and polyphenols content through an electron-transfer (by H^+ transfer) reaction between the sample under examination, in particular compounds with phenolic groups, and the Folin-Ciocalteu reagent. The results are calculated by using external calibration curves, usually in gallic acid, and expressed as mg/g GAE (Gallic Acid Equivalents). Thus, this test evaluates the total phenol compounds by determining the total antioxidant capacity in solution. The in vitro antioxidant activity showed a correlation between total phenols and minor polar compounds, as confirmed by previous studies carried out by comparing different electron transfer reaction assays (e.g., ferric reducing ability of plasma-FRAP, trolox equivalent antioxidant capacity-TEAC and oxygen radical absorbance capacity-ORAC) and in vitro assays on human low-density lipoproteins (LDL) [43,57]. The total phenol and polyphenol content in the examined OFS was 69.186 mg/g GAE.

The assay with DPPH• stable radical gave a measure of the antiradical activity of a sample, expressed as its EC_{50} (amount of sample inhibiting DPPH• activity to 50%). The EC_{50} of the OFS was calculated by measuring the antiradical activity of five different dilutions of the extract according to the procedure described in the "Materials and Methods" section, and calculating the molar concentration in polyphenols of the solution that inhibits the DPPH• activity by 50%. The measured EC_{50} was 0.251 ± 0.009 mg of OFS (3 μg polyphenols).

3.2. In Vivo Study

In the present pilot study, 16 patients with recurrent UTIs, 8 males (mean age 70 ± 2.5 years) and 8 females (mean age 61 ± 1.4 years), were enrolled as the OFS group, and 10 patients with recurrent UTIs, 5 males (mean age 69 ± 1.8 years) and 5 females (mean age 65 ± 2.0 years), were enrolled as the control group (untreated). The epidemiological parameters of the study populations and the evaluation of homogeneity based on gender in the two groups (OFS and control groups) are shown in Table 2.

Table 2. Epidemiological findings of study populations (OFS and control groups) and evaluation of the homogeneity divided according to gender.

	OFS Patients			Control Group		
	Males	**Females**	***p* (ANOVA Test)**	**Males**	**Females**	***p* (ANOVA Test)**
N	8	8		5	5	
Age (years)	70 ± 2.5 [a]	61 ± 1.4 [a]	ns	69 ± 1.8 [a]	65 ± 2.0 [a]	ns
Weight (kg)	74.2 ± 4.6 [a]	73.9 ± 3.5 [a]	ns	73.1 ± 3.9 [a]	73.5 ± 3.4 [a]	ns
BMI (kg/m^2)	26.6 ± 1.8 [a]	26.0 ± 1.7 [a]	ns	26.1 ± 1.9 [a]	25.8 ± 1.8 [a]	ns

[a] Data expressed as mean ± standard deviation; Abbreviations: ns = not significant. OFS = Oral food supplement.

Only five of the eight female treated patients completed the study protocol; three dropouts were recorded in female sex treated patients who complained of side effects in the gastrointestinal tract, such as epigastralgia, nausea and heartburn.

The laboratory parameters (T0 vs. T1) of the OFS group (males and females) are reported in Table 3. Assessment of renal function, monitored by creatinine and e-GRF, did not show statistically significant changes in either OFS subgroup. The evaluation of the inflammation indices showed a statistically significant reduction of ESR in male OFS patients (16.7 ± 2.2 mm/h vs. 11.3 ± 1.5 mm/h, $p = 0.0062$), while the reduction was not statistically significant in female OFS patients. In both genders, no significant reduction in CRP was observed.

Moreover, the urinalysis showed a reduction of leukocytes in the urinary sediment in both OFS subgroups (male: 43.5 (1–450) n/uL vs. 15 ± 5.7 n/uL, $p = 0.0391$; female: 28.5 (1–990) n/uL vs. 7 (1–91) n/uL, $p = 0.0625$). As regards the reduction of the urinary bacterial flora, a significant reduction was observed only in the male OFS subgroup (428 ± 143.4 n/uL vs. 34 (0–450) n/uL, $p = 0.0156$).

The laboratory parameters (T0 vs. T1) of the control group (males and females) are reported in Table 4. No statistically significant differences were shown between T1 and T0.

During the study period (6 weeks), we did not observe any UTI relapse in OFS population, while we detected three UTI relapses in the control group (two cases of *Escherichia.coli* and one case of *Enterococcus faecalis*).

Table 3. Laboratory parameters of two subgroups of OFS patients.

	Male Patients			Female Patients		
	T0	T1	T0 vs. T1	T0	T1	T0 vs. T1
Creatinine (mg/dL)	1.57 ± 0.8 [a]	1.51 ± 0.9 [a]	ns [b]	1.1 ± 0.2 [a]	0.8 ± 0.1 [a]	ns [b]
e-GFR (mL/min/1.73 m^2)	43.0 ± 2.5 [a]	46.55 (36 − 49) [c]	ns [d]	64.5 ± 10.0 [a]	80.0 ± 12.0 [a]	ns [b]
CRP (mg/L)	1.0 (0.5 − 7.5) [c]	0.9 (0.5–3.4) [c]	ns [d]	1.1 (0.3–10.1) [c]	1.5 (1.1–8.4) [c]	ns [d]
ESR (mm/h)	16.7 ± 2.2 [a]	11.3 ± 1.5 [a]	0.0062 [b]	27 ± 7.7 [a]	28.2 ± 5.3 [a]	ns [b]
Urine pH	6 (5.5–7.5) [c]	6.5 (5.5–7.5) [c]	ns [d]	6.3 ± 1.0 [a]	5.6 ± 0.4 [a]	ns [b]
Urinary erythrocytes (n/uL)	10 (0–25) [c]	7 (4–24) [c]	ns [d]	4.5 (1–49) [c]	1 (1–7) [c]	ns [d]
Urinary leukocytes (n/uL)	43.5 (1–450) [c]	15 ± 5.7 [a]	0.0391 [d]	28.5 (1–990) [c]	7 (1–91) [c]	0.0625 [d]
Urinary bacterial flora (n/uL)	428 ± 143.4 [a]	34 (0–450) [c]	0.0156 [d]	553 (36–22,807) [c]	559 (16–61,990) [c]	ns [d]

[a] Data expressed as mean ± standard deviation; [b] Applied test: *t*-test for paired data. [c] Data expressed as a median and the minimum-maximum range is shown in brackets; [d] Applied test: Wilcoxon test; Values of $p \leq 0.05$ are considered statistically significant. Abbreviations: e-GFR, estimated glomerular filtration rate; TC, total-cholesterol; HDL-C, high-density lipoprotein cholesterol; LDL-C, low-density lipoprotein cholesterol; CRP, C-reactive protein; ESR, erythrocyte sedimentation rate; PTH, parathyroid hormone; ns, not significant.

Table 4. Laboratory parameters of control group divided in two subgroups according to gender.

	Male Patients			Female Patients		
	T0	T1	T0 vs. T1	T0	T1	T0 vs. T1
Creatinine (mg/dL)	1.60 ± 0.7 [a]	1.59 ± 0.8 [a]	ns [b]	1.2 ± 0.1 [a]	1.1 ± 0.2 [a]	ns [b]
e-GFR (mL/min/1.73 m^2)	43.4 ± 2.7 [a]	43.6 ± 2.8 [a]	ns [b]	47.4 ± 9.5 [a]	52.6 ± 11.2 [a]	ns [b]
CRP (mg/L)	1.1 (0.6–6.8) [c]	1.0 (0.4–3.2) [c]	ns [d]	1.2 (0.3–10.5) [c]	1.3 (1.0–7.6) [c]	ns [d]
ESR (mm/h)	11.4 ± 1.9 [a]	11.0 ± 1.6 [a]	ns [b]	15.7 ± 5.1 [a]	14.0 ± 5.2 [a]	ns [b]
Urine pH	6.2 ± 1.1 [a]	6.5 ± 1.2 [a]	ns [b]	6.2 ± 1.0 [a]	5.9 ± 0.6 [a]	ns [b]
Urinary erythrocytes (n/uL)	9 (4–25) [c]	7 (5–21) [c]	ns [d]	5 (1–30) [c]	2 (1–9) [c]	ns [d]
Urinary leukocytes (n/uL)	33.5 (1–45) [c]	36.0 ± 5.6 [a]	ns [d]	20.5 (1–70) [c]	22.9 (1–50) [c]	ns [d]
Urinary bacterial flora (n/uL)	428 ± 143.4 [a]	430 ± 140.4 [a]	ns [b]	543 (46–21,700) [c]	552 (20–22,980) [c]	ns [d]

[a] Data expressed as mean ± standard deviation; [b] Applied test: *t*-test for paired data. [c] Data expressed as a median and the minimum-maximum range is shown in brackets; [d] Applied test: Wilcoxon test; Values of $p \leq 0.05$ are considered statistically significant. Abbreviations: e-GFR, estimated glomerular filtration rate; TC, total-cholesterol; HDL-C, high-density lipoprotein cholesterol; LDL-C, low-density lipoprotein cholesterol; CRP, C-reactive protein; ESR, erythrocyte sedimentation rate; PTH, parathyroid hormone; ns, not significant.

The anthropometric parameters and the body composition assessment of OFS group were reported in Table 5, while those of control group were reported in Table 6. After six weeks of OFS treatment, no statistically significant differences were highlighted in either group.

Table 5. Body composition assessment of two subgroups of OFS patients.

	Male Patients			Female Patients		
	T0	T1	T0 vs. T1	T0	T1	T0 vs. T1
Weight (kg)	74.2 ± 4.6 [a]	74.8 ± 4.6 [a]	ns [b]	73.9 ± 3.5 [a]	73.0 ± 3.9 [a]	ns [b]
BMI (kg/m^2)	26.6 ± 1.85 [a]	28.8 ± 1.8 [a]	ns [b]	26.0 ± 1.7 [a]	26.0 ± 1.7 [a]	ns [b]
Resistance (ohm)	493.7 ± 21.3 [a]	480 ± 12.7 [a]	ns [b]	566.7 ± 28.6 [a]	531.0 ± 21.8 [a]	ns [b]
Reactance (ohm)	46.7 ± 2.8 [a]	44.0 ± 0.5 [a]	ns [b]	42.0 ± 2.4 [a]	50.8 ± 2.0 [a]	ns [b]
Phase angle (°)	5.4 ± 0.3 [a]	5.3 ± 0.2 [a]	ns [b]	4.3 ± 0.3 [a]	4.8 ± 0.2 [a]	ns [b]
Hydration status:						
TBW (%)	56.2 ± 1.9 [a]	56.8 ± 1.6 [a]	ns [b]	50.8 ± 1,1 [a]	48.7 ± 2.0 [a]	ns [b]
ICW (%)	51.2 ± 1.8 [a]	50.7 ± 1.3 [a]	ns [b]	45.3 (29.9–51.6) [c]	47.7 ± 1.3 [a]	ns [d]
ECW (%)	48.8 ± 1.8 [a]	49.3 ± 1.1 [a]	ns [b]	54.7 (48.4–55.7) [c]	52.3 ± 1.3 [a]	ns [d]

[a] Data expressed as mean ± standard deviation; [b] Applied test: *t*-test for paired data; [c] Data expressed as a median and the minimum-maximum range is shown in brackets; [d] Applied test: Wilcoxon test; Values of $p \leq 0.05$ are considered statistically significant. Abbreviations: BMI, body mass index; TBW, total body water; ICW, intra cell water; ECW, extra cell water.

At the end of the study, we observed a statistically significant decrease in oxidative stress monitored by FORT (261.4 ± 26.3 vs. 160 (160–250), p = 0.00391), and an increase in antioxidant defenses monitored by FORD (0.88 ± 0.1 vs. 1.43 ± 0.03, p = 0.0030) in male OFS patients, as reported in Table 7. We also observed a statistically significant increase in

FORT (268 ± 46.2 vs. 312 ± 46.1, p = 0.0172) in female OFS patients. Oxidative parameters did not show statistically significant differences in the control group (Table 8).

Table 6. Body composition assessment of control group divided into two subgroups according to gender.

	Male Patients			Female Patients		
	T0	T1	T0 vs. T1	T0	T1	T0 vs. T1
Weight (kg)	73.1 ± 3.9 [a]	73.2 ± 4.3 [a]	ns [b]	73.5 ± 3.4 [a]	73.4 ± 3.0 [a]	ns [b]
BMI (kg/m^2)	26.1 ± 1.9 [a]	26.8 ± 1.7 [a]	ns [b]	25.8 ± 1.8 [a]	25.6 ± 1.7 [a]	ns [b]
Resistance (ohm)	489.7 ± 13.3 [a]	490 ± 11.6 [a]	ns [b]	570.7 ± 27.5 [a]	531.0 ± 21.8 [a]	ns [b]
Reactance (ohm)	45.9 ± 2.6 [a]	45.3 ± 1.7 [a]	ns [b]	43.1 ± 2.5 [a]	45.4 ± 3.4 [a]	ns [b]
Phase angle (°)	5.5 ± 0.2 [a]	5.4 ± 0.2 [a]	ns [b]	4.4 ± 0.4 [a]	4.5 ± 0.3 [a]	ns [b]
Hydration status:						
TBW (%)	56.2 ± 1.9 [a]	56.8 ± 1.6 [a]	ns [b]	50.8 ± 1,1 [a]	48.7 ± 2.0 [a]	ns [b]
ICW (%)	50.6 ± 1.8 [a]	50.8 ± 1.7 [a]	ns [b]	47.2 ± 1.5 [a]	48.1 ± 1.4 [a]	ns [b]
ECW (%)	49.4 ± 1.9 [a]	49.2 ± 1.3 [a]	ns [b]	52.8 ± 1.5 [a]	51.9 ± 1.3 [a]	ns [b]

[a] Data expressed as mean ± standard deviation; [b] Applied test: t-test for paired data; Values of $p \leq 0.05$ are considered statistically significant. Abbreviations: BMI, body mass index; TBW, total body water; ICW, intra cell water; ECW, extra cell water.

Table 7. Oxidative stress and antioxidant defense mechanism efficiency assessment of the OFS group.

	Male Patients			Female Patients		
	T0	T1	T0 vs. T1	T0	T1	T0 vs. T1
FORT (U)	261.4 ± 26.3 [a]	160 (160–250) [c]	p = 0.00391 [d]	268 ± 46.2 [a]	312 ± 46.1 [a]	p = 0.0172 [b]
FORD (mmol/L Trolox)	0.88 ± 0.1 [a]	1.43 ± 0.03 [a]	p = 0.0030 [b]	1.29 ± 0.2 [a]	1.37 ± 0.1 [a]	ns [b]

[a] Data expressed as mean ± standard deviation; [b] Applied test: t-test for paired data; [c] Data expressed as a median and the minimum-maximum range is shown in brackets; [d] Applied test: Wilcoxon test; Values of $p \leq 0.05$ are considered statistically significant. Abbreviations: FORT, Free Oxygen Radical Test; FORD, Free Oxygen Radical Defense; ns, not significant.

Table 8. Oxidative stress and antioxidant defense mechanism efficiency assessment of the control group.

	Male Patients			Female Patients		
	T0	T1	T0 vs. T1	T0	T1	T0 vs. T1
FORT (U)	271.5 ± 27.4 [a]	269.2 ± 30.3 [a]	ns [b]	259.1 ± 51.3 [a]	260.4 ± 40.6 [a]	ns [b]
FORD (mmol/L Trolox)	0.71 ± 0.3 [a]	1.10 ± 0.5 [a]	ns [b]	1.34 ± 0.1 [a]	1.45 ± 0.7 [a]	ns [b]

[a] Data expressed as mean ± standard deviation; [b] Applied test: t-test for paired data; Values of $p \leq 0.05$ are considered statistically significant. Abbreviations: FORT, Free Oxygen Radical Test; FORD, Free Oxygen Radical Defense; ns, not significant.

We did not reveal any statistically significant differences in the patients' lifestyle, monitored by PREDIMED questionnaire and IPAQ, as shown in Table 9.

Table 9. PREDIMED and IPAQ questionnaires of study population.

	PREDIMED		
	T0	T1	p (McNemar's Test)
Minimal adherence (%)	0	0	ns
Average adherence (%)	52.2	56.5	ns
Maximal adherence (%)	47.8	43.5	ns
	IPAQ		
	T0	**T1**	**p (McNemar's Test)**
Inactive (%)	65.2	60.8	ns
Sufficiently active (%)	34.8	39.2	ns
Very active (%)	0	0	ns

Abbreviation: ns, not significant.

After six weeks, the psychological aspect was also assessed through the administration of the SCL-90R questionnaire. A statistically significant reduction in anxiety and depression spheres (and to a lesser extent in somatization) was observed in the male OFS subgroup. In female OGS subjects, this result was less evident. The results of SCL-90R of the OFS group are shown in Figure 3A–C, while the results of SCL-90R of the untreated group are shown in Figure 4A–C. The untreated group showed a slight worsening of the depressive and anxiety spheres, probably due to the chronic course of the disease.

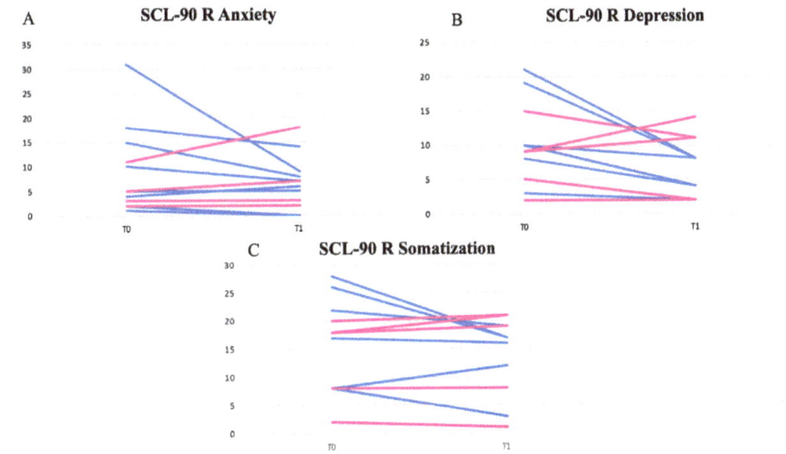

Figure 3. Results of anxiety (**A**), depression (**B**) and somatization (**C**) spheres of SCL-90 questionnaires of OFS group. Legend: blue lines for male patients; pink lines for female patients.

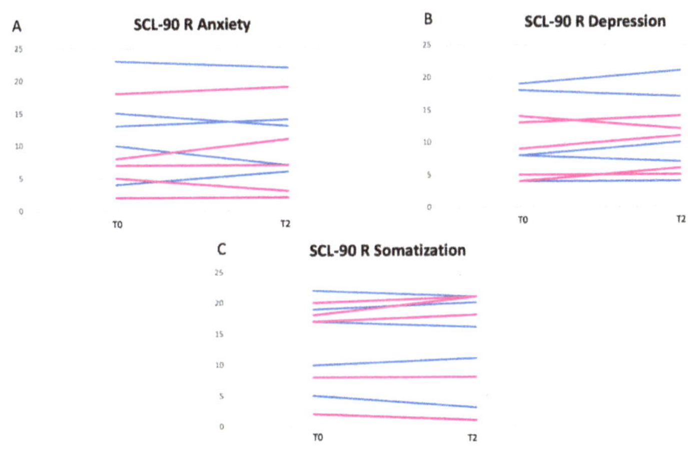

Figure 4. Results of anxiety (**A**), depression (**B**) and somatization (**C**) spheres of SCL-90 questionnaires of untreated group. Legend: blue lines for male patients; pink lines for female patients.

4. Discussion

The hydrolysable tannins from Sweet Chestnut, in combination with the anthocyanosides and proanthocyanidins from Cranberry, seem to exert an antimicrobial action in a gender-dependent manner, useful for countering recurrent UTIs.

The monomers and aromatic acids derived from proanthocyanidins (the main phenolic metabolites of tannins) are found in the urine. Their absorption occurs at the gastrointestinal level, through the gut microbiota [26].

Ellagic acid and gallic acid are metabolites of proanthocyanidins and are found in the plasma, where they undergo conjugation processes with methyl, glucuronyl and sulphate groups and then they are excreted in the urine [58]. One hour after its oral intake, ellagic acid can be detected in the plasma.

A study by Seeram et al. [59] on 18 healthy subjects evaluated the presence of ellagic acid in the urine after pomegranate juice consumption. The authors highlighted that ellagic acid was not present in the plasma the previous day, while it was found both on the day and the day after the juice consumption in 24 h urine. Specifically, the gallic acid metabolites such as dimethylellagic acid glucuronide (DMEAG) and hydroxy-6H-benzopyran-6-one derivatives (urolithins) were detected in plasma in both free and conjugated form. Among the metabolites, the most important was urolithin A-glucuronide, which persisted in the urine for 48 h. This laid the basis for deducing that the daily intake of anthocyanidins and hydrolysable tannins for prolonged periods can exert an antimicrobial action in the urinary apparatus, counteracting relapses of UTIs.

In our pilot study, we administered Chestnut tannins and anthocyanins as OFS to evaluate their possible antimicrobial action on UTIs. This action was confirmed by the significant reduction in the urinary bacterial flora, in leukocyturia and in ESR, in male OFS patients. Cranberry standardized extracts and juices, rich in anthocyanidin/proanthocyanidin, demonstrated interesting effects in UTI prevention, as anthocyanidin/proanthocyanidin inhibit the adherence of P-fimbriated *Escherichia coli* to eukaryotic cells [32]. More recent studies have also demonstrated the efficacy of products containing Cranberry in inhibiting the in vitro growth of *Escherichia coli* strains, whereas the same products showed a lower activity compared to other pathogens like *Klebsiella pneumoniae*, *Enterococcus faecalis*, *Pseudomonas aeruginosa* and *Proteus mirabilis* [35–38,60].

Recent studies showed that Sweet Chestnut extracts and hydrolysable tannins present an antimicrobial activity against different pathogens. This finding suggests interesting and sustainable applications for food or feed safety and agronomics products [61–66]. In the biomedical field, tannic acid, tested in vitro both individually and in combination with fusidic acid on three strains of *Staphylococcus aureus* resistant to methycillin, showed a synergistic effect in preventing of additional adaptive mutations in the bacteria [39]. The chemical 1-methoxy-2,3-digalloylglucose, in mixture in the two anomeric forms, was tested both alone and in sub-inhibitory concentrations in combination with amphotericin B on *Candida albicans*, *Candida glabrata* and *Issatchenkia orientalis*, showing a strong synergistic activity [67].

A study on some terpenoids and on a wide variety of polyphenols, in particular hydrolysable tannins, selected among the representative molecules of natural extracts well known for their antibacterial properties, has confirmed their effectiveness against *Helicobacter pylori* of many of the tested compounds, in particular of hydrolysable tannins with MIC_{50}s in plate between 6.25 and 50 μg/mL [33]. These results, together with those reported above, suggest the possibility of combining hydrolysable tannins from Sweet Chestnut and anthocyanosides/proanthocyanosides from Cranberry, for obtaining products with a wider spectrum of antimicrobial action and with possible synergistic effects. The above-reported in vitro and in vivo results, obtained with active compounds from Cranberry and Sweet Chestnut on microorganisms responsible for UTIs, by both our and other research groups, led us to the innovative formulation of the OFS object of the present pilot study.

In the male OFS subgroup in our study, a statistically significant improvement of the parameters related to oxidative stress (FORT and FORD) was also described, according to the high antioxidant activity and the low EC_{50} antiradical activity, measured using in vitro assays with Folin-Ciocalteu reactive and stable radical DPPH·, respectively.

Such results can be ascribed to the presence of *Serenoa repens* present in the OFS in addition to Chestnut tannins and anthocyanosides. *Serenoa repens* (also commonly called saw palmetto) is a ripe berry of the North American dwarf-palm, traditionally used as treatment for the main male urogenital disturbances. Previous studies have highlighted that *Serenoa repens* has an antispastic, anti-edema, anti-proliferative and anti-androgenic effect [68]. Moreover, *Serenoa repens* extract, in particular its free fatty acid, such as lauric acid and linoleic acid, seems to exert an anti-inflammatory action through inhibition the cyclooxygenase activity, 5-lipoxygenase pathway and pro-inflammatory cytokines biosynthesis [59–62].

However, these promising effects did not seem to be visible in female treated patients. Furthermore, in these subgroups, the OFS seems to have had side effects in the gastrointestinal tract. The OFS tested, due to the high content of *Serenoa repens*, could have induced nausea, vomiting and other minor gastrointestinal symptoms in female subjects as a result of overdose with respect to their body weight [69]. The presence of this substance could explain the reason we observed cases of dropout in female treated subjects. Three female treated subjects, in fact, did not complete the study due to reported gastrointestinal disorders, leading us to hypothesize that the amount of *Serenoa repens* present in the OFS represented an overdose.

Furthermore, in male OFS patients, we observed a reduction in somatization, anxiety and depression state at the end of the study (after six weeks of OFS assumption). These results seem to be in line with the results obtained on the improvement of urinary symptoms in male subgroups. Gender differences have been reported for polyphenols and other bioactive compounds, related to their biotransformation, bioavailability, pharmacodynamics and pharmacokinetics. The metabolism of polyphenolic compounds takes place via gut microbiota and via endogenous enzymes such as cytochrome P450 mono-oxygenases in a gender-dependent manner. The specific targets of these bioactive compounds can be differently expressed in the different genders. Their renal excretion, which is the main excretion route, presents sex differences [70–74].

Thus, the gender is a variable that must be carefully considered. From this perspective, and according to the results of this pilot study, the research will continue with the experimentation of two different OFSs, specific for gender: the OFS (described above) for male patients and a newly formulated OFS, avoiding the use of *Serenoa repens* extract, for female patients.

5. Conclusions

The psychophysical distress induced by UTI recurrence, the possible negative effects related to repeated antibiotics treatment, and the possibility of antibiotic-resistance, have led to interest in finding a natural OFS designed to counteract recurrent UTIs and improve the quality of life of patients. The preliminary results of our pilot study demonstrate the possible therapeutic and preventive efficacy of a natural OFS based on polyphenols, specifically based on Sweet Chestnut tannins, in UTI recurrent patients. However, these results seem to be referable only to male patients. The significant side effects associated with the poor ability to reduce microbial flora in female patients raises the problem of finding a natural OFS able to counteract UTIs in female subjects. To confirm the results obtained for the parameters related to oxidative stress, inflammatory status and gender-dependent antimicrobial activity, a future study should be planned on a larger sample of patients selected by gender, also including a placebo group not treated with tannins. This study should take into account gender differences, formulating two OFS that differ with respect to the presence of *Serenoa repens* extract, which is not completely tolerated by the female population.

Author Contributions: Conceptualization, A.N., N.D.D. and A.R.; methodology, F.D.D., M.C., M.D.L., A.P.Z. and G.M.; investigation, M.D.L., A.P.Z. and G.M.; data curation, F.D.D. and M.C.; writing—original draft preparation, A.N. and A.R.; writing—review and editing, N.D.D.; supervision, A.N. and A.R.; All authors have read and agreed to the published version of the manuscript.

Funding: This research was funded by Tuscany Region, PSR 2014-2020—PS-GO project CAST-AMI-BEN—Use, innovative transformation and enhancement of the Amiata Mount Chestnut PGI in the food, nutraceutical and wellness sectors. Project leader Gruppo Mauro Saviola srl.

Institutional Review Board Statement: The study was conducted according to the guidelines of the Helsinki declaration by the PTV Independent Ethics Committee (Protocol registration number 78/18).

Informed Consent Statement: Informed consent was obtained from all subjects involved in the study.

Data Availability Statement: Data available on request due to privacy restrictions. The data presented in this study are available on request from the corresponding author.

Acknowledgments: We are indebted to Nadia Consalvo and Berardicurti Tiziana for nursing assistance. The authors acknowledge Chiara Cassiani and the Industrial PhD in Economics and Management PON "R&I" 2014–2020.

Conflicts of Interest: The authors declare no conflict of interest.

References

1. Geerlings, S.E. Clinical presentations and epidemiology of urinary tract infections. *Microbiol. Spectr.* **2016**, *4*. [CrossRef]
2. Tandogdu, Z.; Wagenlehner, F.M. Global epidemiology of urinary tract infections. *Curr. Opin. Infect. Dis.* **2016**, *29*, 73–79. [CrossRef]
3. Tan, C.W.; Chlebicki, M.P. Urinary tract infections in adults. *Singap. Med. J.* **2016**, *57*, 485–490. [CrossRef]
4. McLaughlin, S.P.; Carson, C.C. Urinary tract infections in women. *Med. Clin. N. Am.* **2004**, *88*, 417–429. [CrossRef]
5. Flores-Mireles, A.L.; Walker, J.N.; Caparon, M.; Hultgren, S.J. Urinary tract infections: Epidemiology, mechanisms of infection and treatment options. *Nat. Rev. Microbiol.* **2015**, *13*, 269–284. [CrossRef]
6. Bono, M.J.; Reygaert, W.C. Urinary tract infection. In *StatPearls*; StatPearls Publishing: Treasure Island, FL, USA, 2020.
7. Scholes, D.; Hawn, T.R.; Roberts, P.L.; Li, S.S.; Stapleton, A.E.; Zhao, L.P.; Stamm, W.E.; Hooton, T.M. Family history and risk of recurrent cystitis and pyelonephritis in women. *J. Urol.* **2010**, *184*, 564–569. [CrossRef]
8. Yamaji, R.; Friedman, C.R.; Rubin, J.; Suh, J.; Thys, E.; McDermott, P.; Hung-Fan, M.; Riley, L.W. A population-based surveillance study of shared genotypes of escherichia coli isolates from retail meat and suspected cases of urinary tract infections. *mSphere* **2018**, *3*. [CrossRef]
9. Groen, J.; Pannek, J.; Castro Diaz, D.; Del Popolo, G.; Gross, T.; Hamid, R.; Karsenty, G.; Kessler, T.M.; Schneider, M.; t'Hoen, L.; et al. Summary of European Association of Urology (EAU) guidelines on neuro-urology. *Eur. Urol.* **2016**, *69*, 324–333. [CrossRef]
10. Foxman, B. The epidemiology of urinary tract infection. *Nat. Rev. Urol.* **2010**, *7*, 653–660. [CrossRef]
11. Little, P.; Merriman, R.; Turner, S.; Rumsby, K.; Warner, G.; Lowes, J.A.; Smith, H.; Hawke, C.; Leydon, G.; Mullee, M.; et al. Presentation, pattern, and natural course of severe symptoms, and role of antibiotics and antibiotic resistance among patients presenting with suspected uncomplicated urinary tract infection in primary care: Observational study. *BMJ* **2010**, *340*, b5633. [CrossRef]
12. Wagenlehner, F.M.; Vahlensieck, W.; Bauer, H.W.; Weidner, W.; Piechota, H.J.; Naber, K.G. Prevention of recurrent urinary tract infections. *Minerva Urol. Nefrol.* **2013**, *65*, 9–20. [PubMed]
13. Pannek, J. Prevention of recurrent urinary tract infections in neurourology. *Eur. Urol. Focus* **2020**, *6*, 817–819. [CrossRef] [PubMed]
14. Bermingham, S.L.; Ashe, J.F. Systematic review of the impact of urinary tract infections on health-related quality of life. *BJU Int.* **2012**, *110*, E830–E836. [CrossRef] [PubMed]
15. Wagenlehner, F.; Wullt, B.; Ballarini, S.; Zingg, D.; Naber, K.G. Social and economic burden of recurrent urinary tract infections and quality of life: A patient web-based study (GESPRIT). *Expert Rev. Pharm. Outcomes Res.* **2018**, *18*, 107–117. [CrossRef] [PubMed]
16. Noce, A.; Fabrini, R.; Dessi, M.; Bocedi, A.; Santini, S.; Rovella, V.; Pastore, A.; Tesauro, M.; Bernardini, S.; Di Daniele, N.; et al. Erythrocyte glutathione transferase activity: A possible early biomarker for blood toxicity in uremic diabetic patients. *Acta Diabetol.* **2014**, *51*, 219–224. [CrossRef] [PubMed]
17. Bocedi, A.; Noce, A.; Rovella, V.; Marrone, G.; Cattani, G.; Iappelli, M.; De Paolis, P.; Iaria, G.; Sforza, D.; Gallu, M.; et al. Erythrocyte glutathione transferase in kidney transplantation: A probe for kidney detoxification efficiency. *Cell Death Dis.* **2018**, *9*, 288. [CrossRef]
18. Di Renzo, L.; Gualtieri, P.; Romano, L.; Marrone, G.; Noce, A.; Pujia, A.; Perrone, M.A.; Aiello, V.; Colica, C.; De Lorenzo, A. Role of personalized nutrition in chronic-degenerative diseases. *Nutrients* **2019**, *11*. [CrossRef]
19. Romani, A.; Bernini, R.; Noce, A.; Urciuoli, S.; Di Lauro, M.; Pietroboni Zaitseva, A.; Marrone, G.; Di Daniele, N. Potential beneficial effects of extra virgin olive oils characterized by high content in minor polar compounds in nephropathic patients: A pilot study. *Molecules* **2020**, *25*. [CrossRef]
20. Di Daniele, N. The role of preventive nutrition in chronic non-communicable diseases. *Nutrients* **2019**, *11*. [CrossRef]
21. Romani, A.; Ieri, F.; Urciuoli, S.; Noce, A.; Marrone, G.; Nediani, C.; Bernini, R. Health effects of phenolic compounds found in extra-virgin olive oil, by-products, and leaf of *Olea europaea* L. *Nutrients* **2019**, *11*. [CrossRef]
22. Cory, H.; Passarelli, S.; Szeto, J.; Tamez, M.; Mattei, J. The role of polyphenols in human health and food systems: A mini-review. *Front. Nutr.* **2018**, *5*, 87. [CrossRef]
23. Silva, R.F.M.; Pogacnik, L. Polyphenols from food and natural products: Neuroprotection and safety. *Antioxidants* **2020**, *9*. [CrossRef] [PubMed]
24. Noce, A.; Bocedi, A.; Campo, M.; Marrone, G.; Di Lauro, M.; Cattani, G.; Di Daniele, N.; Romani, A. A Pilot study of a natural food supplement as new possible therapeutic approach in chronic kidney disease patients. *Pharmaceuticals* **2020**, *13*. [CrossRef]

25. Noce, A.; Marrone, G.; Di Lauro, M.; Urciuoli, S.; Pietroboni Zaitseva, A.; Wilson Jones, G.; Di Daniele, N.; Romani, A. Cardiovascular protection of nephropathic male patients by oral food supplements. *Cardiovasc. Ther.* **2020**, *2020*, 1807941. [CrossRef] [PubMed]
26. Serrano, J.; Puupponen-Pimia, R.; Dauer, A.; Aura, A.M.; Saura-Calixto, F. Tannins: Current knowledge of food sources, intake, bioavailability and biological effects. *Mol. Nutr. Food Res.* **2009**, *53* (Suppl. S2), S310–S329. [CrossRef]
27. Pizzi, A. Tannins: Prospectives and actual industrial applications. *Biomolecules* **2019**, *9*. [CrossRef]
28. Widsten, P.; Cruz, C.D.; Fletcher, G.C.; Pajak, M.A.; McGhie, T.K. Tannins and extracts of fruit byproducts: Antibacterial activity against foodborne bacteria and antioxidant capacity. *J. Agric. Food Chem.* **2014**, *62*, 11146–11156. [CrossRef]
29. Redondo, L.M.; Chacana, P.A.; Dominguez, J.E.; Fernandez Miyakawa, M.E. Perspectives in the use of tannins as alternative to antimicrobial growth promoter factors in poultry. *Front. Microbiol.* **2014**, *5*, 118. [CrossRef] [PubMed]
30. De Vasconcelos, M.C.; Bennett, R.N.; Rosa, E.A.; Ferreira-Cardoso, J.V. Composition of European chestnut (Castanea sativa Mill.) and association with health effects: Fresh and processed products. *J. Sci. Food Agric.* **2010**, *90*, 1578–1589. [CrossRef]
31. Silva, V.; Falco, V.; Dias, M.I.; Barros, L.; Silva, A.; Capita, R.; Alonso-Calleja, C.; Amaral, J.S.; Igrejas, G.; Ferreira, I.C.F.R.; et al. Evaluation of the phenolic profile of castanea sativa mill. by-products and their antioxidant and antimicrobial activity against multiresistant bacteria. *Antioxidants* **2020**, *9*. [CrossRef]
32. Ohnishi, R.; Ito, H.; Kasajima, N.; Kaneda, M.; Kariyama, R.; Kumon, H.; Hatano, T.; Yoshida, T. Urinary excretion of anthocyanins in humans after cranberry juice ingestion. *Biosci. Biotechnol. Biochem.* **2006**, *70*, 1681–1687. [CrossRef] [PubMed]
33. Funatogawa, K.; Hayashi, S.; Shimomura, H.; Yoshida, T.; Hatano, T.; Ito, H.; Hirai, Y. Antibacterial activity of hydrolyzable tannins derived from medicinal plants against Helicobacter pylori. *Microbiol. Immunol.* **2004**, *48*, 251–261. [CrossRef]
34. Buzzini, P.; Arapitsas, P.; Goretti, M.; Branda, E.; Turchetti, B.; Pinelli, P.; Ieri, F.; Romani, A. Antimicrobial and antiviral activity of hydrolysable tannins. *Mini Rev. Med. Chem.* **2008**, *8*, 1179–1187. [CrossRef] [PubMed]
35. Kontiokari, T.; Sundqvist, K.; Nuutinen, M.; Pokka, T.; Koskela, M.; Uhari, M. Randomised trial of cranberry-lingonberry juice and Lactobacillus GG drink for the prevention of urinary tract infections in women. *BMJ* **2001**, *322*, 1571. [CrossRef]
36. Guay, D.R. Cranberry and urinary tract infections. *Drugs* **2009**, *69*, 775–807. [CrossRef]
37. Ledda, A.; Belcaro, G.; Dugall, M.; Riva, A.; Togni, S.; Eggenhoffner, R.; Giacomelli, L. Highly standardized cranberry extract supplementation (Anthocran(R)) as prophylaxis in young healthy subjects with recurrent urinary tract infections. *Eur. Rev. Med. Pharmacol. Sci.* **2017**, *21*, 389–393.
38. Coppini, C.; Gelinski, J.M.L.N.; Frighetto, M. Cranberry juice inhibit bacterial pathogens associated to urinary tract infection. *J. Sci. Res. Rep.* **2020**, *26*. [CrossRef]
39. Kyaw, B.M.; Lim, C.S.; Wei, Z. Tannic acid as phytochemical potentiator for antibiotic resistance adaptation. *APCBEE Procedia* **2013**, *7*, 175–181.
40. Campo, M.; Pinelli, P.; Romani, A. Hydrolyzable tannins from sweet chestnut fractions obtained by a sustainable and eco-friendly industrial process. *Nat. Prod. Commun.* **2016**, *11*, 409–415. [CrossRef]
41. Lucarini, M.; Durazzo, A.; Romani, A.; Campo, M.; Lombardi-Boccia, G.; Cecchini, F. Bio-Based compounds from grape seeds: A biorefinery approach. *Molecules* **2018**, *23*. [CrossRef]
42. Ninfali, P.; Mea, G.; Giorgini, S.; Rocchi, M.; Bacchiocca, M. Antioxidant capacity of vegetables, spices and dressings relevant to nutrition. *Br. J. Nutr.* **2005**, *93*, 257–266. [CrossRef] [PubMed]
43. Romani, A.; Lapucci, C.; Cantini, C.; Ieri, F.; Mulinacci, N.; Visioli, F. Evolution of minor polar compounds and antioxidant capacity during storage of bottled extra virgin olive oil. *J. Agric. Food Chem.* **2007**, *55*, 1315–1320. [CrossRef] [PubMed]
44. Heimlera, D.; Vignolini, P.; Dini, M.G.; Vincieri, F.F.; Romani, A. Antiradical activity and polyphenol composition of local Brassicaceae edible varieties. *Food Chem.* **2006**, *99*, 464–469. [CrossRef]
45. Ikizler, T.A.; Burrowes, J.D.; Byham-Gray, L.D.; Campbell, K.L.; Carrero, J.-J.; Chan, W.; Fouque, D.; Friedman, A.N.; Ghaddar, S.; Goldstein-Fuchs, D.J.; et al. KDOQI Nutrition in CKD Guideline Work Group. KDOQI clinical practice guideline for nutrition in CKD: 2020 update. *Am. J. Kidney Dis.* **2020**, *76*. [CrossRef] [PubMed]
46. Cesarone, M.R.; Belcaro, G.; Carratelli, M.; Cornelli, U.; De Sanctis, M.T.; Incandela, L.; Barsotti, A.; Terranova, R.; Nicolaides, A. A simple test to monitor oxidative stress. *Int. Angiol.* **1999**, *18*, 127–130.
47. Lewis, N.A.; Newell, J.; Burden, R.; Howatson, G.; Pedlar, C.R. Critical difference and biological variation in biomarkers of oxidative stress and nutritional status in athletes. *PLoS ONE* **2016**, *11*, e0149927. [CrossRef] [PubMed]
48. Lohman, T.G.; Roche, A.F.; Reynaldo Martorell, H.M. *Anthropometric Standardization Reference Manual*; Human Kinetics Books: Champaign, IL, USA, 1988.
49. Bellizzi, V.; Scalfi, L.; Terracciano, V.; De Nicola, L.; Minutolo, R.; Marra, M.; Guida, B.; Cianciaruso, B.; Conte, G.; Di Iorio, B.R. Early changes in bioelectrical estimates of body composition in chronic kidney disease. *J. Am. Soc. Nephrol.* **2006**, *17*, 1481–1487. [CrossRef]
50. Martinez-Gonzalez, M.A.; Garcia-Arellano, A.; Toledo, E.; Salas-Salvado, J.; Buil-Cosiales, P.; Corella, D.; Covas, M.I.; Schroder, H.; Aros, F.; Gomez-Gracia, E.; et al. A 14-item Mediterranean diet assessment tool and obesity indexes among high-risk subjects: The PREDIMED trial. *PLoS ONE* **2012**, *7*, e43134. [CrossRef]
51. Wanner, M.; Probst-Hensch, N.; Kriemler, S.; Meier, F.; Autenrieth, C.; Martin, B.W. Validation of the long international physical activity questionnaire: Influence of age and language region. *Prev. Med. Rep.* **2016**, *3*, 250–256. [CrossRef]

52. Prunas, A.; Sarno, I.; Preti, E.; Madeddu, F.; Perugini, M. Psychometric properties of the Italian version of the SCL-90-R: A study on a large community sample. *Eur. Psychiatry* **2012**, *27*, 591–597. [CrossRef]
53. Hardt, J.; Gerbershagen, H.U.; Franke, P. The symptom check-list, SCL-90-R: Its use and characteristics in chronic pain patients. *Eur. J. Pain* **2000**, *4*, 137–148. [CrossRef]
54. Xiang, J.X. On two-sample McNemar test. *J. Biopharm. Stat.* **2016**, *26*, 217–226. [CrossRef]
55. Wu, X.; Prior, R.L. Systematic identification and characterization of anthocyanins by HPLC-ESI-MS/MS in common foods in the United States: Fruits and berries. *J. Agric. Food Chem.* **2005**, *53*, 2589–2599. [CrossRef]
56. Brown, P.N.; Shipley, P.R. Determination of anthocyanins in cranberry fruit and cranberry fruit products by high-performance liquid chromatography with ultraviolet detection: Single-laboratory validation. *J. AOAC Int.* **2011**, *94*, 459–466. [CrossRef] [PubMed]
57. Huang, D.; Ou, B.; Prior, R.L. The chemistry behind antioxidant capacity assays. *J. Agric. Food Chem.* **2005**, *53*, 1841–1856. [CrossRef]
58. Cerda, B.; Espin, J.C.; Parra, S.; Martinez, P.; Tomas-Barberan, F.A. The potent in vitro antioxidant ellagitannins from pomegranate juice are metabolised into bioavailable but poor antioxidant hydroxy-6H-dibenzopyran-6-one derivatives by the colonic microflora of healthy humans. *Eur. J. Nutr.* **2004**, *43*, 205–220. [CrossRef]
59. Seeram, N.P.; Henning, S.M.; Zhang, Y.; Suchard, M.; Li, Z.; Heber, D. Pomegranate juice ellagitannin metabolites are present in human plasma and some persist in urine for up to 48 hours. *J. Nutr.* **2006**, *136*, 2481–2485. [CrossRef] [PubMed]
60. Howell, A.B. Cranberry proanthocyanidins and the maintenance of urinary tract health. *Crit. Rev. Food Sci. Nutr.* **2002**, *42*, 273–278. [CrossRef] [PubMed]
61. Bargiacchi, E.; Bellotti, P.; Costa, G.; Miele, S.; Pinelli, P.; Romani, A.; Zambelli, P.; Scardigli, A. Use of Sweet Chestnut Tannin Extract as an Antioxidant, Antimicrobial Additive and to Reduce Nitrosamines and Mycotoxins. U.S. Patent Application No. 14/615,615, 2014.
62. Biancalani, C.; Cerboneschi, M.; Tadini-Buoninsegni, F.; Campo, M.; Scardigli, A.; Romani, A.; Tegli, S. Global analysis of type three secretion system and quorum sensing inhibition of pseudomonas savastanoi by polyphenols extracts from vegetable residues. *PLoS ONE* **2016**, *11*, e0163357. [CrossRef]
63. Hoque, M.; Akanda, A.; Miah, M.; Bhuiyan, M.; Miah, M.; Begum, F. In vitro screening of fungicides and tannins against fungal pathogens of jujube fruits. *Progress. Agric.* **2016**, *27*, 154–161. [CrossRef]
64. Agnolucci, M.; Daghio, M.; Mannelli, F.; Secci, G.; Cristani, C.; Palla, M.; Giannerini, F.; Giovannetti, M.; Buccioni, A. Use of chitosan and tannins as alternatives to antibiotics to control mold growth on PDO Pecorino Toscano cheese rind. *Food Microbiol.* **2020**, *92*, 103598. [CrossRef]
65. Messini, A.; Buccioni, A.; Minieri, S.; Mannelli, F.; Mugnai, L.; Comparini, C.; Venturi, M.; Viti, C.; Pezzanti, A.; Rapaccini, S. Effect of chestnut tannin extract (Castanea sativa Miller) on the proliferation of Cladosporium cladosporioides on sheep cheese rind during the ripening. *Int. Dairy J.* **2017**, *66*, 6–12. [CrossRef]
66. Kõrge, K.; Bajić, M.; Likozar, B.; Novak, U. Active chitosan–chestnut extract films used for packaging and storage of fresh pasta. *Int. J. Food Sci. Technol.* **2020**, *55*, 3043–3052. [CrossRef]
67. Romani, A.; Menichetti, S.; Arapitsas, P.; Nativi, C.; Turchetti, B.; Buzzini, P. O-Methylglucogalloyl esters: Synthesis and evaluation of their antimycotic activity. *Bioorg. Med. Chem. Lett.* **2005**, *15*, 4000–4003. [CrossRef]
68. Koch, E. Extracts from fruits of saw palmetto (Sabal serrulata) and roots of stinging nettle (Urtica dioica): Viable alternatives in the medical treatment of benign prostatic hyperplasia and associated lower urinary tracts symptoms. *Planta Med.* **2001**, *67*, 489–500. [CrossRef] [PubMed]
69. Avins, A.L.; Lee, J.Y.; Meyers, C.M.; Barry, M.J.; Group, C.S. Safety and toxicity of saw palmetto in the CAMUS trial. *J. Urol.* **2013**, *189*, 1415–1420. [CrossRef]
70. Campesi, I.; Marino, M.; Cipolletti, M.; Romani, A.; Franconi, F. Put "gender glasses" on the effects of phenolic compounds on cardiovascular function and diseases. *Eur. J. Nutr.* **2018**, *57*, 2677–2691. [CrossRef]
71. Campesi, I.; Romani, A.; Franconi, F. The sex-gender effects in the road to tailored botanicals. *Nutrients* **2019**, *11*. [CrossRef] [PubMed]
72. Kelly, G.E.; Nelson, C.; Waring, M.A.; Joannou, G.E.; Reeder, A.Y. Metabolites of dietary (soya) isoflavones in human urine. *Clin. Chim. Acta* **1993**, *223*, 9–22. [CrossRef]
73. Gradolatto, A.; Basly, J.P.; Berges, R.; Teyssier, C.; Chagnon, M.C.; Siess, M.H.; Canivenc-Lavier, M.C. Pharmacokinetics and metabolism of apigenin in female and male rats after a single oral administration. *Drug Metab. Dispos.* **2005**, *33*, 49–54. [CrossRef] [PubMed]
74. Wruss, J.; Lanzerstorfer, P.; Huemer, S.; Himmelsbach, M.; Mangge, H.; Hoglinger, O.; Weghuber, D.; Weghuber, J. Differences in pharmacokinetics of apple polyphenols after standardized oral consumption of unprocessed apple juice. *Nutr. J.* **2015**, *14*, 32. [CrossRef] [PubMed]

nutrients

MDPI

Article

Higher Intakes of Potassium and Magnesium, but Not Lower Sodium, Reduce Cardiovascular Risk in the Framingham Offspring Study

R. Taylor Pickering , M. Loring Bradlee, Martha R. Singer and Lynn L. Moore *

Department of Medicine/Preventive Medicine & Epidemiology, Boston University School of Medicine, Boston, MA 02118, USA; rtpicker@bu.edu (R.T.P.); lbradlee@bu.edu (M.L.B.); msinger@bu.edu (M.R.S.)
* Correspondence: llmoore@bu.edu

Abstract: We explored the dose-response relations of sodium, potassium, magnesium and calcium with cardiovascular disease (CVD) risk in the Framingham Offspring Study, as well as the combined effects of these minerals. Analyses included 2362 30–64 year-old men and women free of CVD at baseline. Cox proportional-hazards models were used estimate adjusted hazard ratios (HR) and 95% confidence intervals (CIs) for mineral intakes and incident CVD. Cox models with restricted cubic spline functions were used to examine dose-response relations, adjusting for confounding by age, sex, body mass index, dietary fiber intake, and time-varying occurrence of hypertension. Lower sodium intake (<2500 vs. ≥3500 mg/d) was not associated with a lower risk of CVD. In contrast, potassium intake ≥3000 (vs. <2500) mg/d was associated with a 25% lower risk (95% CI: 0.59, 0.95), while magnesium intake ≥320 (vs. <240) mg/d led to a 34% lower risk (95% CI: 0.51, 0.87) of CVD. Calcium intake ≥700 (vs. <500) mg/d was associated with a non-statistically significant 19% lower risk. Restricted cubic spline curves showed inverse dose-response relations of potassium and magnesium with CVD risk, but no such associations were observed for sodium or calcium. These results highlight the importance of potassium and magnesium to cardiovascular health.

Keywords: cardiovascular disease; sodium; potassium

Citation: Pickering, R.T.; Bradlee, M.L.; Singer, M.R.; Moore, L.L. Higher Intakes of Potassium and Magnesium, but Not Lower Sodium, Reduce Cardiovascular Risk in the Framingham Offspring Study. *Nutrients* **2021**, *13*, 269. https://doi.org/10.3390/nu13010269

Received: 10 December 2020
Accepted: 16 January 2021
Published: 19 January 2021

1. Introduction

The need for population-wide salt reduction has been a topic of considerable debate for years. The World Health Organization and others suggest that universal salt reduction may lead to significant public health gains [1]. While there is general agreement that sodium reduction will lead to some reduction in blood pressure, there is considerable disagreement about whether this action will lower cardiovascular risk and mortality. It has even been suggested by some that there may be harm in strict salt reduction guidelines for some people due to unanticipated effects of sodium reduction on other pathways related to cardiovascular disease (CVD) occurrence [2]. As a result, the need for additional research examining the association between lower sodium intakes and CVD risk in the general population of healthy adults as well as high-risk segments of the population has been identified. A 2018 systematic review of several clinical trials, however, examined the efficacy of lower sodium intakes among individuals with prevalent heart failure and concluded that there was insufficient evidence to support salt reduction as a strategy for reducing incident cardiovascular events or mortality in that population [3].

The independent and combined effects of other minerals on CVD risk are of vital interest as well. Potassium is generally thought to have beneficial effects on cardiovascular health through its effects on vascular tone, although this effect may not be fully realized due to widespread under-consumption of potassium among Americans [4]. Evidence for an inverse dose-response relation between potassium and incident CVD is growing [5] as well as evidence for a role of potassium in mediating salt sensitivity [6]. In addition, both calcium and magnesium have been thought to impact cardiovascular health. However, evidence from clinical trials of calcium supplementation is generally inconclusive, with most studies showing no effect on CVD risk [7]. Magnesium is involved in blood pressure and metabolic regulation and may be important to the prevention and management of CVD risk, although evidence on this topic is somewhat sparse and inconclusive [8,9]. Finally, there is also little evidence on the cardiovascular effects of dietary sodium in combination with intakes of calcium and magnesium. Such data will help to expand the evidence base informing future *Dietary Guidelines* [10].

The overall goal of this study is to examine the dose-response relations between sodium, potassium, magnesium, and calcium and risk of CVD in the prospective Framingham Offspring Study. A secondary aim is to evaluate the combined effects of dietary sodium and these other minerals on incident CVD risk.

2. Materials and Methods

The Framingham Offspring Study (FOS) began in 1972 with the enrollment of 5124 offspring (and spouses) of the original Framingham Heart Study cohort. Data related to medical history, cardiometabolic risk factors, lifestyle habits, psychosocial factors, and physical functioning were collected at repeated examination visits occurring at intervals of approximately four years. For these analyses, subjects with complete data on diet, CVD outcomes, and all confounders of interest who were between the ages of 30 and 64 years at examination visit three (when diet was first assessed) were eligible for inclusion. Those missing data or who were less than age 30 at exam 3 were included starting at exam 5 (the next time when dietary intake was assessed).

Of the original 5124 subjects, the following individuals were excluded from the analyses: 283 who died prior to the baseline dietary assessment, 324 outside of the requisite age range (<30 or ≥65 years), 497 who failed to attend exams 3–5, 77 with prevalent cancer at baseline (except non-melanoma skin cancer), 1040 with missing food diaries, 11 with a body mass index (BMI) <18.5 kg/m^2, and 1 with missing blood pressure. As has been done in previous studies, we excluded 389 with implausibly high or low values for energy intake (men reporting usual energy intake <1200 or >4000 kilocalories (kcals)/d, women reporting energy intakes <1000 or >3500 kcals/d [11,12]), or subjects reporting more than 20% of energy intake from alcohol. Finally, 140 with prevalent CVD were excluded, leaving 2362 subjects for these analyses. The original study protocols were approved by the Institutional Review Board of the Boston University Medical Center and all subjects provided written informed consent.

2.1. Dietary Intake

Diet was assessed using two sets of three-day diet records in the third and fifth examination cycles of the study (1984–1988 and 1991–1995). Nearly 75% of participants completed the requested records, with most providing all days of dietary data for a total of approximately 16,000 days of dietary records. Each set of three-day diet records included two weekdays and one weekend day. All participants were instructed in the completion of the diet records by a trained study nutritionist. The participants were instructed to write down everything that they ate or drank for meals and snacks in a 24-h period. They were asked to document portion sizes, specific brand names, cooking methods, and recipes (for home-cooked dishes) as well as salt added at the table. The sources of any take-out foods were also recorded. Fats and seasonings including salt that were added during cooking were also recorded. Two-dimensional food models were used to aid in the estimation of correct portion sizes. The study nutritionist reviewed the dietary records and debriefed the participants as needed for clarification before entering the dietary data into the Nutrition Data System (NDS) of the University of Minnesota [13].

The NDS program provides detailed nutrient composition data for all food items including sodium, potassium, magnesium, and calcium. Total intake for each nutrient from all food sources is then summed over each day. Since adult dietary intakes tend to be stable over time [14] and to reduce random variability in intake, average intakes of sodium, potassium, magnesium, and calcium as well as other nutrients were calculated as a mean across all available days of dietary records.

2.2. CVD Outcomes

All Framingham participants were monitored on an on-going basis for CVD events including fatal and nonfatal myocardial infarction, unstable angina (defined as ischemic episodes with reversible ST-segment changes), heart failure, and ischemic or hemorrhagic stroke. A panel of three investigators adjudicated all possible CVD events and diagnoses using standard long-standing Framingham guidelines. Atherosclerotic cardiovascular disease (ASCVD) cases excluded intracerebral or subarachnoid hemorrhage, ischemic cardiomyopathy, and congestive heart failure.

2.3. Potential Confounders

A wide range of CVD risk factors were routinely assessed in Framingham. Factors explored as potential confounders in these analyses included age, sex, height, education level, BMI, physical activity, cigarette smoking, alcohol intake, a wide range of individual dietary factors, a Dietary Approaches to Stop Hypertension (DASH) eating pattern score, prevalent hypertension and use of anti-hypertensive medications, as well as time-varying development of hypertension or use of anti-hypertensive medications. Blood pressure was measured twice at each examination following a standardized protocol using a mercury-column sphygmomanometer and an appropriate-sized cuff. Modified JNC-7 criteria for defining prevalent hypertension at baseline as well as incident hypertension during follow up have been previously described and, briefly, include those with either a mean systolic blood pressure (SBP) $\geq$ 140 mm Hg, diastolic blood pressure (DBP) $\geq$ 90 mm Hg at 2 consecutive exams, or both, or taking blood pressure lowering medication at any exam, or either an SBP $\geq$ 160, DBP $\geq$ 95, or both, at a single exam) [15].

Height and weight were measured at each exam visit using a standard balance beam scale with a stadiometer. Education was determined by self-report and categorized as less than college vs. some college or more. For those who smoked, the number of cigarettes smoked per day at each exam was recorded. Physical activity was derived from a standardized Framingham questionnaire in which the self-reported hours usually spent per day in sleep, sedentary, light, moderate, and heavy physical activity were determined. Moderate and vigorous activities were weighted for energy expenditure determined by estimated oxygen consumption and summed to obtain an estimate of usual moderate/vigorous physical activity [16]. Alcohol intake (g/d) was based on self-report of usual consumption of beer, wine, and spirits.

2.4. Statistical Analysis

Mineral intakes are often expressed per 1000 kilocalories of intake. However, energy intake is frequently misreported, typically by under-reporting intakes of foods and beverages that are perceived to be less healthy [17], and this under-reporting is differential by body size [18]. As a result, we chose to normalize the intakes of sodium, potassium, magnesium, and calcium for body weight, as a means of accounting for differences in energy intake and overall body size, using the residuals from linear regression models [19]. Each mineral was regressed on body weight; residuals for each subject were then added to the median intakes of that mineral in the FOS population to express the weight-adjusted mineral intakes on the original scale.

For some analyses, mineral intakes were categorized on the basis of dietary recommendations for that mineral [20] as well as power considerations and the sensitivity of the results to changes in the cutoff values. For magnesium and calcium, the current Recommended Dietary Allowance (RDA) values informed the categorization of these variables. However, for sodium or potassium, the lack of an established RDA led to the use of adequate intake (AI) levels to inform the categorizations [20]. Since only 15.8%

of men and 27.8% of women met the AI of <2300 mg/d level for sodium, we chose the following categories to enhance statistical power: <2500, 2500 to <3500, and ≥3500 mg/d. The current AI for potassium is ≥3400 mg/d for men and ≥2600 mg/d for women, with 25.5% of men and 44.9% of women meeting these guidelines. To capture the effects of low potassium consumption, we classified intake for these analyses as <2500, 2500 to <3000, and ≥3000 mg/d. To evaluate the combined effects of adequate intakes of both sodium and potassium, we dichotomized intakes initially based on the established AI levels but since only 1.7% of subjects met the guidelines for both nutrients, we chose to define inadequate intake (referent group) as a sodium level ≥2500 mg/d with a potassium intake of <2500 mg/d, a category comprising approximately 23% of subjects. We similarly chose categories for magnesium and calcium based on the RDA values, power considerations, and sensitivity analyses. Only 10.6% of men and 20.2% of women met the RDA for magnesium (≥420 mg/d for men; ≥320 mg/d for women) while only 21.8% of men and 7.8% of women met the calcium RDA (≥1000 mg/d for men and women 31–50 years of age; ≥1200 mg/d for women ≥51 years of age). The categories for magnesium were <240, 240 to <320, and ≥320 mg/d while those for calcium were <500, 500 to <700, and ≥700 mg/d to enhance statistical power.

Cox proportional hazards models were used to estimate the risk of total CVD and ASCVD associated with categories of intake for each of the four minerals. Only those factors that were found to be actual confounders (as defined by a 5% change or more in the overall effect estimate when included in the model) of the relation between mineral intake and CVD were retained in the final models. Final models for risk of CVD included age, sex, BMI, dietary fiber (for sodium models), and prevalent and time-varying occurrence of hypertension as confounders of the effects. There was no confounding by height, education levels, cigarette smoking, physical activity levels, alcohol intake, energy intake, a DASH eating score, or energy-adjusted macronutrient intakes. Certain dietary variables such as dietary fiber were strongly collinear with some minerals and therefore not included in the final models. To evaluate the dose-response relation between mineral consumption (on a continuous scale) and risk of CVD, we used Cox proportional hazards models with restricted cubic spline functions. Three knots at the 25th, 50th, and 75th percentiles were used, with the 25th percentile serving as the reference point. All analyses were carried out with SAS version 9.4. Figures were created using GraphPad Prism version 8.0 (www.graphpad.com).

3. Results

Table 1 describes the baseline characteristics of the subjects according to weight-adjusted sodium and potassium intakes. Subjects with the lowest intakes of sodium were slightly older, had a higher BMI, were more frequently female, and had lower education levels. They also tended to have lower intakes of potassium, magnesium, and calcium. Those with lower intakes of potassium had a higher BMI, lower education levels, and more were frequently female. There was little to no association between cigarette smoking and sodium or potassium intakes. Notably, those with higher intakes of both sodium and potassium had higher intakes of fruits and vegetables and dairy foods.

Table 2 shows the rates and hazard ratios for CVD and ASCVD associated with intake of each of the four minerals. During the 41,170 person-years of total follow-up (median follow-up time of 19.7 years) there were 404 cases of incident CVD, 367 of which were atherosclerotic in nature. The majority of subjects in this study had moderate sodium intakes (median intake 2927 mg/d). Within the range of usual intake in Framingham, there was no association between sodium and risk of CVD or ASCVD. In contrast, both potassium and magnesium intakes were inversely associated with risks of CVD and ASCVD. For example, those consuming ≥3000 mg/d (vs. <2500 mg/d of potassium had a 25% lower risk of total CVD (hazard ratio (HR) = 0.75; 95% CI: 0.59, 0.95) and a 28% reduction in risk of ASCVD. Consumption of at least 320 mg/d of magnesium (vs. <240 mg/d) was associated with a 34% reduction in risk of total CVD and a 38% reduction in risk of ASCVD. Calcium intake ≥500 mg/d was associated with a 17–19% non-statistically significant (95% CIs: 0.65–1.07 and 0.63–1.08) decreased risks of CVD and ASCVD, respectively.

To assess dose-response relations between minerals and CVD, we utilized Cox proportional hazards models with restricted cubic spline functions. In Figure 1, panels A–D show the dose-response relations between each of the four minerals and risk of total CVD over 12 years of follow-up. In these analyses, only potassium and magnesium were inversely associated with cardiovascular risk in a dose-dependent manner.

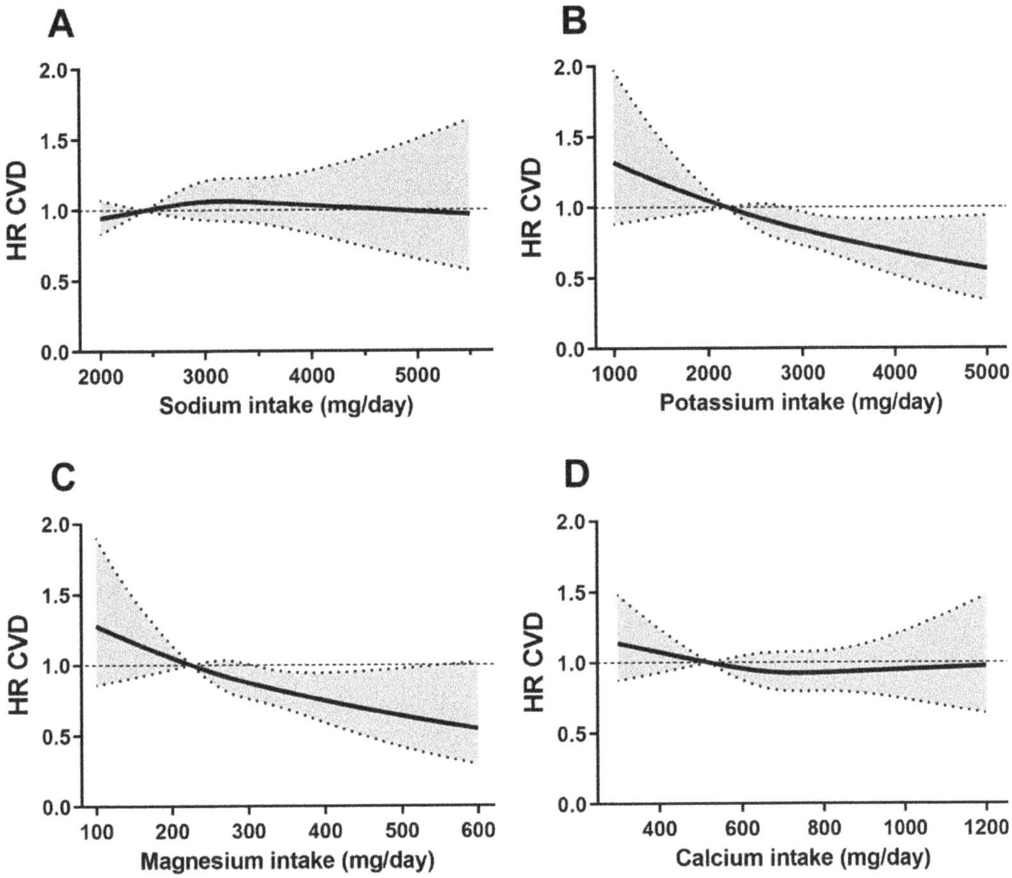

Figure 1. Dose-response relations between mineral intakes and incidence of cardiovascular disease over 12 years of follow-up. Separate dose-response assessments for (**A**) sodium, (**B**) potassium, (**C**) magnesium, and (**D**) calcium with risk of cardiovascular disease (CVD) over 12 years of follow-up using restricted cubic spline analyses. All models were adjusted for age, sex, body mass index (BMI), and time-varying occurrence of hypertension. Dotted lines represent 95% confidence bands. Abbreviations: CVD, Cardiovascular Disease; HR, Hazard Ratio.

Table 1. Baseline subject characteristics by category of sodium and potassium intakes in the Framingham Offspring Study.

	Sodium Intake (mg/d) [a]				Potassium Intake [a]			
	<2500	2500 to <3500	≥3500		<2500	2500 to <3000	≥3000	
	Mean (SD)			*p*-Value	Mean (SD)			*p*-Value
n	661	1018	683		954	650	758	
Age, mean (SD), years	50.5 (8.5)	48.8 (8.7)	47.3 (8.9)	<0.0001	48.7 (8.8)	48.9 (8.7)	49.1 (8.8)	0.58
BMI mean (SD), kg/m^2	27.0 (5.1)	25.5 (4.1)	25.7 (4.0)	<0.0001	26.4 (5.0)	25.7 (4.2)	25.6 (3.8)	0.0001
Cigarettes/d, mean (SD)	4.8 (10.5)	5.0 (11.3)	6.0 (12.3)	0.1004	5.4 (11.0)	5.2 (11.7)	5.0 (11.5)	0.74
Alcohol, mean (SD) gm/d	10.2 (16.9)	10.5 (15.3)	14.0 (18.7)	<0.0001	10.1 (15.8)	11.9 (19.1)	12.7 (16.1)	0.0045
PA, mean (SD), MET score	12.0 (7.5)	11.9 (7.7)	13.3 (9.2)	0.0010	11.8 (7.7)	12.6 (8.2)	12.7 (8.6)	0.0427
SBP, mean (SD), mm Hg	125.1 (16.9)	122.0 (16.2)	120.9 (15.0)	<0.0001	123.0 (16.7)	121.5 (15.3)	122.8 (16.2)	0.15
DBP, mean (SD), mm Hg	78.9 (9.5)	77.5 (9.6)	77.1 (9.3)	0.0019	77.9 (9.4)	77.5 (9.5)	77.8 (9.5)	0.71
Energy, mean (SD), kcals/d	1518 (334)	1870 (422)	2358 (493)	<0.0001	1591 (361)	1945 (431)	2290 (525)	<0.0001
Sodium mean (SD), mg/d [a]	2103 (264)	2973 (276)	4213 (635)	<0.0001	2686 (712)	3154 (813)	3538 (959)	<0.0001
Potassium, mean (SD), mg/d [a]	2389 (645)	2707 (664)	3143 (731)	<0.0001	2067 (295)	2746 (144)	3595 (514)	<0.0001
Magnesium, mean (SD), mg/d [a]	243 (68)	282 (78)	334 (92)	<0.0001	218 (43)	286 (41)	373 (80)	<0.0001
Calcium, mean (SD), mg/d [a]	575 (205)	730 (253	874 (288)	<0.0001	574 (193)	733 (222)	918 (286)	<0.0001
Fiber, mean (SD), gm/d	13.9 (5.5)	15.6 (5.60)	18.6 (6.4)	<0.0001	12.1 (3.6)	15.9 (4.2)	21.0 (6.3)	<0.0001
Fruits and veg, mean (SD), servings/d	2.9 (1.4)	3.0 (1.3)	3.3 (1.4)	<0.0001	2.2 (0.8)	3.1 (1.0)	4.1 (1.4)	<0.0001
Dairy, mean (SD), servings/d	1.0 (0.7)	1.4 (0.8)	1.7 (0.9)	<0.0001	1.0 (0.6)	1.4 (0.7)	1.8 (0.9)	<0.0001
Male (column %)	199 (30.1)	408 (40.1%)	446 (65.3%)	<0.0001	324 (34.0%)	286 (44.0%)	443 (58.4%)	<0.0001
College or more (column %)	183 (27.7)	383 (37.6%)	258 (37.8%)	0.0001	288 (30.2%)	238 (36.6%)	298 (39.3%)	<0.0001

Abbreviations: PA, physical activity; SBP, systolic blood pressure; DBP, diastolic blood pressure; veg, vegetables. [a] Weight adjusted mineral intake.

Table 2. Rates and Hazard Ratios for Risk of CVD Associated with Categories of Mineral intake in the Framingham Offspring Study.

		Risk of Total CVD			Risk of ASCVD		
			Unadjusted	Adjusted [b,c]		Unadjusted	Adjusted [b,c]
Minerals [a]	*n*	**I/1000 py**	**HR (95% CI)**	**HR (95% CI)**	**I/1000 py**	**HR (95% CI)**	**HR (95% CI)**
Sodium (mg/d)							
Sodium <2500	661	9.81	1.00 (ref)	1.00 (ref)	8.73	1.00 (ref)	1.00 (ref)
Sodium 2500 to <3500	1018	9.90	1.04 (0.82, 1.31)	1.12 (0.88, 1.43)	9.04	1.04 (0.82, 1.33)	1.11 (0.86, 1.43)
Sodium ≥3500	683	9.68	1.03 (0.79, 1.34)	1.06 (0.80, 1.39)	8.79	1.01 (0.77, 1.33)	1.02 (0.76, 1.36)
Potassium (mg/d)							
Potassium <2500	954	10.46	1.00 (ref)	1.00 (ref)	9.39	1.00 (ref)	1.00 (ref)
Potassium 2500 to <3000	650	9.73	0.90 (0.71, 1.15)	0.85 (0.67, 1.08)	9.09	0.95 (0.74, 1.21)	0.87 (0.68, 1.12)
Potassium ≥3000	758	9.09	0.86 (0.68, 1.09)	0.75 (0.59, 0.95)	8.09	0.85 (0.67, 1.09)	0.72 (0.56, 0.93)
Magnesium (mg/d)							
Magnesium <240	758	11.05	1.00 (ref)	1.00 (ref)	10.08	1.00 (ref)	1.00 (ref)
Magnesium 240 to <320	929	9.69	0.85 (0.68, 1.07)	0.81 (0.64, 1.02)	8.75	0.85 (0.67, 1.07)	0.78 (0.61, 0.99)
Magnesium ≥320	675	8.64	0.77 (0.60, 1.00)	0.66 (0.51, 0.87)	7.77	0.76 (0.58, 1.00)	0.62 (0.47, 0.83)
Calcium (mg/d)							
Calcium <500	516	11.97	1.00 (ref)	1.00 (ref)	10.76	1.00 (ref)	1.00 (ref)
Calcium 500 to <700	719	9.31	0.77 (0.59, 1.01)	0.82 (0.63, 1.08)	8.41	0.78 (0.59, 1.03)	0.81 (0.61, 1.08)
Calcium ≥700	1127	9.20	0.75 (0.59, 0.96)	0.83 (0.65, 1.07)	8.37	0.76 (0.59, 0.98)	0.81 (0.63, 1.06)

Abbreviations: CVD, cardiovascular disease; ASCVD, atherosclerotic CVD; py, person years; CI, confidence interval; I/1000 py, incidence per 1000 person years; ref, reference. [a] Weight adjusted mineral intakes; [b] hazard ratios for all minerals adjusted for age, sex, BMI, and prevalent hypertension (a time varying covariate); [c] hazard ratios for sodium also adjusted for dietary fiber.

Finally, to determine whether the intakes of potassium, magnesium, or sodium on CVD risk were modified by the daily intake of sodium, we explored the combined intakes of these minerals (Table 3). Here, we dichotomized and then cross-classified intakes of sodium (higher vs. lower) with other minerals. The referent group for each of these analyses was chosen to be that group with the highest expected risk of CVD (i.e., higher sodium plus lower intakes of potassium, magnesium, or calcium). Compared with the referent category, those with higher potassium intakes (when sodium intake was also high) had a 27% lower risk (95% CI: 0.57, 0.93) of CVD while higher potassium intake combined with lower sodium intakes was associated with a 31% lower risk (95% CI: 0.48, 0.99). Higher magnesium intake (≥240 mg/d), regardless of sodium intake, was associated with substantially lower risks of total CVD (HR = 0.72, 95% CI: 0.56, 0.94; HR = 0.61, 95% CI: 0.42, 0.88 among those with higher and lower sodium intakes, respectively). Finally, there were no statistically significant reductions in risk of CVD or ASCVD associate with a lower sodium intake alone.

Table 3. Rates and Hazard Ratios for Risk of CVD Associated with Categories of Mineral Intakes in the Framingham Offspring Study.

		Unadjusted			Adjusted [b]		
Mineral Intakes (mg/d) [a]	*n*	**I/1000 py**	**HR**	**95% CI**	**I/1000 py**	**HR**	**95% CI**
Na ≥ 2500, K < 2500	532	11.18	1.00	ref	10.14	1.00	ref
Na ≥ 2500, K ≥ 2500	1169	9.21	0.73	0.57, 0.93	8.41	0.73	0.57, 0.94
Na < 2500, K < 2500	422	9.55	0.78	0.57, 1.07	8.43	0.80	0.57, 1.11
Na < 2500, K ≥ 2500	239	10.27	0.69	0.48, 0.99	9.28	0.70	0.48, 1.02
Na ≥ 2500, Mg < 240	399	11.41	1.00	ref	10.48	1.00	ref
Na ≥ 2500, Mg ≥ 240	1302	9.34	0.72	0.56, 0.94	8.48	0.70	0.53, 0.92
Na < 2500, Mg < 240	359	10.64	0.85	0.61, 1.19	9.61	0.87	0.61, 1.24
Na < 2500, Mg ≥ 240	302	8.88	0.61	0.42, 0.88	7.73	0.59	0.40, 0.86
Na ≥ 2500, Ca < 700	737	10.41	1.00	ref	9.50	1.00	ref
Na ≥ 2500, Ca ≥ 700	964	9.37	0.96	0.76, 1.20	8.52	0.94	0.73, 1.19
Na < 2500, Ca < 700	498	10.35	0.95	0.72, 1.25	9.15	0.96	0.71, 1.28
Na < 2500, Ca ≥ 700	163	8.19	0.72	0.46, 1.12	7.47	0.74	0.46, 1.18

Abbreviations: CVD, cardiovascular disease;; CI, confidence interval; I/1000 py, incidence per 1000 person years; Na, sodium; K, potassium; Mg, magnesium; Ca, calcium; ref, reference. [a] Weight-adjusted mineral intakes; [b] hazard ratios for all minerals adjusted for age, sex, BMI, and prevalent hypertension (a time varying covariate).

4. Discussion

In these analyses, we first sought to determine the dose-response relations between sodium, potassium, magnesium, and calcium and risk of CVD. In these analyses, sodium intake was not associated with risk of CVD at the levels consumed by this generally healthy community-based population of adults. Both potassium and magnesium were consistently inversely associated with the risk of incident CVD in a dose-dependent manner. While potassium consumption at or above 3000 mg/d (vs. <2500 mg/d) was linked with at least a 25% reduction in risk of both total and atherosclerotic CVD, we also observed that risks declined steadily throughout the distribution for intakes up to 5000 mg/d. Similarly, CVD risks declined steadily with intakes of magnesium ranging from 100 to 600 mg/d. Higher intakes of potassium and magnesium were both associated with reduced risks of CVD regardless of sodium intake while lower intakes of sodium had no independent beneficial effects on CVD risk.

The association between dietary sodium intake and CVD risk is controversial [21]. There have been several prospective studies evaluating the relation between urinary sodium and incident CVD or CVD mortality. One study of subjects in their mid-40s who were overweight and tended to have elevated blood pressure levels had a higher risk of non-fatal CVD associated with increasing sodium intakes [22]. Thus, it is possible that CVD risk may be different among individuals with prevalent obesity and high blood pressure. However, our results are consistent with those from the Health, Aging, and Body Composition (Health ABC) study which found no association between sodium intake and incident CVD even after controlling for prevalent hypertension [23]. Since hypertension may be part of the causal pathway to CVD, we also ran all statistical models excluding those with prevalent and time-varying hypertension from the models and the results were virtually identical. Some studies have observed a non-linear (i.e., J-shaped) relation between urinary sodium and CVD risk. A 2014 meta-analysis of largely prospective cohort studies concluded that lower sodium intakes (<2645 mg/d) were associated with higher risks of all-cause mortality as well as CVD incidence than were more moderate intakes (2645–4945 mg/d) [24]. In Framingham, we had too few subjects with sodium intakes above 5000 mg/d to evaluate intakes at that level. Further, only 15.8% of men and 27.8% of women met the current dietary guidelines for sodium of less than 2300 mg/d, limiting our assessment of risk at this intake level.

It is possible that the lack of a beneficial effect of lower sodium intake on cardiovascular risk in this and other studies may be due to other adverse effects associated with reducing dietary sodium intake. Some randomized clinical trials targeting dietary sodium reduction have shown unintended results—that is, increases in renin, aldosterone, catecholamines, total cholesterol, and triglycerides [5,25]. Since all of these effects are linked with higher risks of heart disease and death, these findings could explain the absence of a beneficial effect of lowering sodium intakes.

Analyses from the Prevention of Renal and Vascular End-Stage Disease (PREVEND) study found that every additional 598 mg of potassium was associated with a 13% reduction in CVD risk [26]. These results are consistent with a 2011 meta-analysis in which every additional 966 mg/d of potassium was associated with a 26% lower risk of total CVD and a 21% lower risk of stroke [27]. Our results in Framingham were similar. We found that for every additional 600 mg/d of potassium consumed, CVD risk declined by 12% (data not shown).

The association between dietary magnesium and risk of CVD dates back to early epidemiologic observations in which water "hardness," a measure of calcium and magnesium content, was inversely associated with CVD risk [28]. Early observations from the Atherosclerosis Risk in Communities (ARIC) study showed that those with prevalent hypertension, CVD, or diabetes had lower serum magnesium levels [29]. In the European Prospective Investigation into Cancer (EPIC)-Norfolk Study, dietary magnesium was also inversely associated with risk of stroke [30]. Our results for CVD are consistent with these earlier findings.

In recent years, concerns about calcium supplements and elevated cardiovascular risk [31] have led to declines in usage. However, dietary calcium may have very different effects than supplements. Data from the large Melbourne Collaborative Cohort Study suggest inverse linear associations between dietary calcium and risk of incident CVD and stroke [32]. However, a meta-analysis of prospective observational studies found that the lowest CVD mortality was at intakes of approximately 800 mg/d, restricted cubic spline analyses in that study suggested that the relation between calcium and CVD mortality may be U-shaped [33]. In the current Framingham analyses, calcium intakes above 500 mg/d were associated with nearly a 20% (non-statistically significant) reduction in risk of CVD. We found no particular benefit of higher total calcium intakes.

The mechanisms by which these dietary minerals may be associated with CVD are complex. For sodium, it seems likely that the effects on blood pressure are mechanistically different for individuals who are salt sensitive and those who are not. Salt sensitivity has been acknowledged for many years but the mechanisms underlying this phenomenon are incompletely understood [34]. Since salt sensitivity varies markedly by race, the largely northern European Caucasian ancestry of the Framingham Study participants suggests that salt sensitivity levels are probably somewhat low in this cohort.

There are a number of mechanisms by which potassium may reduce cardiovascular risk. First, sufficient potassium intake promotes negative sodium balance by inducing sodium excretion [35]. Increased plasma potassium levels also have beneficial effects on endothelial cells, thereby reducing vascular stiffness and enhancing nitric oxide-mediated vasodilation [36,37]. In addition to these independent effects on blood pressure and other cardiovascular outcomes, potassium has been shown to suppress the effects of sodium on blood pressure among salt sensitive individuals [6]. Magnesium regulates numerous biochemical processes, many of which control blood glucose, blood pressure, and inflammation [38]. It also acts as a calcium antagonist and inhibits coagulation. Magnesium deficiency has been associated with increased oxidative stress [39].

It is important to note that since potassium and magnesium are strongly correlated with one another and strongly associated with consumption of fruits and vegetables and dairy products (and hence, a DASH dietary pattern), it is difficult to separate the effects of these minerals from one another and their underlying food sources. In Framingham, 29% of potassium intake and 21% of magnesium intake was derived from vegetables, while 19% of potassium and 11% of magnesium came from fruit consumption and 16–17% of both potassium and magnesium from dairy intake. In summary, these data from the FOS confirm that dietary potassium and magnesium have important roles in the prevention of CVD. This finding may provide additional support for the value of a DASH eating pattern in the reduction of CVD risk. Finally, sodium intake had no independent effect on risk of CVD and dietary calcium had only weak beneficial effects.

There are important strengths and limitations to the present study. One limitation is that we had very few individuals with sodium intakes above 5000 mg/d (n = 88, 3.7%) or below 2000 mg/d (n = 210, 8.9%), preventing accurate estimation of the dose-response relations at extremes of the distribution. Nonetheless, we found no adverse effect of sodium intake on CVD risk at the usual levels of intake (mean intake = 2977 mg/d) in this population. Further, we were unable to examine the potential role of kidney dysfunction in this study due to having too few individuals with creatinine measures at baseline. However, when we examined the proportion of individuals with renal failure at exam 7, when nearly all individuals had serum creatinine levels, the proportions were not different across sodium or potassium intake categories. Additionally, we did not consider the effects of supplemental calcium in this study. The FOS population is largely of European descent, limiting the generalizability of these results to more racially-diverse populations. Finally, self-reported dietary records are an imperfect measurement of salt intake, and may be subject to reporting bias and underreporting of sodium intake among high-risk individuals. There also may be non-differential error in reporting of the intakes of these minerals, which would result in estimates of effect that were biased towards the null. There are many important strengths of this study. Even though we did not have urinary biomarkers for mineral intakes, the dietary recalls have been shown to have stronger correlations with more objective measures of sodium and potassium intake than have other methods of dietary assessment [40]. Additionally, the FOS's well-characterized population, carefully-adjudicated cardiovascular outcomes, and thorough measurement of a wide range of potential confounding factors allow for more accurate assessment of outcomes and control of confounding.

5. Conclusions

Our data provide no evidence for sodium reduction in a healthy population as a means of reducing risk of CVD in a population with moderate sodium intake. It does however support the importance of increasing potassium and magnesium for the purpose of reducing cardiovascular risk. Interventions focused on the promotion of potassium- and magnesium-rich foods in the diet may be effective targets for reducing the occurrence of CVD. To improve generalizability and expand these findings, future studies should examine broader ranges of intakes in more ethnically diverse cohorts.

Author Contributions: Conceptualization, R.T.P. and L.L.M.; methodology, R.T.P., L.L.M., and M.R.S.; formal analysis, M.R.S. and R.T.P.; data curation, M.R.S.; writing—original draft preparation, R.T.P., M.L.B., and L.L.M.; writing—review and editing, R.T.P., M.L.B., and L.L.M.; supervision,

L.L.M.; funding acquisition, L.L.M. All authors have read and agreed to the published version of the manuscript.

Funding: This research was funded by the National Heart, Lung and Blood Institute (NHBLI; grant numbers 5T32HL125232, N01-HC-25195 and HHSN268201500001I) and a grant from the National Dairy Council, grant number 2497.

Institutional Review Board Statement: This study was conducted according to the Declaration of Helsinki and Approved by the Institutional Review Board of the Boston University Medical Center (IRB H-32132).

Informed Consent Statement: Informed Consent was obtained from all subjects involved in the study.

Data Availability Statement: Restrictions apply to the availability of these data. Data was obtained from the Framingham Heart Study and can be requested at https://framinghamheartstudy.org/fhs-for-researchers/data-available-overview/.

Conflicts of Interest: The authors declare no conflict of interest and the funders had no role in the design of the study; in the collection, analyses, or interpretation of data; in the writing of the manuscript, or in the decision to publish the results.

References

1. Beaglehole, R.; Bonita, R.; Horton, R.; Adams, C.; Alleyne, G.; Asaria, P.; Baugh, V.; Bekedam, H.; Billo, N.; Casswell, S.; et al. Priority actions for the non-communicable disease crisis. *Lancet* **2011**, *377*, 1438–1447. [CrossRef]
2. IOM (Institute of Medicine). *Sodium Intake in Populations: Assessment of Evidence*; The National Academies Press: Washington, DC, USA, 2013. [CrossRef]
3. Mahtani, K.R.; Heneghan, C.; Onakpoya, I.; Tierney, S.; Aronson, J.K.; Roberts, N.; Hobbs, F.D.R.; Nunan, D. Reduced salt intake for heart failure: A systematic review. *JAMA Intern. Med.* **2018**, *178*, 1693–1700. [CrossRef] [PubMed]
4. Bailey, R.L.; Parker, E.A.; Rhodes, D.G.; Goldman, J.D.; Clemens, J.C.; Moshfegh, A.J.; Thuppal, S.V.; Weaver, C.M. Estimating sodium and potassium intakes and their ratio in the American diet: Data from the 2011–2012 NHANES. *J. Nutr.* **2016**, *146*, 745–750. [CrossRef] [PubMed]
5. O'Donnell, M.; Mente, A.; Rangarajan, S.; McQueen, M.J.; Wang, X.; Liu, L.; Yan, H.; Lee, S.F.; Mony, P.; Devanath, A.; et al. Urinary sodium and potassium excretion, mortality, and cardiovascular events. *N. Engl. J. Med.* **2014**, *371*, 612–623. [CrossRef] [PubMed]
6. Morris, R.C.; Sebastian, A.; Forman, A.; Tanaka, M.; Schmidlin, O. Normotensive salt sensitivity: Effects of race and dietary potassium. *Hypertension* **1999**, *33*, 18–23. [CrossRef] [PubMed]
7. Chung, M.; Tang, A.M.; Fu, Z.; Wang, D.D.; Newberry, S.J. Calcium intake and cardiovascular disease risk: An updated systematic review and meta-analysis. *Ann. Intern. Med.* **2016**, *165*, 856–866. [CrossRef]
8. Kass, L.; Weekes, J.; Carpenter, L. Effect of magnesium supplementation on blood pressure: A meta-analysis. *Eur. J. Clin. Nutr.* **2012**, *66*, 411–418. [CrossRef]
9. Veronese, N.; Watutantrige-Fernando, S.; Luchini, C.; Solmi, M.; Sartore, G.; Sergi, G.; Manzato, E.; Barbagallo, M.; Maggi, S.; Stubbs, B. Effect of magnesium supplementation on glucose metabolism in people with or at risk of diabetes: A systematic review and meta-analysis of double-blind randomized controlled trials. *Eur. J. Clin. Nutr.* **2016**, *70*, 1354–1359. [CrossRef]
10. US Department of Health and Human Services and US Department of Agriculture. *2015–2020 Dietary Guidelines for Americans*, 8th ed.; 2015. Available online: https://health.gov/dietaryguidelines/2015/guidelines/ (accessed on 1 February 2020).
11. Hu, F.B.; Stampfer, M.J.; Manson, J.E.; Rimm, E.; Colditz, G.A.; Rosner, B.A.; Hennekens, C.H.; Willett, W.C. Dietary fat intake and the risk of coronary heart disease in women. *N. Engl. J. Med.* **1997**, *337*, 1491–1499. [CrossRef]
12. Joshipura, K.J.; Ascherio, A.; Manson, J.E.; Stampfer, M.J.; Rimm, E.B.; Speizer, F.E.; Hennekens, C.H.; Spiegelman, D.; Willett, W.C. Fruit and vegetable intake in relation to risk of ischemic stroke. *JAMA* **1999**, *282*, 1233–1239. [CrossRef]
13. Schakel, S.F.; Sievert, Y.A.; Buzzard, I.M. Sources of data for developing and maintaining a nutrient database. *J. Am. Diet. Assoc.* **1988**, *88*, 1268–1271. [PubMed]
14. Dunn, J.E.; Liu, K.; Greenland, P.; Hilner, J.E.; Jacobs, D.R., Jr. Seven-year tracking of dietary factors in young adults: The CARDIA study. *Am. J. Prev. Med.* **2000**, *18*, 38–45. [CrossRef]
15. Moore, L.L.; Visioni, A.J.; Qureshi, M.M.; Bradlee, M.L.; Ellison, R.C.; D'Agostino, R. Weight loss in overweight adults and the long-term risk of hypertension: The Framingham study. *Arch. Intern. Med.* **2005**, *165*, 1298–1303. [CrossRef] [PubMed]
16. Kannel, W.B.; Sorlie, P. Some health benefits of physical activity. The Framingham Study. *Arch. Intern. Med.* **1979**, *139*, 857–861. [CrossRef]
17. Lafay, L.; Mennen, L.; Basdevant, A.; Charles, M.A.; Borys, J.M.; Eschwege, E.; Romon, M. Does energy intake underreporting involve all kinds of food or only specific food items? Results from the Fleurbaix Laventie Ville Sante (FLVS) study. *Int. J. Obes. Relat. Metab. Disord.* **2000**, *24*, 1500–1506. [CrossRef]

18. Poppitt, S.D.; Swann, D.; Black, A.E.; Prentice, A.M. Assessment of selective under-reporting of food intake by both obese and non-obese women in a metabolic facility. *Int. J. Obes. Relat. Metab. Disord.* **1998**, *22*, 303–311. [CrossRef]
19. Brown, C.C.; Kipnis, V.; Freedman, L.S.; Hartman, A.M.; Schatzkin, A.; Wacholder, S. Energy adjustment methods for nutritional epidemiology: The effect of categorization. *Am. J. Epidemiol.* **1994**, *139*, 323–338. [CrossRef]
20. National Academies of Sciences Engineering and Medicine (U.S.). Food and Nutrition Board. Dietary reference intakes for sodium and potassium. In *Consensus Study Report of the National Academies of Sciences, Engineering, Medicine*; Stallings, V.A., Quirk, M., Oria, M., Eds.; The National Academies Press: Washington, DC, USA, 2019; p. 1.
21. Alderman, M.H.; Cohen, H.W. Dietary sodium intake and cardiovascular mortality: Controversy resolved? *Am. J. Hypertens.* **2012**, *25*, 727–734. [CrossRef]
22. Tuomilehto, J.; Jousilahti, P.; Rastenyte, D.; Moltchanov, V.; Tanskanen, A.; Pietinen, P.; Nissinen, A. Urinary sodium excretion and cardiovascular mortality in Finland: A prospective study. *Lancet* **2001**, *357*, 848–851. [CrossRef]
23. Kalogeropoulos, A.P.; Georgiopoulou, V.V.; Murphy, R.A.; Newman, A.B.; Bauer, D.C.; Harris, T.B.; Yang, Z.; Applegate, W.B.; Kritchevsky, S.B. Dietary sodium content, mortality, and risk for cardiovascular events in older adults: The Health, Aging, and Body Composition (Health ABC) Study. *JAMA Intern. Med.* **2015**, *175*, 410–419. [CrossRef]
24. Graudal, N.; Jürgens, G.; Baslund, B.; Alderman, M.H. Compared with usual sodium intake, low- and excessive-sodium diets are associated with increased mortality: A meta-analysis. *Am. J. Hypertens.* **2014**, *27*, 1129–1137. [CrossRef] [PubMed]
25. Graudal, N.A.; Hubeck-Graudal, T.; Jurgens, G. Effects of low-sodium diet vs. high-sodium diet on blood pressure, renin, aldosterone, catecholamines, cholesterol, and triglyceride (Cochrane Review). *Am. J. Hypertens.* **2012**, *25*, 1–15. [CrossRef] [PubMed]
26. Kieneker, L.M.; Gansevoort, R.T.; de Boer, R.A.; Brouwers, F.P.; Feskens, E.J.; Geleijnse, J.M.; Navis, G.; Bakker, S.J.; Joosten, M.M.; Group, P.S. Urinary potassium excretion and risk of cardiovascular events. *Am. J. Clin. Nutr.* **2016**, *103*, 1204–1212. [CrossRef] [PubMed]
27. D'Elia, L.; Barba, G.; Cappuccio, F.P.; Strazzullo, P. Potassium intake, stroke, and cardiovascular disease a meta-analysis of prospective studies. *J. Am. Coll. Cardiol.* **2011**, *57*, 1210–1219. [CrossRef] [PubMed]
28. Comstock, G.W. Water hardness and cardiovascular diseases. *Am. J. Epidemiol.* **1979**, *110*, 375–400. [CrossRef]
29. Ma, J.; Folsom, A.R.; Melnick, S.L.; Eckfeldt, J.H.; Sharrett, A.R.; Nabulsi, A.A.; Hutchinson, R.G.; Metcalf, P.A. Associations of serum and dietary magnesium with cardiovascular disease, hypertension, diabetes, insulin, and carotid arterial wall thickness: The ARIC study. Atherosclerosis Risk in Communities Study. *J. Clin. Epidemiol.* **1995**, *48*, 927–940. [CrossRef]
30. Bain, L.K.; Myint, P.K.; Jennings, A.; Lentjes, M.A.; Luben, R.N.; Khaw, K.T.; Wareham, N.J.; Welch, A.A. The relationship between dietary magnesium intake, stroke and its major risk factors, blood pressure and cholesterol, in the EPIC-Norfolk cohort. *Int. J. Cardiol.* **2015**, *196*, 108–114. [CrossRef]
31. Reid, I.R.; Birstow, S.M.; Bolland, M.J. Calcium and cardiovascular disease. *Endocrinol. Metab.* **2017**, *32*, 339–349. [CrossRef]
32. Khan, B.; Nowson, C.A.; Daly, R.M.; English, D.R.; Hodge, A.M.; Giles, G.G.; Ebeling, P.R. Higher dietary calcium intakes are associated with reduced risks of fractures, cardiovascular events, and mortality: A prospective cohort study of older men and women. *J. Bone Miner. Res.* **2015**, *30*, 1758–1766. [CrossRef]
33. Wang, X.; Chen, H.; Ouyang, Y.; Liu, J.; Zhao, G.; Bao, W.; Yan, M. Dietary calcium intake and mortality risk from cardiovascular disease and all causes: A meta-analysis of prospective cohort studies. *BMC Med.* **2014**, *12*, 158. [CrossRef]
34. Elijovich, F.; Weinberger, M.H.; Anderson, C.A.; Appel, L.J.; Bursztyn, M.; Cook, N.R.; Dart, R.A.; Newton-Cheh, C.H.; Sacks, F.M.; Laffer, C.L.; et al. Salt sensitivity of blood pressure: A scientific statement from the American Heart Association. *Hypertension* **2016**, *68*, e7–e46. [CrossRef] [PubMed]
35. Gallen, I.W.; Rosa, R.M.; Esparaz, D.Y.; Young, J.B.; Robertson, G.L.; Batlle, D.; Epstein, F.H.; Landsberg, L. On the mechanism of the effects of potassium restriction on blood pressure and renal sodium retention. *Am. J. Kidney Dis.* **1998**, *31*, 19–27. [CrossRef] [PubMed]
36. Oberleithner, H.; Callies, C.; Kusche-Vihrog, K.; Schillers, H.; Shahin, V.; Riethmuller, C.; Macgregor, G.A.; de Wardener, H.E. Potassium softens vascular endothelium and increases nitric oxide release. *Proc. Natl. Acad. Sci. USA* **2009**, *106*, 2829–2834. [CrossRef] [PubMed]
37. Penton, D.; Czogalla, J.; Loffing, J. Dietary potassium and the renal control of salt balance and blood pressure. *Pflug. Arch.* **2015**, *467*, 513–530. [CrossRef]
38. Nielsen, F.H. Magnesium deficiency and increased inflammation: Current perspectives. *J. Inflamm. Res.* **2018**, *11*, 25–34. [CrossRef]
39. Gröber, U.; Schmidt, J.; Kisters, K. Magnesium in prevention and therapy. *Nutrients* **2015**, *7*, 8199–8226. [CrossRef]
40. Yuan, C.; Spiegelman, D.; Rimm, E.B.; Rosner, B.A.; Stampfer, M.J.; Barnett, J.B.; Chavarro, J.E.; Rood, J.C.; Harnack, L.J.; Sampson, L.K.; et al. Relative Validity of Nutrient Intakes Assessed by Questionnaire, 24-Hour Recalls, and Diet Records as Compared With Urinary Recovery and Plasma Concentration Biomarkers: Findings for Women. *Am. J. Epidemiol.* **2018**, *187*, 1051–1063. [CrossRef]

nutrients

Review

Role of Cachexia and Fragility in the Patient Candidate for Cardiac Surgery

Calogera Pisano [1,*,†], Daniele Polisano [2,†], Carmela Rita Balistreri [3,†], Claudia Altieri [1], Paolo Nardi [1], Fabio Bertoldo [1], Daniele Trombetti [1], Laura Asta [1], Maria Sabrina Ferrante [1], Dario Buioni [1], Calogero Foti [2,‡] and Giovanni Ruvolo [1,‡]

1 Department of Cardiac Surgery, Tor Vergata University Hospital, 00133 Rome, Italy; claudia.altieri@ptvonline.it (C.A.); pa.nardi4@libero.it (P.N.); fabio.bertoldo@uniroma2.it (F.B.); daniele.trombetti@famigliatrombetti.it (D.T.); astalaura92@gmail.com (L.A.); fesab882011@libero.it (M.S.F.); docyuk@libero.it (D.B.); giovanni.ruvolo@uniroma2.it (G.R.)

2 Physical and Rehabilitation Medicine, Tor Vergata University of Rome, 00133 Rome, Italy; daniele.polisano94@gmail.com (D.P.); foti@med.uniroma2.it (C.F.)

3 Department of Biomedicine, Neuroscience and Advanced Diagnostics (Bi.N.D.), University of Palermo, 90133 Palermo, Italy; carmelarita.balistreri@unipa.it

* Correspondence: lindapisano82@gmail.com; Tel.: +39-328-329-7692; Fax: +39-(06)-2090-3538

† These authors contributed equally to this work.

‡ These authors contributed equally to this work.

Abstract: Frailty is the major expression of accelerated aging and describes a decreased resistance to stressors, and consequently an increased vulnerability to additional diseases in elderly people. The vascular aging related to frail phenotype reflects the high susceptibility for cardiovascular diseases and negative postoperative outcomes after cardiac surgery. Sarcopenia can be considered a biological substrate of physical frailty. Malnutrition and physical inactivity play a key role in the pathogenesis of sarcopenia. We searched on Medline (PubMed) and Scopus for relevant literature published over the last 10 years and analyzed the strong correlation between frailty, sarcopenia and cardiovascular diseases in elderly patient. In our opinion, a right food intake and moderate intensity resistance exercise are mandatory in order to better prepare patients undergoing cardiac operation.

Keywords: frailty; vascular aging; age related syndrome; sarcopenia; malnutrition

Citation: Pisano, C.; Polisano, D.; Balistreri, C.R.; Altieri, C.; Nardi, P.; Bertoldo, F.; Trombetti, D.; Asta, L.; Ferrante, M.S.; Buioni, D.; et al. Role of Cachexia and Fragility in the Patient Candidate for Cardiac Surgery. *Nutrients* **2021**, *13*, 517. https://doi.org/10.3390/nu13020517

Academic Editor: Annalisa Noce

Received: 20 December 2020

Accepted: 29 January 2021

Published: 5 February 2021

Publisher's Note: MDPI stays neutral with regard to jurisdictional claims in published maps and institutional affiliations.

1. Introduction

The concept of frailty was first evidenced in the 1979 [1] and entered in the common medical language, thanks to recognized value in predicting the risk to many chronic diseases in old population, evidencing the marked differences in the two sexes (especially in female people), with respect to the traditional risk factors for these diseases, and in facilitating (or precisely quantifying) the increase of health age-related deficits. Nevertheless, its definition remains uncertain, although three researchers have advanced some major proposals: (1) Fried [2] defines frailty as the process that decreases the physiological reserves and results in a major vulnerability to stressors (pathologies, surgery); (2) Rockwood [3] describes it as the result of the presence of adverse variables in old people, including those of cardiovascular nature (i.e., hypertension, heart attack and arrhythmia); (3) Gobbens [4] suggests that damages in the psychosocial sphere of an individual cause some adverse effects to the health. Currently, advances in the field propose frailty as major phenotype of accelerated aging characterized by a multiorgan dysfunction and/or significantly associated with an increased vulnerability to diverse diseases (multimorbidity) in elderly people [5]. Sarcopenia can be considered a biological substrate of physical frailty [6]. Muscle loss typically begins in the fifth decade of life and proceeds at a rate of decline of 0.8% years [7]. Epidemiological data suggest a wide variability in the prevalence of sarcopenia, depending on the type of population studied, sex, age and diagnostic criteria used. The prevalence of sarcopenia is between 7.5% and 77.6% [8]. There are numerous factors responsible for this muscle loss: the aging process, genetic susceptibility, environmental factors, such as suboptimal diet, prolonged bed rest, sedentary lifestyle, chronic diseases and drugs [9,10]. In most cases the etiology of sarcopenia is multifactorial and sarcopenia is considered PRIMARY (age-related) when the only obvious cause is aging [11–14]. Malnutrition plays a key role in the pathogenesis of sarcopenia and frailty. The malnutrition refers to an imbalance condition of protein or other nutrient imbalance, responsible for negative effects on body composition, physical function and clinical outcome [15]. Although, malnutrition is not inevitably associated with the aging process. Numerous causes can contribute to a decline in nutritional status: anorexia, edentulism, dysgeusia, dysphagia, motor and visual disability represent physiological and physical causes that can compromise an adequate intake of nutrients [16]. We will see in this narrative review the correlation between frailty, sarcopenia and malnutrition in the management of the elderly patient. At the same time, we proposed a right food intake in order to better prepare patients undergoing cardiac surgery.

2. Materials and Methods

2.1. Data Sources and Search Strategy

Current literature investigating frailty, sarcopenia and malnutrition is analyzed and contextualized in this review. Specifically, research was conducted on Medline (Pubmed) and Scopus. To review recent studies on frailty, sarcopenia, malnutrition, and cardiovascular disease, we selected scientific papers published in English 10 years since the European Working Group on Sarcopenia in Older People (EWGSOP) was published in 2010 [17]. We used the search term frailty, sarcopenia, malnutrition, cachexia, cardiovascular disease, mortality and morbidity, cardiac surgery.

2.2. Study Selection

2.2.1. Inclusion Criteria

The inclusion criteria for the included studies in this review were as follows: (1) assessment of frailty and sarcopenic patients; (2) inclusion of both gender and all races; (3) examination of the impact of undernutrition, sarcopenia and frailty on clinical outcomes; (4) frailty evaluation; (5) evaluation of muscle strength and/or muscle mass for diagnosing sarcopenia; (6) evaluation of the correlation between frailty/sarcopenia and cardiovascular diseases; (7) identification of frailty biomarkers in predicting vascular aging and cardiovascular disease; (8) morbidity and mortality in frailty patients underwent cardiac surgery; (9) application of a specific dietary intake in order to prevent sarcopenia in cardiac surgery patients.

2.2.2. Exclusion Criteria

Editorial, case report, letters to editor, and conference abstracts were excluded from this review.

3. Frailty Definition and Quantification

Two main models have been proposed for the frailty evaluation: the phenotypic (primary frailty) or the deficits accumulation model (secondary frailty). Different instruments have been proposed for measuring frailty. Of note are the data from the Cardiovascular Health Study [18] that evidenced in about 25% of older participants signs of frailty without either multiple comorbidities or disabilities (physiological ageing). In this context, frailty has been defined as "primary frailty", with a phenotypic presentation involving the decline in physical functions and psychological status, without taking into consideration associated diseases or pathological conditions. In measuring the primary frailty, the Fried's phenotype frailty index has been widely adopted [2]. It derived from an analysis of five health factors: nutrition; physical exhaustion; low energy expenditure (or inactive status); mobility and muscular strength. Deterioration in one of these examined factors was scored as 1 if present or 0 if absent, giving a potential score spanning from 0 to 5.

The phenotypic model permitted to classify three groups of individuals: robust (no deterioration); pre-frail (one or two altered factors); or frail (three or more altered factors). This classification was independently correlated with outcomes, such as survival, falls, disability, and institutionalization. The secondary frailty considered the accumulation of multiple deficits, including symptoms, signs, disabilities, pathological conditions, and abnormal laboratory values. Furthermore, its evaluation was based on frailty index based on the defects accumulation's model [19]. Every deficit has been coded as binary (1 or 0) or ordinal (0, 0.5, 1), consequently the frailty index was the sum of the deficit's values divided by the total number of deficits listed. This approach evidenced an important issue on measurement of frailty based on phenotypic or deficits accumulation model, that has revealed it is complex and time consuming. Alternative and easier instruments have been, subsequently, proposed for fragility assessment both in general population or in clinical practice [20]. For frailty estimation in general population, two simple scales and multifaceted tools requiring comprehensive geriatric assessment (CGA) have been introduced. Among the scales, the most important tool adopted is the Edmonton frailty scale (EFS) [21]. It examines nine domains of frailty (cognition, general health status, functional independence, social support, medication usage, nutrition, mood, continence, functional performance). The results have been reported in a scale ranging from 0 to 17 values and the participants have been conventionally classified into three categories. A higher score has been associated with a higher degree of frailty. The same nine domains of frailty have been also assessed by using a specific tool requiring comprehensive geriatric assessment. For example, this is the case of the Mini-Mental State Examination (MMSE) [22] or the Geriatric Depression Scale (GDS) [23,24]. In contrast, the functional performance has been detected by using the Handgrip strength test [25,26]: handgrip measurement is assessed on the dominant hand using a Jamar dynamometer adhering to the standardized protocol recommended by the American Society of Hand Therapists and the average value of the handgrip in the two genders is used to define the scores. Thus, a lower score than 30 kg for man and lower than 20 kg for women is considered weak [27]. The valuation of the nutritional status has been performed using the Mini Nutritional Assessment (MNA) [28]. The MNA is composed of 18 items divided in four categories: anthropometric assessment, general state, dietary assessment, and self-assessment. A score ≥24 points indicates a good nutritional status. A score from 17 to 23.5 points is an indicator of a risk of malnutrition, while a score ≤17 points indicates malnutrition. In the appraisal of the general health status, the assessment of the sarcopenia [29], osteoporosis [30], and serum albumin [31] is important. Beside the instruments used to estimate fragility in general population, it is necessary to mention the main tests used in hospital environment. Among these the *SHARE-FI scale* [32] (Survey on Health, Ageing and Retirement in Europe Frailty Instrument) is one of the most counted. This instrument is based on the first wave of the Survey of Health, Ageing and Retirement in Europe, a large population-based survey (n = 31,115) conducted in 2004–2005 in 12 European countries. It measures five variables approximating Fried's frailty definition (exhaustion, weight loss, weakness, slowness, and low activity) with different coefficients for men and women. Another important scale is the SPPB [33] (Short Physical Performance Battery). This battery analyzes physical performance features as 4-m gait speed, balance capacity and sit-to-stand time. The lower extremity performance is a long-term mortality predictor independent of NYHA class and ejection fraction in elderly hospitalized patients. The EFT [34] (Essential Frailty Instrument) is a specific scale used in cardiac surgery. This scale includes biological, physical and mental state thus it may identify subclinical frailty; in fact, its features are albumin and hemoglobin blood values, sit-to-stand time and MMSE test. In the context of cardiac surgery two important risk scores have been introduced in order to predict 30-day mortality and 1-year mortality: respectively, the Comprehensive assessment of frailty [35] (CAF) and the Frailty predicts death One years after Elective Cardiac Surgery Test (FORECAST) [36]. The CAF is composed of different items to quantify the physical performance and coordinative abilities of the patients in addition to scores that are already used to define frailty in medicine.

In addition, several laboratory tests result as creatinine and FEV_1. The FORECAST is a simple version of the CAF with a higher predictive power. It is composed of those five test items (chair rise, weak, stair, clinical frailty scale from the Canadian Study of Health and Aging, serum creatinine level).

4. Sarcopenia as Biological Substrate of Physical Frailty

Sarcopenia is considered the biological substrate of physical frailty. The prevalence of sarcopenia is higher in male with low body max index (BMI) [37,38]. Sarcopenia is a common condition in the elderly but can also be seen in younger patients. It is defined *primary* or age related when no cause is highlighted, other than the aging process. It is considered *secondary* when one or more causes are identifiable and in this case it is called activity-related, disease-related, nutrition-related [39]. Sarcopenia is a syndrome characterized by the progressive and generalized loss of mass, muscle strength, physical performance, which leads to an increased risk of disability, poor quality of life, falls, numerous complications, and death [40]. Muscle trophism is a consequence of a balance between anabolic triggers (insulin, physical exercise, amino acids, adrenaline, testosterone) and catabolic triggers (cortisol, catecholamines, glucagon, cytokines, intense exercise) [41]. In the elderly, the catabolic state is associated with the normal aging process, which becomes predominant when particular conditions of comorbidity are concomitant. In these cases, the muscle mass suffers the effects of the general catabolic state in which the body is found. Several factors contribute to the pathophysiology of sarcopenia [42]. In particular the main factor are reduction of sex hormone levels, reduction of growth hormone levels, increased production of cytokines, interleukin-1 (IL-1), interleukin-6 (IL-6), Tumor Necrosis Factor-Alpha (TNF-alpha), alteration of the cellular redox-status, neuromuscular changes, physical inactivity, and malnutrition [43,44]. Drugs can also play a protective or causative role in the development of sarcopenia. Statins, sulfonylureas, glinides have a potential harmful effect on muscle metabolism; while ACE inhibitors, allopurinol, Vitamin D play a protective role on muscle function [45,46]. The muscle is formed of different types of muscle fibers: slow fibers (type I) and fast fibers (types IIa and IIb). With aging, especially in sarcopenic patients, there is a reduction in the diameter of muscle fibers, as well as a progressive loss of fast fibers which results in a reduction in strength, coordination of movements, and walking speed. This happens because the lost fast muscle fibers are replaced by slow fibers. Given the dynamic nature of neuromuscular remodeling, it has been seen that the muscle of the elderly subject under certain triggers maintains the ability to respond and to adapt to the new state. It has been shown that even lifestyle alone can greatly influence the development of muscle mass. This means that effective therapeutic intervention could be applied in order to reverse the processes that lead to sarcopenia [47,48].

5. Sarcopenia Diagnosis

The simultaneous presence of muscle mass loss associated with reduced muscle strength or physical performance is recommended for the diagnosis of sarcopenia. There are various methods for the assessment of the sarcopenia.

The muscle mass usually is calculated using the Impedancemetry [49]. This is a valid and recognized alternative to more complex and expensive methods, such as magnetic resonance imaging (MRI), dual X-ray absorptiometry (DEXA) and computed tomography (CT). The exam lasts a few minutes, is absolutely painless, safe and allows you to know the body areas in terms of fat, lean tissue and water content. It is based on the principle that tissues full of water and electrolytes offer less resistance to the passage of an electric current than adipose tissue. The result is then compared with the reference values obtained according to normalization formulas for race, age, sex, body weight. Muscle strength is measured through the Handgrip (dynamometer), a simple tool that evaluates the force developed by gripping the hand; usually three tests are performed, of which the best is chosen [50]. The result is compared with threshold values calculated according to age, sex and BMI. Physical performance can be assessed through the quick and easy walking speed test [51]. The main symptoms related to sarcopenia are muscle weakness and fatigue. This concept does not only concern the bedridden people but also the person who has functional autonomy. This condition does not only concern the thin and undernourished patient, but even obese patient with increased body max index that have a reduction in muscle mass. This scenario is called sarcopenic obesity [52]. Sarcopenic obesity is related to an increased cardiovascular risk, due to the unfavorable metabolic effects of the increased visceral adipose component. The muscle tissue is one of the major contributors to the peripheral action of insulin on the uptake of circulating glucose. The sarcopenic patient also has a condition of insulin resistance which can contribute to establishing and maintaining harmful metabolic circulation.

6. Role of Frailty and Sarcopenia in Cardiovascular Disease

Cardiovascular disease (CVD), both clinical and subclinical, has been proposed as one of the pathological conditions associated with frailty [53]. The Women's Health Initiative Observational Study has been the first and largest study to confirm that CVD was a risk factor for developing frailty [54]. The relationship between these conditions indicates a common pathophysiological mechanism characterized by an abnormal inflammatory response, resulting in an increase in inflammatory markers, leading to chronic inflammation [55]. On the other hand, chronic inflammation is the main cause of endothelial dysfunction that leads to onset of cardiovascular diseases. The association between endothelial dysfunction and frailty confirms the role of CVD in frailty. This suggests its relevance since the early stages of vascular dysfunction (in case of only functional impairment) are apparent [56]. The endothelial dysfunction has been associated with a lot of cardiovascular risk factors: hypertension, hypercholesterolemia, diabetes mellitus, obesity, smoking. Ricci et al. [57] in a population-based study assessed that frail and pre-frail older people corresponded to a substantial proportion of those with greater CVD risk factor. In particular, diabetes mellitus (DM) seems to be the most prevalent CVD risk factor in frail and pre-frail older people. It has been demonstrated that DM is one of the strongest risk factors for atherosclerosis and, consequently, diabetic individuals present an increased risk, 3–4 times higher, of developing CVD and a double risk of mortality when compared to general population. Cacciatore et al. [58] in a 12-year survival analysis study, showed that frail individuals were 2.6 times more likely to have a complication related to DM, regarding age, sex, and number of years living with this pathology. The relationship between DM and frailty seems to be influenced by the sarcopenia. The muscle impairment in diabetic people is the result of fat infiltration in the muscle tissue, higher insulin resistance levels, the increased levels of cytokines, and reduction in motor and plates. Other important CVD risk factors in older people are hypertension and smoking, regardless of the frailty classification. No relationship has been found between frailty, obesity and waist circumference. Anyway, the concept of sarcopenic obesity behind the BMI has been introduced: frail people are characterized by sarcopenia and fat infiltration of muscle. This increased fat tissue allows the production of proinflammatory cytokines and mediators, such as interleukin-6 and C-reactive protein, which induces a state of chronic inflammation present in the frailty syndrome [59]. Therefore, it is interesting to note that aging per se induces endothelial dysfunction, in absence of cardiovascular risk factor and CVD related to increased oxidative stress and proinflammatory profile [60].

7. Crosstalk between Frailty and Cardiovascular Diseases in Molecular Mechanisms Level

Recent literature data suggest considering the pathophysiological mechanisms involved in the development or progression of a frailty status, for identifying frailty biomarkers [61,62]. In the context of vascular aging related to frail phenotype, several mechanisms are strongly associated with the onset of cardiovascular disease [63]. For example, inflammation is the predominant mechanism in vascular aging that induces the activation of endothelial and vascular smooth muscle cells and the migration in the wall of leukocytes. They may evolve in atherosclerosis condition or in other degenerative pathological age-related conditions. According a recent review, there are 44 most important biomarkers related to frailty. They propose a core panel of 19 high priority markers and an expanded panel with 22 medium priority markers [64]. In addition, three low priority markers are reported. These markers might be assembled in different groups according the mechanism in which they are involved: inflammation, mitochondria and apoptosis, calcium homeostasis, fibrosis, neuromuscular junction and neurons, cytoskeleton and hormones, and others. Alterations in immune system seem to be one of the most important triggers related to vascular aging. According Monti et al. [65], the aging process is related to a systemic increase in proinflammatory mediators from various sources. In addition, aging induces important changes in immune cell phenotypes and function, called "immunosenescence". It is characterized by a shift from lymphoid to myeloid differentiation was described for B and T cell populations. Equally, there is a change in the function and receptor signaling (i.e., Toll-like receptors and RAGE) in monocytes, macrophages, dentritic cells, and neutrophils. Moreover, immune cells go through "immunosenescence" process [66]. These cells change their surface marker expression, reduce the production of reactive oxygen species (ROS) and their migration capacity, increase the production of proinflammatory over anti-inflammatory cytokines. All these events induce the release of inflammatory molecules that might be used as vascular aging and fragility biomarkers. The most important inflammatory markers are CD14 antigen also known as myeloid cell-specific leucine rich glycoprotein [67]; CX3CL1 (C-X3-C motif chemokine ligand 1, aka fractalkine) [68,69]; pentraxin [70,71]; sVCAM (soluble vascular cell adhesion molecule 1/soluble intercellular adhesion molecule 1) [72,73]; IL-6 (interleukin 6) [74,75]; CXCL 10 (C-X-C motif chemokine 10) [76]; defensins (a large family of antimicrobial and cytotoxic peptides involved in host defense and in immunomodulation) [77]. Among these seven potential inflammatory biomarkers for frailty, three seem to have high priority (IL-6, CXCL10, CX3CL1), three medium priority (pentraxin, sVCAM/sICAM, defensin), and one low priority candidate (CD14) [64]. The other group of frailty biomarkers is related to the impairment of mitochondrial function and apoptosis typical of several ageing disorders [78]. Among these the most important are GDF15 (growth differentiation factor 15 or myomitokine) [79]; FNDC5 (fibronectin type III domain containing 5) [80]; Vimentin (type III intermediate filament protein) [81]; APP (amyloid precursor protein beta) [82]; LDH (lactate dehydrogenase) [83]. From the five markers in the mitochondria and apoptosis category, the profile of GDF15, FNDC5 and vimentin are considered high priority biomarkers [64]. In addition, because in aged people the calcium homeostasis is usually altered, changes in calcium signaling and/or binding proteins have been proven to be effective markers of cellular and tissue dysfunction in these patients. The three "calcium homeostasis" biomarkers of fragility are S100B (S 100 calcium binding protein B) [84]; SMP30 (senescence-marker protein 30) [85]; calreticulin (a multifunctional protein initially identified as a Ca^{2+} storage protein) [86]. Both regucalcin and calreticum reached high priority. Another important group of biomarkers is related to fibrotic changes that various tissues show with age. One of the most important factors involved in the fibrosis is the transforming growth factor-ß (TGF-ß) [87]. A higher concentration of TGF-ß seems to be related to various diseases associated to age and fragility such as atherosclerosis, acute and chronic liver and kidney disease, autoimmunity, osteoarthritis, and neurodegenerative diseases [88]. Other fibrosis markers related to the TGF-ß pathway activation are PAI-19 (plasminogen activator inhibitor1) [89]; PLAU (urokinase plasminogen activator) [90]; MMP7 (matrix metalloproteinases 7) [91]; TGM2 (transglutaminase 2) [92]; THBS2 (thrombospondin 2) [93]; AGT100 (angiotensinogen) [94]. Yet, with the aging syndrome there is a damage of the cell cytoskeleton that leads to hormones dysregulation [95]. Consequently, a lot of hormones could likely be used as frailty biomarkers, such as the GH (growth hormone), IGF (insulin-like growth factor-1), FGF23 (fibroblast growth factor 23), resistin, adiponectine, leptin, and ghrelin.

8. Clinical Impact of Frailty and Sarcopenia

Frailty is emerging as a new and more specific predictor of morbidity and mortality in patients with CVD [96] (acute myocardial infarction, heart failure, heart valve disease).

On the other hand, frailty seems to be a more accurate perioperative risk score than those currently used in patients underwent cardiac surgery and transcatheter aortic valve replacement. Recently, Graham et al. [97] analyzed the prognostic significance of frailty measure by the EFS in patients ≥65 years of age admitted to the hospital with acute coronary syndrome. Patients with higher EFS scores (i.e., more frailty) were older, had greater comorbidity, and were less suitable for revascularization. An EFS ≥7 was related to a longer duration of hospitalization and mortality compared with those with an EFS score ≤3. In a cohort of older patients hospitalized with non-ST-elevation myocardial infarction, Ekerstand et al. [98] assessed that frailty, quantified by the clinical frailty score (CFS), was an independent predictor of major adverse events (death, reinfarction, revascularization, major bleeding, stroke or renal replacement therapy, rehospitalization) at 1 month. In a follow up, Sanchis et al. [99] analyzed the relationship between clinical factors and laboratory parameters (i.e., inflammation, coagulation activation, hormonal dysregulation, nutritional status, kidney and cardiac function), frailty and a composite of death/myocardial infarction among survivors of ACS at the time of hospital discharge. Four clinical variables (age ≥75 years, female sex, ischemic heart disease, heart failure) and three laboratory variables (hemoglobin ≤125 g/L, vitamin D level ≤9 mg/dL, cystatin Clevel ≥1.2 mg/dL) had a predictor power similar to that of the Fried criteria for the composite outcome. Frailty might also have important prognostic implications also in patients with heart failure (HF). About 18–54% of patients with heart failure (HF) are frail [100]. A baseline frail state was found to independently predict incident HF [101]. Frailty was associated with higher likelihood of hospitalization for HF decompensation and 1-year mortality [102]. On the other hand, the presence of frailty at the time of left ventricular assist device implantation in patients with end stage-HF was shown to be associated with longer recovery time, a risk for rehospitalization and mortality [103]. Frailty has been increasingly recognized to have significance in predicting the risk of perioperative complications, resource use, and outcomes after cardiac surgery. In a lot of studies, frailty seems to improve and outperform conventional perioperative risk scores for predicting adverse outcomes [104]. A recent systematic review showed a significant association with postoperative mortality and major cardiac and cerebrovascular events (MACCE) and frailty [105]. In addition, Sudermann et al. [35] assessed a moderate correlation of the frailty with the EuroSCORE and the Society of Thoracic Surgeons (STS) scores to predict mortality in patients aged ≥74 years who were referred to cardiac surgery. In conditions where new scale is used (SHARE-FI scale), frailty seems to be better than the EuroSCORE II in predicting 1-year mortality. The role of frailty evaluation in predicting morbidity and mortality after cardiac surgery has been studied in different types of surgical procedures: aortic valve replacement [106], mitral valve surgery [107], coronary artery bypass surgery [108]. Finally, the most important application of the frailty evaluation in terms of prognostic factor is the transcatheter aortic valve replacement (TAVR) field [109]. In fact, patients undergoing TAVR are generally older, have multimorbidities and are frail. Stortecky et al. evaluated the association between preoperative multidimensional geriatric assessment (MGA) and 30-day and 1-year risk of MACCE and mortality among 100 patients undergoing TAVR. Nearly all domains of the MGA evaluated showed association with MACCE and death [110].

9. Dietary Intake to Prevent Sarcopenia in Patients Undergoing Cardiac Surgery

The cardiac operation is a moment of stress for ill patients. Metabolic demands and muscle breakdown are accelerated by bedrest and poor oral intake causing an important loss of muscle mass. This is particularly deleterious for frail older patients, who have lower reserves of muscle mass and strength [111]. Frailty increases the age-related changes in protein and muscle metabolism by increasing the rate of protein catabolism and decreasing the response to anabolic factors. A correct protein intake is necessary in particular in ill patients [112] (Table 1). The American Society for Parenteral and Enteral Nutrition (ASPEN) advised a protein intake of 2.0 g per kilogram of body weight per day (g/kg/d) [112]. Instead the European Society for Parenteral and Enteral Nutrition (ESPEN) recommends 1.5 g/kg/d [113]. On the other hand, these guidelines recommended aggressive postoperative nutritional support, including early enteral nutrition when necessary to meet the postoperative caloric and protein needs. In fact, in the 6 weeks after cardiac surgery, older adults lose on average 5% of their body mass, and this increased the risk of readmission in the hospital [114]. A study of patients admitted to the surgical intensive care unit showed that those with a postoperative protein deficit were less likely to be discharged home [115]. A multinational study revealed that consuming close to the recommended protein intake was associated with 60-day survival and ventilator free days [116]. A prospective interventional study demonstrated that aggressive protein supplementation was associated with a 66% reduction in infectious complications in the surgical intensive care unit [117]. A retrospective study of 1007 postsurgical patients at eight hospitals found that those with sufficient protein intake, defined as >60% of the recommended protein intake, had decreased length of stay and hospital costs [118]. The Nutrition Care in Canadian Hospitals (NCCH) study showed that surgical patients who ate less than half of the provided food had signs of malnutrition and increased length of stay [119]. To improve the nutrition of patient candidates for surgery, one of the core components of the ERAS program is a recommendation to liberally prescribe oral nutritional supplements in the pre- and postoperative periods [120]. Exercise plays a key role in the prevention and treatment of sarcopenia and today it is the most effective approach. Through the stimulus given by physical activity, numerous pathways are activated at the muscle level that converge towards anabolic pathways, with positive consequences on trophism and muscle quality. In particular, it is the moderate intensity resistance exercises that produce the most results. Intense exercise does not bring further benefits, if not actually harmful [121].

Table 1. Dietary protein intake and sarcopenia.

Study Type	Duration of the Study	Material and Methods	References	Main Findings
Review	The articles were conducted from January 2010 until April 2015	The first group, containing eight articles, discussed protein or amino acid supplementation alone on sarcopenia. The second group, containing six articles, discussed exercise alone on sarcopenia. The last group, containing six articles, discussed both protein or amino acid supplementation and exercise on sarcopenia	Naseeb MA et al. [122]	Protein intakes should exceed the current recommended dietary allowance RDA (0.8 g/kg body mass per day)
The European Society for Clinical Nutrition and Metabolism (ESPEN) hosted a Workshop on Protein Requirements in the Elderly		Healthy older people for older people who are malnourished or at risk of malnutrition because they have acute or chronic illness. Older adults above 65 years	Deutz NE et al. [121]	Higher protein intakes in older adults in relation to the current protein RDA: a 25–50% increase for healthy individuals, a 50–90% increase those suffering from acute or chronic disease, and greater than 50% increase above the RDA for those experiencing severe illness or injury
Observational and cross-sectional study	Noninstitutionalized participants from the 2005-2014 National Health and Nutrition Examination Survey	Data from 11,680 adults were categorized into 51–60 years (n = 4016), 61–70 years (n = 3854), and 71 years and older (n = 3810) for analysis	Krok-Schoen JL et al. [123]	Over 30% failed to meet the current protein RDA
Parallel-group randomized trial, protein consumption at the current RDA or twice the RDA (2RDA) affects skeletal muscle mass and physical function in elderly men	Before treatment and after 10 wk of intervention	29 men aged > 70 y (mean $\pm$ SD) body mass index (in kg/m^2): 28.3 $\pm$ 4.2	Mitchell CJ et al. [124]	Increasing protein intake to twice RDA (1.6 g/kg per day) resulted in significant gains in lean tissue mass in healthy older men

Table 1. *Cont.*

Study Type	Duration of the Study	Material and Methods	References	Main Findings
The current evidence related to dietary protein intake and muscle health in elderly adults	/	Elderly population of the United State	Baum JI et al. [125]	The consumption of dietary protein consistent with the upper end of the AMDRs (as much as 30–35% of total caloric intake) may prove to be beneficial, although practical limitations may make this level of dietary protein intake difficult The consumption of high-quality proteins that are easily digestible and contain a high proportion of EAAs lessens the urgency of consuming diets with an extremely high protein content
A meta-analysis of randomized controlled trials to investigate effect of whey protein supplementation on long- and short-term appetite	/	Eight publications met inclusion criteria, five records were on short-term and three records on long-term appetite	Mollahosseini M et al. [126]	Increasing daily protein intake to twice the RDA translates to an 80 kg older adult consuming about 130 protein daily. Given that protein increases satiety in a dose-dependent manner
Review		Seniors over 50 with reduced protein intake	Paddon-Jones D et al. [127]	Results from muscle protein anabolism, appetite regulation and satiety research support the contention that meeting a protein threshold (approximately 30 g/meal) represents a promising strategy for middle-aged and older adults concerned with maintaining muscle mass while controlling body fat
A multicenter, randomized, double-blinded, controlled trial with evaluation the effects of two high-quality oral nutritional supplements (ONS) differing in amount and type of key nutrients in older adult men and women	A 24-week intervention period with two energy-rich (330 kcal) ONS treatment groups	Malnourished and sarcopenic men and women, 65 years and older (n = 330)	Cramer JT et al. [128]	The recommendation to increase protein intake while simultaneously maintaining, and in many cases increasing, energy intake can present a protein paradox. Dietary supplementation strategies to increase protein intake may unintentionally result in partial energy redistribution, which may negatively affect both protein and energy intake
Experimental protocol that compared the stimulatory role of leucine, BCAA and EAA ingestion on anabolic signaling following exercise	/	Eight healthy male volunteers with mean (± E) age 27 ± 2 yr, body weight 84 ± 3 kg, height 181 ± 3 cm, and maximal leg strength 430± 13 kg	Moberg M et al. [129]	The capacity of resistance exercise to sensitize muscle to the anabolic potential of dietary protein is primarily achieved through a timely supply of EAA
All trials were single-blind, randomized, and counterbalanced	All laboratory visits were separated by a minimum of 7 days	Seven (n = 7) younger (18–45 years; four males, three females) and seven (n = 7) older (60–80 years; four males, three females) volunteers	Lees MJ et al. [130]	The ingestion of a novel, gel-based, leucine-enriched EAA supplement results in substantial aminoacidemia and anabolic signaling in younger and older individuals. This formulation can augment dietary protein consumption, intracellular anabolic signaling, and aminoacidemia in older adults without deleterious effects on appetite and subsequent energy intake

Table 1. *Cont.*

Study Type	Duration of the Study	Material and Methods	References	Main Findings
A systematic review of interventional evidence was performed through the use of a random-effects meta-analysis model	The effect of dietary protein supplementation during prolonged (>6 wk) resistance-type exercise training	680 subjects	Cermak NM et al. [131]	Gains in lean tissue mass were of greater magnitude in both younger and older adults when combining resistance training with protein supplementation vs. resistance training alone. Increases in type II muscle fiber cross-sectional area in older adults following resistance training
A systematic review, meta-analysis and meta-regression	Only randomized controlled trials with RET ≥6 weeks in duration and dietary protein supplementation	49 studies with 1863 participants	Morton RW et al. [132]	Protein supplementation augmented muscle growth during resistance training when habitual dietary protein intakes were, on average, below 1.6 g/kg body mass per day in younger adults. However, the impact of protein supplementation on muscle mass was reduced with advancing age
Available studies linking protein intake with physical function and health parameters in elderly 80 years old or older	/	Elderly cohorts including very old participants aged 80 years and older	Franzke B et al. [133]	The amino acid composition of a given protein source can influence the extent and amplitude of postprandiam MPS, and induce varying patterns of aminoacidemia
Studies assessing the relation between dietary protein intake and indexes of muscle mass, physical function, distribution, and muscle mass and function	/	Persons aged > 80 y sarcopenic	Traylor DA et al. [134]	The leucine content of a given protein source is particularly important in attenuating declines in a muscle mass when consumed alongside other essential amino acid EAA
Clinical trials anabolic response to essential amino acid plus whey protein composition is greater than whey protein alone in young healthy adults	/	16 healthy male and females. Characteristics: age, body weight, body mass index, lean body mass, fat mass	Park S et al. [135]	Provision of ample dietary EAA and leucine are necessary to support a skeletal muscle anabolic response in older adults. Nutritional supplementation with EAA and leucine alongside meals containing suboptimal protein content (i.e., breakfast and lunch) could assist older adults in achieving their per meal protein intake
Study time-dependent concordance and discordance between human muscle protein synthesis and mTORC1 signaling	Recruitment for healthy young men (n = 8; 21 ± 2 y old (mean ± SEM); body mass index (in kg/m^2): 22.9 ± 0.9) began in January 2008	Eight postabsorptive healthy men (≈21 y of age) were studied during 8.5 h of primed continuous infusion of [1,2-$^{13}C_2$]leucine with intermittent quadriceps biopsies for determination of muscle protein synthesis MPS and anabolic signaling	Atherton PJ et al. [136]	When skeletal muscle is refractory to the anabolic effects of leucine during the postprandial 'muscle-full' period, it would be prudent that protein-based snacks or supplements are administered between meals when additional nutritional supplementation is required to reach their daily protein goal
Protocol study that demonstrates the refractoriness of muscle to nutrient-led anabolic stimulation in the postprandial period	/	Healthy, recreationally active older males (n = 16, 70.3 ± 2.6 years, BMI 25.5 ± 1.8 (mean ± SD) were recruited by mail and local advertising	Mitchell WK et al. [137]	Supplements may be most likely to be effective when taken in between meals, perhaps in the form of low dose EAA mixtures, rather than leucine alone; the efficacy of which may be limited in the absence of exogenous EAA to promote whole body and skeletal muscle net balance

10. Conclusions

Frailty is the major expression of a decreased resistance to stressors, and consequently an increased vulnerability to additional diseases in elderly people. Sarcopenia can be considered a biological substrate of physical frailty. The vascular aging related to frail phenotype reflects the high susceptibility for cardiovascular diseases and negative postoperative outcomes after cardiac surgery. For this reason, a frail phenotype is a risk factor of mortality and morbidity for several diseases. A lot of biomarkers have been identified as expression of the crosstalk between frailty and cardiovascular diseases in molecular mechanisms level.

Malnutrition plays a key role in the pathogenesis of sarcopenia and frailty. Malnutrition is defined as an imbalance condition of protein or other nutrient imbalance, responsible for negative effects on body composition, physical function and clinical outcome. An optimal nutrition status and protein intake, as well as a moderate intensity resistance exercise are the key points. Accordingly, in the future the role of a specialized team-workers will be very important in the management of cardiac surgery patients. Cardiac surgeons, cardiologists, geriatricians, physiatrists, and dieticians should work in a complementary way in order to better prepare patients undergoing cardiac surgery and reduce postoperative mortality and morbidity.

Author Contributions: Conceptualization C.P., D.P., C.F., G.R.; methodology, C.P., C.R.B.; software, C.P., D.P., C.R.B.; validation: C.F., G.R.; formal analysis: C.P., D.P., C.R.B.; investigation, C.A., D.P., P.N., F.B., D.T., L.A., M.S.F., D.B.; resources: C.P.; data curation, C.P., D.P.; writing—original draft preparation, C.P., D.P., C.R.B.; writing—review and editing, C.F., G.R.; visualization, C.P., C.R.B.; supervision, G.R., C.F.; project administration, C.P. All authors have read and agreed to the published version of the manuscript.

Funding: This research received no external funding.

Institutional Review Board Statement: The study was conducted according to the guidelines of the Declaration of Helsinki, and approved by the Ethics Committee of Tor Vergata University (protocol code 134/19 and date of approval 17 July 2019).

Informed Consent Statement: Not applicable.

Data Availability Statement: Not applicable.

Conflicts of Interest: The authors declare no conflict of interest.

References

1. Malaguarnera, M.; Vacante, M.; Frazzetto, P.M.; Motta, M. What is the frailty in elderly? Value and significance of the multidimensional assessments. *Arch. Gerontol. Geriatr.* **2013**, *56*, 23–26. [CrossRef]
2. Fried, L.P.; Tangen, C.M.; Walston, J.; Newman, A.B.; Hirsch, C.; Gottdiener, J.; Seeman, T.; Tracy, R.; Kop, W.J.; Burke, G.; et al. Frailty in older adults: Evidence for a phenotype. *J. Gerontol. A Biol. Sci. Med. Sci.* **2001**, *56*, M146–M156. [CrossRef]
3. Rockwood, K.; Mitnitski, A. Frailty defined by deficit accumulation and geriatric medicine defined by Frailty. *Clin. Geriatr. Med.* **2011**, *27*, 17–26. [CrossRef]
4. Gobbens, R.J.; Luijkx, K.G.; Wijnen-Sponselee, M.T.; Schols, J.M. In search of an integral conceptual definition of frailty: Opinions of experts. *J. Am. Med. Dir. Assoc.* **2010**, *11*, 338–343. [CrossRef]
5. Chang, S.F.; Yang, R.S.; Lin, T.C.; Chiu, S.C.; Chen, M.L.; Lee, H.C. The discrimination of using the short physical performance battery to screen frailty for community-dwelling older people. *J. Nurs. Scholarsh.* **2014**, *46*, 207–215. [CrossRef] [PubMed]
6. Petermann-Rocha, F.; Gray, S.R.; Pell, J.P.; Ho, F.K.; Celis-Morales, C. The joint association of sarcopenia and frailty with incidence and mortality healthoutcomes: A prospective study. *Clin. Nutr.* **2020**. [CrossRef] [PubMed]
7. Bose, C.; Alves, I.; Singh, P.; Palade, P.T.; Carvalho, E.; Børsheim, E.; Jun, S.R.; Cheema, A.; Boerma, M.; Awasthi, S.; et al. Sulforaphane prevents age-associated cardiac and muscular dysfunction through Nrf2 signaling. *Aging Cell* **2020**, *19*, e13261. [CrossRef]
8. Inoue, T.; Maeda, K.; Nagano, A.; Shimizu, A.; Ueshima, J.; Murotani, K.; Sato, K.; Tsubaki, A. Undernutrition, Sarcopenia, and Frailty in Fragility Hip Fracture: Advanced Strategies for Improving Cliniccal Outcomes. *Nutrients* **2020**, *12*, 3743. [CrossRef] [PubMed]
9. Bielecka-Dabrowa, A.; Ebner, N.; Dos Santos, M.R.; Ishida, J.; Hasenfuss, G.; von Haehling, S. Cachexia, muscle wasting, and frailty in cardiovascular disease. *Eur. J. Heart Fail.* **2020**. [CrossRef] [PubMed]

10. Krysztofiak, H.; Wleklik, M.; Migaj, J.; Dudek, M.; Uchmanowicz, I.; Lisiak, M.; Kubiela, G.; Straburzyńska-Migaj, E.; Lesiak, M.; Kałużna-Oleksy, M. Cardiac Cachexia: A Well- Known but Challenging Complication of Heart Failure. *Clin. Interv. Aging* **2020**, *15*, 2041–2051. [CrossRef] [PubMed]
11. Yin, J.; Lu, X.; Qian, Z.; Xu, W.; Zhou, X. New insights into the pathogenesis and treatment of sarcopenia in chronic heart failure. *Theranostics* **2019**, *9*, 4019–4029. [CrossRef]
12. Lena, A.; Anker, M.S.; Springer, J. Muscle Wasting and Sarcopenia in Heart Failure- The Current State of Science. *Int. J. Mol. Sci.* **2020**, *21*, 6549. [CrossRef]
13. Abe, K.; Yano, T.; Katano, S.; Ohori, K.; Ishigo, T.; Kouzu, H.; Moniwa, N.; Miura, T. Reply to the comments on "Utility of the sarcopenia index for assessment of muscle mass and nutritional status in patients with chronic heart failure: Comparison with anthropometric parameters". *Geriatr. Gerontol. Int.* **2020**, *20*, 998–999. [CrossRef] [PubMed]
14. Li, X.; Xu, F.; Hu, L.; Fang, H.; An, Y. Revisiting: "prevalence of and factorsassociated with sarcopenia among multi-ethnic ambulatory older Asians with type 2 diabetes mellitus in a primary care setting". *BMC Geriatr.* **2020**, *20*, 415. [CrossRef] [PubMed]
15. Reginster, J.Y.; Beaudart, C.; Al-Daghri, N.; Avouac, B.; Bauer, J.; Bere, N.; Bruyère, O.; Cerreta, F.; Cesari, M.; Rosa, M.M.; et al. Update on the ESCEO recommendation for the conduct of clinical trials for drugs aiming at the treatment of sarcopenia in older adults. *Aging Clin. Exp. Res.* **2020**. [CrossRef] [PubMed]
16. Izumida, T.; Imamura, T. Sarcopenia in patients with cardiovascular disease. *J. Cardiol.* **2020**, *76*, 636. [CrossRef]
17. Cruz-Jentoft, A.J.; Baeyens, J.P.; Bauer, J.M.; Boirie, Y.; Cederholm, T.; Landi, F.; Martin, F.C.; Michel, J.P.; Rolland, Y.; Schneider, S.M.; et al. Sarcopenia: European consensus on definition and diagnosis. *Age Ageing* **2010**, *39*, 412–423. [CrossRef] [PubMed]
18. Afilalo, J.; Alexander, K.P.; Mack, M.J.; Maurer, M.S.; Green, P.; Allen, L.A.; Popma, J.J.; Ferrucci, L.; Forman, D.E. Frailty assessment in the cardiovascular care of older adults. *J. Am. Coll. Cardiol.* **2014**, *63*, 747–762. [CrossRef] [PubMed]
19. Rockwood, K.; Mitnitski, A.; Howlett, S.E. Frailty: Scaling from Cellular Deficit Accumulation? *Interdiscip. Top. Gerontol. Geriatr.* **2015**, *41*, 1–14. [CrossRef]
20. Vigorito, C.; Abreu, A.; Ambrosetti, M.; Belardinelli, R.; Corrà, U.; Cupples, M.; Davos, C.H.; Hoefer, S.; Iliou, M.C.; Schmid, J.P.; et al. Frailty and cardiac rehabilitation: A call to action from the EAPC Cardiac Rehabilitation Section. *Eur. J. Prev. Cardiol.* **2017**, *24*, 577–590. [CrossRef]
21. He, Y.; Li, L.W.; Hao, Y.; Sim, E.Y.; Ng, K.L.; Lee, R.; Lim, M.S.; Poopalalingam, R.; Abdullah, H.R. Assessment of predictive validity and feasibility of Edmonton Frail Scale in identifying postoperative complications among elderly patients: A prospective observational study. *Sci. Rep.* **2020**, *10*, 14682. [CrossRef]
22. Saczynski, J.S.; Inouye, S.K.; Guess, J.; Jones, R.N.; Fong, T.G.; Nemeth, E. The Montreal Cognitive Assessment: Creating a Crosswalk with the Mini-Mental State Examination. *J. Am. Geriatr. Soc.* **2015**, *63*, 2370–2374. [CrossRef] [PubMed]
23. Mitchell, A.J.; Bird, V.; Rizzo, M.; Meader, N. Diagnostic validity and added value of the Geriatric Depression Scale for depression in primary care: A meta- analysis of GDS30 and GDS15. *J. Affect. Disord.* **2010**, *125*, 10–17. [CrossRef] [PubMed]
24. Hollocks, M.J.; Lawrence, A.J.; Brookes, R.L.; Barrick, T.R.; Morris, R.G.; Husain, M.; Markus, H.S. Differential relationships between apathy and depression with white matter microstructural changes and functional outcomes. *Brain* **2015**, *138*, 3803–3815. [CrossRef] [PubMed]
25. Ramírez-Vélez, R.; Sáez de Asteasu, M.L.; Martínez-Velilla, N.; Zambom-Ferraresi, F.; García-Hermoso, A.; Izquierdo, M. Handgrip Strength as a Complementary Test for Mobility Limitations Assessment in Acutely Hospitalized Oldest Old. *Rejuvenation Res.* **2021**. [CrossRef] [PubMed]
26. Landi, F.; Calvani, R.; Martone, A.M.; Salini, S.; Zazzara, M.B.; Candeloro, M.; Coelho-Junior, H.J.; Tosato, M.; Picca, A.; Marzetti, E. Normative values of muscle strength across ages in a 'real world' population: Results from the longevity check-up 7+ project. *J. Cachexia Sarcopenia Muscle* **2020**, *11*, 1562–1569. [CrossRef]
27. Valdes, K.; Blausey, J.; Campbell, J.; Duran, R.; Giles, A.; Matthys, C.; Miesner, S.; Schroeder, B.; Smolyansky, D. Certified hand therapists membership in the American Society of Hand Therapists: A survey study. *J. Hand Ther.* **2020**. [CrossRef]
28. Schrader, E.; Baumgartel, C.; Gueldenzoph, H.; Stehle, P.; Uter, W.; Sieber, C.C.; Volkerf, D. Nutritional status according to Mini Nutritional Assessment is related to functional status in geriatric patients—Independent of health status. *J. Nutr. Health Aging* **2014**, *18*, 257–263. [CrossRef]
29. Kirk, B.; Phu, S.; Brennan-Olsen, S.L.; Bani Hassan, E.; Duque, G. Associations between osteoporosis, the severity of sarcopenia and fragility fractures in community- dwelling older adults. *Eur. Geriatr. Med.* **2020**, *11*, 443–450. [CrossRef]
30. Tsourdi, E.; Nees, J.A.; Hofbauer, L.C. Osteoporosis in the geriatric population. *Dtsch. Med. Wochenschr.* **2020**, *145*, 728–732.
31. Seng, W.R.; Belani, M.H.; Ramason, R.; Naidu, G.; Doshi, H.K. Functional Improvement in Geriatric Hip Fractures: Does Vitamin D Deficiency Affect the Functional Outcome of Patients With Surgically Treated Intertrochanteric Hip Fractures. *Geriatr. Orthop. Surg. Rehabil.* **2015**, *6*, 186–191. [CrossRef]
32. Romero-Ortuno, R.; Walsh, C.D.; Lawlor, B.A.; Kenny, R.A. A frailty instrument for primary care: Findings from the Survey of Health, Ageing and Retirement in Europe (SHARE). *BMC Geriatr.* **2010**, *10*, 57. [CrossRef]
33. Lauretani, F.; Ticinesi, A.; Gionti, L.; Prati, B.; Nouvenne, A.; Tana, C.; Meschi, T.; Maggio, M. Short-Physical Performance Battery (SPPB) score is associated with falls in older outpatients. *Aging Clin. Exp. Res.* **2019**, *31*, 1435–1442. [CrossRef]
34. Afilalo, J.; Lauck, S.; Kim, D.H.; Lefèvre, T.; Piazza, N.; Lachapelle, K.; Martucci, G.; Lamy, A.; Labinaz, M.; Peterson, M.D.; et al. Frailty in Older Adults Undergoing Aortic Valve Replacement: The FRAILTY-AVR Study. *J. Am. Coll. Cardiol.* **2017**, *70*, 689–700. [CrossRef]

35. Sündermann, S.; Dademasch, A.; Praetorius, J.; Kempfert, J.; Dewey, T.; Falk, V.; Mohr, F.W.; Walther, T. Comprehensive assessment of frailty for elderly high-risk patients undergoing cardiac surgery. *Eur. J. Cardiothorac. Surg.* **2011**, *39*, 33–37. [CrossRef]
36. Sündermann, S.; Dademasch, A.; Rastan, A.; Praetorius, J.; Rodriguez, H.; Walther, T.; Mohr, F.W.; Falk, V. One-year follow-up of patients undergoing elective cardiac surgery assessed with the Comprehensive Assessment of Frailty test and its simplified form. *Interact. Cardiovasc. Thorac. Surg.* **2011**, *13*, 119–123; discussion 123. [CrossRef]
37. Cao, L.; Morley, J.E. Sarcopenia Is Recognized as an Independent Condition by an International Classification of Disease, Tenth Revision, Clinical Modification (ICD-10-CM) Code. *J. Am. Med. Dir. Assoc.* **2016**, *17*, 675–677. [CrossRef]
38. Anker, S.D.; Morley, J.E.; von Haehling, S. Welcome to the ICD-10 code for sarcopenia. *J. Cachexia Sarcopenia Muscle* **2016**, *7*, 512–514. [CrossRef]
39. Yamada, M.; Nishiguchi, S.; Fukutani, N.; Tanigawa, T.; Yukutake, T.; Kayama, H.; Aoyama, T.; Arai, H. Prevalence of sarcopenia in community-dwelling Japanese older adults. *J. Am. Med. Dir. Assoc.* **2013**, *14*, 911–915. [CrossRef]
40. Edwards, M.H.; Dennison, E.M.; Aihie Sayer, A.; Fielding, R.; Cooper, C. Osteoporosis and sarcopenia in older age. *Bone* **2015**, *80*, 126–130. [CrossRef]
41. Ham, D.J.; Börsch, A.; Lin, S.; Thürkauf, M.; Weihrauch, M.; Reinhard, J.R.; Delezie, J.; Battilana, F.; Wang, X.; Kaiser, M.S.; et al. The neuromuscular junction is a focal point of mTORC1 signaling in sarcopenia. *Nat. Commun.* **2020**, *11*, 4510. [CrossRef]
42. Bano, G.; Trevisan, C.; Carraro, S.; Solmi, M.; Luchini, C.; Stubbs, B.; Manzato, E.; Sergi, G.; Veronese, N. Inflammation and sarcopenia: A systematic review and meta-analysis. *Maturitas* **2017**, *96*, 10–15. [CrossRef]
43. Thoma, A.; Lightfoot, A.P. NF-kB and Inflammatory Cytokine Signalling: Role in Skeletal Muscle Atrophy. *Adv. Exp. Med. Biol.* **2018**, *1088*, 267–279.
44. Rong, Y.D.; Bian, A.L.; Hu, H.Y.; Ma, Y.; Zhou, X.Z. Study on relationship between elderly sarcopenia and inflammatory cytokine IL-6, anti-inflammatory cytokine IL-10. *BMC Geriatr.* **2018**, *18*, 308. [CrossRef] [PubMed]
45. Gurcay, E.; Kara, M. Is sarcopenia primarily an age-related or renin-angiotensin system-related disorder? *Geriatr. Gerontol. Int.* **2020**, *20*, 997. [CrossRef]
46. Ekiz, T.; Kara, M.; Özcan, F.; Ricci, V.; Özçakar, L. Sarcopenia and COVID-19: A Manifold Insight on Hypertension and the Renin Angiotensin System. *Am. J. Phys. Med. Rehabil.* **2020**, *99*, 880–882. [CrossRef]
47. Bao, Z.; Cui, C.; Chow, S.K.; Qin, L.; Wong, R.M.Y.; Cheung, W.H. AChRs Degeneration at NMJ in Aging-Associated Sarcopenia-A Systematic Review. *Front. Aging Neurosci.* **2020**, *12*, 5978. [CrossRef] [PubMed]
48. Uezumi, A.; Ikemoto-Uezumi, M.; Zhou, H.; Kurosawa, T.; Yoshimoto, Y.; Nakatani, M.; Hitachi, K.; Yamaguchi, H.; Wakatsuki, S.; Araki, T.; et al. Mesenchymal Bmp3b expression maintains skeletal muscle integrity and decreases in age-related sarcopenia. *J. Clin. Investig.* **2021**, *131*, e139617. [CrossRef]
49. Marini, E.; Buffa, R.; Saragat, B.; Coin, A.; Toffanello, E.D.; Berton, L.; Manzato, E.; Sergi, G. The potential of classic and specific bioelectrical impedance vector analysis for the assessment of sarcopenia and sarcopenic obesity. *Clin. Interv. Aging* **2012**, *7*, 585–591. [CrossRef]
50. Bahat, G.; Kilic, C.; Altinkaynak, M.; Akif Karan, M. Comparison of standard "versus" population-specific handgrip strength cut-off points in the detection of probable sarcopenia after launch of EWGSOP2. *Aging Male* **2021**, *12*, 1–6. [CrossRef] [PubMed]
51. Zanker, J.; Patel, S.; Blackwell, T.; Duchowny, K.; Brennan-Olsen, S.; Cummings, S.R.; Evans, W.J.; Orwoll, E.S.; Scott, D.; Vogrin, S.; et al. Walking Speed and Muscle Mass Estimated by the D_3-Creatine Dilution Method Are Important Components of Sarcopenia Associated With Incident Mobility Disability in Older Men: A Classification and Regression Tree Analysis. *J. Am. Med. Dir. Assoc.* **2020**, *21*, 1997–2002. [CrossRef]
52. Poggiogalle, E.; Parrinello, E.; Barazzoni, R.; Busetto, L.; Donini, L.M. Therapeutic strategies for sarcopenic obesity: A systematic review. *Curr. Opin. Clin. Nutr. Metab. Care* **2021**, *24*, 33–41. [CrossRef]
53. Rahman, A.; Jafry, S.; Jeejeebhoy, K.; Nagpal, A.D.; Pisani, B.; Agarwala, R. Malnutrition and Cachexia in Heart Failure. *JPEN J. Parenter. Enter. Nutr.* **2016**, *40*, 475–486. [CrossRef]
54. Woods, N.F.; LaCroix, A.Z.; Gray, S.L.; Aragaki, A.; Cochrane, B.B.; Brunner, R.L.; Masaki, K.; Murray, A.; Newman, A.B. Women's Health Initiative. Frailty: Emergence and consequences in women aged 65 and older in the Women's Health Initiative Observational Study. *J. Am. Geriatr. Soc.* **2005**, *53*, 1321–1330, Erratum in **2017**, *65*, 1631–1632. [CrossRef] [PubMed]
55. Gary, R. Evaluation of frailty in older adults with cardiovascular disease: Incorporating physical performance measures. *J. Cardiovasc. Nurs.* **2012**, *27*, 120–131. [CrossRef]
56. Alonso-, C.; Carcaillon, L.; García-García, F.J.; Amor-Andrés, M.S.; El Assar, M.; Rodríguez-Mañas, L. Association between endothelial dysfunction and frailty: The Toledo Study for Healthy Aging. *Age* **2014**, *36*, 495–505. [CrossRef]
57. Ricci, N.A.; Pessoa, G.S.; Ferriolli, E.; Dias, R.C.; Perracini, M.R. Frailty andcardiovascular risk in community-dwelling elderly: A population-based study. *Clin. Interv. Aging* **2014**, *9*, 1677–1685. [CrossRef]
58. Cacciatore, F.; Testa, G.; Galizia, G.; Della-Morte, D.; Mazzella, F.; Langellotto, A.; Pirozzi, G.; Ferro, G.; Gargiulo, G.; Ferrara, N.; et al. Clinical frailty and long-term mortality in elderly subjects with diabetes. *Acta Diabetol.* **2013**, *50*, 251–260. [CrossRef]
59. Hubbard, R.E.; Lang, I.A.; Llewellyn, D.J.; Rockwood, K. Frailty, body mass index, and abdominal obesity in older people. *J. Gerontol. A Biol. Sci. Med. Sci.* **2010**, *65*, 377–381. [CrossRef]
60. Angulo, J.; Vallejo, S.; El Assar, M.; García-Septiem, J.; Sánchez-Ferrer, C.F.; Rodríguez-Mañas, L. Age-related differences in the effects of α and γ peroxisome proliferator-activated receptor subtype agonists on endothelial vasodilation in human microvessels. *Exp. Gerontol.* **2012**, *47*, 734–740. [CrossRef]

61. Liberale, L.; Montecucco, F.; Tardif, J.C.; Libby, P.; Camici, G.G. Inflamm-ageing: The role of inflammation in age-dependent cardiovascular disease. *Eur. Heart J.* **2020**, *41*, 2974–2982. [CrossRef]
62. Whitehead, J.C.; Hildebrand, B.A.; Sun, M.; Rockwood, M.R.; Rose, R.A.; Rockwood, K.; Howlett, S.E. A clinical frailty index in aging mice: Comparisons with frailty index data in humans. *J. Gerontol. Ser. A Biol. Sci. Med. Sci.* **2014**, *69*, 621–632. [CrossRef] [PubMed]
63. Lopez-Otin, C.; Blasco, M.A.; Partridge, L.; Serrano, M.; Kroemer, G. The hallmarks of aging. *Cell* **2013**, *153*, 1194–1217. [CrossRef] [PubMed]
64. Cardoso, A.L.; Fernandes, A.; Aguilar-Pimentel, J.A.; de Angelis, M.H.; Guedes, J.R.; Brito, M.A.; Ortolano, S.; Pani, G.; Athanasopoulou, S.; Gonos, E.S.; et al. Towards frailty biomarkers: Candidates from genes and pathways regulated in aging and age-related diseases. *Ageing Res. Rev.* **2018**, *47*, 214–277. [CrossRef]
65. Monti, D.; Ostan, R.; Borelli, V.; Castellani, G.; Franceschi, C. Inflammaging and human longevity in the omics era. *Mech. Ageing Dev.* **2017**, *165*, 129–138. [CrossRef] [PubMed]
66. Baker, D.J.; Wijshake, T.; Tchkonia, T.; LeBrasseur, N.K.; Childs, B.G.; van de Sluis, B.; Kirkland, J.L.; van Deursen, J.M. Clearance of p16Ink4a-positive senescent cells delays ageing-associated disorders. *Nature* **2011**, *479*, 232–236. [CrossRef]
67. Yoshida, N.; Yamamoto, H.; Shinke, T.; Otake, H.; Kuroda, M.; Terashita, D.; Takahashi, H.; Sakaguchi, K.; Hirota, Y.; Emoto, T.; et al. Impact of CD14(+ +)CD16(+) monocytes on plaque vulnerability in diabetic and non-diabetic patients with asymptomatic coronary artery disease: A cross-sectional study. *Cardiovasc. Diabetol.* **2017**, *16*, 96. [CrossRef]
68. Verschoor, C.P.; Johnstone, J.; Millar, J.; Parsons, R.; Lelic, A.; Loeb, M.; Bramson, J.L.; Bowdish, D.M. Alterations to the frequency and function of peripheral blood monocytes and associations with chronic disease in the advanced-age, frail elderly. *PLoS ONE* **2014**, *9*, e104522. [CrossRef]
69. Shah, R.; Matthews, G.J.; Shah, R.Y.; McLaughlin, C.; Chen, J.; Wolman, M.; Master, S.R.; Chai, B.; Xie, D.; Rader, D.J.; et al. Serum Fractalkine (CX3CL1) and Cardiovascular Outcomes and Diabetes: Findings From the Chronic Renal Insufficiency Cohort (CRIC) Study. *Am. J. Kidney Dis.* **2015**, *66*, 266–273. [CrossRef]
70. Anuurad, E.; Enkhmaa, B.; Gungor, Z.; Zhang, W.; Tracy, R.P.; Pearson, T.A.; Kim, K.; Berglund, L. Age as a modulator of inflammatory cardiovascular risk factors. *Arterioscler. Thromb. Vasc. Biol.* **2011**, *31*, 2151–2156. [CrossRef]
71. Pavanello, S.; Stendardo, M.; Mastrangelo, G.; Bonci, M.; Bottazzi, B.; Campisi, M.; Nardini, M.; Leone, R.; Mantovani, A.; Boschetto, P. Inflammatory long pentraxin 3 is associated with leukocyte telomere length in night-shift workers. *Front. Immunol.* **2017**, *8*, 516. [CrossRef]
72. Tobias, D.K.; Akinkuolie, A.O.; Chandler, P.D.; Lawler, P.R.; Manson, J.E.; Buring, J.E.; Ridker, P.M.; Wang, L.; Lee, I.M.; Mora, S. Markers of inflammation and in- cident breast Cancer risk in the women's health study. *Am. J. Epidemiol.* **2018**, *187*, 705–716. [CrossRef] [PubMed]
73. Wang, J.; Ma, R.; Sharma, A.; He, M.; Xue, J.; Wu, J.; Dun, B.; Li, G.; Wang, X.; Ji, M.; et al. Inflammatory serum proteins are severely altered in metastatic gastric adenocarcinoma patients from the Chinese population. *PLoS ONE* **2015**, *10*, e0123985. [CrossRef] [PubMed]
74. Tung, B.T.; Rodriguez-Bies, E.; Talero, E.; Gamero-Estevez, E.; Motilva, V.; Navas, P.; Lopez-Lluch, G. Anti-inflammatory effect of resveratrol in old mice liver. *Exp. Gerontol.* **2015**, *64*, 1–7. [CrossRef]
75. Zhuang, P.Y.; Zhang, K.W.; Wang, J.D.; Zhou, X.P.; Liu, Y.B.; Quan, Z.W.; Shen, J. Effect of TALEN-mediated IL-6 knockout on cell proliferation, apoptosis, invasion and anti-cancer therapy in hepatocellular carcinoma (HCC-LM3) cells. *Oncotarget* **2017**, *8*, 77915–77927. [CrossRef]
76. Trott, D.W.; Lesniewski, L.A.; Donato, A.J. Selected life-extending interventions reduce arterial CXCL10 and macrophage colony-stimulating factor in aged mouse arteries. *Cytokine* **2017**, *96*, 102–106. [CrossRef]
77. Jin, T.; Sun, Z.; Chen, X.; Wang, Y.; Li, R.; Ji, S.; Zhao, Y. Serum human Beta- Defensin-2 is a possible biomarker for monitoring response to JAK inhibitor in psoriasis patients. *Dermatology* **2017**, *233*, 164–169. [CrossRef]
78. Szczepanowska, K.; Trifunovic, A. Origins of mtDNA mutations in ageing. *Essays Biochem.* **2017**, *61*, 325–337. [PubMed]
79. Corre, J.; Hebraud, B.; Bourin, P. Concise review: Growth differentiation factor 15 in pathology: A clinical role? *Stem Cells Transl. Med.* **2013**, *2*, 946–952. [CrossRef]
80. Fujita, Y.; Taniguchi, Y.; Shinkai, S.; Tanaka, M.; Ito, M. Secreted growth differentiation factor 15 as a potential biomarker for mitochondrial dysfunctions in aging and age-related disorders. *Geriatr. Gerontol. Int.* **2016**, *16* (Suppl. S1), 17–29. [CrossRef] [PubMed]
81. Dogan, S.; Ray, A.; Cleary, M.P. The influence of different calorie restriction protocols on serum pro-inflammatory cytokines, adipokines and IGF-I levels in female C57BL6 mice: Short term and long term diet effects. *Meta Gene* **2017**, *12*, 22–32. [CrossRef]
82. Bayes-Genis, A.; Barallat, J.; de Antonio, M.; Domingo, M.; Zamora, E.; Vila, J.; Subirana, I.; Gastelurrutia, P.; Pastor, M.C.; Januzzi, J.L.; et al. Bloodstream amyloid-beta (1-40) peptide, cognition, and outcomes in heart failure. *Rev. Esp. Cardiol.* **2017**, *70*, 924–932. [CrossRef] [PubMed]
83. Diem, S.; Kasenda, B.; Spain, L.; Martin-Liberal, J.; Marconcini, R.; Gore, M.; Larkin, J. Serum lactate dehydrogenase as an early marker for outcome in patients treated with anti-PD-1 therapy in metastatic melanoma. *Br. J. Cancer* **2016**, *114*, 256–261. [CrossRef]
84. Bluhm, B.; Laffer, B.; Hirnet, D.; Rothermundt, M.; Ambree, O.; Lohr, C. Normal cerebellar development in S100B-deficient mice. *Cerebellum* **2015**, *14*, 119–127. [CrossRef]

85. Yamaguchi, M. Regulatory role of regucalcin in heart calcium signaling: Insight into cardiac failure (Review). *Biomed. Rep.* **2014**, *2*, 303–308. [CrossRef]
86. Fischer, C.R.; Mikami, M.; Minematsu, H.; Nizami, S.; Goo Lee, H.; Stamer, D.; Patel, N.; Yu Soung, D.; Back, J.H.; Song, L.; et al. Calreticulin inhibits inflammation-induced osteoclastogenesis and bone resorption. *J. Orthop. Res.* **2017**, *35*, 2658–2666. [CrossRef]
87. Eming, S.A.; Martin, P.; Tomic-Canic, M. Wound repair and regeneration: Mechanisms, signaling, and translation. *Sci. Transl. Med.* **2014**, *6*, 265–266. [CrossRef] [PubMed]
88. Zeisberg, M.; Kalluri, R. Cellular mechanisms of tissue fibrosis. 1. Common and organ-specific mechanisms associated with tissue fibrosis. *Am. J. Physiol. Cell Physiol.* **2013**, *304*, C216–C225. [CrossRef]
89. Ghosh, A.K.; Rai, R.; Park, K.E.; Eren, M.; Miyata, T.; Wilsbacher, L.D.; Vaughan, D.E. A small molecule inhibitor of PAI-1 protects against doxorubicin-induced cellular senescence. *Oncotarget* **2016**, *7*, 72443–72457. [CrossRef]
90. Sudol, M. From Rous sarcoma virus to plasminogen activator, src oncogene and cancer management. *Oncogene* **2011**, *30*, 3003–3010. [CrossRef] [PubMed]
91. Musial, K.; Bargenda, A.; Zwolinska, D. Urine matrix metalloproteinases and their extracellular inducer EMMPRIN in children with chronic kidney disease. *Ren. Fail.* **2015**, *37*, 980–984. [CrossRef]
92. Szondy, Z.; Korponay-Szabo, I.; Kiraly, R.; Sarang, Z.; Tsay, G.J. Transglutaminase 2 in human diseases. *BioMedicine* **2017**, *7*, 15. [CrossRef]
93. Kimura, Y.; Izumiya, Y.; Hanatani, S.; Yamamoto, E.; Kusaka, H.; Tokitsu, T.; Takashio, S.; Sakamoto, K.; Tsujita, K.; Tanaka, T.; et al. High serum levels of thrombospondin-2 correlate with poor prognosis of patients with heart failure with preserved ejection fraction. *Heart Vessels* **2016**, *31*, 52–59. [CrossRef]
94. Miranda, A.S.; Simoes, E.S.A.C. Serum levels of angiotensin converting enzyme as a biomarker of liver fibrosis. *World J. Gastroenterol.* **2017**, *23*, 8439–8442. [CrossRef]
95. Maggio, M.; Cattabiani, C.; Lauretani, F.; Ferrucci, L.; Luci, M.; Valenti, G.; Ceda, G. The concept of multiple hormonal dysregulation. *Acta Bio-Medica Atenei Parm.* **2010**, *81* (Suppl. S1), 19–29.
96. Rajabali, N.; Rolfson, D.; Bagshaw, S.M. Assessment and Utility of Frailty Measures in Critical Illness, Cardiology, and Cardiac Surgery. *Can. J. Cardiol.* **2016**, *32*, 1157–1165. [CrossRef]
97. Graham, M.M.; Galbraith, P.D.; O'Neill, D.; Rolfson, D.B.; Dando, C.; Norris, C.M. Frailty and outcome in elderly patients with acute coronary syndrome. *Can. J. Cardiol.* **2013**, *29*, 1610–1615. [CrossRef] [PubMed]
98. Ekerstad, N.; Swahn, E.; Janzon, M.; Alfredsson, J.; Löfmark, R.; Lindenberger, M.; Carlsson, P. Frailty is independently associated with short-term outcomes for elderly patients with non-ST-segment elevation myocardial infarction. *Circulation* **2011**, *124*, 2397–2404. [CrossRef] [PubMed]
99. Sanchis, J.; Núñez, E.; Ruiz, V.; Bonanad, C.; Fernández, J.; Cauli, O.; García-Blas, S.; Mainar, L.; Valero, E.; Rodríguez-Borja, E.; et al. Usefulness of Clinical Data and Biomarkers for the Identification of Frailty After Acute Coronary Syndromes. *Can. J. Cardiol.* **2015**, *31*, 1462–1468. [CrossRef]
100. Jha, S.R.; Ha, H.S.; Hickman, L.D.; Hannu, M.; Davidson, P.M.; Macdonald, P.S.; Newton, P.J. Frailty in advanced heart failure: A systematic review. *Heart Fail. Rev.* **2015**, *20*, 553–560. [CrossRef] [PubMed]
101. Khan, H.; Kalogeropoulos, A.P.; Georgiopoulou, V.V.; Newman, A.B.; Harris, T.B.; Rodondi, N.; Bauer, D.C.; Kritchevsky, S.B.; Butler, J. Frailty and risk for heart failure in older adults: The health, aging, and body composition study. *Am. Heart J.* **2013**, *166*, 887–894. [CrossRef]
102. Lupón, J.; González, B.; Santaeugenia, S.; Altimir, S.; Urrutia, A.; Más, D.; Díez, C.; Pascual, T.; Cano, L.; Valle, V. Prognostic implication of frailty and depressive symptoms in an outpatient population with heart failure. *Rev. Esp. Cardiol.* **2008**, *61*, 835–842. [CrossRef]
103. Farhat, J.S.; Velanovich, V.; Falvo, A.J.; Horst, H.M.; Swartz, A.; Patton, J.H., Jr.; Rubinfeld, I.S. Are the frail destined to fail? Frailty index as predictor of surgical morbidity and mortality in the elderly. *J. Trauma Acute Care Surg.* **2012**, *72*, 1526–1530; discussion 1530–1531. [CrossRef] [PubMed]
104. Kim, S.W.; Han, H.S.; Jung, H.W.; Kim, K.I.; Hwang, D.W.; Kang, S.B.; Kim, C.H. Multidimensional frailty score for the prediction of postoperative mortality risk. *JAMA Surg.* **2014**, *149*, 633–640. [CrossRef]
105. Sepehri, A.; Beggs, T.; Hassan, A.; Rigatto, C.; Shaw-Daigle, C.; Tangri, N.; Arora, R.C. The impact of frailty on outcomes after cardiac surgery: A systematic review. *J. Thorac. Cardiovasc. Surg.* **2014**, *148*, 3110–3117. [CrossRef]
106. Bruno, R.R.; Wolff, G.; Wernly, B.; Kelm, M.; Jung, C. Frailty Assessment in Patients Undergoing Aortic Valve Replacement: Be Quick and Be Sure. *JACC Cardiovasc. Interv.* **2020**, *13*, 1965–1967. [CrossRef] [PubMed]
107. Iyengar, A.; Goel, N.; Kelly, J.J.; Han, J.; Brown, C.R.; Khurshan, F.; Chen, Z.; Desai, N. Effects of Frailty on Outcomes and 30-day Readmissions After Surgical Mitral Valve Replacement. *Ann. Thorac. Surg.* **2020**, *109*, 1120–1126. [CrossRef]
108. Tran, D.T.T.; Tu, J.V.; Dupuis, J.Y.; Bader Eddeen, A.; Sun, L.Y. Association of Frailty and Long-Term Survival in Patients Undergoing Coronary Artery Bypass Grafting. *J. Am. Heart Assoc.* **2018**, *7*, e009882. [CrossRef]
109. Lindman, B.R.; Patel, J.N. Multimorbidity in Older Adults with Aortic Stenosis. *Clin. Geriatr. Med.* **2016**, *32*, 305–314. [CrossRef]
110. Stortecky, S.; Schoenenberger, A.W.; Moser, A.; Kalesan, B.; Jüni, P.; Carrel, T.; Bischoff, S.; Schoenenberger, C.M.; Stuck, A.E.; Windecker, S.; et al. Evaluation of multidimensional geriatric assessment as a predictor of mortality and cardiovascular events after transcatheter aortic valve implantation. *JACC Cardiovasc. Interv.* **2012**, *5*, 489–496. [CrossRef] [PubMed]

111. Deutz, N.E.; Pereira, S.L.; Hays, N.P.; Oliver, J.S.; Edens, N.K.; Evans, C.M.; Wolfe, R.R. Effect of β-hydroxy-β-methylbutyrate (HMB) on lean body mass during 10 days of bed rest in older adults. *Clin. Nutr.* **2013**, *32*, 704–712. [CrossRef]
112. Hoffer, L.J.; Bistrian, B.R. Appropriate protein provision in critical illness: A systematic and narrative review. *Am. J. Clin. Nutr* **2012**, *96*, 591–600. [CrossRef] [PubMed]
113. McClave, S.A.; Taylor, B.E.; Martindale, R.G.; Warren, M.M.; Johnson, D.R.; Braunschweig, C.; McCarthy, M.S.; Davanos, E.; Rice, T.W.; Cresci, G.A.; et al. Guidelines for the provision and assessment of nutrition support therapy in the adult critically ill patient: Society of critical care medicine (SCCM) and American society for parenteral and enteral nutrition (A.S.P.E.N.). *JPEN J. Parenter. Enter. Nutr.* **2016**, *40*, 159–211. [CrossRef]
114. Singer, P.; Berger, M.M.; Van den Berghe, G.; Biolo, G.; Calder, P.; Forbes, A.; Griffiths, R.; Kreyman, G.; Leverve, X.; Pichard, C.; et al. ESPEN Guidelines on Parenteral Nutrition: Intensive care. *Clin. Nutr.* **2009**, *28*, 387–400. [CrossRef] [PubMed]
115. Ljungqvist, O.; Scott, M.; Fearon, K.C. Enhanced recovery after surgery: A review. *JAMA Surg.* **2017**, *152*, 292–298. [CrossRef]
116. Yeh, D.D.; Fuentes, E.; Quraishi, S.A.; Cropano, C.; Kaafarani, H.; Lee, J.; King, D.R.; DeMoya, M.; Fagenholz, P.; Butler, K.; et al. Adequate Nutrition May Get You Home: Effect of Caloric/Protein Deficits on the Discharge Destination of Critically Ill Surgical Patients. *JPEN J. Parenter. Enter. Nutr.* **2016**, *40*, 37–44. [CrossRef]
117. Elke, G.; Wang, M.; Weiler, N.; Day, A.G.; Heyland, D.K. Close to recom- mended caloric and protein intake by enteral nutrition is associated with better clinical outcome of critically ill septic patients: Secondary analysis of a large international nutrition database. *Crit. Care* **2014**, *18*, R29. [CrossRef] [PubMed]
118. Yeh, D.D.; Cropano, C.; Quraishi, S.A.; Fuentes, E.; Kaafarani, H.M.; Lee, J.; Chang, Y.; Velmahos, G. Implementation of an Aggressive Enteral Nutrition Protocol and the Effect on Clinical Outcomes. *Nutr. Clin. Pract.* **2017**, *32*, 175–181. [CrossRef]
119. Allard, J.P.; Keller, H.; Jeejeebhoy, K.N.; Laporte, M.; Duerksen, D.R.; Gramlich, L.; Payette, H.; Bernier, P.; Vesnaver, E.; Davidson, B.; et al. Malnutrition at Hospital Admission-Contributors and Effect on Length of Stay: A Prospective Cohort Study From the Canadian Malnutrition Task Force. *JPEN J. Parenter. Enter. Nutr.* **2016**, *40*, 487–497. [CrossRef]
120. Slieker, J.; Frauche, P.; Jurt, J.; Addor, V.; Blanc, C.; Demartines, N.; Hübner, M. Enhanced recovery ERAS for elderly: A safe and beneficial pathway in colorectal surgery. *Int. J. Colorectal. Dis.* **2017**, *32*, 215–221. [CrossRef]
121. Deutz, N.E.; Bauer, J.M.; Barazzoni, R.; Biolo, G.; Boirie, Y.; Bosy-Westphal, A.; Cederholm, T.; Cruz-Jentoft, A.; Krznariç, Z.; Nair, K.S.; et al. Protein intake and exercise for optimal muscle function with aging: Recommendations from the ESPEN Expert Group. *Clin. Nutr.* **2014**, *33*, 929–936. [CrossRef]
122. Naseeb, M.A.; Volpe, S.L. Protein and exercise in the prevention of sarcopenia and aging. *Nutr. Res.* **2017**, *40*, 1–20. [CrossRef]
123. Krok-Schoen, J.L.; Archdeacon Price, A.; Luo, M.; Kelly, O.J.; Taylor, C.A. Low Dietary Protein Intakes and Associated Dietary Patterns and Functional Limitations in an Aging Population: A NHANES analysis. *J. Nutr. Health Aging* **2019**, *23*, 338–347. [CrossRef] [PubMed]
124. Mitchell, C.J.; Milan, A.M.; Mitchell, S.M.; Zeng, N.; Ramzan, F.; Sharma, P.; Knowles, S.O.; Roy, N.C.; Sjödin, A.; Wagner, K.H.; et al. The effects of dietary protein intake on appendicular lean mass and muscle function in elderly men: A 10-wk randomized controlled trial. *Am. J. Clin. Nutr.* **2017**, *106*, 1375–1383. [CrossRef] [PubMed]
125. Baum, J.I.; Kim, I.Y.; Wolfe, R.R. Protein Consumption and the Elderly: What Is the Optimal Level of Intake? *Nutrients* **2016**, *8*, 359. [CrossRef] [PubMed]
126. Mollahosseini, M.; Shab-Bidar, S.; Rahimi, M.H.; Djafarian, K. Effect of whey protein supplementation on long and short term appetite: A meta-analysis of randomized controlled trials. *Clin. Nutr. ESPEN* **2017**, *20*, 34–40. [CrossRef]
127. Paddon-Jones, D.; Leidy, H. Dietary protein and muscle in older persons. *Curr. Opin. Clin. Nutr. Metab. Care* **2014**, *17*, 5–11. [CrossRef]
128. Cramer, J.T.; Cruz-Jentoft, A.J.; Landi, F.; Hickson, M.; Zamboni, M.; Pereira, S.L.; Hustead, D.S.; Mustad, V.A. Impacts of High-Protein Oral Nutritional Supplements Among Malnourished Men and Women with Sarcopenia: A Multicenter, Randomized, Double-Blinded, Controlled Trial. *J. Am. Med. Dir. Assoc.* **2016**, *17*, 1044–1055. [CrossRef]
129. Moberg, M.; Apró, W.; Ekblom, B.; van Hall, G.; Holmberg, H.C.; Blomstrand, E. Activation of mTORC1 by leucine is potentiated by branched-chain amino acids and even more so by essential amino acids following resistance exercise. *Am. J. Physiol. Cell Physiol.* **2016**, *310*, C874–C884. [CrossRef] [PubMed]
130. Lees, M.J.; Wilson, O.J.; Webb, E.K.; Traylor, D.A.; Prior, T.; Elia, A.; Harlow, P.S.; Black, A.D.; Parker, P.J.; Harris, N.; et al. Novel Essential Amino Acid Supplements Following Resistance Exercise Induce Aminoacidemia and Enhance Anabolic Signaling Irrespective of Age: A Proof-of-Concept Trial. *Nutrients* **2020**, *12*, 2067. [CrossRef]
131. Cermak, N.M.; Res, P.T.; de Groot, L.C.; Saris, W.H.; van Loon, L.J. Protein supplementation augments the adaptive response of skeletal muscle to resistance- type exercise training: A meta-analysis. *Am. J. Clin. Nutr.* **2012**, *96*, 1454–1464. [CrossRef]
132. Morton, R.W.; Murphy, K.T.; McKellar, S.R.; Schoenfeld, B.J.; Henselmans, M.; Helms, E.; Aragon, A.A.; Devries, M.C.; Banfield, L.; Krieger, J.W.; et al. A systematic review, meta-analysis and meta-regression of the effect of protein supplementation on resistance training-induced gains in muscle mass and strength in healthy adults. *Br. J. Sports Med.* **2018**, *52*, 376–384. [CrossRef] [PubMed]
133. Franzke, B.; Neubauer, O.; Cameron-Smith, D.; Wagner, K.H. Dietary Protein, Muscle and Physical Function in the Very Old. *Nutrients* **2018**, *10*, 935. [CrossRef]
134. Traylor, D.A.; Gorissen, S.H.M.; Phillips, S.M. Perspective: Protein Requirements and Optimal Intakes in Aging: Are We Ready to Recommend More Than the Recommended Daily Allowance? *Adv. Nutr.* **2018**, *9*, 171–182. [CrossRef]

135. Park, S.; Church, D.D.; Azhar, G.; Schutzler, S.E.; Ferrando, A.A.; Wolfe, R.R. Anabolic response to essential amino acid plus whey protein composition is greater than whey protein alone in young healthy adults. *J. Int. Soc. Sports Nutr.* **2020**, *17*, 9. [CrossRef] [PubMed]
136. Atherton, P.J.; Etheridge, T.; Watt, P.W.; Wilkinson, D.; Selby, A.; Rankin, D.; Smith, K.; Rennie, M.J. Muscle full effect after oral protein: Time-dependent concordance and discordance between human muscle protein synthesis and mTORC1 signaling. *Am. J. Clin. Nutr.* **2010**, *92*, 1080–1088. [CrossRef] [PubMed]
137. Mitchell, W.K.; Phillips, B.E.; Hill, I.; Greenhaff, P.; Lund, J.N.; Williams, J.P.; Rankin, D.; Wilkinson, D.J.; Smith, K.; Atherton, P.J. Human skeletal muscle is refractory to the anabolic effects of leucine during the postprandial muscle-full period in older men. *Clin. Sci.* **2017**, *131*, 2643–2653. [CrossRef]

nutrients

Review

Effects of Caloric Restriction Diet on Arterial Hypertension and Endothelial Dysfunction

Nicola Di Daniele [1,*], Giulia Marrone [1,2], Manuela Di Lauro [1], Francesca Di Daniele [1,2], Daniela Palazzetti [1], Cristina Guerriero [1] and Annalisa Noce [1]

1 UOC of Internal Medicine-Center of Hypertension and Nephrology Unit, Department of Systems Medicine, University of Rome Tor Vergata, Via Montpellier 1, 00133 Rome, Italy; giul.marr@gmail.com (G.M.); dilauromanuela@gmail.com (M.D.L.); francesca.didaniele@gmail.com (F.D.D.); daniela.palazzetti96@gmail.com (D.P.); cristinaguerriero@hotmail.it (C.G.); annalisa.noce@uniroma2.it (A.N.)

2 School of Applied Medical, Surgical Sciences, University of Rome Tor Vergata, Via Montpellier 1, 00133 Rome, Italy

* Correspondence: didaniele@med.uniroma2.it; Tel.: +39-062090-2982; Fax: +39-062090-3362

Abstract: The most common manifestation of cardiovascular (CV) diseases is the presence of arterial hypertension (AH), which impacts on endothelial dysfunction. CV risk is associated with high values of systolic and diastolic blood pressure and depends on the presence of risk factors, both modifiable and not modifiable, such as overweight, obesity, physical exercise, smoking, age, family history, and gender. The main target organs affected by AH are the heart, brain, vessels, kidneys, and eye retina. AH onset can be counteracted or delayed by adopting a proper diet, characterized by a low saturated fat and sodium intake, a high fruit and vegetable intake, a moderate alcohol consumption, and achieving and maintaining over time the ideal body weight. In this review, we analyzed how a new nutritional approach, named caloric restriction diet (CRD), can provide a significant reduction in blood pressure values and an improvement of the endothelial dysfunction. In fact, CRD is able to counteract aging and delay the onset of CV and neurodegenerative diseases through the reduction of body fat mass, systolic and diastolic values, free radicals production, and oxidative stress. Currently, there are few studies on CRD effects in the long term, and it would be advisable to perform observational studies with longer follow-up.

Keywords: arterial hypertension; endothelial dysfunction; organ damage; caloric restriction diet; intermittent fasting

Citation: Di Daniele, N.; Marrone, G.; Di Lauro, M.; Di Daniele, F.; Palazzetti, D.; Guerriero, C.; Noce, A. Effects of Caloric Restriction Diet on Arterial Hypertension and Endothelial Dysfunction. *Nutrients* **2021**, *13*, 274. https://doi.org/10.3390/nu13010274

Received: 20 December 2020
Accepted: 15 January 2021
Published: 19 January 2021

1. Introduction

As is well known, high blood pressure (BP) is one of the most important public health problems worldwide. High BP values may be caused by elevated cardiac output, enhancement of peripheral vascular resistance, or by a combination of both, and they negatively impact on the average life expectancy [1,2]. In fact, it is estimated that the prevalence of arterial hypertension (AH) among adults will increase from 26.4% to 29.2% in the year 2025. Currently, the World Health Organization (WHO) states that it affects one in four men and one in five women, meaning more than 1 billion people [3]. Therefore, cardiovascular (CV) diseases are one of the main causes of death in the word, together with newly arising infectious causes such as SARS-CoV-2 [4]. The most frequent CV diseases are myocardial infarction and stroke [5]. The affected target organs by AH are heart, brain, kidneys, vessels, and eye retina [3]. According to the degree of AH, clinicians should provide the most appropriate pharmacological and non-pharmacological therapy, as well [6]. The latter include undertake a healthy lifestyle, practice habitual physical exercise, avoid smoke, follow a diet with low-sodium intake (<100 mEq/day), and lose weight (in case of overweight or obese subjects). If lifestyle changes are not able to normalize BP, it is mandatory to begin a pharmacological treatment in order to restore BP values in the range of normality [7].

Currently, it is speculated that one of the possible nutritional strategies useful for the management of AH is a caloric restriction diet (CRD) [8,9].

Numerous studies have shown that eating habits are able to modify CV risk factors [10–12]. These impact on endothelial function, favoring the inflammatory processes underlying atherosclerosis [13]. In physiological conditions, the vascular endothelium maintains its tone through the release of signaling molecules with vasodilator (such as nitric oxide- (NO)) and vasoconstrictor (such as angiotensin II) action [14]. The endothelial dysfunction occurs when there is an abnormal production of reactive oxygen species (ROS) of pro-inflammatory cytokines, such as interleukin (IL)-1 and tumor necrosis factor (TNF)-α, and a decrease release of NO [15]. This condition triggers the atherosclerosis process [14]. For this reason, it is important, if not essential, undertake a nutritional treatment in patients with high BP. In this review, we focused on the possible beneficial effect of CRD on the BP control, highlighting the main antihypertensive mechanisms exerted by this nutritional approach.

2. Definition, Classification, and Management of Arterial Hypertension

AH is defined as a "condition characterized by increased blood pressure in the blood vessels" and when it becomes too high, and persists over time, it can damage the arteries and organs, leading the heart to a greater cardiac output [16]. Therefore, a condition in which systolic BP (SBP) is higher than 140 mm Hg and/or diastolic BP (DBP) is more than 90 mm Hg is defined as AH, as reported by the latest 2018 European Society of Cardiology (ESC) and European Society of Hypertension guidelines (ESH) [17].

Specifically, BP is optimal when SPB values are <120 mm Hg and DBP values are <80 mm Hg, and it is normal when SPB values are between 120 and 129 mm Hg and DBP values are between 80 and 84 mm Hg. BP is considered high-normal for SBP values between 130 and 139 mm Hg and for DPB values between 85 and 89 mm Hg. It defines grade 1 AH when SPB values are between 140 and 159 mm Hg and/or DPB values are between 90 and 99 mm Hg; grade 2 AH when SPB values are between 160 and 179 mm Hg and/or DPB values are between 100 and 109 mm Hg; grade 3 AH when SPB values are ≥180 mm Hg and/or DPB values are ≥110 mm Hg [17].

High BP can cause left ventricular hypertrophy and then heart failure, renal failure (nephroangiosclerosis), accelerated atherosclerosis, and hypertensive retinopathy (Figure 1). AH arises from the combination of genetic and environmental factors. Therefore, it should be useful to identify predisposed patients and instruct them to lifestyle changes. Such lifestyle changes concern to the adoption of healthy diet, in order to avoid high alcohol consumption, to promote regular physical activity, to maintain the normal BW, to stop smoking and to avoid passive smoking [18]. In Italy, it is estimated that there are about 15 million hypertensive subjects; in fact, 28.3% of the population is affected by AH [19,20]. Several clinical trials have shown that a lifestyle change can delay or prevent AH onset in people who are not hypertensive, delay or prevent drug therapy in subjects with grade I AH, contribute to the BP reduction in hypertensive individuals already in pharmacological therapy, and reduce the number and dosage of antihypertensive drugs [21].

The wide availability of drugs offers the possibility to obtain a fast-hypotensive effect and to act positively on the mechanisms that predispose to CV events.

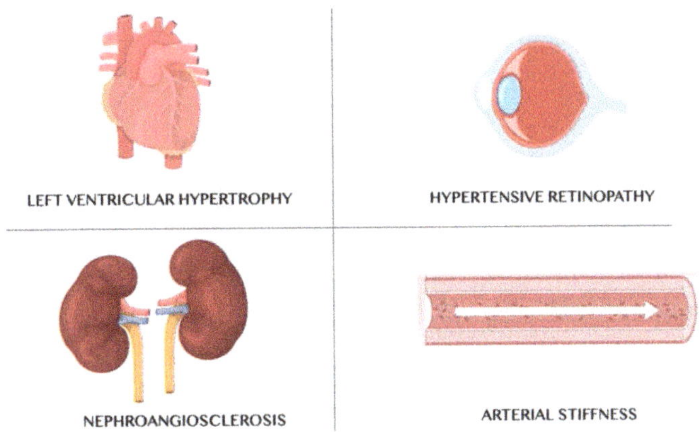

Figure 1. Main target organs of arterial hypertension.

3. Arterial Hypertension and Endothelial Alterations

AH is the most important risk factor for CV diseases [22]. In fact, vascular alterations (such as endothelial dysfunction) are a pathophysiological response mechanism to the development of organ damage, and they should be taken into account for the global CV risk assessment. The endothelium plays a pivotal role, and it consists of 1.2 billion cells, with a weight exceeding 1.5 kg and an area of 400 m^2 [23]. It forms a thin cellular lamina in direct contact with the bloodstream, representing the innermost layer of the vessel wall. The most important functions performed by endothelium are the modulation of the inflammation, the regulation of the vasomotor tone, the promotion and inhibition of cell proliferation and the modulation of the coagulative cascade [24–26]. The main physiological mediator of endothelium is NO, but it displays also other important functions at the level of the central nervous and immune systems. Endothelial cells produce NO through the enzyme NO-synthase (NOS), which transforms L-arginine amino acid into citrulline [27]. The activity of NOS is stimulated by numerous mediators, such as bradykinin and acetylcholine, or by mechanical forces, mainly "shear stress" [27]. NO is a volatile gas that has few seconds half-life and that, spreading towards the vessel wall smooth muscle cells, causes the release of cyclic guanosine-monophosphate (cGMP) with the consequent reduction of intracellular calcium [28].

The term "endothelial dysfunction" identifies a pathophysiological condition characterized by anatomically intact endothelial cells, but when stimulated, instead of solely determining the production of NO, activates in parallel the production of ROS, which causes the degradation of NO [29]. Although the endothelial dysfunction is mainly caused by increased destruction of NO, it may also depend on its reduced production due to L-arginine substrate deficiency, or by vasoconstriction induced by factors derived from cyclooxygenase [23].

The endothelial damage is instead represented by the destruction of endothelial cells, where the regeneration of these cells is difficult to achieve [30]. The BP increase causes an enhancement in the production of superoxide and a decrease in the bioavailability of NO [31]. There is also another mechanism involving the renin–angiotensin–aldosterone system (RAAS). The angiotensin conversion enzyme (ACE) acts on the endothelium by converting angiotensin I to angiotensin II, an endocrine vasoactive peptide. The latter is an active protein able to induce vasoconstriction through the calcium-dependent myosin phosphorylation with the contraction of arterial smooth cells, inducing an enhancement of BP levels. Moreover, angiotensin II stimulates in the kidneys the sodium reabsorption which, in turn, induces water retention. Angiotensin II promotes also the production of endothelin, a class of proteins with paracrine/vasoconstrictive action, causing an increase in blood pressure [32,33]. Endothelin is synthetized in the endothelial cells through two different pathways. The first is named "constitutive", and it is characterized by a continuous release of endothelin from macrovesicles which, in turn, interact with their own receptors, maintaining the vascular tone. The second is defined as a "regulated" mechanism and is activated by external physiological or pathophysiological stimuli. In fact, the endothelin that is released by the endothelial cells through this system, interacts with two types of receptors, namely, ET_A, placed on the smooth muscle layer of the vessel, and, to a lesser extent, with ET_B, which mediate the vasoconstrictive action [34].

Normally, there is a balance between vasoconstrictive and vasodilating substances in the bloodstream, but in the case of AH, the bioavailability of endothelin can be increased in parallel with a reduction in the bioavailability of NO. Angiotensin conversion enzyme inhibitors (ACE-I) and angiotensin receptor blockers (ARB) represent the first-line therapies for AH [7,17]. Dysfunctional endothelium shows a poor vasomotor function, which leads to an elevated BP at rest [35]. To date, it is speculated that there is a correlation between AH and endothelial dysfunction, but the issue is still unresolved due to the few studies in this regard [36–39].

4. Methods

PubMed, Web of Science, and Scopus online libraries were searched up until November 2020 in order to assess the most interesting and recent evidences about CRD and AH. The search was performed manually, using a combination of MeSH terms and keywords such as "caloric restriction diet", "arterial hypertension", "endothelial dysfunction", and "dietary protocol". All the studies were in the English language.

5. Caloric Restriction Diet

The CRD innovative approach consists of a chronic reduction in daily caloric intake of about 25–30% compared to the normal caloric intake, without any exclusion of food groups [40]. Although this regimen is not standardized, numerous studies show its effectiveness. Currently, according to the Calorie Restriction Society, subjects who follow a self-imposed CRD regimen are characterized by an increase in life expectancy. This regime consists of a caloric restriction with a daily consumption lower than 1800 kcal for an average period of 15 years and with an energy intake 30% lower than a group of individuals (homogeneous for age, gender, and socio-economic status) who consumed a Western diet model [40,41].

The first animal study to assess the beneficial effects of the CRD was carried out in rats in the 1900s [42]. Following studies showed that restricting food intake in mice delayed their growth but also extended their lifespan by to two times. Compared to a group that did not follow any restrictions, the CRD mice group showed anti-aging effects [43]. The most reliable hypothesis of the anti-aging effect of CRD is associated with reduced oxidative stress (OS). During the cellular respiration, oxygen is converted into ROS. ROS are able to react with macromolecules within cells causing the process of cellular aging. Therefore, CRD was initially designed for its antioxidant effect and consequently for its anti-aging action [44]. At this regard, numerous studies have been carried out, although contrasting results have been obtained. In fact, in some mice, without the superoxide dismutase enzyme, there was not increase in the aging, despite the enhanced of oxidative damage [45,46]. Animal studies on CRD showed that initial food restriction followed by alternating fasting acted positively on glycemic control, body weight (BW) reduction, insulin sensitivity, and BP control [47]. Anisimov and Bartke confirmed that serum values of glycemia and insulin were lower under CRD treatment [48]. In this context, the key role exerted by CRD is due to the decrease in insulin and insulin-like growth factor-1 (IGF-1) release and to the insulin sensitivity enhancement, observed both in rodents and in monkeys [49,50]. The results obtained in these studies confirm that the aging-related pathologies' arose later in the CRD group compared to *ad libitum* diet group [51].

In recent years, many theories have been developed to explain the CRD beneficial effects. Most of them focused their attention on the sirtuin family. Sirtuins are proteins nicotinamide dinucleotide (NAD+)-dependent deacylases able to prevent some diseases and to modulate several aspects of cellular aging, promoting DNA integrity through the maintenance of the normal degree of chromatin condensation and repairing promptly DNA damage [52–55]. In mammals, sirtuin1 (SIRT1) cytoplasmatic protein acts on the control of cell cycle, particularly on the apoptosis and on the mitochondrial metabolism [56]. Several studies showed that sirtuins levels after CRD increased in some specific tissues such as the gut, brain, kidney, adipose tissue, and skeletal muscle [57–59]. In a study conducted by Meyer et al., the authors demonstrated a significant effect of CRD on cardiovascular aging [60]. Twenty-five healthy subjects (mean age 53 $\pm$ 12 years) that followed CRD were matched with 25 subjects homogeneous for age and gender (control group) that followed a typical Western diet. All enrolled subjects underwent to diastolic function evaluation by transmitral flow, Doppler tissue imaging, and model-based image processing (MBIP) of E waves and inflammatory status assessment thanks to C-reactive protein (CRP), TNF-α, and transforming growth factor-beta1 (TGF-β1) blood sampling. Results showed that parameters such as Doppler flow diastolic function mean indexes, BP, CRP, TNF-α, and TGF-β1 were significantly lower in the CRD subject, compared to the control group. The results of this study highlighted that CRD is able to ameliorate cardiac function, slowing the cardiac-aging process and decreasing systemic inflammation and BP values. Furthermore, it is confirmed that CRD increases life expectancy.

Following a CRD, it is possible counteract aging in all living forms, lengthening both the median and the maximum duration of life, and delaying over time the appearance of CV and neurodegenerative diseases [61,62].

Human's life span has increased considerably due to the improvement of hygienic conditions and greater availability of pharmacological therapies. Taking into account the idea that the caloric restriction prolongs the life span, the mechanism of action by which this is possible, is not fully understood yet. Recent studies have shown that CRD can determine the damaged DNA repair and decrease fat mass, SBP and DBP values, the production of free radicals. The results obtained from the CRD can occur quickly, but they can mitigate in case of its suspension [63].

6. Caloric Restriction Diet and Arterial Hypertension

The CRD would seem to exert a beneficial effect against AH (Table 1) and for this reason represents a useful tool for its clinical management (Figure 2).

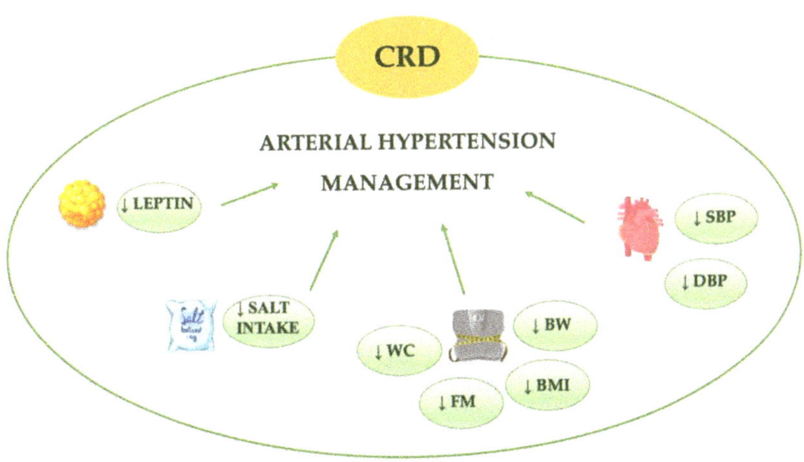

Figure 2. Beneficial effects of caloric restriction diet (CRD) on arterial hypertension. BMI, body mass index; BW, body weight; DBP, diastolic blood pressure; FM, fat mass; SBP, systolic blood pressure; WC, waist circumference, ↓ decrease.

Table 1. Dietary approaches in the arterial hypertension management.

Nutritional Treatment	Type of the Study	Authors	Year	Organ/System/Metabolic Target	Duration of Intervention	Impact of Intervention
CRD	Animal study	Zanetti et al. [64]	2010	• Vascular endothelial • Oxidative stress	3 weeks	↓ Oxidative stress ↓ iNOS ↓ Total nitrite ↓ Calcium-independent NOS activity ↑ SOD activity
CRD	Animal study	Rippe et al. [65]	2010	• Vascular endothelial • Oxidative stress	8 weeks	↓ Oxidative stress ↓ Blood glucose ↓ TG ↑ NO bioavailability ↑ Arterial expression of SIRT1
CRD	Animal study	Donato et al. [66]	2013	• Oxidative stress • Cardiovascular	30–31 months	↓ NADPH oxidase activity ↓ Oxidative stress ↓ TG ↓ BW ↑ SOD activity ↑ Catalase
CRD	Animal study	Kobara et al. [67]	2015	• Cardiovascular Oxidative stress	4 weeks	↓ ROS ↓ Cardiac hypertrophy and fibrosis
CRD	Animal study	Waldman et al. [68]	2018	• Cardiovascular • Glucose profile	4 weeks	↓ Oxidative stress ↓ Inflammation ↑ SIRT1
CRD	Animal study	An et al. [69]	2020	• Cardiovascular Iron metabolism	12 weeks	↓ Cardiac hypertrophy and fibrosis ↓ Cardiac inflammation ⇅ Cellular regulation of iron homeostasis
CRD	Human study	Wadden et al. [70]	1998	• Leptin • Body composition	40 weeks	↓ BW ↓ FM ↓ Serum leptin

Table 1. *Cont.*

Nutritional Treatment	Type of the Study	Authors	Year	Organ/System/Metabolic Target	Duration of Intervention	Impact of Intervention
CRD	Human study	Nakano et al. [71]	2001	• Glucose profile • ANS • Body composition	2 weeks	↓ BMI ↓ BW ↓ TG ↓ HOMA-index ↓ SBP ↓ DBP
CRD	Human study	Facchini et al. [72]	2003	• Cardiovascular • Body composition	3 weeks	↓ BMI ↓ Heart rate ↑ Parasympathetic activity
CRD	Human study	Das et al. [73]	2007	• Body composition • Lipid profile	12 months	↓ FM ↓ BW ↓ TG ↓ insulin ↓ LDL-c ↓ TC ↑ HDL-c
CRD	Human study	Lecoultre et al. [74]	2011	• Body composition • Leptin	6 months	↓ BW ↓ Mean 24 h circulating leptin ↓ Urinary norepinephrine
CRD	Human study	Stewart et al. [75]	2013	• Body composition • Lipid profile	24 months	↓ LDL-c ↓ TC/HDL-c↓ TG ↓ DBP ↓ BW ↓ BMI ↓ FM ↓ MSS ↑ IS
CRD	Human study	Ruggenenti et al. [76]	2017	• Renal function • Glucose profile • Body composition	6 months	↓ WC ↓ BW ↓ BMI ↓ BG ↓ HbA_{1c}

Table 1. *Cont.*

Nutritional Treatment	Type of the Study	Authors	Year	Organ/System/Metabolic Target	Duration of Intervention	Impact of Intervention
CRD	Human study	Most et al. [77]	2018	• Cardiovascular • Glucose profile	24 months	↓ VAT ↓ SAAT ↓ SBP ↓ DBP ↓ TC ↓ LDL-c ↓ IR
CRD	Human study	Kraus et al. [78]	2019	• Body composition • Lipid profile • Glucose profile • Cardiovascular • Lipidic profile	24 months	↓ LDL-c ↓ TC/HDL-c ↓ SBP ↓ DBP ↓ MSS ↑ IS
DASH DIET	Human study	Harsha et al. [79]	1999	• Cardiovascular	8 weeks	↓ SBP ↓ DBP
IF	Human study	Arnason et al. [80]	2017	• Glucose profile • Body composition	6 weeks	↓ BW ↓ BMI
IF	Human study	Erdem et al. [81]	2018	• Cardiovascular	-	↓ SBP ↓ Heart rate ↓ USE
IF	Human study	Furmli et al. [82]	2018	• Glucose profile • Body composition	12 weeks	↓ HbA_{1c} ↓ BW ↓ WC

Table 1. *Cont.*

Nutritional Treatment	Type of the Study	Authors	Year	Organ/System/Metabolic Target	Duration of Intervention	Impact of Intervention
IF	Human study	Wilhelmi de Toledo et al. [83]	2019	• Cardiovascular • Glucose profile • Body composition • Lipidic profile	24 months	↓ SBP ↓ DBP ↓ BW ↓ Abdominal circumference ↓ Blood glucose ↓ TG ↓ LDL-c ↓ HDL-c ↓ TC ↑ Physical and emotional well-being
CRD *vs* IF	Animal study	Magger et al. [84]	2006	• Cardiovascular • Body composition	16 weeks	↓ BW ↓ Heart rate ↓ SBP ↓ DBP ↓ Blood glucose

Abbreviations:ANS, autonomic nervous system; BG, blood glucose; BMI, body mass index; BW, body weight; CRD, caloric restriction diet; DBP, diastolic blood pressure; FM, fat mass; HbA_{1c}, glycated hemoglobin; HDL-c, high-density lipoprotein cholesterol; IF, intermittent fasting; iNOS, inducible nitric oxide synthase; IR, insulin resistance; IS, insulin sensitivity; LDL-c, low density lipoprotein cholesterol; MSS, metabolic syndrome score; NADPH, nicotinamide adenine dinucleotide phosphate; NO, nitric oxide; ROS, reactive oxygen species; SAAT, subcutaneus abdominal adipose tissue; SBP, systolic blood pressure; SIRT1, sirtuin 1; SOD, superoxide dismutase; TC, total cholesterol; TG, triglycerides; USE, urinary sodium excretion; VAT, visceral adipose tissue; WC, waist circumference, ↑ increase, ↓ decrease.

An important study conducted in this regard was the CALERIE (Comprehensive Assessment of Long-term Effects of Reducing Intake of Energy) [85]. The CALERIE study was a randomized controlled trial with a two-year follow up. This study was divided in two phases: CALERIE-1 and CALERIE-2 [73,75]. CALERIE-1 study was performed to assess the possible effects induced by a reduction of 10-30% of caloric intake on body composition parameters and lipid profile after 6 and 12 months in a population of middle-aged non-obese subjects. CALERIE-1 results showed an improvement in lipid and glycemic profile and a reduction in BW and fat mass. CALERIE-2 was the largest multi-center study on CRD, involving three centers, namely the Pennington Biomedical Research Center (Baton Rouge, LA, USA), Tufts University (Boston, MA, USA), and Washington University School of Medicine (St. Louis, MO, USA), and was coordinated by Duke Clinical Research Institute (Durham, NC, USA). A total of 2020 subjects were enrolled randomly with a 2:1 allocation into two subgroups: 145 in the CRD group and 75 in the *ad libitum* group. The CRD group followed 25% caloric restriction for two-years [78]. After two years of diet treatment, cardiometabolic risk factors such as low-density lipoprotein cholesterol (LDL-c), total cholesterol/high-density lipoprotein cholesterol (HDL-c) ratio, SBP and DBP decreased. Moreover, serum biomarkers such as CRP, insulin sensitivity index and metabolic syndrome score were reduced. Moreover, BW was significantly lower in the CRD group when compared to the *ad libitum* group (average weight loss in CRD group was 7,5 kg *vs* average BW increase of 0,1 kg in *ad libitum* group). These data showed that a period of two–years of CRD was able to decrease cardiometabolic risk factors in middle-aged non-obese subjects. For this reason, it is possible to consider CRD as nutritional therapeutic approach to enhance life expectation and reduce the onset of chronic non-communicable diseases such as diabetes mellitus, cancer, chronic kidney disease, and AH, among others [86,87].

Other studies have been conducted to investigate the role of CRD in the control of AH. In particular, a study performed by Most et al. on caloric restriction (25%), with two years follow–up, evaluated the possible reduction of CV risk factors and insulin resistance in non-obese subjects and whether the results obtained were maintained over time or were limited to the period study. The authors showed a significant weight loss associated to a decrease in SBP and DBP and an improvement in other parameters, such as lipid profile and insulin resistance. These improvements, with the exception of insulin sensitivity, appeared to be maintained over time [77].

Another study examined the impact of CRD (25%) for six–month follow-up period in patients affected by type 2 diabetes mellitus, AH, and glomerular hyperfiltration. The results were positive, since the authors observed that the CRD reduced glomerular hyperfiltration and, improved insulin sensitivity and SBP and DBP values, compared to the control the group that followed a standard diet. In general, it is speculated that CRD ameliorates CV risk factors [76]. To explain the mechanism underlying the reduction in BP induced by CRD, it has been hypothesized that it may act through the activation of the autonomic nervous system. This hypothesis was investigated by Nakano et al, who observed a reduction in SBP and DBP in obese hypertensive patients in CRD treatment (800 kcal/day) with normal sodium content for two weeks [71]. In obese individuals, CRD would appear to improve the balance of night activation between the vagal/sympathetic system. In fact, in obese subjects, there is an alteration of the autonomic control of the heart due to a prevalence of the sympathetic component over the parasympathetic component one in the autonomic equilibrium. A subsequent study confirmed the effects of short-term CRD (three weeks) in combination with a high–intensity exercise program on heart rate variability (HRV) in normotensive obese subjects, demonstrating that a short-term program of CRD associated with high intensity exercise can positively change the autonomic profile, leading to a reduction in HRV and an increase of parasympathetic activity [72]. The long-term effects of CRD on autonomic nervous system activity are relevant. In an animal study, long-term CRD intake would appear to be able to slow down age-related functional changes in the autonomic system. In fact, in male rats, CRD causes an increase in HRV of the highest frequency components (biomarkers of parasympathetic nervous system activity) and reduces the low-frequency component of diastolic pressure variability (biomarkers of sympathetic tone) [84].

7. Caloric Restriction Diet and Hypertensive Organ Damages

The damage caused by AH is directed to important target organs and develops essentially as a direct and/or indirect consequence of vascular pathology (Figure 3) [88]. In the majority of patients with high BP, the increased values may not be too severe, and the direct damage may develop after years of the AH onset. As previously mentioned, the organs most affected by AH are the heart, kidneys, eye retina, vessels and brain. The pathogenesis of organ damages is complex and varied [89].

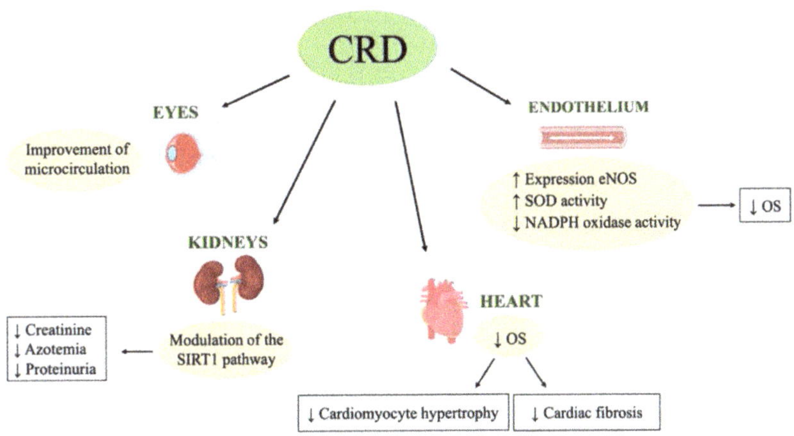

Figure 3. Improvement of caloric restriction diet on hypertensive organ damage. CRD, caloric restriction diet; eNOS, endothelial nitric oxide synthase; NADPH, nicotinamide adenine dinucleotide phosphate hydrogen; OS, oxidative stress; SIRT1, sirtuin 1; SOD, superoxide dismutase, ↑ increase, ↓ decrease.

Increased BP, in addition to the direct damage to the vessel walls, can cause additional pathogenic pathways that include endothelial dysfunction, inflammation, and OS, as well as, all the changes that induce vascular structure alteration, namely arterial rigidity, already underlining to the aging. Therefore, AH is considered to be an accelerated form of cardiovascular aging. Several studies have highlighted a protective action exerted by CRD against hypertensive organ damage.

7.1. Left Ventricular Hypertrophy

Over time, in hypertensive patients, the left ventricular becomes increasingly rigid and diastolic filling is compromised. Left ventricular hypertrophy (LVH) can be described as the thickening of the ventricular wall caused to cope with the overload imposed to the heart for offset peripheral resistances, hence the increase in BP. LVH may occur in both sexes and leads to an increase in either oxygen consumption or energy expenditure of cardiac output [90,91].

It has recently been observed that in hypertensive subjects, LVH is related to hemodynamic and non-hemodynamic mechanisms, which are observed in particular in obese and overweight subjects. In fact, LVH is observed mainly in these patients. The non-hemodynamic mechanisms involve an impairment of lipid metabolism but also the production by adipose tissue of adipokines, such as adiponectin, leptin, and TNF-α [92]. In confirmation of this, recent animal studies have demonstrated that leptin-deficient and leptin-resistant mice exhibit obesity, insulin resistance, and LVH, although the mechanisms that related LVH and alteration of leptin metabolism are not fully understood [93]. Most obese and hypertensive subjects show leptin resistance, and their leptin plasma level is directly related to cardiac hypertrophy [94]. Leptin is involved in heart complications typical of obesity, including AH. It has been shown that an acute increase in leptin does not appear to have any effect on the BP values. On the contrary, a chronic enhancement in leptin increases BP values, stimulating the sympathetic nervous system and simultaneously altering the mechanisms designed to counteract it, such as the natriuresis and the synthesis of NO [95]. Finally, chronic hyperleptinemia appears to be directly related with BP levels [96].

However, leptin may play also a cardioprotective role, related to BW reduction and improvement of myocardial metabolism. Leptin avoids an excessive accumulation of lipids in the heart in obese subjects and inhibits the formation of toxic lipid derivatives, which induce a condition called "cardiac lipotoxicity" [97]. This cardioprotective role exerted by leptin in obese patients, was confirmed only in a study conducted on 1172 black obese women, which revealing an inverse correlation between leptin and LVH severity [98].

CRD is capable of reducing leptin blood levels [74]. This reduction would initially seem to be related to the degree of caloric restriction, but in the long time it appears to be related to the reduction of BW and visceral fat [70]. The CRD seems to exert a cardioprotective action derived from the decrease of OS and inflammation, from the cellular regulation of iron homeostasis and from cardiac remodeling. For these reasons, CRD is definitely a new adjuvant treatment, in combination with pharmacological therapy, for cardiomyopathy in hypertensive patients [99].

Moreover, CRD helps to improve the picture of cardiac hypertrophy through SIRT1 and peroxisome proliferator-activated receptor gamma coactivator 1(PGC1)-α pathways [68]. In fact, as previously reported, CRD would seem to increase the activity of SIRT1, which in turn stimulates PGC1-α, involved in inflammatory signaling pathways. The protective action induced by SIRT1 is related to glucose metabolism as it counteracts the accumulation of cardiac fatty acids, exerting a protective action against cardiotoxicity [68]. Another CRD cardioprotective effect is related to suppression of OS present in AH condition. In fact, recent studies show how ROS production is involved in cardiomyocyte hypertrophy and cardiac fibrosis [100]. Kobara et al., demonstrated that CRD reduces ROS production at the cardiac level, improving the condition of OS and ameliorating cardiac hypertrophy and fibrosis in hypertensive mice, subjected to a 40% reduction in caloric intake compared to the control group, after four weeks [67]. Finally, the last cardioprotective effect induced by CRD is related to the regulation of cellular iron homeostasis. The authors observed a reduction in inflammation, OS, and LVH, associated with a normalization of iron overload in leptin-resistant obese mice subjected to 12 weeks of CRD. This effect would appear to be due to the modulation of gene expression induced by CRD of the genes involved in iron homeostasis at the level of heart tissue [69].

Further studies will be needed for demonstrate if factors such as age and period of following CRD impact on obtained results.

7.2. Kidney Damage—Nephroangiosclerosis

Classically, the term nephroangiosclerosis (NAS) refers to the presence of non-immune-mediated vascular lesions and it is a very common condition in presence of high BP values [101]. The organ damage caused by AH on the renal parenchyma shows up by the increase in plasma creatinine and in early stage by the presence of albuminuria [102]. The classic definition has been revised as the immune cells contribute to determining kidney organ damage. In fact, it has been highlighted how different stimuli, including an increase in angiotensin II production or a diet with a high salt content, can induce a response by the immune system that causes an inflammatory reaction mainly at the level of the organs involved in the control of BP values (such as kidneys, heart and nervous system) [103–105]. This observation is confirmed by the fact that hypertensive patients usually present a low-grade chronic inflammatory state. It has also been shown that a high salt diet can stimulate cells of the immune system, including T lymphocytes, which in turn produce pro-inflammatory cytokines, while the activation of monocytes and macrophages can induce both vasoconstriction and renal sodium retention [106]. The three mechanisms described above, taken together, contribute to producing organ damage at the level of the kidney, such as NAS [107]. In terms of NAS, it is important to reduce BP values and proteinuria since they are related to the progression of the nephropathy and to cardiovascular events [108].

According to a recent meta-analysis that examined 27 studies, CRD appears to exert a protective effect against renal damage induced by AH [109]. This meta-analysis demonstrated that in chronic kidney disease (CKD) rat models undergoing CRD, there was a higher reduction in BP, creatinine, azotemia, and proteinuria, compared to a control group fed *ad libitum* [109]. The authors also demonstrated a survival rate increase at 700-800 days. This nephroprotective effect of CRD would seem to be exerted through the activation of AMP-activated protein kinase (AMPK) and the modulation of the SIRT1 pathways, the latter having been previously described [110]. In particular, the CRD seems to impact, in CKD rat models, on the activity of the AMPK and of the mammalian target of rapamycin (m-TOR) pathways, both implicated in cellular energy metabolism. Increased expression of m-TOR has been shown to accelerate kidney aging, while phosphorylation of AMPK inhibits this process. A short-term, CRD has been seen to induce an up-regulation of AMPK and a down-regulation of m-TOR, resulting in slowing the aging process of renal tubular cells [111]. Therefore, CRD in combination with a reduced sodium intake could be a valid nutritional alternative in the treatment of NAS patients.

In fact, a meta-analysis conducted by D'Elia et al. [112] confirmed the nephroprotective role of sodium restriction, pointing out the relation among dietary sodium restriction and urinary albumin excretion (UAE), a biomarker of CKD progression. The authors considered a total of 11 studies who agreed on the close association between the reduction of UAE induced by a decrease of dietary sodium intake. Moreover, as previously demonstrated in the literature, the authors confirmed that this nutritional approach could ameliorate the therapy based on RAAS-blocking drugs in hypertensive patients. In this perspective, the reduction of UAE induced by low dietary sodium intake could impact on slowing down the progression of CKD, rather than reducing cardiovascular morbidity and mortality.

7.3. Arterial Stiffness

Arterial stiffness (AS) is an important CV risk factor [108], and it induces the rigidity of the arterial wall [113,114]. BP is the main determinant of AS, and the aortic rigidity is accentuated by other concomitant diseases such as diabetes mellitus, metabolic syndrome, CKD, and obesity [115].

The aortic pulsating wave velocity (PWV) correlates with the presence of organ damage and in the heart, an increased AS induces ventricular afterload, ventricular hypertrophy, and reduced coronary perfusion [116]. Sustained increases in BP promote collagen matrix synthesis, causing subsequent vascular thickening and vasal stiffening [117].

The protective role of CRD on AS has been extensively studied. Specifically, CRD seems to exert a protective effect on the endothelium through an antioxidant action and thanks to an increased NO bioavailability [64]. Rippe et al. observed an improvement in carotid artery endothelium-dependent dilation mediated by the enhancer of the expression of endothelial nitric oxide synthase (eNOS) in mice undergoing CRD (approximately 30% of caloric restriction) compared to the control group (fed *ad libitum*) after 8 weeks [65]. Similar results were observed in a subsequent study, where the authors demonstrated a reduction in nicotinamide adenine dinucleotide phosphate hydrogen (NADPH) oxidase activity and an increase in superoxide dismutase (SOD) and in catalase activities, confirming that the CRD has an important antioxidant action, in animal models [66]. The effect on the increase of NO availability would seem to be exerted by the action of CRD on SIRT1, which deacetylases and activates eNOS and increases the expression of antioxidant genes [118].

A recent meta-analysis [119] examined 11 clinical trials on the correlation between dietary salt restriction and AS. In the analyzed studies, an average decrease in dietary salt restriction highlighted a 2.38% reduction of PWV. This data would seem of great interest, but there are some concerns, common to all the studies examined, such as the duration of the intervention (ranging from a minimum of 1 to a maximum of 6 weeks) and the small samples of enrolled subjects in each study.

7.4. *Hypertensive Retinopathy*

Hypertensive retinopathy is an ocular disease involving the arteries and veins of the retina, the optic nerve, and the choroid. The cause of this condition is AH. In subjects that have high BP values, over the time, a modification of the retinal arteries takes place, which tends to shrink, and the retinal veins instead tend to assume a different course, no longer linear. These changes can impair normal vision and lead to the formation of ischemic areas of the retina, with the formation of exudates [120]. In more advanced forms, the vision can be compromised and become blurred or distorted. The diagnosis is made by examining the ocular fund which allows the physician to assess the vision, or to see if there are small bleeding or edema. Depending on what can be observable, it can be possible to evaluate the stage of the disease. When AH is associated with diabetes mellitus, both increase the risk of CV complications and the retinopathy becomes more severe. Since the retina is one of the organs most sensitive to changes in microcirculation, signs of damage to this organ have a high prognostic value for myocardial ischemia, carotid sclerosis and coronary artery damage [121].

The role of CRD on the eye has been extensively investigated, especially with regard to age-related eye diseases [122], but no studies have been conducted to investigate the *in vivo* effect of CRD on hypertensive retinopathy. However, as reported above, it is well known that the control of BP within the normal range, also through CRD, exerts a preventive role against the onset of eye microvascular abnormalities leading to hypertensive retinopathy [123].

8. Secondary AH: Traditional Dietary Nutritional Protocols

To date, secondary AH affects about 5-10% of the hypertensive subjects. In the greater percentage of these cases, it is possible to observe reversible causes [124]. In the population over 65 years of age, the etiology of secondary AH is due to conditions such as renal artery stenosis, CKD, hypothyroidism, hyperaldosteronism, obstructive sleep apnea, and Cushing syndrome [125]. Resistant AH is a condition characterized by high levels of BP (i.e. >140/90 mmHg in the general population and 130/80 mm Hg in diabetic or nephropathic patients) that could not be controlled by medications (for example the combination of three or more antihypertensive drugs belonging to different classes, including diuretics at their maximum tolerated dosage), according to current guidelines [126].

A large number of scientific evidences supports the notion that multiple dietary factors influence BP levels in the general population both in cases of primary and secondary AH [127]. Secondary forms of AH are relatively rare, but their recognition is very important if we consider that some of them can heal permanently after the removal of the cause.

The suspicion of a AH secondary form should raise if (i) AH is resistant to a triple drug therapy that includes a diuretic drug, (ii) AH, characterized by particularly high BP values, diagnosed in young people (about 30 years of age), especially in women, (iii) in case of accelerated or malignant hypertension (renovascular hypertension), (iv) AH becomes suddenly uncontrollable with poly-pharmacological therapy or (v) there is an excessive BP values reduction after administration of RAAS blocking drugs [128]. For the prevention and treatment of secondary AH, as well as for the primary, is used a dietary model similar to the Mediterranean diet called DASH diet is used, which stands for "Dietary approaches to Stop Hypertension". The DASH diet was developed by studies supported by the U.S National Institute of Health (NIH), as a nutritional approach for the treatment of AH that could be of valuable assistance in terms of reaching optimal BP values, with minimal use of antihypertensive drugs (Figure 4) [129]. The purpose of the DASH diet is to restore homeostasis in all AH patients with an incorrect and unhealthy lifestyle. The DASH diet is considered to be the dietary "gold standard" approach by the American Society of Hypertension in reducing multiple CV risk factors, related to primary and secondary AH [130]. The DASH diet is characterized by a high intake of fish, fruit, and vegetables; a drastic reduction of the consumption of foods rich in saturated fats, such as red meat and aged cheeses; and a complete elimination of alcohol and "kitchen salt" assumption. While CRD, previously discussed, focuses on the number of calories consumed, reducing them by 25–30%, the DASH diet mainly focuses on the quality of micro- and macro-nutrients assumed. Numerous studies have shown that the DASH diet induces a greater reduction in BP values than other dietary interventions or physical exercise programs [131].

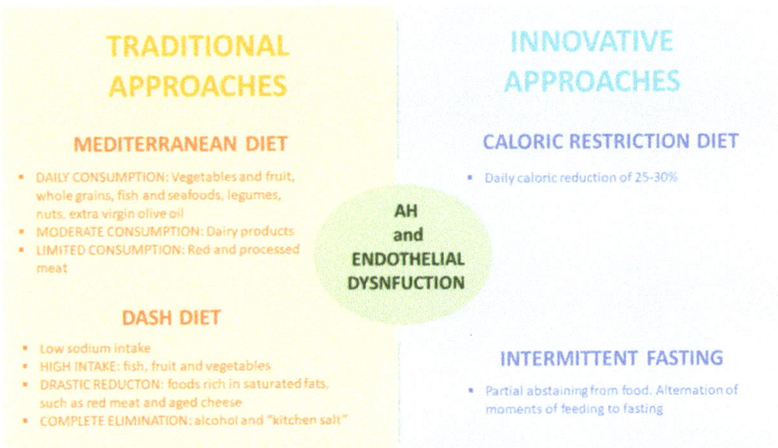

Figure 4. Traditional vs. innovative nutritional approaches for arterial hypertensions and endothelial dysfunction treatment. AH, arterial hypertension.

Most of the studies on salt reduction and weight loss were carried out on middle-aged subjects. In particular, the Trial of Nonpharmacological Interventions in The Elderly (TONE) study showed that, in AH subjects, a moderate salt restriction and weight loss allow the reduction of the number and dosage of antihypertensive drugs [131].

The study carried out by Harsha et al. on 459 adults with SBP values lower than 160 mm Hg and DBP values between 80 and 95 mm Hg demonstrated that the DASH diet reduced BP and represent a nutritional approach to prevent and treat AH [79].

For this reason, the Joint National Committee for Prevention, Treatment and Evaluation of High Blood Pressure strongly recommends following this dietary-nutritional approach to counteract high BP values [132].

Moreover, a recent meta-analysis conducted by D'Elia et al. [133] examined the impact of dietary sodium restriction on central blood pressure (cBP). In fact, if the effects of sodium restriction on peripheral BP are well known, its effects on cBP are scarce. The authors found a statistically significant association between the reduction of BP and central pulse pressure, speculating that sodium restriction also impacts on central BP values. For this reason, a diet with a low sodium intake is a useful tool to counteract the onset and/or the progression of CV disease, especially in normotensive subjects and in prehypertensive patients. This effect supports previous studies that exhort the benefits of a low sodium intake diet for the optimal control of BP values.

9. Other Innovative Nutritional Approaches for Essential AH Treatment

Among the innovative nutritional approaches for AH, considerable interest was aroused by intermittent fasting (IF) [134], which is less restrictive when compared to CRD [135]. IF consists in the assumption of the normal daily caloric quote in a well-defined time gap, daily or weekly [136]. In the IF panorama, two basics scenarios are commonly used. The first one is time-restricted feeding (TRF), and it can be applied in three different alternatives: (i) 16/8, (ii) 18/6, or (iii) 20/4. For example, in the case of "16/8" the first value indicates 16 hours of fast, and 8 indicates the nutritional window. This scheme can be applied to the different variants [136]. An alternative to TRF is a 24-h fasting period (meaning caloric assumption of around 400-600 kcal/day) followed by a 24-h eating period twice or three times *per* week. In this scenario, there are two possible combinations: (i) 5/2 or (ii) 4/3. For instance, in the "5/2" plan caloric restriction is applied for two days, while normal diet is followed for five days [137].

The overview of the mechanisms that have been hypothesized to assess the correlation between IF and human health are the reduction of OS, improvement of cognition function, and anti-inflammatory effects. With regard to the cardioprotective effects, research has observed a reduction of adipose tissue (visceral); an increased concentration of adiponectin; a decrease of leptin and LDL-c; and prevention of CV diseases, particularly AH [138].

According to recent studies, it is assumed that there are mechanisms to assess the correlation between IF and human health [139], namely circadian rhythm (if disturbed, it can cause an energy imbalance and it means an enhanced risk of chronic non-communicable diseases onset [140]) and gut microbiota dysbiosis (if the microbiota is altered there is an impaired gut permeability, with the promotion of systemic inflammation [141–143]). Fasting regiments can also impact on modifiable CV risk factors such as energy intake [144], energy expenditure, [145] and sleep quality [146].

The IF diet has shown positive effects in lowering BP, with a recent study conducted by Erdem et al. [81] evaluating, in 60 subjects, the possible effects on BP values. The obtained results showed that IF significantly reduced 24-hour urinary sodium excretion, being associated with a decrease in SBP ($p < 0.001$) and DBP ($p < 0.039$) values. The authors concluded that this reduction is partly related to the decrease sodium consumption during the IF period.

Another interesting study was conducted by Wilhelmi de Toledo et al. [83] on 1422 subjects with one-year follow-up. IF period was comprised between 4 and 21 days, and the enrolled patients were instructed to perform a moderate-intensity physical exercise. Enrolled patients were grouped into four subgroups of 5, 10, 15, and 20 $\pm$ 2 days of IF. Every meal contained 200–250 kcal on average. The authors observed a decrease of SBP and DBP in patients who followed IF for a longer period of time; however, in this study, the authors did not specify the amount of salt introduced during the IF period and in particular whether there were any changes in its intake respect to the non-fasting period. They speculated that the drop of BP was due to the enhancement of parasympathetic nervous system activity, of norepinephrine excretion by the kidney and natriuretic peptides, and greater insulin sensitivity. Moreover, it has been observed that the positive effects induced by IF were limited to the nutritional intervention time; in fact, when the nutritional treatment was suspended, the BP values returned to their initial values [84]. To understand whether the reduction of the BP values during the IF are related to reduced caloric intake or reduced salt intake, further randomized clinical trials are needed to assess the impact of each variable on BP values.

During IF, the concentration of plasma glucose decreases. The glycogen reserves in the liver are consumed, and at the same time the gluconeogenesis process is activated. Insulin and IGF-1 levels are reduced in the bloodstream, while glucagon levels are increased. During the lipolysis process, fatty acids are released to be converted through beta oxidation, to be released in the blood and used as a source of energy [147]. An IF nutritional plan improves glucose metabolism even in patients with type 2 diabetes mellitus. As reported in a study conducted by Furmili et al. [82], after 12 weeks of almost 850 kcal daily meals, enrolled subjects showed a weight loss with a normalized fasting glycemia, reduced values of glycated hemoglobin (Hba1c) and increased insulin sensitivity [80].

In animal and in human studies, it has been showed that IF is able to improve physical function, moreover, in mice kept in an IF state, showed greater resistance to running. Other parameters such as balance and coordination were also positive [148].

10. Conclusions

In recent years, there has been a growing interest in identifying an alternative food strategy for achieving and maintaining weight loss in overweight people and to counteract the onset of CV disease. The high BP is one of the main risk factors of serious disabling diseases such as stroke, myocardial ischemia, and heart and kidney failure. In all AH forms, from the milder ones to those refractories to the pharmacotherapy, the correct nutritional approach turns out to be effective in order to reduce in a short period of time the BP values until the target range is obtained. The use of CRD, as an innovative nutritional-dietary approach, highlighted a significant reduction in BP values, an improvement in endothelial dysfunction, causing a decrease in metabolic and inflammatory parameters related to chronic non-communicable diseases. The results of the clinical intervention studies do not allow us to reach solid and definitive conclusions about the long-term efficacy and safety of this nutritional approach, especially in terms of the risk of developing side effects and nutritional inadequacy from lack of macro and micronutrients.

Author Contributions: Conceptualization, N.D.D. and A.N.; writing—original draft preparation, G.M., M.D.L., F.D.D., D.P., and C.G.; writing—review and editing, N.D.D. and A.N. All authors have read and agreed to the published version of the manuscript.

Funding: This research received no external funding.

Institutional Review Board Statement: Not applicable.

Informed Consent Statement: Not applicable.

Data Availability Statement: Not applicable.

Conflicts of Interest: The authors declare no conflict of interest.

Abbreviations

ACE	angiotensin conversion enzyme
AH	arterial hypertension
AMPK	AMP-activated protein kinase
ARB	angiotensin receptor blockers
AS	arterial stiffness
BMI	body mass index
BP	blood pressure
BW	body weight
CALERIE	Comprehensive Assessment of Long-term Effects of Reducing Intake of Energy
cBP	central blood pressure
cGMP	cyclic guanosine monophosphate
CKD	chronic kidney disease
CRD	caloric restriction diet
CRP	c-reactive protein
CV	cardiovascular diseases
DBP	diastolic blood pressure
eNOS	endothelial nitric oxide synthase
ESC	European Society of Cardiology
ESH	European Society of Hypertension
HDL-c	high-density lipoprotein cholesterol
HRV	heart rate variability
IF	intermittent fasting
IGF-1	insulin-like growth factor-1
IL	interleukin
LDL-c	low-density lipoprotein cholesterol
LVH	left ventricular hypertrophy

m-TOR	mammalian target of rapamycin
MBIP	model-based image processing
MDRD	modification of diet in real disease study
NAD	nicotinamide dinucleotide
NAS	nephroangiosclerosis
NIH	National Institute of Health
NO	nitric oxide
NOS	nitric oxide-synthase
OS	oxidative stress
PGC1-α	peroxisome proliferator-activated receptor gamma coactivator 1-α
PWV	pulsating wave velocity
RAAS	renin–angiotensin–aldosterone system
ROS	reactive oxygen species
SBP	systolic bloody pressure
SIRT1	sirtuin1
SOD	superoxide dismutase
TGF- β1	transforming growth factor-β1
TNF	tumor necrosis factor
TRF	time-restricted feeding
UAE	urinary albumin excretion
WHO	World Health Organization

References

1. Kitt, J.; Fox, R.; Tucker, K.L.; McManus, R.J. New Approaches in Hypertension Management: A Review of Current and Developing Technologies and Their Potential Impact on Hypertension Care. *Curr. Hypertens. Rep.* **2019**, *21*, 44. [CrossRef] [PubMed]
2. Jordan, J.; Kurschat, C.; Reuter, H. Arterial Hypertension. *Dtsch. Arztebl. Int.* **2018**, *115*, 557–568. [CrossRef]
3. World Healt Organization (WHO). Hypertension. Available online: https://www.who.int/health-topics/hypertension/#tab=tab_1 (accessed on 13 November 2020).
4. Lawes, C.M.; Vander Hoorn, S.; Rodgers, A.; International Society of Hypertension. Global burden of blood-pressure-related disease, 2001. *Lancet* **2008**, *371*, 1513–1518. [CrossRef]
5. Yusuf, S.; Hawken, S.; Ounpuu, S.; Dans, T.; Avezum, A.; Lanas, F.; McQueen, M.; Budaj, A.; Pais, P.; Varigos, J.; et al. Effect of potentially modifiable risk factors associated with myocardial infarction in 52 countries (the INTERHEART study): Case-control study. *Lancet* **2004**, *364*, 937–952. [CrossRef]
6. Volpe, M.; Rosei, E.A.; Ambrosioni, E.; Cottone, S.; Cuspidi, C.; Borghi, C.; de Luca, N.; Fallo, F.; Ferri, C.; Morganti, A.; et al. 2012 consensus document of the Italian Society of Hypertension (SIIA): Strategies to improve blood pressure control in Italy: From global cardiovascular risk stratification to combination therapy. *High Blood Press Cardiovasc. Prev.* **2013**, *20*, 45–52. [CrossRef] [PubMed]
7. Flack, J.M.; Adekola, B. Blood pressure and the new ACC/AHA hypertension guidelines. *Trends Cardiovasc. Med.* **2020**, *30*, 160–164. [CrossRef] [PubMed]
8. Brandhorst, S.; Longo, V.D. Dietary Restrictions and Nutrition in the Prevention and Treatment of Cardiovascular Disease. *Circ. Res.* **2019**, *124*, 952–965. [CrossRef] [PubMed]
9. Kord-Varkaneh, H.; Nazary-Vannani, A.; Mokhtari, Z.; Salehi-Sahlabadi, A.; Rahmani, J.; Clark, C.C.T.; Fatahi, S.; Zanghelini, F.; Hekmatdoost, A.; Okunade, K.; et al. The Influence of Fasting and Energy Restricting Diets on Blood Pressure in Humans: A Systematic Review and Meta-Analysis. *High Blood Press Cardiovasc. Prev.* **2020**, *27*, 271–280. [CrossRef]
10. Ross, R.; Faggiotto, A.; Bowen-Pope, D.; Raines, E. The role of endothelial injury and platelet and macrophage interactions in atherosclerosis. *Circulation* **1984**, *70*, III77–III82.
11. Lerman, A.; Zeiher, A.M. Endothelial function: Cardiac events. *Circulation* **2005**, *111*, 363–368. [CrossRef]
12. Behrendt, D.; Ganz, P. Endothelial function. From vascular biology to clinical applications. *Am. J. Cardiol.* **2002**, *90*, 40L–48L. [CrossRef]
13. Davis, N.; Katz, S.; Wylie-Rosett, J. The effect of diet on endothelial function. *Cardiol. Rev.* **2007**, *15*, 62–66. [CrossRef]
14. Davignon, J.; Ganz, P. Role of endothelial dysfunction in atherosclerosis. *Circulation* **2004**, *109*, III27–III32. [CrossRef]
15. Gonzalez, M.A.; Selwyn, A.P. Endothelial function, inflammation, and prognosis in cardiovascular disease. *Am. J. Med.* **2003**, *115*, 99S–106S. [CrossRef]
16. Sullivan, D. The Effects of Hypertension on the Body. Available online: https://www.healthline.com/health/high-blood-pressure-hypertension/effect-on-body (accessed on 10 November 2020).
17. Williams, B.; Mancia, G.; Spiering, W.; Agabiti Rosei, E.; Azizi, M.; Burnier, M.; Clement, D.L.; Coca, A.; de Simone, G.; Dominiczak, A.; et al. 2018 ESC/ESH Guidelines for the management of arterial hypertension. *Eur. Heart J.* **2018**, *39*, 3021–3104. [CrossRef]
18. Frisoli, T.M.; Schmieder, R.E.; Grodzicki, T.; Messerli, F.H. Beyond salt: Lifestyle modifications and blood pressure. *Eur. Heart J.* **2011**, *32*, 3081–3087. [CrossRef] [PubMed]
19. OsMed. Appropriatezza Prescrittiva e Aderenza alle Terapie. Il Quadro Emerso dagli Indicatori OsMed per Alcune Aree Terapeutiche. Available online: http://www.agenziafarmaco.gov.it/content/appropriatezza-prescrittiva-e-aderenza-alle-terapie-il-quadro-emerso-dagli-indicatori-osmed- (accessed on 15 October 2020).

20. Scuteri, A.; Modestino, A.; Frattari, A.; di Daniele, N.; Tesauro, M. Occurrence of Hypotension in Older Participants. Which 24-hour ABPM Parameter Better Correlate With? *J. Gerontol. A Biol. Sci. Med. Sci.* **2012**, *67*, 804–810. [CrossRef] [PubMed]
21. Li, J.; Zheng, H.; Du, H.B.; Tian, X.P.; Jiang, Y.J.; Zhang, S.L.; Kang, Y.; Li, X.; Chen, J.; Lu, C.; et al. The multiple lifestyle modification for patients with prehypertension and hypertension patients: A systematic review protocol. *BMJ Open* **2014**, *4*, e004920. [CrossRef] [PubMed]
22. Fuchs, F.D.; Whelton, P.K. High Blood Pressure and Cardiovascular Disease. *Hypertension* **2020**, *75*, 285–292. [CrossRef] [PubMed]
23. Versari, D.; Daghini, E.; Virdis, A.; Ghiadoni, L.; Taddei, S. Endothelial dysfunction as a target for prevention of cardiovascular disease. *Diabetes Care* **2009**, *32*, S314–S321. [CrossRef]
24. Sandoo, A.; van Zanten, J.J.; Metsios, G.S.; Carroll, D.; Kitas, G.D. The endothelium and its role in regulating vascular tone. *Open Cardiovasc. Med. J.* **2010**, *4*, 302–312. [CrossRef] [PubMed]
25. Van Hinsbergh, V.W. Endothelium–role in regulation of coagulation and inflammation. *Semin. Immunopathol.* **2012**, *34*, 93–106. [CrossRef]
26. Godo, S.; Shimokawa, H. Endothelial Functions. *Arterioscler. Thromb Vasc. Biol.* **2017**, *37*, e108–e114. [CrossRef] [PubMed]
27. Forstermann, U.; Sessa, W.C. Nitric oxide synthases: Regulation and function. *Eur. Heart J.* **2012**, *33*, 829–837. [CrossRef] [PubMed]
28. Sriram, K.; Laughlin, J.G.; Rangamani, P.; Tartakovsky, D.M. Shear-Induced Nitric Oxide Production by Endothelial Cells. *Biophys. J.* **2016**, *111*, 208–221. [CrossRef] [PubMed]
29. Lee, M.Y.; Griendling, K.K. Redox signaling, vascular function, and hypertension. *Antioxid. Redox Signal.* **2008**, *10*, 1045–1059. [CrossRef]
30. Gimbrone, M.A., Jr.; Garcia-Cardena, G. Endothelial Cell Dysfunction and the Pathobiology of Atherosclerosis. *Circ. Res.* **2016**, *118*, 620–636. [CrossRef]
31. Gryglewski, R.J.; Palmer, R.M.; Moncada, S. Superoxide anion is involved in the breakdown of endothelium-derived vascular relaxing factor. *Nature* **1986**, *320*, 454–456. [CrossRef]
32. Luscher, T.F. Endothelial dysfunction: The role and impact of the renin-angiotensin system. *Heart* **2000**, *84*, i20–i22:discussion i50. [CrossRef]
33. Campia, U.; Tesauro, M.; di Daniele, N.; Cardillo, C. The vascular endothelin system in obesity and type 2 diabetes: Pathophysiology and therapeutic implications. *Life Sci.* **2014**, *118*, 149–155. [CrossRef]
34. Davenport, A.P.; Hyndman, K.A.; Dhaun, N.; Southan, C.; Kohan, D.E.; Pollock, J.S.; Pollock, D.M.; Webb, D.J.; Maguire, J.J. Endothelin. *Pharmacol. Rev.* **2016**, *68*, 357–418. [CrossRef] [PubMed]
35. Flammer, A.J.; Anderson, T.; Celermajer, D.S.; Creager, M.A.; Deanfield, J.; Ganz, P.; Hamburg, N.M.; Luscher, T.F.; Shechter, M.; Taddei, S.; et al. The assessment of endothelial function: From research into clinical practice. *Circulation* **2012**, *126*, 753–767. [CrossRef]
36. Dharmashankar, K.; Widlansky, M.E. Vascular endothelial function and hypertension: Insights and directions. *Curr. Hypertens. Rep.* **2010**, *12*, 448–455. [CrossRef]
37. Dinh, Q.N.; Drummond, G.R.; Sobey, C.G.; Chrissobolis, S. Roles of inflammation, oxidative stress, and vascular dysfunction in hypertension. *Biomed. Res. Int.* **2014**, *2014*, 406960. [CrossRef]
38. Mordi, I.; Mordi, N.; Delles, C.; Tzemos, N. Endothelial dysfunction in human essential hypertension. *J. Hypertens.* **2016**, *34*, 1464–1472. [CrossRef]
39. Yannoutsos, A.; Levy, B.I.; Safar, M.E.; Slama, G.; Blacher, J. Pathophysiology of hypertension: Interactions between macro and microvascular alterations through endothelial dysfunction. *J. Hypertens.* **2014**, *32*, 216–224. [CrossRef] [PubMed]
40. Most, J.; Tosti, V.; Redman, L.M.; Fontana, L. Calorie restriction in humans: An update. *Ageing Res. Rev.* **2017**, *39*, 36–45. [CrossRef] [PubMed]
41. Fontana, L.; Meyer, T.E.; Klein, S.; Holloszy, J.O. Long-term calorie restriction is highly effective in reducing the risk for atherosclerosis in humans. *Proc. Natl. Acad. Sci. USA* **2004**, *101*, 6659–6663. [CrossRef]
42. Osborne, T.B.; Mendel, L.B.; Ferry, E.L. The Effect of Retardation of Growth Upon the Breeding Period and Duration of Life of Rats. *Science* **1917**, *45*, 294–295. [CrossRef]
43. Weindruch, R.; Walford, R.L.; Fligiel, S.; Guthrie, D. The retardation of aging in mice by dietary restriction: Longevity, cancer, immunity and lifetime energy intake. *J. Nutr.* **1986**, *116*, 641–654. [CrossRef]
44. Lee, S.H.; Min, K.J. Caloric restriction and its mimetics. *BMB Rep.* **2013**, *46*, 181–187. [CrossRef]
45. Zhang, Y.; Ikeno, Y.; Qi, W.; Chaudhuri, A.; Li, Y.; Bokov, A.; Thorpe, S.R.; Baynes, J.W.; Epstein, C.; Richardson, A.; et al. Mice deficient in both Mn superoxide dismutase and glutathione peroxidase-1 have increased oxidative damage and a greater incidence of pathology but no reduction in longevity. *J. Gerontol. A Biol. Sci. Med. Sci.* **2009**, *64*, 1212–1220. [CrossRef]
46. Smirnov, A.; Panatta, E.; Lena, A.; Castiglia, D.; di Daniele, N.; Melino, G.; Candi, E. FOXM1 regulates proliferation, senescence and oxidative stress in keratinocytes and cancer cells. *Aging (Albany N. Y.)* **2016**, *8*, 1384–1397. [CrossRef]
47. Singh, R.; Lakhanpal, D.; Kumar, S.; Sharma, S.; Kataria, H.; Kaur, M.; Kaur, G. Late-onset intermittent fasting dietary restriction as a potential intervention to retard age-associated brain function impairments in male rats. *Age* **2012**, *34*, 917–933. [CrossRef] [PubMed]
48. Anisimov, V.N.; Bartke, A. The key role of growth hormone-insulin-IGF-1 signaling in aging and cancer. *Crit. Rev. Oncol. Hematol.* **2013**, *87*, 201–223. [CrossRef]

49. Willcox, D.C.; Willcox, B.J.; Todoriki, H.; Curb, J.D.; Suzuki, M. Caloric restriction and human longevity: What can we learn from the Okinawans? *Biogerontology* **2006**, *7*, 173–177. [CrossRef]
50. Colman, R.J.; Anderson, R.M.; Johnson, S.C.; Kastman, E.K.; Kosmatka, K.J.; Beasley, T.M.; Allison, D.B.; Cruzen, C.; Simmons, H.A.; Kemnitz, J.W.; et al. Caloric restriction delays disease onset and mortality in rhesus monkeys. *Science* **2009**, *325*, 201–204. [CrossRef]
51. Mattison, J.A.; Roth, G.S.; Beasley, T.M.; Tilmont, E.M.; Handy, A.M.; Herbert, R.L.; Longo, D.L.; Allison, D.B.; Young, J.E.; Bryant, M.; et al. Impact of caloric restriction on health and survival in rhesus monkeys from the NIA study. *Nature* **2012**, *489*, 318–321. [CrossRef]
52. Guarente, L. Sir2 links chromatin silencing, metabolism, and aging. *Genes Dev.* **2000**, *14*, 1021–1026.
53. Toiber, D.; Sebastian, C.; Mostoslavsky, R. Characterization of nuclear sirtuins: Molecular mechanisms and physiological relevance. *Handb. Exp. Pharmacol.* **2011**, *206*, 189–224. [CrossRef] [PubMed]
54. Lee, S.H.; Lee, J.H.; Lee, H.Y.; Min, K.J. Sirtuin signaling in cellular senescence and aging. *BMB Rep.* **2019**, *52*, 24–34. [CrossRef] [PubMed]
55. Yamashita, S.; Ogawa, K.; Ikei, T.; Udono, M.; Fujiki, T.; Katakura, Y. SIRT1 prevents replicative senescence of normal human umbilical cord fibroblast through potentiating the transcription of human telomerase reverse transcriptase gene. *Biochem. Biophys. Res. Commun.* **2012**, *417*, 630–634. [CrossRef] [PubMed]
56. Jeong, S.M.; Xiao, C.; Finley, L.W.; Lahusen, T.; Souza, A.L.; Pierce, K.; Li, Y.H.; Wang, X.; Laurent, G.; German, N.J.; et al. SIRT4 has tumor-suppressive activity and regulates the cellular metabolic response to DNA damage by inhibiting mitochondrial glutamine metabolism. *Cancer Cell* **2013**, *23*, 450–463. [CrossRef] [PubMed]
57. Cohen, H.Y.; Miller, C.; Bitterman, K.J.; Wall, N.R.; Hekking, B.; Kessler, B.; Howitz, K.T.; Gorospe, M.; de Cabo, R.; Sinclair, D.A. Calorie restriction promotes mammalian cell survival by inducing the SIRT1 deacetylase. *Science* **2004**, *305*, 390–392. [CrossRef]
58. Chen, D.; Bruno, J.; Easlon, E.; Lin, S.J.; Cheng, H.L.; Alt, F.W.; Guarente, L. Tissue-specific regulation of SIRT1 by calorie restriction. *Genes Dev.* **2008**, *22*, 1753–1757. [CrossRef]
59. Deota, S.; Chattopadhyay, T.; Ramachandran, D.; Armstrong, E.; Camacho, B.; Maniyadath, B.; Fulzele, A.; Gonzalez-de-Peredo, A.; Denu, J.M.; Kolthur-Seetharam, U. Identification of a Tissue-Restricted Isoform of SIRT1 Defines a Regulatory Domain that Encodes Specificity. *Cell Rep.* **2017**, *18*, 3069–3077. [CrossRef]
60. Meyer, T.E.; Kovacs, S.J.; Ehsani, A.A.; Klein, S.; Holloszy, J.O.; Fontana, L. Long-term caloric restriction ameliorates the decline in diastolic function in humans. *J. Am. Coll. Cardiol.* **2006**, *47*, 398–402. [CrossRef]
61. Sohal, R.S.; Weindruch, R. Oxidative stress, caloric restriction, and aging. *Science* **1996**, *273*, 59–63. [CrossRef]
62. Liguori, I.; Russo, G.; Curcio, F.; Bulli, G.; Aran, L.; Della-Morte, D.; Gargiulo, G.; Testa, G.; Cacciatore, F.; Bonaduce, D.; et al. Oxidative stress, aging, and diseases. *Clin. Interv. Aging* **2018**, *13*, 757–772. [CrossRef]
63. Ingram, D.K.; Cutler, R.G.; Weindruch, R.; Renquist, D.M.; Knapka, J.J.; April, M.; Belcher, C.T.; Clark, M.A.; Hatcherson, C.D.; Marriott, B.M.; et al. Dietary restriction and aging: The initiation of a primate study. *J. Gerontol.* **1990**, *45*, B148–B163. [CrossRef]
64. Zanetti, M.; Gortan Cappellari, G.; Burekovic, I.; Barazzoni, R.; Stebel, M.; Guarnieri, G. Caloric restriction improves endothelial dysfunction during vascular aging: Effects on nitric oxide synthase isoforms and oxidative stress in rat aorta. *Exp. Gerontol.* **2010**, *45*, 848–855. [CrossRef] [PubMed]
65. Rippe, C.; Lesniewski, L.; Connell, M.; LaRocca, T.; Donato, A.; Seals, D. Short-term calorie restriction reverses vascular endothelial dysfunction in old mice by increasing nitric oxide and reducing oxidative stress. *Aging Cell* **2010**, *9*, 304–312. [CrossRef] [PubMed]
66. Donato, A.J.; Walker, A.E.; Magerko, K.A.; Bramwell, R.C.; Black, A.D.; Henson, G.D.; Lawson, B.R.; Lesniewski, L.A.; Seals, D.R. Life-long caloric restriction reduces oxidative stress and preserves nitric oxide bioavailability and function in arteries of old mice. *Aging Cell* **2013**, *12*, 772–783. [CrossRef] [PubMed]
67. Kobara, M.; Furumori-Yukiya, A.; Kitamura, M.; Matsumura, M.; Ohigashi, M.; Toba, H.; Nakata, T. Short-Term Caloric Restriction Suppresses Cardiac Oxidative Stress and Hypertrophy Caused by Chronic Pressure Overload. *J. Card. Fail* **2015**, *21*, 656–666. [CrossRef] [PubMed]
68. Waldman, M.; Cohen, K.; Yadin, D.; Nudelman, V.; Gorfil, D.; Laniado-Schwartzman, M.; Kornwoski, R.; Aravot, D.; Abraham, N.G.; Arad, M.; et al. Regulation of diabetic cardiomyopathy by caloric restriction is mediated by intracellular signaling pathways involving 'SIRT1 and PGC-1alpha'. *Cardiovasc. Diabetol.* **2018**, *17*, 111. [CrossRef]
69. An, H.S.; Lee, J.Y.; Choi, E.B.; Jeong, E.A.; Shin, H.J.; Kim, K.E.; Park, K.A.; Jin, Z.; Lee, J.E.; Koh, J.S.; et al. Caloric restriction reverses left ventricular hypertrophy through the regulation of cardiac iron homeostasis in impaired leptin signaling mice. *Sci. Rep.* **2020**, *10*, 7176. [CrossRef]
70. Wadden, T.A.; Considine, R.V.; Foster, G.D.; Anderson, D.A.; Sarwer, D.B.; Caro, J.S. Short- and long-term changes in serum leptin dieting obese women: Effects of caloric restriction and weight loss. *J. Clin. Endocrinol. Metab.* **1998**, *83*, 214–218. [CrossRef]
71. Nakano, Y.; Oshima, T.; Sasaki, S.; Higashi, Y.; Ozono, R.; Takenaka, S.; Miura, F.; Hirao, H.; Matsuura, H.; Chayama, K.; et al. Calorie restriction reduced blood pressure in obesity hypertensives by improvement of autonomic nerve activity and insulin sensitivity. *J. Cardiovasc. Pharmacol.* **2001**, *38*, S69–S74. [CrossRef]
72. Facchini, M.; Malfatto, G.; Sala, L.; Silvestri, G.; Fontana, P.; Lafortuna, C.; Sartorio, A. Changes of autonomic cardiac profile after a 3-week integrated body weight reduction program in severely obese patients. *J. Endocrinol. Invest.* **2003**, *26*, 138–142. [CrossRef]

73. Das, S.K.; Gilhooly, C.H.; Golden, J.K.; Pittas, A.G.; Fuss, P.J.; Cheatham, R.A.; Tyler, S.; Tsay, M.; McCrory, M.A.; Lichtenstein, A.H.; et al. Long-term effects of 2 energy-restricted diets differing in glycemic load on dietary adherence, body composition, and metabolism in CALERIE: A 1-y randomized controlled trial. *Am. J. Clin. Nutr.* **2007**, *85*, 1023–1030. [CrossRef]
74. Lecoultre, V.; Ravussin, E.; Redman, L.M. The fall in leptin concentration is a major determinant of the metabolic adaptation induced by caloric restriction independently of the changes in leptin circadian rhythms. *J. Clin. Endocrinol. Metab.* **2011**, *96*, E1512–E1516. [CrossRef] [PubMed]
75. Stewart, T.M.; Bhapkar, M.; Das, S.; Galan, K.; Martin, C.K.; McAdams, L.; Pieper, C.; Redman, L.; Roberts, S.; Stein, R.I.; et al. Comprehensive Assessment of Long-term Effects of Reducing Intake of Energy Phase 2 (CALERIE Phase 2) screening and recruitment: Methods and results. *Contemp. Clin. Trials* **2013**, *34*, 10–20. [CrossRef] [PubMed]
76. Ruggenenti, P.; Abbate, M.; Ruggiero, B.; Rota, S.; Trillini, M.; Aparicio, C.; Parvanova, A.; Petrov Iliev, I.; Pisanu, G.; Perna, A.; et al. Renal and Systemic Effects of Calorie Restriction in Patients With Type 2 Diabetes With Abdominal Obesity: A Randomized Controlled Trial. *Diabetes* **2017**, *66*, 75–86. [CrossRef] [PubMed]
77. Most, J.; Gilmore, L.A.; Smith, S.R.; Han, H.; Ravussin, E.; Redman, L.M. Significant improvement in cardiometabolic health in healthy nonobese individuals during caloric restriction-induced weight loss and weight loss maintenance. *Am. J. Physiol. Endocrinol. Metab.* **2018**, *314*, E396–E405. [CrossRef]
78. Kraus, W.E.; Bhapkar, M.; Huffman, K.M.; Pieper, C.F.; Krupa Das, S.; Redman, L.M.; Villareal, D.T.; Rochon, J.; Roberts, S.B.; Ravussin, E.; et al. 2 years of calorie restriction and cardiometabolic risk (CALERIE): Exploratory outcomes of a multicentre, phase 2, randomised controlled trial. *Lancet Diabetes Endocrinol.* **2019**, *7*, 673–683. [CrossRef]
79. Harsha, D.W.; Lin, P.H.; Obarzanek, E.; Karanja, N.M.; Moore, T.J.; Caballero, B.; DASH Collaborative Research Group. Dietary Approaches to Stop Hypertension: A summary of study results. *J. Am. Diet Assoc.* **1999**, *99*, S35–S39. [CrossRef]
80. Arnason, T.G.; Bowen, M.W.; Mansell, K.D. Effects of intermittent fasting on health markers in those with type 2 diabetes: A pilot study. *World J. Diabetes* **2017**, *8*, 154–164. [CrossRef]
81. Erdem, Y.; Ozkan, G.; Ulusoy, S.; Arici, M.; Derici, U.; Sengul, S.; Sindel, S.; Erturk, S.; Turkish Society of, H.; Renal, D. The effect of intermittent fasting on blood pressure variability in patients with newly diagnosed hypertension or prehypertension. *J. Am. Soc. Hypertens.* **2018**, *12*, 42–49. [CrossRef]
82. Furmli, S.; Elmasry, R.; Ramos, M.; Fung, J. Therapeutic use of intermittent fasting for people with type 2 diabetes as an alternative to insulin. *BMJ Case Rep.* **2018**, *2018*. [CrossRef]
83. Wilhelmi de Toledo, F.; Grundler, F.; Bergouignan, A.; Drinda, S.; Michalsen, A. Safety, health improvement and well-being during a 4 to 21-day fasting period in an observational study including 1422 subjects. *PLoS ONE* **2019**, *14*, e0209353. [CrossRef]
84. Mager, D.E.; Wan, R.; Brown, M.; Cheng, A.; Wareski, P.; Abernethy, D.R.; Mattson, M.P. Caloric restriction and intermittent fasting alter spectral measures of heart rate and blood pressure variability in rats. *FASEB J.* **2006**, *20*, 631–637. [CrossRef] [PubMed]
85. Rickman, A.D.; Williamson, D.A.; Martin, C.K.; Gilhooly, C.H.; Stein, R.I.; Bales, C.W.; Roberts, S.; Das, S.K. The CALERIE Study: Design and methods of an innovative 25% caloric restriction intervention. *Contemp. Clin. Trials* **2011**, *32*, 874–881. [CrossRef] [PubMed]
86. Di Renzo, L.; Gualtieri, P.; Romano, L.; Marrone, G.; Noce, A.; Pujia, A.; Perrone, M.A.; Aiello, V.; Colica, C.; de Lorenzo, A. Role of Personalized Nutrition in Chronic-Degenerative Diseases. *Nutrients* **2019**, *11*, 1707. [CrossRef] [PubMed]
87. Di Daniele, N. The Role of Preventive Nutrition in Chronic Non-Communicable Diseases. *Nutrients* **2019**, *11*, 1074. [CrossRef] [PubMed]
88. Bidani, A.K.; Griffin, K.A. Basic science: Hypertensive target organ damage. *J. Am. Soc. Hypertens.* **2015**, *9*, 235–237; quiz 238. [CrossRef] [PubMed]
89. Mensah, G.A.; Croft, J.B.; Giles, W.H. The heart, kidney, and brain as target organs in hypertension. *Cardiol. Clin.* **2002**, *20*, 225–247. [CrossRef]
90. Aronow, W.S. Hypertension and left ventricular hypertrophy. *Ann. Transl. Med.* **2017**, *5*, 310. [CrossRef]
91. Gallu, M.; Marrone, G.; Legramante, J.M.; de Lorenzo, A.; Di Daniele, N.; Noce, A. Female Sex as a Thromboembolic Risk Factor in the Era of Nonvitamin K Antagonist Oral Anticoagulants. *Cardiovasc. Ther.* **2020**, *2020*, 1743927. [CrossRef]
92. Landecho, M.F.; Tuero, C.; Valenti, V.; Bilbao, I.; de la Higuera, M.; Fruhbeck, G. Relevance of Leptin and Other Adipokines in Obesity-Associated Cardiovascular Risk. *Nutrients* **2019**, *11*, 2664. [CrossRef]
93. Hall, M.E.; Harmancey, R.; Stec, D.E. Lean heart: Role of leptin in cardiac hypertrophy and metabolism. *World J. Cardiol.* **2015**, *7*, 511–524. [CrossRef]
94. Raju, S.V.; Zheng, M.; Schuleri, K.H.; Phan, A.C.; Bedja, D.; Saraiva, R.M.; Yiginer, O.; Vandegaer, K.; Gabrielson, K.L.; O'Donnell, C.P.; et al. Activation of the cardiac ciliary neurotrophic factor receptor reverses left ventricular hypertrophy in leptin-deficient and leptin-resistant obesity. *Proc. Natl. Acad. Sci. USA* **2006**, *103*, 4222–4227. [CrossRef] [PubMed]
95. Beltowski, J. Role of leptin in blood pressure regulation and arterial hypertension. *J. Hypertens.* **2006**, *24*, 789–801. [CrossRef] [PubMed]
96. Bravo, P.E.; Morse, S.; Borne, D.M.; Aguilar, E.A.; Reisin, E. Leptin and hypertension in obesity. *Vasc. Health Risk Manag.* **2006**, *2*, 163–169. [CrossRef] [PubMed]
97. Chiu, H.C.; Kovacs, A.; Ford, D.A.; Hsu, F.F.; Garcia, R.; Herrero, P.; Saffitz, J.E.; Schaffer, J.E. A novel mouse model of lipotoxic cardiomyopathy. *J. Clin. Invest.* **2001**, *107*, 813–822. [CrossRef]

98. Kamimura, D.; Suzuki, T.; Wang, W.; deShazo, M.; Hall, J.E.; Winniford, M.D.; Kullo, I.J.; Mosley, T.H.; Butler, K.R.; Hall, M.E. Higher plasma leptin levels are associated with reduced left ventricular mass and left ventricular diastolic stiffness in black women: Insights from the Genetic Epidemiology Network of Arteriopathy (GENOA) study. *Hypertens. Res.* **2018**, *41*, 629–638. [CrossRef]
99. Abel, E.D.; Litwin, S.E.; Sweeney, G. Cardiac remodeling in obesity. *Physiol. Rev.* **2008**, *88*, 389–419. [CrossRef]
100. Maulik, S.K.; Kumar, S. Oxidative stress and cardiac hypertrophy: A review. *Toxicol. Mech. Methods* **2012**, *22*, 359–366. [CrossRef]
101. Cianci, R.; Barbano, B.; Martina, P.; Gigante, A.; Polidori, L.; Lai, S.; Ascoli, G.; de Francesco, I.; di Donato, D.; Fuiano, G.; et al. Nephroangiosclerosis and its pharmacological approach. *Curr. Vasc. Pharmacol.* **2011**, *9*, 238–243. [CrossRef]
102. Yamanouchi, M.; Hoshino, J.; Ubara, Y.; Takaichi, K.; Kinowaki, K.; Fujii, T.; Ohashi, K.; Mise, K.; Toyama, T.; Hara, A.; et al. Clinicopathological predictors for progression of chronic kidney disease in nephrosclerosis: A biopsy-based cohort study. *Nephrol. Dial. Transplant.* **2019**, *34*, 1182–1188. [CrossRef]
103. O'Donnell, M.; Mente, A.; Yusuf, S. Sodium intake and cardiovascular health. *Circ. Res.* **2015**, *116*, 1046–1057. [CrossRef]
104. Wade, B.; Petrova, G.; Mattson, D.L. Role of immune factors in angiotensin II-induced hypertension and renal damage in Dahl salt-sensitive rats. *Am. J. Physiol. Regul. Integr. Comp. Physiol.* **2018**, *314*, R323–R333. [CrossRef] [PubMed]
105. Noce, A.; Marrone, G.; di Lauro, M.; Urciuoli, S.; Pietroboni Zaitseva, A.; Wilson Jones, G.; di Daniele, N.; Romani, A. Cardiovascular Protection of Nephropathic Male Patients by Oral Food Supplements. *Cardiovasc. Ther.* **2020**, *2020*, 1807941. [CrossRef]
106. Mattson, D.L.; James, L.; Berdan, E.A.; Meister, C.J. Immune suppression attenuates hypertension and renal disease in the Dahl salt-sensitive rat. *Hypertension* **2006**, *48*, 149–156. [CrossRef] [PubMed]
107. Lu, X.; Crowley, S.D. Inflammation in Salt-Sensitive Hypertension and Renal Damage. *Curr. Hypertens. Rep.* **2018**, *20*, 103. [CrossRef] [PubMed]
108. Hildebrand, A.M.; Garg, A.X. Blood pressure targets in chronic kidney disease: Does proteinuria dictate how low we go? *CMAJ* **2013**, *185*, 941–942. [CrossRef]
109. Xu, X.M.; Cai, G.Y.; Bu, R.; Wang, W.J.; Bai, X.Y.; Sun, X.F.; Chen, X.M. Beneficial Effects of Caloric Restriction on Chronic Kidney Disease in Rodent Models: A Meta-Analysis and Systematic Review. *PLoS ONE* **2015**, *10*, e0144442. [CrossRef] [PubMed]
110. Wang, S.Y.; Cai, G.Y.; Chen, X.M. Energy restriction in renal protection. *Br. J. Nutr.* **2018**, *120*, 1149–1158. [CrossRef] [PubMed]
111. Dong, D.; Cai, G.Y.; Ning, Y.C.; Wang, J.C.; Lv, Y.; Hong, Q.; Cui, S.Y.; Fu, B.; Guo, Y.N.; Chen, X.M. Alleviation of senescence and epithelial-mesenchymal transition in aging kidney by short-term caloric restriction and caloric restriction mimetics via modulation of AMPK/mTOR signaling. *Oncotarget* **2017**, *8*, 16109–16121. [CrossRef]
112. D'Elia, L.; Rossi, G.; Schiano di Cola, M.; Savino, I.; Galletti, F.; Strazzullo, P. Meta-Analysis of the Effect of Dietary Sodium Restriction with or without Concomitant Renin-Angiotensin-Aldosterone System-Inhibiting Treatment on Albuminuria. *Clin. J. Am. Soc. Nephrol.* **2015**, *10*, 1542–1552. [CrossRef]
113. Cheung, Y.F. Arterial stiffness in the young: Assessment, determinants, and implications. *Korean Circ. J.* **2010**, *40*, 153–162. [CrossRef]
114. Schinzari, F.; Iantorno, M.; Campia, U.; Mores, N.; Rovella, V.; Tesauro, M.; di Daniele, N.; Cardillo, C. Vasodilator responses and endothelin-dependent vasoconstriction in metabolically healthy obesity and the metabolic syndrome. *Am. J. Physiol. Endocrinol. Metab.* **2015**, *309*, E787–E792. [CrossRef] [PubMed]
115. Noce, A.; Fabrini, R.; Dessi, M.; Bocedi, A.; Santini, S.; Rovella, V.; Pastore, A.; Tesauro, M.; Bernardini, S.; di Daniele, N.; et al. Erythrocyte glutathione transferase activity: A possible early biomarker for blood toxicity in uremic diabetic patients. *Acta Diabetol.* **2014**, *51*, 219–224. [CrossRef] [PubMed]
116. Covic, A.; Gusbeth-Tatomir, P.; Goldsmith, D.J. Arterial stiffness in renal patients: An update. *Am. J. Kidney Dis.* **2005**, *45*, 965–977. [CrossRef] [PubMed]
117. Sun, Z. Aging, arterial stiffness, and hypertension. *Hypertension* **2015**, *65*, 252–256. [CrossRef] [PubMed]
118. Martens, C.R.; Seals, D.R. Practical alternatives to chronic caloric restriction for optimizing vascular function with ageing. *J. Physiol.* **2016**, *594*, 7177–7195. [CrossRef]
119. D'Elia, L.; Galletti, F.; La Fata, E.; Sabino, P.; Strazzullo, P. Effect of dietary sodium restriction on arterial stiffness: Systematic review and meta-analysis of the randomized controlled trials. *J. Hypertens.* **2018**, *36*, 734–743. [CrossRef]
120. Henderson, A.D.; Bruce, B.B.; Newman, N.J.; Biousse, V. Hypertension-related eye abnormalities and the risk of stroke. *Rev. Neurol. Dis.* **2011**, *8*, 1–9.
121. Chatterjee, S.; Chattopadhyay, S.; Hope-Ross, M.; Lip, P.L. Hypertension and the eye: Changing perspectives. *J. Hum. Hypertens.* **2002**, *16*, 667–675. [CrossRef]
122. Kawashima, M.; Ozawa, Y.; Shinmura, K.; Inaba, T.; Nakamura, S.; Kawakita, T.; Watanabe, M.; Tsubota, K. Calorie restriction (CR) and CR mimetics for the prevention and treatment of age-related eye disorders. *Exp. Gerontol.* **2013**, *48*, 1096–1100. [CrossRef]
123. Katsi, V.; Marketou, M.; Vlachopoulos, C.; Tousoulis, D.; Souretis, G.; Papageorgiou, N.; Stefanadis, C.; Vardas, P.; Kallikazaros, I. Impact of arterial hypertension on the eye. *Curr. Hypertens. Rep.* **2012**, *14*, 581–590. [CrossRef]
124. Viera, A.J.; Neutze, D.M. Diagnosis of secondary hypertension: An age-based approach. *Am. Fam. Phys.* **2010**, *82*, 1471–1478.
125. Charles, L.; Triscott, J.; Dobbs, B. Secondary Hypertension: Discovering the Underlying Cause. *Am. Fam. Phys.* **2017**, *96*, 453–461.

126. Calhoun, D.A.; Jones, D.; Textor, S.; Goff, D.C.; Murphy, T.P.; Toto, R.D.; White, A.; Cushman, W.C.; White, W.; Sica, D.; et al. Resistant hypertension: Diagnosis, evaluation, and treatment: A scientific statement from the American Heart Association Professional Education Committee of the Council for High Blood Pressure Research. *Circulation* **2008**, *117*, e510–e526. [CrossRef] [PubMed]
127. Siddiqui, M.A.; Mittal, P.K.; Little, B.P.; Miller, F.H.; Akduman, E.I.; Ali, K.; Sartaj, S.; Moreno, C.C. Secondary Hypertension and Complications: Diagnosis and Role of Imaging. *Radiographics* **2019**, *39*, 1036–1055. [CrossRef] [PubMed]
128. Acelajado, M.C.; Calhoun, D.A. Resistant hypertension, secondary hypertension, and hypertensive crises: Diagnostic evaluation and treatment. *Cardiol. Clin.* **2010**, *28*, 639–654. [CrossRef] [PubMed]
129. Filippou, C.D.; Tsioufis, C.P.; Thomopoulos, C.G.; Mihas, C.C.; Dimitriadis, K.S.; Sotiropoulou, L.I.; Chrysochoou, C.A.; Nihoyannopoulos, P.I.; Tousoulis, D.M. Dietary Approaches to Stop Hypertension (DASH) Diet and Blood Pressure Reduction in Adults with and without Hypertension: A Systematic Review and Meta-Analysis of Randomized Controlled Trials. *Adv. Nutr.* **2020**, *11*, 1150–1160. [CrossRef]
130. Troyer, J.L.; Racine, E.F.; Ngugi, G.W.; McAuley, W.J. The effect of home-delivered Dietary Approach to Stop Hypertension (DASH) meals on the diets of older adults with cardiovascular disease. *Am. J. Clin. Nutr.* **2010**, *91*, 1204–1212. [CrossRef]
131. Fu, J.; Liu, Y.; Zhang, L.; Zhou, L.; Li, D.; Quan, H.; Zhu, L.; Hu, F.; Li, X.; Meng, S.; et al. Nonpharmacologic Interventions for Reducing Blood Pressure in Adults With Prehypertension to Established Hypertension. *J. Am. Heart Assoc.* **2020**, *9*, e016804. [CrossRef]
132. The sixth report of the Joint National Committee on prevention, detection, evaluation, and treatment of high blood pressure. *Arch. Intern. Med.* **1997**, *157*, 2413–2446. [CrossRef]
133. D'Elia, L.; La Fata, E.; Giaquinto, A.; Strazzullo, P.; Galletti, F. Effect of dietary salt restriction on central blood pressure: A systematic review and meta-analysis of the intervention studies. *J. Clin. Hypertens.* **2020**, *22*, 814–825. [CrossRef]
134. Malinowski, B.; Zalewska, K.; Wesierska, A.; Sokolowska, M.M.; Socha, M.; Liczner, G.; Pawlak-Osinska, K.; Wicinski, M. Intermittent Fasting in Cardiovascular Disorders-An Overview. *Nutrients* **2019**, *11*, 673. [CrossRef]
135. Barnosky, A.R.; Hoddy, K.K.; Unterman, T.G.; Varady, K.A. Intermittent fasting vs daily calorie restriction for type 2 diabetes prevention: A review of human findings. *Transl. Res.* **2014**, *164*, 302–311. [CrossRef] [PubMed]
136. Johnstone, A. Fasting for weight loss: An effective strategy or latest dieting trend? *Int. J. Obes.* **2015**, *39*, 727–733. [CrossRef] [PubMed]
137. Harvie, M.; Howell, A. Potential Benefits and Harms of Intermittent Energy Restriction and Intermittent Fasting Amongst Obese, Overweight and Normal Weight Subjects-A Narrative Review of Human and Animal Evidence. *Behav. Sci.* **2017**, *7*, 4. [CrossRef] [PubMed]
138. Aly, S.M. Role of intermittent fasting on improving health and reducing diseases. *Int. J. Health Sci.* **2014**, *8*, V–VI. [CrossRef]
139. Patterson, R.E.; Sears, D.D. Metabolic Effects of Intermittent Fasting. *Annu. Rev. Nutr.* **2017**, *37*, 371–393. [CrossRef] [PubMed]
140. Eckel-Mahan, K.L.; Patel, V.R.; de Mateo, S.; Orozco-Solis, R.; Ceglia, N.J.; Sahar, S.; Dilag-Penilla, S.A.; Dyar, K.A.; Baldi, P.; Sassone-Corsi, P. Reprogramming of the circadian clock by nutritional challenge. *Cell* **2013**, *155*, 1464–1478. [CrossRef] [PubMed]
141. Turnbaugh, P.J.; Ley, R.E.; Mahowald, M.A.; Magrini, V.; Mardis, E.R.; Gordon, J.I. An obesity-associated gut microbiome with increased capacity for energy harvest. *Nature* **2006**, *444*, 1027–1031. [CrossRef] [PubMed]
142. Noce, A.; Marrone, G.; Di Daniele, F.; Ottaviani, E.; Wilson Jones, G.; Bernini, R.; Romani, A.; Rovella, V. Impact of Gut Microbiota Composition on Onset and Progression of Chronic Non-Communicable Diseases. *Nutrients* **2019**, *11*, 1073. [CrossRef]
143. Noce, A.; Tarantino, A.; Tsague Djoutsop, C.; Erald, V.; de Lorenzo, A.; di Daniele, N. Gut Microbioma Population: An Indicator Really Sensible to Any Change in Age, Diet, Metabolic Syndrome, and Life-Style. *Mediat. Inflam.* **2014**, *2014*. [CrossRef]
144. Chowdhury, E.A.; Richardson, J.D.; Tsintzas, K.; Thompson, D.; Betts, J.A. Effect of extended morning fasting upon ad libitum lunch intake and associated metabolic and hormonal responses in obese adults. *Int. J. Obes.* **2016**, *40*, 305–311. [CrossRef] [PubMed]
145. Hatori, M.; Vollmers, C.; Zarrinpar, A.; DiTacchio, L.; Bushong, E.A.; Gill, S.; Leblanc, M.; Chaix, A.; Joens, M.; Fitzpatrick, J.A.; et al. Time-restricted feeding without reducing caloric intake prevents metabolic diseases in mice fed a high-fat diet. *Cell Metab.* **2012**, *15*, 848–860. [CrossRef] [PubMed]
146. Yamaguchi, M.; Uemura, H.; Katsuura-Kamano, S.; Nakamoto, M.; Hiyoshi, M.; Takami, H.; Sawachika, F.; Juta, T.; Arisawa, K. Relationship of dietary factors and habits with sleep-wake regularity. *Asia Pac. J. Clin. Nutr.* **2013**, *22*, 457–465. [CrossRef] [PubMed]
147. Camandola, S.; Mattson, M.P. Brain metabolism in health, aging, and neurodegeneration. *EMBO J.* **2017**, *36*, 1474–1492. [CrossRef]
148. Mattson, M.P.; Longo, V.D.; Harvie, M. Impact of intermittent fasting on health and disease processes. *Ageing Res. Rev.* **2017**, *39*, 46–58. [CrossRef]

nutrients

MDPI

Article

Dietary Patterns and Progression of Impaired Kidney Function in Japanese Adults: A Longitudinal Analysis for the Fukushima Health Management Survey, 2011–2015

Enbo Ma [1,2,3,*], Tetsuya Ohira [1,2,3], Seiji Yasumura [3,4], Hironori Nakano [2,3], Eri Eguchi [2,3], Makoto Miyazaki [1,3], Mitsuaki Hosoya [1,3,5], Akira Sakai [3,6], Atsushi Takahashi [3,7], Hiromasa Ohira [7], Junichiro Kazama [3,8], Michio Shimabukuro [3,9], Hirooki Yabe [3,10], Masaharu Maeda [3,11], Hitoshi Ohto [3] and Kenji Kamiya [3,12]

1 Health Promotion Centre, Fukushima Medical University, Fukushima 960-1295, Japan; teoohira@fmu.ac.jp (T.O.); m-miya@fmu.ac.jp (M.M.); mhosoya@fmu.ac.jp (M.H.)
2 Department of Epidemiology, Fukushima Medical University School of Medicine, Fukushima 960-1295, Japan; h-nakano@fmu.ac.jp (H.N.); e-eguchi@fmu.ac.jp (E.E.)
3 Radiation Medical Science Centre for Fukushima Health Management Survey, Fukushima Medical University, Fukushima 960-1295, Japan; yasumura@fmu.ac.jp (S.Y.); sakira@fmu.ac.jp (A.S.); junior@fmu.ac.jp (A.T.); jjkaz@fmu.ac.jp (J.K.); shima01@fmu.ac.jp (M.S.); hyabe@fmu.ac.jp (H.Y.); masagen@fmu.ac.jp (M.M.); hit-ohto@fmu.ac.jp (H.O.); kkamiya@fmu.ac.jp (K.K.)
4 Department of Public Health, Fukushima Medical University School of Medicine, Fukushima 960-1295, Japan
5 Department of Paediatrician, Fukushima Medical University School of Medicine, Fukushima 960-1295, Japan
6 Department of Radiation Life Sciences, Fukushima Medical University School of Medicine, Fukushima 960-1295, Japan
7 Department of Gastroenterology, Fukushima Medical University School of Medicine, Fukushima 960-1295, Japan; h-ohira@fmu.ac.jp
8 Department of Nephrology and Hypertension, Fukushima Medical University School of Medicine, Fukushima 960-1295, Japan
9 Department of Diabetes, Endocrinology, and Metabolism, Fukushima Medical University School of Medicine, Fukushima 960-1295, Japan
10 Department of Neuropsychiatry, Fukushima Medical University School of Medicine, Fukushima 960-1295, Japan
11 Department of Disaster Psychiatry, Fukushima Medical University School of Medicine, Fukushima 960-1295, Japan
12 Research Institute for Radiation Biology and Medicine, Hiroshima University, Hiroshima 734-8553, Japan
* Correspondence: mae@fmu.ac.jp; Tel.: +81-24-547-1788; Fax: +81-24-547-1789

Citation: Ma, E.; Ohira, T.; Yasumura, S.; Nakano, H.; Eguchi, E.; Miyazaki, M.; Hosoya, M.; Sakai, A.; Takahashi, A.; Ohira, H.; et al. Dietary Patterns and Progression of Impaired Kidney Function in Japanese Adults: A Longitudinal Analysis for the Fukushima Health Management Survey, 2011–2015. *Nutrients* **2021**, *13*, 168. https://doi.org/10.3390/nu13010168

Received: 11 December 2020
Accepted: 5 January 2021
Published: 7 January 2021

Publisher's Note: MDPI stays neutral with regard to jurisdictional claims in published maps and institutional affiliations.

Abstract: To investigate associations between dietary patterns and the risk of impaired kidney function, we analyzed data from 14,732 participants (40–89 years) who completed the baseline diet questionnaire of The Fukushima Health Management Survey in 2011. The incidence of chronic kidney disease (CKD) (estimated glomerular filtration rate (eGFR) <60 mL/min/1.73 m^2 or proteinuria (≥1+ by dipstick test)) and annual changes in eGFR were assessed from 2012 to 2015. Three major dietary patterns were identified. The adjusted cumulative incidence ratio of the highest vs. lowest tertile of a vegetable diet scores was 0.90 (95% confidence interval (CI): 0.82, 1.00) for eGFR < 60 mL/min/1.73 m^2, 0.68 (95% CI: 0.52, 0.90) for proteinuria, and 0.88 (95% CI: 0.80, 0.97) for CKD (*P* for trend = 0.031, 0.007, and 0.005, respectively). The incident risk of CKD in the highest tertile of juice diet scores was 18% higher than the lowest tertile. The odds ratio of the highest vs. lowest tertile of vegetable diet scores was 0.85 (95% CI: 0.75, 0.98) in the rapidly decreasing eGFR group (*P* for trend = 0.009). We did not observe significant associations for the meat dietary pattern. A Japanese vegetable diet could reduce the risk of developing impaired kidney function and CKD.

Keywords: chronic kidney disease; dietary pattern; eGFR; Fukushima Health Management Survey; proteinuria; trajectory analysis

1. Introduction

In 2017, the global prevalence of chronic kidney disease (CKD) was 9.1% (697.5 million cases), and deaths from CKD were 1.2 million, which was an increase of 41.5% from 1990 [1]. Approximately 7.6% (1.4 million) of deaths from cardiovascular disease (CVD) could be attributed to impaired kidney function. Patients with CKD have similar mortality risks and medical costs compared to those with CVD [2,3]. Between 1990 and 2017, CVD mortality declined by 30.4%; however, a decline in CKD mortality was not observed [1].

CKD is rarely diagnosed in the early stages, and the health-related quality of life in patients decreases with CKD progression [4]. Patients with CKD, especially those with dialysis treatment, are at increased risk of osteoporosis and suffering from bone pain and fractures [5]. Most patients with CKD 3 to 5 stages show an increased serum parathyroid hormone level [6]. In CKD-mineral and bone disorders, circulating factors released from vessels may affect bone metabolism, while the bone disease may increase vascular calcification, highlighting the mortality risk [6,7].

In kidney function impairment, the persistent low-grade inflammation has a prominent feature, and gut microbiota dysbiosis is a source of microinflammation [8]. Studies indicated that the most harmful uremic toxins produced by the gut microbiota are protein-bound and are recalcitrant to removal by dialysis [9]; the function of uremic toxins such as indoxyl sulfate may play roles in CVD development through altered monocytes activation, intensified inflammatory process, and augmented oxidative stress [10,11]. In addition, CKD and Type 2 diabetes have similar negative effects on both intestinal microbiota and function [12]. A recent animal study reported an increase in concentrations of uremic toxins in serum associated with a stronger impairment in cognition and higher permeability of the blood–brain barrier [13].

For the prevention of CKD, dietary control is among the modifiable factors receiving increasing attention. Dietary balance between acid-producing foods and alkaline-producing foods is important [14]. Given than some nutrients are highly correlated or interactive, studies on dietary patterns focusing on multiple food groups provide results resembling actual eating behaviors more than those studies based on a single dietary product or nutrient [15–17].

Dietary patterns such as the Mediterranean diet (higher intake of fruits, vegetables, legumes, cereals, and fish) [18–20], the Dietary Approaches to Stop Hypertension (DASH)-style diet (higher intake of fruits, vegetables, and whole grains) [21], and other diets [22] have been reported to be associated with a lower risk of estimated glomerular filtration rate (eGFR) in healthy adults in Western countries. Only a few studies on dietary patterns associated with loss of kidney function have been performed in Asian populations [23–25].

In Japan, CKD mortality is the eighth leading cause of death [1]. The Great East Japan Earthquake in March 2011 affected the health status of the residents in Fukushima, with increased cardiometabolic (e.g., overweight/obesity, hypertension, diabetes mellitus, and dyslipidemia) and possibly CKD risks [26,27]. In a previous study, we elucidated that a vegetable diet was inversely associated with cardiometabolic risks [28]. However, the association between dietary patterns and CKD risk has not been intensively investigated in general populations in Japan.

In this study, we investigated the associations between dietary patterns identified by the Fukushima Health Management Survey (FHMS) in 2011 and the risks of developing impaired kidney function and CKD in the follow-up years until 2015 in Fukushima, Japan.

2. Materials and Methods

2.1. Study Participants

The FHMS was initiated in 2011 after the Great East Japan Earthquake. The target population comprised 210,189 residents living in evacuation zones along with the radiation disclosure areas [29]. The FHMS included a self-administered questionnaire on socio-economic and demographic information, medical history and clinical treatment, lifestyle behaviors, and a food frequency questionnaire (FFQ) (n = 88,613). We used data from individuals aged between 40 and 90 years who had completed both the FFQ and a comprehensive health checkup in the 2011 fiscal year (n = 28,602). Partial participants who completed the health checkup in 2014 and 2015 were also utilized in the previous study [28].

2.2. Dietary Intake Assessment

We used a short-form FFQ with 19 food items to determine the participant's food intake during the 6 months preceding the survey date. The FFQ was a validated and modified version of the Hiroshima and Nagasaki Life Span Study [30]. In the validation study of the original FFQ, the frequency of food intake as measured by the FFQ was moderately correlated with food intake as measured by the 24-h recall records; for example, the Spearmen correlation coefficient of fruit, milk, miso soup, beef/pork, rice, and bread was between 0.14 and 0.34 [30]. The food items included non-juice fruits, non-juice vegetables (green vegetables, red and orange vegetables, and light-colored vegetables), fruit juice, vegetable juice, meat (chicken, beef/pork, and ham/sausages), soybean product (fermented soybean, soy milk, miso soup, tofu, and boiled beans), fish (raw and cooked), dairy (milk, yogurt, and lactobacillus drinks), rice, and bread. We asked the participants how often they consumed individual food items, with 6 response choices for frequency: none, <1 time/week, 1–2 times/week, 3–4 times/week, 5–6 times/week, or every day.

2.3. End-Point Determination

We retrieved data on participants with at least 1 comprehensive health checkup conducted between the 2012 and 2015 fiscal years. Urinalysis by the dipstick method was conducted for a single spot urine specimen. The results of proteinuria testing were recorded based on the guidelines of the Japanese Committee for Clinical Laboratory Standards (http://jccls.org/) [27]. Serum creatinine was assayed using the enzymatic method. We calculated the eGFR using the Modification of Diet in Renal Disease formula recommended by the Japanese Society of Nephrology [31]: eGFR (mL/min/1.73 m^2) = 194 $\times$ Cr$^{-1.094}$ $\times$ age$^{-0.287}$ $\times$ 0.739 (if female). We defined CKD as an impaired kidney function with an eGFR < 60 mL/min/1.73 m^2 and/or proteinuria ($\geq$1+ by dipstick test) [32].

2.4. Other Variables

Concerning covariates at baseline in 2011, we classified education status as "less or more than vocational university"; smoking history as "never, former, or current"; and alcohol consumption as "never, occasional, or regular". We grouped physical activity into "none, 1 time/week, 2–4 times/week, or every day" and resident status at post-earthquake into "living in a shelter or temporary house, an apartment or rental house, or at relatives" or own house". We assessed the mental health status of the participants with the Japanese version of the Kessler Psychological Distress Scale (K6). Scores ranged from 0 to 24, and we defined nonspecific distress as corresponding to a K6 score of $\geq$13 [33].

Blood pressure was measured with a standard sphygmomanometer or an automated device by medical staff at local institutes. We defined the participants' cardiometabolic factors as follows: overweight as body mass index (BMI) $\geq$25 kg/m^2; hypertension as systolic blood pressure (SBP) $\geq$140 mm Hg, diastolic blood pressure as $\geq$90 mm Hg, or the use of antihypertensive medication; and diabetes mellitus as fasting plasma glucose $\geq$126 mg/dL, hemoglobin A1c (HbA1c) $\geq$6.5% or the use of insulin or other medications. Participants who met the following criteria were diagnosed with dyslipidemia: hypo-high-density lipoprotein (HDL) cholesterolemia (as HDL-C < 40 mg/dL), hyper-low-density lipoprotein (LDL) cholesterolemia (as LDL-C $\geq$ 140 mg/dL), or hypertriglyceridemia (as high triglycerides $\geq$ 150 mg/dL).

2.5. Statistical Analysis

In the FFQs, we excluded pregnant women, participants with more than 3 missing FFQ answers [34], those with underlying kidney disease or abnormal kidney function, and those with missing estimated glomerular filtration rate (eGFR). In all, 14,732 individuals underwent the baseline survey and a health checkup in 2011 and at least 1 health checkup in the follow-up years (Figure 1).

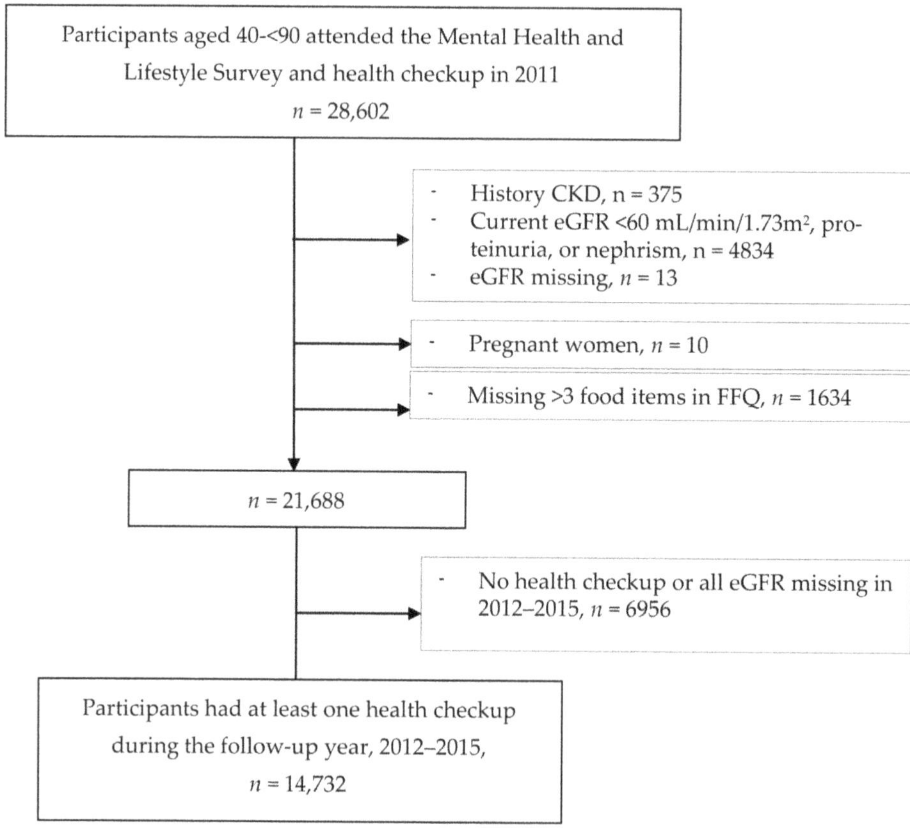

Figure 1. Flow diagram of study participants, Fukushima Health Management Survey, 2011–2015. CKD, chronic kidney disease; eGFR, estimated glomerular filtration rate; FFQ, food frequency questionnaire.

For the remaining participants, those who did not answer some dietary questions (12.4% missed 1 and 4.2% missed 2 in the FFQ), we replaced the missing values with the sex-specific median value of that food item frequency [34]. For the intake frequency of each food item, we used the daily midpoint for the frequency category; e.g., we assessed "3–4 times/week" as 0.5 times/day [34].

We derived dietary patterns from food items using the principal component method of factor analysis. To achieve a simpler structure with greater interpretability, a varimax rotation was performed. We selected factor numbers mainly according to eigenvalues > 1.5, scree plots, and factor interpretability concerning food items, with absolute factor loadings $\geq$ 0.3 to account for each component [35]. Among 5 factors that satisfied the criteria, a 3-factor solution appeared to describe the most meaningfully distinctive dietary patterns of the study population. We labeled derived dietary patterns as "vegetables", "juice", and "meat" based on food items with high factor loadings on each pattern. Eigenvalues of the vegetable, juice, and meat patterns were 4.02, 1.72, and 1.55, respectively. The cumulative variance explained was 38.4%. Cronbach's alpha coefficient for each dietary pattern indicated higher internal reliability of these measures: 0.79 for vegetables, 0.80 for juice, and 0.81 for meat. Given we derived almost the same factors from the factor analysis by sex, we only reported dietary patterns for total participants. We assigned each participant a pattern-specific score, which we calculated as the sum of the products of the factor loading coefficients and the standardized intake of food items. The factor scores reflect how closely a participant's diet resembles each identified pattern, with higher scores representing closer resemblance [21]. We categorized the dietary pattern scores into tertiles for further analysis.

We calculated the annual eGFR change rate by the differences between baseline and final measurements and divided them by the follow-up time [36]. In the trajectory analysis for eGFR changes over time, the participants were classified into 3 groups: the group in which the eGFR level decreased most was defined as the early decliner group [37].

To compare baseline characteristics across groups, we used the chi-squared test for categorical data and an analysis of variance Kruskal–Wallis test (for median) and F test (for means) for continuous variables. We used the Poisson regression model to estimate the cumulative incidence ratio (CIR) between means of dietary pattern scores (with the first tertile as the reference) and the risk of impaired kidney function in the follow-up years. We applied the multiple linear regression model to measure the annual eGFR decline rate associated with dietary patterns. Finally, we performed a polytomous logistic regression analysis to examine the association between dietary pattern scores and groups of eGFR decliners, using the moderate lowering eGFR group as a reference [36]. We input the tertiles of each dietary pattern score simultaneously in the model, and the associations with outcomes were adjusted for age (continuous) and sex (Model 1); further adjustments were made for education, smoking history, alcohol consumption, physical activity, resident status, distress scale, BMI, diabetes, hypertension, and dyslipidemia by categories as mentioned above (Model 2). We selected these variables for adjustment based on previous publications for FHMS and clinical relevance [26–28,34,38,39]. Tests for trend were performed using median pattern scores in the tertile categories as continuous variables. Given that age is a strong factor in kidney function progression, we also added an age square for adjustment in modeling when both the age and age square were statistically significant. We performed a sensitivity analysis for participants who attended all the follow-up health checkups and for participants without cardiometabolic risk factors.

We analyzed all the data using SAS statistical software version 9.4 for Windows (SAS Institute, Cary, NC, USA). All p-values reported were 2-sided, and $p < 0.05$ was considered statistically significant.

3. Results

Table 1 shows three independent dietary patterns with a minor overlapping of food group loadings. The vegetable diet pattern includes vegetables (white, green, red, and yellow vegetables), fish, fruits, bean products (tofu, fermented beans, boiled beans, and miso soup), and rice. The juice dietary pattern includes vegetable juice, fruit juice, yogurt, soymilk, fruits, milk, boiled beans, bread, and red/yellow vegetables. The meat dietary pattern includes chicken, beef/pork, and ham/sausage, and bread.

Table 2 shows the social and demographic characteristics and health checkups of participants at the baseline. Comparing with men, women had higher education, less current smokers, less current alcohol drinkers, and higher dietary pattern scores but less frequent physical activity and more depression. The health checkup conditions were better in women than men, except for the LDL level. The elderly were more likely to follow vegetable and juice diets, not a meat diet. Those with higher consumption of the juice and meat diets were likely to have higher education and to exercise frequently. Current smokers and alcohol drinkers were less likely to follow vegetable and juice diets, with an inverse tendency of following a meat diet. Residents in temporary houses or shelters were likely to have lower consumption of vegetables (Table S1).

The SBP level, hypertension proportion, and fasting blood glucose level were higher in participants with a higher intake of vegetables but lower in participants with a higher intake of meat. Furthermore, participants with hyper-LDL-C showed an inverse tendency in consumption of a vegetable and juice diet. Participants with higher triglyceride levels declined along the ascendant tertiles of all the dietary patterns.

Table 1. Factor loadings of dietary patterns identified by principal component method of factor analysis, FHMS, 2011 (n = 14,732).

Food Groups	Vegetable	Juice	Meat
White vegetables	**0.69**	0.14	0.22
Green vegetables	**0.65**	0.20	0.18
Tofu	**0.64**	0.13	0.06
Miso soup	**0.63**	−0.12	−0.09
Red/yellow vegetables	**0.63**	**0.30**	0.25
Fish	**0.51**	0.06	0.23
Fermented beans	**0.48**	0.14	−0.13
Fruit	**0.45**	0.41	0.01
Boiled beans	**0.38**	0.37	0.08
Rice	**0.34**	−0.22	−0.05
Vegetable juice	−0.02	**0.71**	0.004
Fruit juice	−0.01	**0.68**	0.08
Yogurt	0.22	**0.53**	0.004
Soybean milk	0.08	**0.40**	−0.04
Bread	−0.23	**0.35**	**0.31**
Milk	0.19	**0.34**	0.06
Beef/pork	0.15	−0.05	**0.74**
Ham/sausage	−0.01	0.07	**0.69**
Chicken	0.16	0.04	**0.68**
% Variance explained	3.26	2.21	1.81

FHMS, Fukushima Health Management Survey. Loadings with an absolute value more than 0.30 are shown in bold.

Table 2. Characteristics of participants at baseline in 2011, FHMS (n = 14,732).

	All	Men (n = 5964)	Women (n = 8768)	p Value
Age (years), mean (SD)	61.4 (10.0)	62.6 (9.9)	60.5 (9.9)	<0.001
Education ≥ vocational university, %	20.9	19.3	22.1	<0.001
Current smoker, %	13.3	24.9	5.4	<0.001
Current alcohol drinking, %	45.3	71.8	27.2	<0.001
Physical activity ≥ 2 times/week, %	42	45.1	39.9	<0.001
Distress scale ≥ 13, %	14.6	11.1	16.9	<0.001
Live at shelter/temporary house, %	39.8	39.1	40.3	0.094
BMI (kg/m^2), mea(SD)	23.7 (3.4)	24.2 (3.1)	23.3 (3.5)	<0.001
BMI ≥ 25 kg/m^2, %	32.2	38.2	28.1	<0.001
Hypertension, %	50.5	58.1	45.4	<0.001
SBP (mmHg), mean (SD)	131 (15.8)	133.5 (15.0)	129.3 (16.1)	<0.001
DBP (mmHg), mean (SD)	78.6 (10.1)	80.6.0 (9.9)	77.2 (10)	<0.001
Fast blood glucose (mg/dL), median (IQR)	97 (90, 105)	100 (93, 110)	95 (89, 102)	<0.001
Fast blood glucose ≥ 126 mg/dl, %	7	10	4.9	<0.001
HbA1C1 ≥ 6.5%, %	6.6	9.1	5	<0.001
LDL-C (mg/dL), mean (SD)	126.8 (31.7)	122.6 (31.9)	129.7 (31.3)	<0.001
LDL-C ≥ 140 mg/dL, %	33.2	29.1	36	<0.001
HDL-C (mg/dL), mean (SD)	60.8 (15.2)	56.1 (14.5)	64 (14.9.0)	<0.001
HDL-C < 40 mg/dL, %	5.7	9.8	2.8	<0.001
Triglycerides (mg/dL), median (IQR)	97 (69, 136)	106 (75, 152)	91 (66, 126)	<0.001
Triglycerides ≥ 150 mg/dL, %	19.5	25.9	15.1	<0.001
eGFR, mL/min/1.73 m^2, median (IQR)	74 (67, 82)	73 (67, 82)	74 (68, 82)	<0.001
Vegetable pattern score, median (IQR)	0.01 (−0.68, 0.73)	−0.09 (−0.77, 0.65)	0.08 (−0.61, 0.78)	<0.001
Juice/milk pattern score, median (IQR)	−0.17 (−0.69, 0.47)	−0.33 (−0.84, 0.29)	−0.06 (−0.58, 0.58)	<0.001
Meat pattern score, median (IQR)	−0.21 (−0.67, 0.46)	−0.31 (−0.71, 0.34)	−0.14 (−0.63, 0.54)	<0.001

BMI, body mass index; DBP, diastolic blood pressure; eGFR, estimated glomerular filtration rate; FHMS, Fukushima Health Management Survey; HDL-C, high-density lipoprotein cholesterolemia; IQR, interquartile; LDL-C, low-density lipoprotein cholesterolemia; SBP, systolic blood pressure; SD, standard deviation.

The mean (standard deviation) eGFR at baseline was 75.7 (11.0) mL/min/1.73 m^2; then, it gradually declined to 70.8 mL/min/1.73 m^2 in 2015 (2.7 $\pm$ 1.2 follow-up years) (Table 3). Participants with eGFR 60–90 mL/min/1.73 m^2 decreased from 89.1% in 2011 to 78.7% in 2015; in contrast, participants with eGFR < 60 mL/min/1.73 m^2 increased to 14.9% in 2015. Compared with impaired eGFR level, proteinuria occurred much less frequently, showing a slight increase over time. The mean annual changes in eGFR rate during the follow-up years declined.

Table 3. Kidney function in participants at follow-up years, 2012–2015, FHMS.

	2011		**2012**		**2013**		**2014**		**2015**		***p* Value**
	(*n* = 14,732)		**(*n* = 10,999)**		**(*n* = 9597)**		**(*n* = 8713)**		**(*n* = 8477)**		
eGFR (mL/min/1.73 m^2), mean (SD)	75.7	(11.0)	73.9	(11.9)	72.3	(11.5)	71.0	(11.5)	70.8	(11.7)	<0.001
eGFR (mL/min/1.73 m^2) category, *n* (%)											<0.001
<60	0		889	(8.1)	1060	(11.1)	1195	(13.7)	1262	(14.9)	
60–90	13,131	(89.1)	9035	(82.1)	7777	(81.2)	6981	(80.1)	6673	(78.7)	
≥90	1601	(10.9)	1075	(9.8)	733	(7.7)	536	(6.1)	541	(6.4)	
Proteinuria											0.049
Negative	14,602	(99.4)	10,771	(97.9)	9358	(97.7)	8528	(97.8)	8254	(97.4)	
Trace	91	(0.6)	123	(1.1)	100	(1.0)	79	(0.9)	102	(1.2)	
Positive	0		95	(0.9)	104	(1.1)	97	(1.1)	115	(1.4)	
eGFR < 60 mL/min/1.73 m^2 or proteinuria, *n* (%)	0		973	(8.8)	1143	(11.9)	1270	(14.6)	1350	(15.9)	<0.001

	2011–2012		**2012–2013**		**2013–2014**		**2014–2015**		***p* Value**
	(*n* = 10,999)		**(*n* = 7342)**		**(*n* = 6612)**		**(*n* = 6337)**		
Annual change of eGFR (mL/min/1.73 m^2 per year), mean (SD)	−1.8	(7.3)	−1.2	(6.9)	−1.3	(6.2)	−0.3	(6.0)	<0.001
Annual change category, *n* (%)									<0.001
<−30%	6589	(59.9)	3954	(41.3)	3715	(42.6)	3034	(35.8)	
−30—< 15%	666	(6.1)	507	(5.3)	457	(5.2)	479	(5.6)	
≥15%	3744	(34.0)	2881	(30.1)	2440	(28.0)	2824	(33.3)	

eGFR, estimated glomerular filtration rate; FHMS, Fukushima Health Management Survey; SD, standard deviation.

Table 4 shows the associations between dietary pattern scores and the risk of impaired kidney function. In Model 2, the highest vs. lowest tertile of a vegetable diet had a 10–32% reduced risk of development of eGFR < 60 mL/min/1.73 m^2 and/or proteinuria with significant decreasing trends. The highest vs. lowest tertile of a juice diet increased the risk of developing impaired kidney function by 18–19% with significant increasing trends. No significant associations between a meat diet and impaired kidney function or CKD were observed.

Changes of eGFR in participants in the follow-up years were divided into three groups by trajectory analysis (Figure 2). We classified eGFR variance as an increasing group (20.3%), a moderate decline group (64.6%), and a rapid decline group (15.1%). The trajectory of the increasing group and the rapid decline group were approximately quadratic curves, whereas that of the moderate decline group was an almost straight line. Among the rapid decline group, most participants had higher risk categories, such as hypertension (14.7% vs. 12.6% nonhypertension), diabetes (17.9% vs. 13.1% nondiabetes), and living at temporary house or shelter (15.4% vs. 12.5% living in other places) (Table S2). For the yearly declines in eGFR, the highest vs. lowest tertile of a vegetable diet could prevent 15% of participants from rapid decline compared with the moderate decline eGFR group (Table 5).

Table 4. Cumulative incidence ratios (95% confidence intervals) of impaired kidney function among dietary patterns, 2012–2015, FHMS.

		eGFR < 60 (mL/min/1.73 m^2)		Proteinuria		eGFR < 60 (mL/min/ 1.73 m^2) or Proteinuria	
		CIR [a]	95% CI	CIR [a]	95% CI	CIR [a]	95% CI
Vegetable							
Model 1	T1 (lowest)	1.00	Referent	1.00	Referent	1.00	Referent
	T2	0.97	0.88, 1.07	0.80	0.61, 1.04	0.95	0.87, 1.05
	T3	0.89	0.81, 0.98	0.67	0.51, 0.88	0.87	0.79, 0.95
	P for trend	0.013		0.005		0.001	
Model 2	T1 (lowest)	1.00	Referent	1.00	Referent	1.00	Referent
	T2	0.98	0.89, 1.08	0.80	0.62, 1.04	0.96	0.88, 1.06
	T3	0.90	0.82, 1.00	0.68	0.52, 0.90	0.88	0.80, 0.97
	P for trend	0.031		0.007		0.005	
Juice							
Model 1	T1 (lowest)	1.00	Referent	1.00	Referent	1.00	Referent
	T2	1.08	0.98, 1.19	0.97	0.74, 1.26	1.07	0.97, 1.17
	T3	1.20	1.09, 1.32	1.08	0.83, 1.41	1.19	1.09, 1.30
	P for trend	<0.001		0.543		<0.001	
Model 2	T1 (lowest)	1.00	Referent	1.00	Referent	1.00	Referent
	T2	1.07	0.97, 1.18	0.94	0.72, 1.22	1.05	0.96, 1.15
	T3	1.19	1.08, 1.31	1.04	0.79, 1.36	1.18	1.08, 1.29
	P for trend	<0.001		0.738		<0.001	
Meat							
Model 1	T1 (lowest)	1.00	Referent	1.00	Referent	1.00	Referent
	T2	0.97	0.89, 1.06	1.00	0.77, 1.31	0.97	0.89, 1.06
	T3	0.96	0.88, 1.06	1.17	0.90, 1.52	0.98	0.90, 1.08
	P for trend	0.459		0.214		0.809	
Model 2	T1 (lowest)	1.00	Referent	1.00	Referent	1.00	Referent
	T2	0.97	0.89, 1.07	1.02	0.78, 1.33	0.98	0.90, 1.06
	T3	0.98	0.89, 1.07	1.20	0.92, 1.55	1.00	0.92, 1.09
	P for trend	0.695		0.158		0.898	

CIR, cumulative incidence ratio; eGFR, estimated glomerular filtration rate; FHMS, Fukushima Health Management Survey; T, tertile. [a] Poisson regression, Model 1: adjusted for age (continuous), age^2 (continuous) and sex; Model 2: model 1+ smoking history (never/former/current), alcohol drinking (never/occasional/regular), education (<occasional university/≥occasional university), physical activity (none/1 time per week/2–4 times per week/every day), distress scale (K6<13/≥13), residence (temporary house or shelter/others), overweight (no/yes), diabetes (no/yes), hypertension (no/yes), hyper-low-density lipoprotein cholesterolemia (no/yes), hypo-high-density lipoprotein cholesterolemia (no/yes), and hypertriglyceridemia (no/yes).

In the sensitivity analysis for participants with all available kidney function data in the follow-up years (n = 4440), we observed similar significant associations between each dietary pattern and impaired kidney function or CKD. In the multivariable adjustment models, the CIR of CKD in the highest vs. lowest tertile was 0.83 (95% CI: 0.71, 0.97) (*P* for trend = 0.006) for the vegetable diet; 1.16 (95% CI: 1.00, 1.34) (*P* for trend = 0.038) for the juice diet; and 0.95 (95% CI: 0.83, 1.10) (*P* for trend = 0.507) for the meat diet.

We repeated the regression analysis for participants with normal BMI, without hypertension, diabetes, or hyperlipidaemia; we observed similar significant associations between the vegetable and the juice dietary patterns and impaired kidney function (Tables S3–S5). However, the significant associations (CIRs) in participants with a normal BMI or without hypertension were attenuated to nonsignificance for the vegetable pattern. Other results in participants with hypertension had similar significant associations (CIRs) with CKD as in the main analysis. Participants with overweight and dyslipidemia showed similar significant associations (CIRs) with CKD for the juice pattern as in the main analysis, whereas the significant associations in participants with diabetes were attenuated to nonsignificance (data not shown).

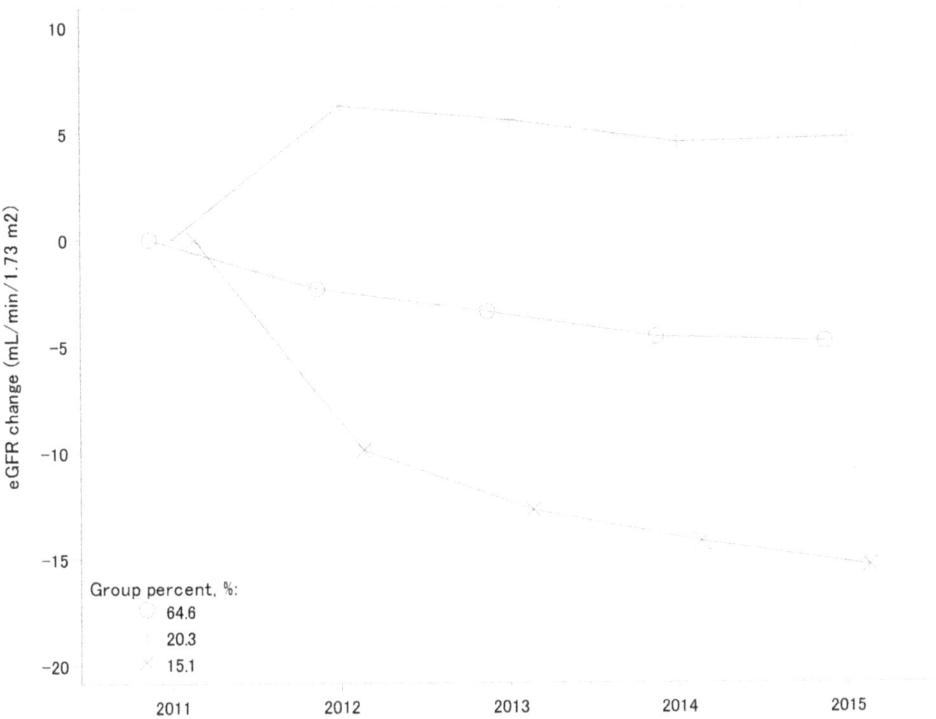

Figure 2. Trajectory groups of eGFR change over time, Fukushima Health Management Survey, 2011–2015. Solid lines were observed trajectory groups and short-dash lines indicate the model estimates. eGFR, estimated glomerular filtration rate.

Table 5. Associations between dietary patterns and annual eGFR change, rising eGFR, and decreasing eGFR groups, 2012–2015, FHMS.

		Annual Change in eGFR ($mL/min/1.73\ m^2$ Per Year)		Increasing eGFR		Rapid Decline in eGFR	
		β [a]	95% CI	OR [b]	95% CI	OR [b]	95% CI
Vegetable							
Model 1	T1 (lowest)	0	Referent	1.00	Referent	1.00	Referent
	T2	0.24	0.03, 0.44	0.96	0.86, 1.06	0.86	0.76, 0.98
	T3	0.27	0.06, 0.49	0.95	0.85, 1.06	0.83	0.73, 0.94
	P for trend	0.012		0.4		0.006	
Model 2	T1 (lowest)	0	Referent	1.00	Referent	1.00	Referent
	T2	0.23	0.02, 0.43	0.96	0.86, 1.06	0.88	0.77, 1.00
	T3	0.26	0.04, 0.47	0.94	0.84, 1.06	0.85	0.75, 0.98
	P for trend	0.019		0.422		0.009	
Juice							
Model 1	T1 (lowest)	0	Referent	1.00	Referent	1.00	Referent
	T2	−0.08	−0.28, 0.12	0.92	0.82, 1.02	1.11	0.98, 1.26
	T3	0.08	−0.12, 0.28	0.92	0.83, 1.03	0.99	0.87, 1.13
	P for trend	0.553		0.204		0.836	
Model 2	T1	1.00	Referent	1.00	Referent	1.00	Referent
	T2	−0.08	−0.28, 0.12	0.92	0.83, 1.03	1.10	0.98, 1.25
	T3	0.07	−0.13, 0.28	0.94	0.85, 1.05	1.00	0.88, 1.14
	P for trend	0.607		0.284		0.937	

Table 5. *Cont.*

		Annual Change in eGFR (mL/min/1.73 m^2 Per Year)		Increasing eGFR		Rapid Decline in eGFR	
		β [a]	95% CI	OR [b]	95% CI	OR [b]	95% CI
Meat							
Model 1	T1 (lowest)	0	Referent	1.00	Referent	1.00	Referent
	T2	−0.06	−0.26, 0.14	0.93	0.83, 1.03	1.04	0.92, 1.18
	T3	0.04	−0.16, 0.24	1.07	0.96, 1.19	1.08	0.95, 1.23
	P for trend	0.62		0.099		0.256	
Model 2	T1 (lowest)	0	Referent	1.00	Referent	1.00	Referent
	T2	−0.07	−0.27, 0.13	0.92	0.83, 1.02	1.04	0.92, 1.18
	T3	0.03	−0.18, 0.23	1.06	0.96, 1.18	1.09	0.96, 1.23
	P for trend	0.732		0.095		0.176	

eGFR, estimated glomerular filtration rate; FHMS, Fukushima Health Management Survey; T, tertile. [a] Multiple linear regression, Model 1: adjusted for age (continuous) and sex; Model 2: model 1+ smoking history (never/former/current), alcohol drinking (never/occasional/regular), education (<occasional university/≥occasional university), physical activity (none/1 time per week/ 2–4 times per week/every day), distress (no/yes), residence (temporary house or shelter/others), overweight (no/yes), diabetes (no/yes), hypertension (no/yes), hyper-low-density lipoprotein cholesterolemia (no/yes), hyper-low-density lipoprotein cholesterolemia (no/yes), and hypertriglyceridemia (no/yes). [b] Polytomous logistic regression, covariates were the same as the analysis for the overall change of eGFR, with adding age^2 and baseline eGFR level for an adjustment (continuous).

4. Discussion

In this large population-based prospective study, we observed significant inverse associations between the vegetable dietary pattern and risks of impaired kidney function or CKD, and significant positive associations between the juice dietary pattern and eGFR < 60 mL/min/1.73 m^2 or CKD. We also elucidated that the intake of a vegetable diet was inversely associated with annual eGFR changes, particularly in the rapid decline group.

The three dietary patterns identified in this study were similar to those in other Japanese studies [40–43]; particularly, the vegetable pattern had the most similar characteristics of high intake in the reproducible healthy/prudent patterns, i.e., the combination of vegetables, fish, fruits, bean products, and rice, in the typical Japanese diet [44–46]. In the Nurses' Health Study in the USA, the DASH-style diet was inversely associated with eGFR decline ≥30% [21]. In a Swedish population, the medium and high adherents to the Mediterranean Diet were 23% and 42% less likely, respectively, to have CKD compared with the low adherents [19]. A recent meta-analysis has reported that the adherence to a healthy dietary pattern (rich in whole grains, vegetables, fruit, legumes, nuts, and fish, and a lower intake of red and processed meats, sodium, and sugar-sweetened beverages) was associated with lower odds of incident CKD and albuminuria [47]. Our study results support these findings and add evidence to these favorable associations in a general Japanese population.

The anti-inflammatory mechanism underlying the beneficial effects of these diets could help reduce the incidence of chronic diseases, improving cardiometabolic profiles [11,18,48] and reducing metabolic acidosis [12,49]. A diet high in cereal fiber was protective against the development of moderate CKD among older adults [50]. High polyunsaturated fatty acid intake has been considered to be renal-protective [50,51]. Japanese and Mediterranean diets similarly feature seafood, vegetables, and fruits, as well as the soybean and soy products that are popular among the Japanese [52,53]. The Japanese dietary pattern is reportedly associated with the intake of antioxidant vitamins, minerals, dietary fiber, and omega-3 fatty acids [46,53]. Dietary intakes of omega-3 marine polyunsaturated fatty acids [54], soy, and isoflavones [55] were inversely associated with the incident risk of ischemic heart disease. Soybeans, a major source of plant protein, are associated with lower CVD mortality [56]. Moreover, the low consumption of sweeteners and the high consumption of green tea could be related to obesity status [57]. Therefore, the adoption of a healthy dietary lifestyle, lowering inflammatory markers, and/or dietary acid loading as mediators of cardiometabolic factors, e.g., preventing or stopping the vicious cycle between the gut microbiota and the cardiovascular/renal systems, might better preserve renal function, thus decreasing the morbidity and mortality of CVD and CKD [10,11,19,21,47,58].

We observed a significant positive association between the juice dietary pattern and eGFR < 60 mL/min/1.73 m^2 and CKD. These results were similar to an Iranian study in which the dietary pattern of high fat and sugar was associated with a significant 46% increased risk of incident CKD [24]. Our results were also similar to the Nurses' Health Study, in which the Western diet showed a positive association with microalbuminuria [21]. This result might be attributable to the higher intake of dairy products and sugar-based juices, such as the high-dairy [43] and bread [59] dietary patterns. The juice pattern in this study can be considered the reverse of the traditional Japanese staple food pattern [59]. The significant relationships between dietary patterns and inflammatory markers such as C-reactive protein could explain the direct association of the juice pattern with proteinuria [14,21].

High saturated fatty acid intake can deteriorate kidney function by affecting plasma creatinine [21]. The Western diet is rich in advanced glycation end products, formed when food is processed at increased temperatures, which is independently associated with the GFR [60]. Of note, the low saturated fat (meat) and high omega-3 polyunsaturated fat (fish) in the Japanese diet contribute to the low prevalence of hypercholesterolemia [61]. Serum levels of advanced glycation end products might be reduced through changes in diet [60] or in cooking practices. This reduction could explain the nonsignificant associations between the meat diet and the eGFR decline in this study, which was similar to those found in the Northern Manhattan Study [18].

In the subgroup analysis, significant associations for the vegetable pattern did not remain in participants without overweight or hypertension. This result might be due to the fact that the prevalence of cardiometabolic risk was high in this population; more than 50% of the participants had hypertension at baseline. Thus, those with hypertension, whether in treatment or not, were on a more cautious diet [62] and might have increased their consumption of vegetables [42]. High salt intake is an established risk factor for kidney function decline, mainly through its adverse effect on blood pressure and vascular health. Vegetables and fruit are high in potassium, and high potassium intake could also positively affect kidney function [14]. Hypertension might also be considered partially mediated by obesity [63] or sodium [41]; however, we did not have data on sodium or the sodium–potassium ratio for analysis. Additionally, although participants living in temporary houses or shelters had lower vegetable pattern scores, higher juice pattern scores, and the same meat pattern scores compared with other residents, we observed similar significant associations when stratified by post-disaster residence.

One of the strengths of this study was that it had a large sample with repeated longitudinal measurements of kidney function, and the associations measured were more robust than in simple cross-sectional surveys. Further, we applied not only the eGFR cut-off level and proteinuria as the outcomes for measuring the associations but also yearly changes based on the trajectory analysis; both showed prominent significant results.

This study had a few limitations. First, FHMS response rates remained at ≈27%; thus, the representativeness of the results might not be generalizable to the entire prefecture or the country's population. Second, we could not compute the food amounts or nutrient amounts with energy adjustment, which would have helped us better elucidate the underlying mechanisms [63]. Meanwhile, a total of 19 food group items might not be able to determine correlations between specific foods [43]. Third, we computed the dietary pattern scores arising from FFQ surveys at baseline; thus, we could not clarify whether changes in the residence, employment, or dietary habits of participants during follow-up years impacted the associations measured [18,47]. Finally, the presence of residual confounding was possible, as in any observational study [27].

5. Conclusions

Our study suggests that the vegetable dietary pattern could be inversely associated with risk of impaired kidney function, including lower eGFR and proteinuria, whereas the juice dietary pattern could be positively associated with the risk of impaired kidney function. Continuous promotion of a balanced diet, particularly a vegetable diet rich in traditional Japanese foods, might be necessary to prevent the progression of impaired kidney function, thus reducing the burden of CKD with aging. Further validation studies or randomized controlled trials on these associations are needed, examining how individual foods influence kidney function.

Supplementary Materials: The following are available online at https://www.mdpi.com/2072-6643/13/1/168/s1, Table S1: Characteristics of participants at baseline according to tertiles of dietary pattern scores, FHMS, 2011 (n = 14, 732). Table S2: Characteristics of participants in groups of eGFR progression, 2012–2015, FHMS. Table S3: Cumulative incidence ratios (95% confidence intervals) of impaired kidney function (eGFR < 60 mL/min/1.73 m^2) among dietary patterns in participants without cardiometabolic risk at baseline, 2012–2015, FHMS. Table S4: Associations (coefficients) between dietary patterns and the annual change of eGFR (mL/min/1.73 m^2) in participants without cardiometabolic risk at baseline, 2012–2015, FHMS. Table S5: Odds ratios (95% confidence intervals) of the eGFR rapid decline group among dietary patterns in participants without cardiometabolic risk at baseline, 2012–2015, FHMS.

Author Contributions: E.M. designed the study, analyzed data, and drafted the manuscript; T.O., S.Y., H.Y., M.M. (Masaharu Maeda), H.O. (Hitoshi Ohto), and K.K. directed and supervised the field survey activities; H.N. cleaned and prepared the database; E.E., M.M. (Makoto Miyazaki), M.H., A.S., A.T., H.O. (Hiromasa Ohira), J.K., and M.S. provided quality assurance and control of data collection. Each author contributed important intellectual content during manuscript drafting or revisions and approved the final manuscript. All authors have read and agreed to the published version of the manuscript.

Funding: This work was supported by the National Health Fund for Children and Adults Affected by the Nuclear Incident for the design and conduction of the study.

Institutional Review Board Statement: This study was approved by the Committee for Ethics at Fukushima Medical University, Japan (nos. 1316, 1319, and 29064).

Informed Consent Statement: Informed consent was obtained from all individual participants included in the study.

Data Availability Statement: The data presented in this study are available on request from the corresponding author. The data are not publicly available due to the privacy of participants from the radiation disaster areas.

Acknowledgments: The findings and conclusions of this article are solely the responsibility of the authors and do not represent the official views of the Fukushima Prefecture government.

Conflicts of Interest: The authors declare there is no conflict of interests.

References

1. GBD Chronic Kidney Disease Collaboration. Global, regional, and national burden of chronic kidney disease, 1990–2017: A systematic analysis for the Global Burden of Disease Study 2017. *Lancet* **2020**, *395*, 709–733. [CrossRef]
2. Tonelli, M.; Muntner, P.; Lloyd, A.; Manns, B.J.; Klarenbach, S.; Pannu, N.; James, M.T.; Hemmelgarn, B.R.; Alberta Kidney Disease Network. Risk of coronary events in people with chronic kidney disease compared with those with diabetes: A population-level cohort study. *Lancet* **2012**, *380*, 807–814. [CrossRef]
3. Muka, T.; Imo, D.; Jaspers, L.; Colpani, V.; Chaker, L.; van der Lee, S.J.; Mendis, S.; Chowdhury, R.; Bramer, W.M.; Falla, A.; et al. The global impact of non-communicable diseases on healthcare spending and national income: A systematic review. *Eur. J. Epidemiol.* **2015**, *30*, 251–277. [CrossRef] [PubMed]
4. Cooper, J.T.; Lloyd, A.; Sanchez, J.J.G.; Sorstadius, E.; Briggs, A.; McFarlane, P. Health related quality of life utility weights for economic evaluation through different stages of chronic kidney disease: A systematic literature review. *Health Qual. Life Outcomes* **2020**, *18*, 310. [CrossRef]
5. Daya, N.; Voskertchian, A.; Schneider, A.L.C.; Ballew, S.; McAdams DeMarco, M.; Coresh, J.; Appel, L.J.; Selvin, E.; Grams, M.E. Kidney Function and Fracture Risk: The Atherosclerosis Risk in Communities (ARIC) Study. *Am. J. Kidney Dis.* **2016**, *67*, 218–226. [CrossRef]

6. Pimentel, A.; Urena-Torres, P.; Zillikens, M.C.; Bover, J.; Cohen-Solal, M. Fractures in patients with CKD-diagnosis, treatment, and prevention: A review by members of the European Calcified Tissue Society and the European Renal Association of Nephrology Dialysis and Transplantation. *Kidney Int.* **2017**, *92*, 1343–1355. [CrossRef]
7. Evenepoel, P.; Opdebeeck, B.; David, K.; D'Haese, P.C. Bone-Vascular Axis in Chronic Kidney Disease. *Adv. Chronic. Kidney Dis.* **2019**, *26*, 472–483. [CrossRef]
8. Mihai, S.; Codrici, E.; Popescu, I.D.; Enciu, A.M.; Albulescu, L.; Necula, L.G.; Mambet, C.; Anton, G.; Tanase, C. Inflammation-Related Mechanisms in Chronic Kidney Disease Prediction, Progression, and Outcome. *J. Immunol. Res.* **2018**, *2018*, 2180373. [CrossRef]
9. Graboski, A.L.; Redinbo, M.R. Gut-Derived Protein-Bound Uremic Toxins. *Toxins* **2020**, *12*. [CrossRef]
10. Kaminski, T.W.; Pawlak, K.; Karbowska, M.; Mysliwiec, M.; Pawlak, D. Indoxyl sulfate—#he uremic toxin linking hemostatic system disturbances with the prevalence of cardiovascular disease in patients with chronic kidney disease. *BMC Nephrol.* **2017**, *18*, 35. [CrossRef]
11. Felizardo, R.J.F.; Watanabe, I.K.M.; Dardi, P.; Rossoni, L.V.; Camara, N.O.S. The interplay among gut microbiota, hypertension and kidney diseases: The role of short-chain fatty acids. *Pharmacol. Res.* **2019**, *141*, 366–377. [CrossRef] [PubMed]
12. Sabatino, A.; Regolisti, G.; Cosola, C.; Gesualdo, L.; Fiaccadori, E. Intestinal Microbiota in Type 2 Diabetes and Chronic Kidney Disease. *Curr. Diabete Rep.* **2017**, *17*, 16. [CrossRef] [PubMed]
13. Bobot, M.; Thomas, L.; Moyon, A.; Fernandez, S.; McKay, N.; Balasse, L.; Garrigue, P.; Brige, P.; Chopinet, S.; Poitevin, S.; et al. Uremic Toxic Blood-Brain Barrier Disruption Mediated by AhR Activation Leads to Cognitive Impairment during Experimental Renal Dysfunction. *J. Am. Soc. Nephrol.* **2020**, *31*, 1509–1521. [CrossRef] [PubMed]
14. Jhee, J.H.; Kee, Y.K.; Park, J.T.; Chang, T.I.; Kang, E.W.; Yoo, T.H.; Kang, S.W.; Han, S.H. A Diet Rich in Vegetables and Fruit and Incident CKD: A Community-Based Prospective Cohort Study. *Am. J. Kidney Dis.* **2019**, *74*, 491–500. [CrossRef] [PubMed]
15. Hu, F.B. Dietary pattern analysis: A new direction in nutritional epidemiology. *Curr. Opin. Lipidol.* **2002**, *13*, 3–9. [CrossRef]
16. Borges, C.A.; Rinaldi, A.E.; Conde, W.L.; Mainardi, G.M.; Behar, D.; Slater, B. Dietary patterns: A literature review of the methodological characteristics of the main step of the multivariate analyzes. *Rev. Bras. Epidemiol.* **2015**, *18*, 837–857. [CrossRef]
17. Wesolowska, E.; Jankowska, A.; Trafalska, E.; Kaluzny, P.; Grzesiak, M.; Dominowska, J.; Hanke, W.; Calamandrei, G.; Polanska, K. Sociodemographic, Lifestyle, Environmental and Pregnancy-Related Determinants of Dietary Patterns during Pregnancy. *Int. J. Environ. Res. Public Health* **2019**, *16*. [CrossRef]
18. Khatri, M.; Moon, Y.P.; Scarmeas, N.; Gu, Y.; Gardener, H.; Cheung, K.; Wright, C.B.; Sacco, R.L.; Nickolas, T.L.; Elkind, M.S. The association between a Mediterranean-style diet and kidney function in the Northern Manhattan Study cohort. *Clin. J. Am. Soc. Nephrol.* **2014**, *9*, 1868–1875. [CrossRef]
19. Huang, X.; Jimenez-Moleon, J.J.; Lindholm, B.; Cederholm, T.; Arnlov, J.; Riserus, U.; Sjogren, P.; Carrero, J.J. Mediterranean diet, kidney function, and mortality in men with CKD. *Clin. J. Am. Soc. Nephrol.* **2013**, *8*, 1548–1555. [CrossRef]
20. Cai, Q.; Dekker, L.H.; Bakker, S.J.L.; de Borst, M.H.; Navis, G.J. Dietary Patterns Based on Estimated Glomerular Filtration Rate and Kidney Function Decline in the General Population: The Lifelines Cohort Study. *Nutrients* **2020**, *12*. [CrossRef]
21. Lin, J.; Fung, T.T.; Hu, F.B.; Curhan, G.C. Association of dietary patterns with albuminuria and kidney function decline in older white women: A subgroup analysis from the Nurses' Health Study. *Am. J. Kidney Dis.* **2011**, *57*, 245–254. [CrossRef] [PubMed]
22. Smyth, A.; Griffin, M.; Yusuf, S.; Mann, J.F.; Reddan, D.; Canavan, M.; Newell, J.; O'Donnell, M. Diet and Major Renal Outcomes: A Prospective Cohort Study. The NIH-AARP Diet and Health Study. *J. Ren. Nutr.* **2016**, *26*, 288–298. [CrossRef] [PubMed]
23. Shi, Z.; Taylor, A.W.; Riley, M.; Byles, J.; Liu, J.; Noakes, M. Association between dietary patterns, cadmium intake and chronic kidney disease among adults. *Clin. Nutr.* **2018**, *37*, 276–284. [CrossRef] [PubMed]
24. Asghari, G.; Momenan, M.; Yuzbashian, E.; Mirmiran, P.; Azizi, F. Dietary pattern and incidence of chronic kidney disease among adults: A population-based study. *Nutr. Metab.* **2018**, *15*, 88. [CrossRef]
25. Kurniawan, A.L.; Hsu, C.Y.; Rau, H.H.; Lin, L.Y.; Chao, J.C. Association of kidney function-related dietary pattern, weight status, and cardiovascular risk factors with severity of impaired kidney function in middle-aged and older adults with chronic kidney disease: A cross-sectional population study. *Nutr. J.* **2019**, *18*, 27. [CrossRef]
26. Ohira, T.; Nakano, H.; Nagai, M.; Yumiya, Y.; Zhang, W.; Uemura, M.; Sakai, A.; Hashimoto, S.; Fukushima Health Management Survey Group. Changes in Cardiovascular Risk Factors After the Great East Japan Earthquake. *Asia Pac. J. Public Health* **2017**, *29*, 47S–55S. [CrossRef]
27. Hayashi, Y.; Nagai, M.; Ohira, T.; Satoh, H.; Sakai, A.; Ohtsuru, A.; Hosoya, M.; Kawasaki, Y.; Suzuki, H.; Takahashi, A.; et al. The impact of evacuation on the incidence of chronic kidney disease after the Great East Japan Earthquake: The Fukushima Health Management Survey. *Clin. Exp. Nephrol.* **2017**, *21*, 995–1002. [CrossRef]
28. Ma, E.; Ohira, T.; Sakai, A.; Yasumura, S.; Takahashi, A.; Kazama, J.; Shimabukuro, M.; Nakano, H.; Okazaki, K.; Maeda, M.; et al. Associations between Dietary Patterns and Cardiometabolic Risks in Japan: A Cross-Sectional Study from the Fukushima Health Management Survey, 2011–2015. *Nutrients* **2020**, *12*. [CrossRef]
29. Yasumura, S.; Hosoya, M.; Yamashita, S.; Kamiya, K.; Abe, M.; Akashi, M.; Kodama, K.; Ozasa, K. Study protocol for the Fukushima Health Management Survey. *J. Epidemiol.* **2012**, *22*, 375–383. [CrossRef]
30. Sauvaget, C.; Allen, N.; Hayashi, M.; Spencer, E.; Nagano, J. Validation of a food frequency questionnaire in the Hiroshima/Nagasaki Life Span Study. *J. Epidemiol.* **2002**, *12*, 394–401. [CrossRef]

31. Matsuo, S.; Imai, E.; Horio, M.; Yasuda, Y.; Tomita, K.; Nitta, K.; Yamagata, K.; Tomino, Y.; Yokoyama, H.; Hishida, A.; et al. Revised equations for estimated GFR from serum creatinine in Japan. *Am. J. Kidney Dis.* **2009**, *53*, 982–992. [CrossRef] [PubMed]
32. Kidney Disease: Improving Global Outcomes (KDIGO) Diabetes Work Group. KDIGO 2020 Clinical Practice Guideline for Diabetes Management in Chronic Kidney Disease. *Kidney Int.* **2020**, *98*, S1–S115. [CrossRef] [PubMed]
33. Kessler, R.C.; Barker, P.R.; Colpe, L.J.; Epstein, J.F.; Gfroerer, J.C.; Hiripi, E.; Howes, M.J.; Normand, S.L.; Manderscheid, R.W.; Walters, E.E.; et al. Screening for serious mental illness in the general population. *Arch. Gen. Psychiatry* **2003**, *60*, 184–189. [CrossRef] [PubMed]
34. Zhang, W.; Ohira, T.; Abe, M.; Kamiya, K.; Yamashita, S.; Yasumura, S.; Ohtsuru, A.; Masaharu, M.; Harigane, M.; Horikoshi, N.; et al. Evacuation after the Great East Japan Earthquake was associated with poor dietary intake: The Fukushima Health Management Survey. *J. Epidemiol.* **2017**, *27*, 14–23. [CrossRef] [PubMed]
35. Luger, E.; Aspalter, R.; Luger, M.; Longin, R.; Rieder, A.; Dorner, T.E. Changes of dietary patterns during participation in a web-based weight-reduction programme. *Public Health Nutr.* **2016**, *19*, 1211–1221. [CrossRef] [PubMed]
36. Kobayashi, A.; Matsuzawa, T.; Hosoya, T.; Yoshida, S. Sulfoxide synthesis from sulfinate esters under Pummerer-like conditions. *Chem. Commun.* **2020**, *56*, 5429–5432. [CrossRef]
37. Burckhardt, P.; Nagin, D.S.; Padman, R. Multi-Trajectory Models of Chronic Kidney Disease Progression. *AMIA Annu. Symp. Proc.* **2016**, *2016*, 1737–1746.
38. Nagai, M.; Ohira, T.; Takahashi, H.; Nakano, H.; Sakai, A.; Hashimoto, S.; Yasumura, S.; Abe, M.; Fukushima Health Management Survey. Impact of evacuation onstrends in the prevalence, treatment, and control of hypertension before and after a disaster. *J. Hypertens* **2018**, *36*, 924–932. [CrossRef]
39. Bazi, T.; Takahashi, S.; Ismail, S.; Bo, K.; Ruiz-Zapata, A.M.; Duckett, J.; Kammerer-Doak, D. Prevention of pelvic floor disorders: International urogynecological association research and development committee opinion. *Int. Urogynecol. J.* **2016**, *27*, 1785–1795. [CrossRef]
40. Maruyama, K.I.H.; Date, C.; Kikuchi, S.; Watanabe, Y.; Wada, Y.; Inaba, Y.; Tamakoshi, A. JACC Study Group. Dietary patterns and risk of cardiovascular deaths among middle-aged Japanese: JACC Study. *Nutr. Metab. Cardiovasc. Dis.* **2013**, *23*, 8. [CrossRef]
41. Okada, E.; Takahashi, K.; Nakamura, K.; Ukawa, S.; Takabayashi, S.; Nakamura, M.; Sasaki, S.; Tamakoshi, A.; Takimoto, H. Dietary patterns and abnormal glucose tolerance among Japanese: Findings from the National Health and Nutrition Survey, 2012. *Public Health Nutr.* **2019**, *22*, 2460–2468. [CrossRef] [PubMed]
42. Sadakane, A.; Tsutsumi, A.; Gotoh, T.; Ishikawa, S.; Ojima, T.; Kario, K.; Nakamura, Y.; Kayaba, K. Dietary patterns and levels of blood pressure and serum lipids in a Japanese population. *J. Epidemiol.* **2008**, *18*, 58–67. [CrossRef]
43. Tomata, Y.; Sugiyama, K.; Kaiho, Y.; Honkura, K.; Watanabe, T.; Zhang, S.; Sugawara, Y.; Tsuji, I. Dietary Patterns and Incident Dementia in Elderly Japanese: The Ohsaki Cohort 2006 Study. *J. Gerontol A Biol. Sci. Med. Sci.* **2016**, *71*, 1322–1328. [CrossRef] [PubMed]
44. Murakami, K.; Shinozaki, N.; Fujiwara, A.; Yuan, X.; Hashimoto, A.; Fujihashi, H.; Wang, H.C.; Livingstone, M.B.E.; Sasaki, S. A Systematic Review of Principal Component Analysis-Derived Dietary Patterns in Japanese Adults: Are Major Dietary Patterns Reproducible Within a Country? *Adv. Nutr.* **2019**, *10*, 237–249. [CrossRef] [PubMed]
45. Nanri, A.; Yoshida, D.; Yamaji, T.; Mizoue, T.; Takayanagi, R.; Kono, S. Dietary patterns and C-reactive protein in Japanese men and women. *Am. J. Clin. Nutr.* **2008**, *87*, 1488–1496. [CrossRef]
46. Ito, T.; Kawakami, R.; Tanisawa, K.; Miyawaki, R.; Ishii, K.; Torii, S.; Suzuki, K.; Sakamoto, S.; Muraoka, I.; Oka, K.; et al. Dietary patterns and abdominal obesity in middle-aged and elderly Japanese adults: Waseda Alumni's Sports, Exercise, Daily Activity, Sedentariness and Health Study (WASEDA'S Health Study). *Nutrition* **2019**, *58*, 149–155. [CrossRef]
47. Bach, K.E.; Kelly, J.T.; Palmer, S.C.; Khalesi, S.; Strippoli, G.F.M.; Campbell, K.L. Healthy Dietary Patterns and Incidence of CKD: A Meta-Analysis of Cohort Studies. *Clin. J. Am. Soc. Nephrol.* **2019**, *14*, 1441–1449. [CrossRef]
48. Babio, N.; Bullo, M.; Salas-Salvado, J. Mediterranean diet and metabolic syndrome: The evidence. *Public Health Nutr.* **2009**, *12*, 1607–1617. [CrossRef]
49. Goraya, N.; Simoni, J.; Jo, C.H.; Wesson, D.E. A comparison of treating metabolic acidosis in CKD stage 4 hypertensive kidney disease with fruits and vegetables or sodium bicarbonate. *Clin. J. Am. Soc. Nephrol.* **2013**, *8*, 371–381. [CrossRef]
50. Gopinath, B.; Harris, D.C.; Flood, V.M.; Burlutsky, G.; Brand-Miller, J.; Mitchell, P. Carbohydrate nutrition is associated with the 5-year incidence of chronic kidney disease. *J. Nutr.* **2011**, *141*, 433–439. [CrossRef]
51. Lauretani, F.; Maggio, M.; Pizzarelli, F.; Michelassi, S.; Ruggiero, C.; Ceda, G.P.; Bandinelli, S.; Ferrucci, L. Omega-3 and renal function in older adults. *Curr. Pharm. Des.* **2009**, *15*, 4149–4156. [CrossRef] [PubMed]
52. Yamori, Y.; Sagara, M.; Arai, Y.; Kobayashi, H.; Kishimoto, K.; Matsuno, I.; Mori, H.; Mori, M. Soy and fish as features of the Japanese diet and cardiovascular disease risks. *PLoS ONE* **2017**, *12*, e0176039. [CrossRef]
53. Tsugane, S. Why has Japan become the world's most long-lived country: Insights from a food and nutrition perspective. *Eur. J. Clin. Nutr.* **2020**. [CrossRef] [PubMed]
54. Iso, H.; Kobayashi, M.; Ishihara, J.; Sasaki, S.; Okada, K.; Kita, Y.; Kokubo, Y.; Tsugane, S.; Group, J.S. Intake of fish and n3 fatty acids and risk of coronary heart disease among Japanese: The Japan Public Health Center-Based (JPHC) Study Cohort I. *Circulation* **2006**, *113*, 195–202. [CrossRef]

55. Kokubo, Y.; Iso, H.; Ishihara, J.; Okada, K.; Inoue, M.; Tsugane, S.; Group, J.S. Association of dietary intake of soy, beans, and isoflavones with risk of cerebral and myocardial infarctions in Japanese populations: The Japan Public Health Center-based (JPHC) study cohort I. *Circulation* **2007**, *116*, 2553–2562. [CrossRef]
56. Budhathoki, S.; Sawada, N.; Iwasaki, M.; Yamaji, T.; Goto, A.; Kotemori, A.; Ishihara, J.; Takachi, R.; Charvat, H.; Mizoue, T.; et al. Association of Animal and Plant Protein Intake With All-Cause and Cause-Specific Mortality in a Japanese Cohort. *JAMA Intern. Med.* **2019**, *179*, 1509–1518. [CrossRef]
57. Hruby, A.; Manson, J.E.; Qi, L.; Malik, V.S.; Rimm, E.B.; Sun, Q.; Willett, W.C.; Hu, F.B. Determinants and Consequences of Obesity. *Am. J. Public Health* **2016**, *106*, 1656–1662. [CrossRef]
58. Onal, E.M.; Afsar, B.; Covic, A.; Vaziri, N.D.; Kanbay, M. Gut microbiota and inflammation in chronic kidney disease and their roles in the development of cardiovascular disease. *Hypertens Res.* **2019**, *42*, 123–140. [CrossRef]
59. Nanri, H.; Nakamura, K.; Hara, M.; Higaki, Y.; Imaizumi, T.; Taguchi, N.; Sakamoto, T.; Horita, M.; Shinchi, K.; Tanaka, K. Association between dietary pattern and serum C-reactive protein in Japanese men and women. *J. Epidemiol.* **2011**, *21*, 122–131. [CrossRef]
60. Semba, R.D.; Ferrucci, L.; Fink, J.C.; Sun, K.; Beck, J.; Dalal, M.; Guralnik, J.M.; Fried, L.P. Advanced glycation end products and their circulating receptors and level of kidney function in older community-dwelling women. *Am. J. Kidney Dis.* **2009**, *53*, 51–58. [CrossRef]
61. Iso, H. Lifestyle and cardiovascular disease in Japan. *J. Atheroscler. Thromb.* **2011**, *18*, 83–88. [CrossRef] [PubMed]
62. Tormo, M.J.; Navarro, C.; Chirlaque, M.D.; Barber, X.; Cancer, E.G.o.S.E.P.I.o. Is there a different dietetic pattern depending on self-knowledge of high blood pressure? *Eur. J. Epidemiol.* **2000**, *16*, 963–971. [CrossRef] [PubMed]
63. Iwasaki, Y.; Arisawa, K.; Katsuura-Kamano, S.; Uemura, H.; Tsukamoto, M.; Kadomatsu, Y.; Okada, R.; Hishida, A.; Tanaka, K.; Hara, M.; et al. Associations of Nutrient Patterns with the Prevalence of Metabolic Syndrome: Results from the Baseline Data of the Japan Multi-Institutional Collaborative Cohort Study. *Nutrients* **2019**, *11*. [CrossRef] [PubMed]

nutrients MDPI

Review

The Impact of Whole Grain Intake on Gastrointestinal Tumors: A Focus on Colorectal, Gastric, and Esophageal Cancers

Valentina Tullio †, Valeria Gasperi *,†, Maria Valeria Catani ‡ and Isabella Savini ‡

Department of Experimental Medicine, Tor Vergata University of Rome, 00133 Rome, Italy; valentinatullio.nu@gmail.com (V.T.); catani@uniroma2.it (M.V.C.); savini@uniroma2.it (I.S.)
* Correspondence: gasperi@med.uniroma2.it; Tel.: +39-06-72596465
† Equally first authors.
‡ Equally senior authors.

Abstract: Cereals are one of staple foods in human diet, mainly consumed as refined grains. Nonetheless, epidemiological data indicate that whole grain (WG) intake is inversely related to risk of type 2 diabetes, cardiovascular disease, and several cancer types, as well as to all-cause mortality. Particularly responsive to WG positive action is the gastrointestinal tract, daily exposed to bioactive food components. Herein, we shall provide an up-to-date overview on relationship between WG intake and prevention of gastrointestinal tumors, with a particular focus on colorectal, stomach, and esophagus cancers. Unlike refined counterparts, WG consumption is inversely associated with risk of these gastrointestinal cancers, most consistently with the risk of colorectal tumor. Some WG effects may be mediated by beneficial constituents (such as fiber and polyphenols) that are reduced/lost during milling process. Beside health-promoting action, WGs are still under-consumed in most countries; therefore, World Health Organization and other public/private stakeholders should cooperate to implement WG consumption in the whole population, in order to reach nutritionally effective intakes.

Keywords: dietary fiber; esophagus; stomach and colorectal cancer; nutrition; polyphenols; refined grains; whole grains

Citation: Tullio, V.; Gasperi, V.; Catani, M.V.; Savini, I. The Impact of Whole Grain Intake on Gastrointestinal Tumors: A Focus on Colorectal, Gastric, and Esophageal Cancers. *Nutrients* **2021**, *13*, 81. https://doi.org/10.3390/nu13010081

Received: 13 November 2020
Accepted: 25 December 2020
Published: 29 December 2020

1. Introduction

Cereals, plant species belonging to the *Poaceae* family, are grasses producing edible grains (wheat, corn, rice, oats, barley, rye, millet, teff, sorghum, canary seed, triticale, and Job's tears). Usually, the term also includes grains from non-herbaceous plants, known as pseudocereals (quinoa, buckwheat, amaranth, and wild rice), which have a composition similar to "real" grains [1–3].

Rice, maize, and wheat are the most common farmed cereals with a production of 2646 million tons in 2018–2019 [4]. As a primary source of carbohydrates, cereals provide about 60% of food energy worldwide and are mostly consumed as refined grains (RGs). However, health effects of cereals are mainly attributed to whole grains (WGs), and therefore governmental authorities are increasingly encouraging WG consumption.

Scientific interest in health properties of WGs began in the late 1970s, when the surgeon Denis Parsons Burkitt, noting the difference in disease incidence in rural Africa and the UK, brought together data coming from several disciplines and launched the dietary fiber hypothesis; he and other researchers, indeed, noted that a diet highly refined and lacking WG foods might be involved in several diseases, including coronary heart disease, obesity, diabetes, dental caries, as well as in some cancer types, such as gastric and colon tumors [5]. Since then, both epidemiological and interventional studies have reported potential health effects of unrefined grains [6–14]. Nonetheless, most of the population does not consume WGs, much likely due to several factors, including lack of nutritional education programs, low variety and palatability of WG-containing products, poor identification, and high purchase costs of WG foods. In addition, national dietary guidelines generally provide qualitative statements, such as "choose WG versions/varieties" or "increase WG intake", and only few countries provide quantitative recommendations: i.e., 48 g/day (corresponding to 3 servings/day) in USA [15], ≥75 g/day in Denmark [16], and 70–90 g/day in Norway and Sweden [17,18]. In addition, among WG consumers, daily intake of unrefined grains is still below recommended levels, except for few countries, such as Denmark [19]. As emerged from National Health and Nutrition Examination Survey (NHANES, 2001–2012), mean WG intakes are 15.52 g/day for adults and 11.84 g/day for children in USA, and less than 8.0 and 1.0% of adults and children, respectively, meets WG recommendations [20]. Low mean intakes have also been reported by National Adults Nutrition Survey (NANS, 2008–2010), National Children's Food Survey (NCFS, 2003–2004) and National Teens' Food Survey (NTFS, 2005–2006) in Ireland: 27.8 g/day for adults (only 19% satisfies recommendations) [21] and 18.5–23.2 g/day for children/adolescents (just 17–39% met recommendations) [22]. Similar findings have been reported for Australia [23] and UK [24], while in other countries, WG intakes are even lower. For example, in France, as emerged from Comportements et Consommations Alimentaires en France survey (CCAF, 2009–2010), averages are 4.7 g/day for adults/older adults and 4.1 g/day for children/adolescents [25]. Finally, the Italian National Food Consumption Survey (INRAN-SCAI, 2005–2006) reported average values of 3.7 g/day for adults/older adults and 2.1 g/day for children/adolescents [26]; as recently emerged from the Italian Nutrition and Health Survey (INHES, 2010–2013), only 27.2% of adults and 21.9% of children/adolescents consume WG foods (mainly bread) at least once per week [27].

2. Whole Grains: An Overview

2.1. Definition

Each country or organization uses different WG definitions that are constantly updated [1,28,29]. The widely used International Definition was provided in 1999 by American Association for Cereal Chemists (AACC), which states: "whole grain shall consist of the intact, ground, cracked or flaked caryopsis, whose principal anatomical components—the starchy endosperm, germ, and bran—are present in the same relative proportions as they exist in the intact caryopsis" [30]. As unambiguous definition is essential for dietary recommendations, nutritional research, flour manufacturing process and labeling, in 2010 the European HEALTHGRAIN Consortium has developed, in line with AACC, a new definition: "whole grain shall consist of the intact, ground, cracked or flaked kernel after the removal of inedible parts, such as the hull and husk. The principal anatomical components—the starchy endosperm, germ and bran—are present in the same relative proportions as they exist in the intact kernel. Small losses of components—that is, less than 2% of the grain/10% of the bran—that occur through processing methods consistent with safety and quality are allowed". Moreover, due to different composition of WG products, HEALTHGRAIN proposed that a product is labelled as WG food if "one for which the product is made with >30% whole-grain ingredients on a dry-weight basis and more whole-grain ingredients than refined-grain ingredients" [29].

2.2. Chemical Composition of Cereal Kernels

Starchy endosperm, germ, and bran are the three principal fractions in cereal kernels [31]. Endosperm represents the most abundant fraction (constituting over 80% of caryopsis) containing large amounts of starch to supply energy and 75% of needed proteins for plant germination, some fiber, and micronutrients (especially iron, riboflavin, niacin, and thiamin). Germ (i.e., the embryo) occupies a small fraction of the seed (2–3% of kernel); particularly rich in proteins, fiber, and fats, it also contains significant amounts of mineral, B and E vitamins. Bran, consisting of multiple layers (pericarp, spermoderm, and perisperm), is the outer portion of the seed (13–17% of kernel); it contains fiber (more than

50%), proteins, starch (as "contamination" from endosperm), B vitamins, minerals, and several bioactive compounds, such as polyphenols (Figure 1) [31–33].

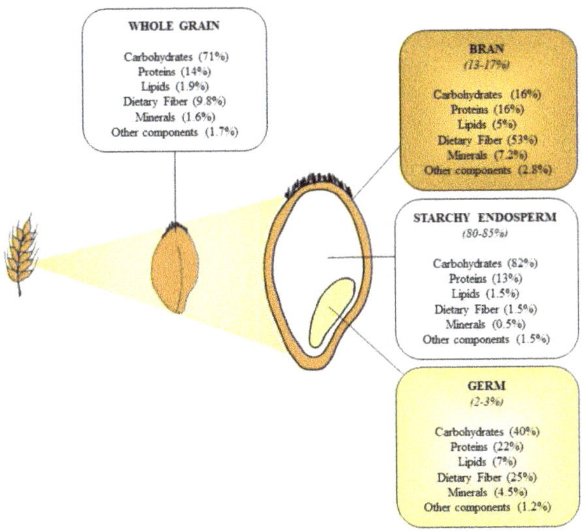

Figure 1. Nutritional composition of wheat kernel. Values are reported as percentage of dry matter.

During milling process of refined flours, bran and germ are removed (and used for food and non-food applications) [33,34]; as a result, RG products contain fewer nutrients than WG counterparts. For example, in refined wheat flour, pantotenic acid, folate, iron and copper content and fiber are reduced, while some vitamins, potassium, magnesium, and manganese are even lost [35]; however, this loss can be compensated by fortifications, such as mandatory folate addition in USA [36].

2.3. *Whole Grains and Health*

As above mentioned, compared to refined counterparts, WG are associated with benefits for human health [30]. Among mechanisms of action, effects on postprandial glycemia, appetite and ad libitum energy intake have been proposed. In a randomized crossover study enrolling twenty young subjects (10 females and 10 males; Body Mass Index (BMI) = 21.7 $\pm$ 2.2 kg/m^2), Kristensen and co-workers [37] reported that, with respect to refined counterparts, WG wheat bread ingestion led to increased satiety and reduced hunger, without modifying energy intake at the subsequent meals. The same group obtained similar results for WG pasta ingestion in overweight/obese (25 < BMI < 40 kg/m^2) subjects [38]. Accordingly, WG food intake appears to be closely associated with reduced risk of obesity: cross-sectional dietary data from NHANES 2001-12 (which included 15,280 children and 29,683 adults) documented that WG intake inversely related to BMI, waist circumference and percentage of overweight/obese individuals [20]. Besides its beneficial role in obesity, WG consumption is closely associated with reduced risk of other chronic diseases, including cardiovascular disease, type II diabetes, metabolic syndrome, and several cancer types [8,10–14].

An important role in chronic diseases seems to be played by gut microbiota, whose composition is influenced not only by genetics and age, but also by diet [39]. A strong link between microbiota composition and food intake exists, as a consequence of long-term dietary habits [40]. In particular, high consumption of WG, vegetables and fruits is associated with greater microbial variety, while diet rich in RG and fats and low in fiber is associated with lower biodiversity [41,42]. An observational study has shown that high adherence to the Mediterranean diet (MD), a typical eating pattern of the Mediterranean basin characterized by high consumption of cereals, fruits, vegetables, and legumes, was associated with increased levels of anti-inflammatory compounds (such as short chain fatty acids, SCFAs) in fecal samples and reduced atherogenic compounds (such as trimethylamine *N*-oxide) in urine samples [43]. As MD recommends daily consumption of cereals, preferably as unrefined grains [44–46], it is conceivable that WGs cooperate with fruits and vegetables to change microbiota composition. Accordingly, in a randomized controlled, six-week trial, high WG consumption displayed better positive effects than high RG consumption, in terms of gut microbiota and immune responses [47].

In this context, it should be underlined that, unlike RG eaters, WG consumers generally follow health and diet recommendations and ingest few, if any, non-recommended, indulgent foods. For this reason, studies on WG intake must take into account all confounders, in order to remove potential bias from data.

3. Dietary Fiber and Polyphenols as Functional Compounds in Whole Grains

Fiber and polyphenols (or phenolic compounds) are the main dietary bioactive compounds studied for prevention of chronic diseases; they have different chemical structures, physical and biological properties, and ability to activate distinct metabolic pathways [48,49].

The definition of dietary fiber is constantly evolving and, although AACC has proposed that it is: "the remnants of the edible part of plants or analogous carbohydrates that are resistant to digestion and absorption in the human small intestine, with complete or partial fermentation in the large intestine" [50], the more recently accepted definition is that provided by Codex Alimentarius (CAC), i.e., "carbohydrate polymers with ten or more monomeric units, which are not hydrolyzed by the endogenous enzymes in the small intestine of humans" [51].

In WG, dietary fiber primarily derives from the outer portion of cereal kernel (although it can also be found in endosperm of some grains, like wheat and barley) and mainly consists of non-starch polysaccharides, distinguished by fermentability to SCFAs, solubility in water, viscosity, and cation exchange capacity [52,53]. Cellulose, galactomannans, xylans, xyloglucans, and lignin are part of insoluble dietary fiber, while arabinoxylans, arabinogalactans, β-glucans, and pectins of soluble dietary fiber [52]. Among cereals, wheat, triticale, and rye are rich in arabinoxylans, while oats and barley mainly contain β-glucans [54] (Table 1). As it will be discussed, high dietary fiber intake improves intestinal health, increases satiety, and reduces risk of some chronic diseases, including cancer [55,56].

Table 1. Content of polyphenols and fiber in some whole grains commonly consumed worldwide.

Compound [1]	Wheat	Oat	Corn	Rice	Refs
Dietary fiber	9.7–13.1	7.6–10.6	2–7.3	1.4–3.75	[57,58]
Total polyphenols	538	471.7	497.1	421.8	[59]
Total phenolic acids	1342 (75%) [2]	472 (75%) [2]	601 (85%) [2]	197–376 (62%) [2]	[60,61]
Ferulic acid	11.6–870	249.4–1044.9	97–584.0	68.2–301.7	[62]
p-coumaric acid	3.5–293.0	607.3	97.0–584.0	22.8–85.0	[62]
Gallic acid	6.5–195.0	1.7–241.2	0.5–116.5	5.5–115.6	[62]
Caffeic acid	0.5–51.9	3.6–9.2	5.7–24.4	1.0–3.5	[62]
Total flavonoids	95.8–212	n.r.	607.1–1277	94–3274	[63]

[1] All data are expressed as μg/g dry weight, except for dietary fiber, expressed as g/100 g of grain and total polyphenols, expressed as mg of Gallic Acid Equivalent/100 g dry weight. [2] Percentage of bound form. n.r.: not reported by authors.

Polyphenols, secondary metabolites found in plant tissues, are heterogeneous compounds, possessing one or more aromatic rings with one or more hydroxyl groups. Polyphenols can be subdivided into (i) phenolic acids, (ii) flavonoids, (iii) stilbenes, and (iv) lignans [59,64]; they can also be distinguished in soluble (free molecules in cell vacuoles) and insoluble (bound to cell wall elements, such as dietary fiber) compounds [32]. These phytochemicals are important for plant growth, defense, reproduction, and color; consequently, beyond genetics, also environmental factors significantly affect their levels, that vary greatly even between cultivars of the same species. Therefore, it is not possible to establish precise amounts of each compound in different plant-foods, and several polyphenols are still unidentified; therefore, literature data on polyphenol content in plant foods, including grains, is incomplete, difficult to compare and often contradictory. According to available data, WGs seem to contain polyphenol amounts similar to those found in fruits and vegetables, with some highly active phenolic compounds more represented (Table 1) [60–62,65,66]. Because of frequency of consumption [4], it has been estimated that WGs provide for about one-third of total polyphenol dietary intake [67].

Ferulic, *p*-cumaric, vanillic, siryngic, gallic, and caffeic acids are the most common phenolic acids of grains (Table 1) [60–62,65,66,68]. Some of them are present as esters or amides; this is the case of γ-oryzanol, a blend of ferulic acid esters and phytosterols, mostly found in rice [69] and avenanthramides, phenolic amides containing anthranilic acid and hydroxycinnamic acid moieties, exclusively found in oats [70]. Significant differences in phenolic acid amounts exist, depending on grain dimension and species, as well as on fiber type and content; ferulic acid, for example, is more abundant in smaller than in larger grains, and the higher the fiber content, the higher the ferulic acid content (Table 1) [61,71].

As above mentioned, WGs also represent a source of flavonoids, among which there are the two flavones apigenin and luteolin and the two flavanones naringenin and eriodictyol [60,63]. Additionally, anthocyanins have been reported in pigmented varieties of some WGs, such as barley, rice, rye, and wheat [63,72]; finally, among lignans, secoisolariciresinol is present in buckwheat and pinoresinol in oats [73].

Phenolic compounds might play a role in chronic diseases and, due to their antioxidant properties and ability to modulate specific signaling pathways involved in cell survival and death, are particularly beneficial in cancer [74–76]. However, physiological effects of these WG components strictly depend on their bioavailability, in turn influenced by binding to dietary fiber [77–79]. In cereals, most polyphenols (95%) are indeed covalently linked to polysaccharide chains of dietary fiber, mainly arabinoxylans [64]. As a consequence, although dietary fiber properties are generally attributed to non-starch polysaccharides, the "dietary fiber concept" is changing towards the "antioxidant dietary fiber concept" [80]. When gut microbiota ferments fiber, phenolic compounds are released into the intestinal lumen and absorbed by enterocytes. Non-fermented and non-absorbable polyphenols counteract the pro-oxidant effects of ingested foods, by scavenging free radicals [48], and meanwhile they synergize with bacteria-derived SCFAs in modulating cell death and differentiation [64,81]. Furthermore, dietary fiber-polyphenol association can downregulate energy metabolism, nuclear receptor signaling and lipid biosynthesis (via tumor necrosis factor-α and peroxisome proliferator-activated receptor-α), pathways involved not only in obesity, but also in cancer (especially of the gastrointestinal tract) [64,82,83].

4. Whole Grains and Gastrointestinal Cancers: An Overview

According to global cancer statistics, 19.3 million new cancer cases and 10 million all cancer deaths occurred in 2020 worldwide; more than one-third of cancer victims suffered from gastrointestinal tumors [84]. Based on molecular phenotype and histological characteristics, these tumors include cancers affecting upper and lower gastrointestinal tract, as well as salivary gland, liver and bile ducts, gallbladder, and exocrine pancreas [85]. Although a geographic description of cancer- and sex-specific incidence and mortality patterns exists, overall more than 60% of gastrointestinal cancer cases and deaths occurred in Asia, followed by Europe and North America [84].

Clinical management of gastrointestinal cancers remains a major challenge for clinicians, especially because most cases are diagnosed in advanced stages, when treatment options are limited [86]. A variety of etiological factors have been identified; it has been estimated that genetic defects account only for 5–10%, while harmful environmental conditions and unhealthy lifestyle represent 90–95% of risk factors [87]. Consequently, primary and secondary prevention strategies, including promotion of healthy lifestyle aimed at deeply modifying some risk behaviors (e.g., tobacco use, physical inactivity, unhealthy diet, and alcohol abuse), are particularly relevant for reducing cancer risk and outcomes.

Consumption of WGs is strongly recommended for gastrointestinal health. A large body of literature data concerning WG effects on gastrointestinal cancers are available, although WG action is not equal (and even absent) in different gastrointestinal organs. To the best of our knowledge, no epidemiological studies about WG intake and risk of gallbladder and bile duct carcinomas have been published, while only one study demonstrated inverse association between WG (and possibly bran and cereal fiber) intake and risk of hepatocellular carcinoma, the predominant histological form of primary liver cancer [88]. As emerged by a meta-analysis of case-control and cohort studies, high intake of WGs was also associated with reduced risk of pancreatic cancer [89]; nonetheless, lack of more prospective cohort studies prevents to draw robust conclusions.

Similarly, literature data on association between unrefined grains and oral cavity and oropharynx cancers are scarce, not updated and just based on few case-control and cohort studies. Some investigations highlighted that WG intake was favorably related to risk of upper aerodigestive tract cancers [90–95]. Conversely, other studies reported no [96–99] or even positive associations [100–102]. Due to these controversial results, data on WGs and oropharyngeal cancer risk are less consistent than those for other plant-derived foods. Finally, except for a large US prospective cohort study showing a marginal inverse relationship between WG food consumption and small intestinal cancer [103], also data referred to small bowel tumors are sparse and difficult to interpret.

Based on this evidence, we focused on colorectal, gastric, and esophageal tumors, the most diagnosed and severe gastrointestinal cancers, for which investigations are more extensive and continuously updated.

5. Whole Grains and Colorectal Cancer

In 2020, 1.9 million of individuals were diagnosed for colorectal cancer, the second mostly incident cancer and the third leading cause of cancer death worldwide. It has slightly higher incidence in males (1,065,960 cases) than females (865,360 cases) [84]; although incidence (10% of all cancer cases) is decreasing in developed countries, cases are increasing among younger adults, especially in USA [104].

Depending on location (proximal colon, distal colon and rectum), colorectal cancer varies in terms of etiology and sensitivity to specific risk factors [49,105]. Only 1–2% of cases have been associated with ulcerative colitis, Crohn disease and inflammatory conditions [106], while modifiable lifestyle factors, typical of industrialized countries (tobacco smoking, physical inactivity, red/processed meat and alcohol consumption, low intake of fruits and vegetables), are long-established risk factors [107].

In this context, WGs represent protective factors, as high intakes have been associated with significant decrease of cancer risk (Table 2). In a 14-year case-control study, conducted in Northern Italy and including 11,990 patients with several cancer types (among them, 955 colon and 625 rectum tumors), multivariate odd ratios (ORs) for the highest category of WG intake (>3 day per week) were 0.5 (95% CI 0.4–0.6) and 0.6 (95% CI 0.4–0.8) for colon and rectum cancers, respectively [108]. Intriguingly, Um and collaborators found sex-related differences in terms of WG association: the prospective CPS-II Nutrition Cohort study enrolling 50,118 men and 62,031 women (1742 incident colorectal cancer cases during the follow-up) found that the highest vs. lowest quintile of WG intake was associated with 23% and 43% lower risk of colorectal and rectal cancer, respectively, among men, but no association was found for women. Moreover, authors did not find any evidence of increased risk with consumption of RGs, grain-based sweets, or desserts [109]. Several meta-analyses have reinforced potential benefits of WGs against colorectal tumors [110–113]. For example, the World Cancer Research Fund International (WCRF) Continuous Update Project (CUP) has updated the systematic review and meta-analysis (until the end of May 2015) of prospective studies reporting 17% decreased risk for each 90 g/day WG increase [110].

Similarly, Schwingshackl and co-workers found 20% decreased risk of colorectal cancer with increasing WG intake up to 120 g/day [112]. Noteworthy, a recent meta-analysis from Zhang's group found that WG/colorectal cancer association was significant only for sample size ≥500 [113]. Finally, as emerged from a Chinese 10-year follow-up study (enrolling 369 colon cancer subjects) high WG consumption (more than 17 g/day) also appears to be correlated with prognosis and survival rates [114]. Although all these findings highlight the positive role of WGs in cancer onset and/or outcomes, nonetheless no randomized clinical trials have tested the long-term impact of WG consumption on colorectal tumorigenesis up until now.

Table 2. Some epidemiological studies on whole grains/whole grain fiber and colorectal cancer.

	Study Type and Design	Main Findings *	Refs
Whole grains	14-year US prospective population-based case-control study (112,149 participants (1742 CRC) from the Cancer Prevention Study-II Nutrition Cohort 1999–2013) Quintiles of WG intake (g/day): Q1: <19 for men; <18 for women Q5: 116.7–1296 for men; 117.1–1255 for women	Similar WG intake in women (mean: 72.8 g/day; 10th–90th percentile distribution: 10.6–168 g/day) and men (mean: 74.5 g/day; 10th–90th percentile distribution: 9.2–174 g/day) High WG intake associated with low CRC risk among older men, but not women (HR = 0.77, 95% CI 0.61–0.97; $p = 0.03$ for men; HR = 1.10, 95% CI 0.88–1.36; $p = 0.14$ for women; p interaction by sex = 0.01) Men in the highest quintile: 43% reduced risk (HR = 0.57, 95% CI 0.35–0.93, $p = 0.04$) No association of RG with CRC risk	[109]
	10-year prospective population study (369 CRC patients, 154 deaths during the follow-up) Quartiles of WG intake (g/day) Q1: ≤7.1 Q2: 7.1–10.7 Q3: 10.7–17.9 Q4: >17.9	High WG intake associated with risk of mortality (HR_{Q4} vs. $_{Q1}$ = 0.56, 95% CI 0.35–0.89; p for trend 0.05)	[114]
	Meta-analysis of 11 prospective studies for WG consumption and 3 reports for RG consumption	Inverse association between CRC risk and WG intake (RR = 0.88, 95% CI 0.83–0.94, $I^2 = 35\%$, $p = 0.13$) (10 studies with 9223 CRC cases; overall intake range: 0–374 g/day) Each additional daily 30 g of WGs inversely associated with CRC risk (RR = 0.95, 95% CI 0.93–0.97, $I^2 = 58\%$, $p = 0.02$); 20% decreased risk with WG intake up to 120 g/day No association for RG intake (RR = 1.46, 95% CI 0.80–2.67, $I^2 = 71\%$, $p = 0.06$) (900 CRC cases, overall intake range: 15–585 g/day)	[112]
	Meta-analysis of 34 studies of WG intake and risk of digestive tract cancer [CRC: 7 case-control and 10 cohort studies (1,489,581 participants and 19,424 cases)]	Inverse association between CRC risk and WG intake (RR = 0.89, 95% CI 0.84–0.93; $p < 0.001$; $I^2 = 38.2\%$, $p = 0.029$). Positive effects of WGs only in studies with sample size ≥500 (RR = 0.91, 95% CI 0.88–0.94, $p < 0.001$) No statistically significant heterogeneity in women ($I^2 = 0\%$, $p = 0.619$), European ($I^2 = 0\%$, $p = 0.732$), before 2010 publication year ($I^2 = 0\%$, $p = 0.622$) and adjustment for energy ($I^2 = 4.6\%$, $p = 0.399$) studies	[113]

Table 2. *Cont.*

	Study Type and Design	Main Findings *	Refs
	Spanish observational case-control study (308 CRC and 308 controls) Tertiles of WG fiber intake not defined, but referred to Healthy Eating Index for Spanish Diet (HEISD) (T1: 69; T2: 69–74.5; T3: >74.5) and MedDietScore (MDS) (T1: <35; T2: 35–37; T3: >37)	WG intake lower in CRC patients than controls (14.4 ± 19.9 vs. 18.8 ± 23.4 g/day, $p = 0.012$). Inverse association between WG intake and CRC risk ($OR_{T3\ vs.\ T1} = 0.62$, 95% CI 0.39–0.98) Consumption of fiber-containing foods, especially WG, associated with lower CRC risk ($OR_{T3\ vs.\ T1} = 0.65$, 95% CI 0.35–1.21).	[115]
Whole grains/ whole grain fiber	US Prospective NIH-AARP Diet and Health Study (1995–2011) including 478,994 subjects (285,456 men and 193,538 women) cancer free at the beginning; 10,200 incident cases (6712 men and 3488 women) at the end. Quintiles of WG intake (servings/1000 kcal/day) Q1: 0.2 Q2: 0.4 Q3: 0.6 Q4: 0.8 Q5: 1.8 Quintiles of WG fiber intake (g/1000 kcal/day) Q1: 1.7 Q2: 2.5 Q3: 3.2 Q4: 4.0 Q5: 5.7	Positive association for both WGs ($HR_{Q5\ vs.\ Q1} = 0.69$, 95% CI 0.64–0.73; $p < 0.001$) and dietary fiber ($HR_{Q5\ vs.\ Q1} = 0.70$, 95% CI 0.66–0.75; $p < 0.0001$) After adjustment for potential confounders: $HR_{Q5\ vs.\ Q1} = 0.83$ (95% CI 0.78–0.89; $p < 0.001$) for WGs and $HR_{Q5\ vs.\ Q1} = 0.92$ (95% CI 0.86–0.99; $p < 0.03$) for dietary fiber intake. The association remained statistically significant after adjustment for folate ($HR_{Q5\ vs.\ Q1} = 0.84$, 95% CI 0.79–0.90; $p < 0.001$) and dietary fiber intake ($HR_{Q5\ vs.\ Q1} = 0.84$, 95% CI 0.78–0.90; $p < 0.001$) Only fiber from grains was inversely associated with CRC ($HR_{Q5\ vs.\ Q1} = 0.89$, 95% CI 0.83–0.96; $p < 0.001$) No sex-dependence ($p = 0.13$ for interaction)	[116]
	963 US females from Nurses' Health Study cohort (NHS; 1980–2010) and 612 US males from Health Professionals Follow-up Study cohort (HPFS; 1986–2010) diagnosed stage I to III CRC throughout follow-up. Quintiles of WG fiber intake (g/1000 kcal/day) Q1: 1.7 Q2: 2.5 Q3: 3.2 Q4: 4.0 Q5: 5.7	WG intake associated with low CRC-specific mortality (HR per 20 g/day increment = 0.72, 95% CI 0.59–0.88; $p = 0.002$), also after adjusting for fiber intake (HR = 0.77, 95% CI 0.62–0.96; $p = 0.02$), and all-cause mortality (HR = 0.88, 95% CI 0.80–0.97; $p = 0.008$ for trend). Cereal fiber intake associated with low CRC-specific mortality (HR per 5 g/day increment = 0.67, 95% CI 0.50–0.90; $p = 0.007$) and all-cause mortality (HR = 0.78, 95% CI 0.68–0.90; $p < 0.001$). Vegetable fiber associated with low all-cause mortality (HR = 0.83, 95% CI 0.72–0.96; $p = 0.009$), but not CRC-specific mortality (HR = 0.82, 95% CI 0.60–1.13; $p = 0.22$); no association for fruit fiber. Patients with increased fiber intake after diagnosis: lower mortality rate [each 5 g/day increase associated with 18% decrease in CRC-specific mortality (95% CI 7–28%; $p = 0.002$) and 14% decrease in all-cause mortality (95% CI 8–19%; $p = 0.001$)].	[117]

Table 2. *Cont.*

Study Type and Design	Main Findings *	Refs
1902 US females from Nurses' Health Study cohort (NHS; 1980–2012) and 1276 US males from Health Professionals Follow-up Study cohort (HPFS; 1986–2012) diagnosed CRC throughout follow-up. Deciles of total fiber intake (g/day): D1: 9.56 in women; 13.1 in men D10: 24.8 in women; 33.7 in men Deciles of cereal/WG fiber intake (g/day): D1: 1.60/6.54 in women; 2.58/9.70 in men D10: 7.43/39.01 in women; 12.0/58.3 in men Deciles of fruit/vegetable fiber intake (g/day): D1: 1.03/2.79 in women; 1.10/3.30 in men D10: 8.50/9.87 in women; 10.2/13.1 in men	No association between total fiber and CRC risk. No association for fruit or vegetable fiber. Inverse association between cereal fiber intake and CRC risk only in men ($HR_{D10\ vs.\ D1}$ = 0.75, 95% CI 0.57–1.00). Inverse association between intake of WG fiber and risk of CRC only in men ($HR_{D10\ vs.\ D1}$ = 0.72, 95% CI 0.54–0.96).	[118]

* Findings on WG intake per se, after adjusting for confounding factors (e.g., age, sex, education, smoking, dietary habits, alcohol, physical activity, etc.) through multivariate models. CI: Confidence interval; CRC: colorectal cancer; HR: Hazard Rate; OR: Odd ratio; RG: refined grain; RR: Relative Risk; WG: whole grain.

The American Institute for Cancer Research and the World Cancer Research Fund stated that eating at least 90 g/day WG reduces colorectal cancer risk, mainly due to its high fiber content [119]. Among mechanisms involved in WG protective effects, fiber-mediated reduction of fecal transit time, dilution, and removal of carcinogens (especially heterocyclic amines), maintenance of epithelial cell integrity and stimulation of bacterial fermentation (and, therefore, SCFA production that inhibits colon carcinogenesis) can be identified [120,121]. Accordingly, among all fiber containing foods, WGs are most consistently associated with incidence of colorectal cancer. Indeed, two large recent prospective US cohort studies did not find any association for total dietary fiber intake, but when different food sources were examined, lower risk for colorectal tumors was observed only in high cereal (especially unrefined) consumers. Moreover, such association was observed in men, but not in women; this sex-disparity might depend on lower fiber intake registered for women (mean fiber intake of 14 g/day for women and 20.0 g/day for men) [118]. Alegria-Lertxundi and co-workers [115] investigated the relationships between food groups, diet quality and colorectal cancer risk and reported no significant differences of intake between control and patient groups for the majority of food classes, except for lower WG intake (and higher egg consumption) in tumor cases; coherently, the observed protective effects of fiber-containing foods appeared to be mainly ascribed to WGs. A recent, large US cohort analysis (with more than 10,000 incident colorectal cases and more than 15 years of follow-up) further confirmed that fiber from grains, but not from other sources, was associated with lower incidence, especially for distal colon and rectal cancers [116]. High fiber and WG intake after diagnosis also leads to lower death rate, and this positive association again depends on fiber sources, with cereal fiber (especially from WG) showing the strongest link [117]. These data apparently disagree with the European Prospective Investigation into Cancer and Nutrition (EPIC) study that observed a significant lower risk of CRC in higher total fiber consumers [122,123]. Such a discrepancy may depend on less fiber in a typical American diet (with respect to the European one), as well as less proportional contribution of WG foods to total dietary fiber intake; indeed, about 39% of dietary fiber derives from grain foods containing no WGs, but RGs that have few amounts of fiber and are consumed in large quantities [124]. Therefore, further studies are necessary to evaluate dose-response relationship and influence of different fiber sources, taking into account that range of fiber intake widely varies depending on the examined population.

Concerning phenolic compounds, these phytochemicals exert anti-cancer activities in colon-cancer cells, mainly by inducing cell-cycle arrest and apoptosis. Just an example, ferulic and *p*-coumaric acids modulate S and G2/M phase transitions, respectively [125]; the two compounds also inhibit cancer cell proliferation, by inhibiting expression of epidermal growth factor receptor, one of the most relevant biomarkers in colorectal cancer [126], and related mitogenic signaling pathways [127,128]. Likewise, in human colon cancer cells, secoisolariciresinol diglycoside and its metabolites (enterolactone and enterodiol) induce S-phase cell cycle arrest, by modulating key regulatory proteins (cyclin A and cyclin-dependent kinase 4) [129–131]. By possessing estrogenic activity, some flavonoids (such as apigenin, naringenin, luteolin, and eriodictyol) contribute to colon cancer prevention, through activation of estrogen receptor-β in colonocytes [132–135]. Finally, some miRNAs involved in colorectal cancer are sensitive to phenolic compounds: for example, miRNA384 is up-regulated by luteolin, thus resulting in decreased expression levels of pleiotrophin, a cytokine upregulated in colorectal tumors [136–138].

In order to overcome challenges in polyphenol delivery to target tissues, recent studies have attempted to find novel strategies for improving bioavailability and anti-tumor efficacy of these phytochemicals. For example, the novel stable ferulic derivative tributyltin(IV) ferulate has been designed and found to potently exert anti-tumor activity; this synthetic compound, indeed, triggers autophagic cell death through generation of reactive oxygen species and endoplasmic reticulum stress in colon cancer cells [139]. Similarly, a novel nanoparticle system, consisting of encapsulated gallic acid and gum arabic as coating material, has shown promising anti-cancer properties: the formulated nanoparticles, indeed, were selectively internalized by cancer cells, thus exerting potent anti-oxidant and anti-neoplastic effects, as assessed by cytotoxic, migration, and apoptosis assays [140].

6. Whole Grains and Gastric Cancer

Gastric cancer is the fourth cause of tumor-related deaths. Incidence (5.6% of all cancer cases) is higher in males (719,523 cases) than females (369,580 cases); 75.3% of cases occur in Asia, followed by Europe (12.5%) and Latin America and Caribbean (6.2%) [84].

Generally, gastric cancer is classified into non-cardiac gastric cancer, originating from distal regions of stomach, and cardiac gastric cancer, arising near the esophageal-gastric junction [141]; both forms are associated with cigarette smoking and *Helicobacter pylori* infection, while cardiac gastric cancer is also related to other risk factors, including esophageal reflux, Barrett's esophagus, and obesity [142–149].

Among factors affecting cancer onset, dietary habits play an important role [150–152]: salt-preserved foods and smoked meats potentiate carcinogenic effects of *H. pylori* infection [153], whereas fruits, vegetables, and WGs are protective factors [154,155]. A prospective population-based case-control study reported in men, but not in women, a modestly lower risk of stomach cancer with diet patterns high in WGs, only when combined with citrus fruit and vegetables [156]. Several meta-analyses have been published on WG/gastric cancer association, relying primarily or entirely on case-control studies and without dose-response analyses. However, all studies reported that increasing WG consumption was notable in showing a negative association with stomach cancer risk (ranging from 13 to 50% lower risk for highest WG consumers) and/or RG intake generally appeared to be a dose-dependent risk factor (63–65% increment of the risk) (Table 3) [92,157–163]. However, it should be recalled that RG-rich diet is usually poor in WGs (and other dietary fiber sources) and associated with unfavorable lifestyles. Therefore, for gastric cancer, nutritional and lifestyle combination, rather than RG alone, may account for direct associations observed in the studies.

Table 3. Some epidemiological studies on whole grains/whole grain fiber and gastric cancer.

	Study Type and Design	Main Findings *	Refs
Whole grains	Prospective 14-year population-based case-control Cancer Prevention Study [533,391 women (439 deaths for GC) and 436,654 men (910 deaths for GC)] Tertiles of WG intake (days/week): T1: < 1 T2: 1–4 (4.5 for women) T3: > 4 (4.5 for women)	Men: high WG consumption associated with decreased risk only in age-adjusted model ($RR_{T2\ vs.\ T1}$ = 0.87, 95% CI 0.74–1.03; $RR_{T3\ vs.\ T1}$ = 0.77, 95% CI 0.66–0.90; $p < 0.001$), but not in multivariate-adjusted model ($RR_{T2\ vs.\ T1}$ = 0.94, 95% CI 0.79–1.11; $RR_{T3\ vs.\ T1}$ = 0.90, 95% CI 0.77–1.06; $p = 0.17$). More than 4 times/week cold cereal intake related to lower risk with respect to low (<once/week) intake (RR = 0.83, 95% CI 0.68–1.00; $p = 0.03$ for trend). Men with positive family GC history, consuming WG products >4 days/week, showed lower risk (RR = 0.31, 95% CI 0.15–0.64) with respect to men with no family GC history. Women: no association between WGs and GC risk. Women consuming brown rice, whole wheat or barley 6 to 7 times/week were at greater risk of fatal stomach cancer with respect to women with no intake ($RR_{T3\ vs.\ T1}$ = 1.41, 95% CI 1.04–1.91; p for trend = 0.05).	[156]
	Retrospective 10-year hospital-based case-control study (745 GC patients and 3526 controls) Tertiles of WG food intake (simple score of consumption): T1: low T2: intermediate T3: high Tertiles of RG food intake (portions/week): T1: 0–14 T2: 15–21 T3: ≥22	Whole meal consumption negatively correlated with GC risk For WG foods: $OR_{T3\ vs.\ T1}$ = 0.5, 95% CI 0.4–0.7 For RG foods: $OR_{T2\ vs.\ T1}$ = 1.24, 95% CI 1.0–1.5 and $OR_{T3\ vs.\ T1}$ = 1.54, 95% CI 1.2–2.0	[92]
	Retrospective 3-year hospital-based case-control study (143 GC patients and 328 controls) Tertiles of whole-meal bread intake (simple score of consumption): T1: low T2: intermediate T3: high	Whole meal consumption negatively correlated with GC risk $RR_{T2\ vs.\ T1}$ = 1.26, 95% CI 0.79–2.01 $RR_{T3\ vs.\ T1}$ = 0.48, 95% CI 0.28–0.82	[160]
	Meta-analysis of 5 hospital-based case-control, 4 population-based case-control and 2 prospective cohort studies (2920 GC cases and 527,256 controls)	WG consumption inversely related to GC in Europe (OR = 0.72, 95% CI 0.19–1.24) and America (OR = 0.61, 95% CI 0.38–0.85), both in hospital-based case-control (OR = 0.50, 95% CI 0.35–0.65) and cohort (OR = 0.61, 95% CI 0.38–0.85) studies	[158]

Table 3. *Cont.*

	Study Type and Design	Main Findings *	Refs
	Meta-analysis of 34 studies of WG intake and risk of digestive tract cancer [GC: 9 case-control and 2 cohort studies (1,021,955 participants and 8274 GC cases)]	WG consumption: 36% decrease in GC risk (RR = 0.64, 95% CI 0.53–0.79; $p < 0.001$), with a significant heterogeneity (I^2 = 78.2%, $p = 0.001$) WG intake was a protective factor for case-control (RR = 0.55, 95% CI 0.41–0.74; $p < 0.001$) and European (RR = 0.64, 95% CI 0.53–0.79; $p < 0.001$) studies No significant association in cohort (RR = 0.89, 95% CI 0.78–1.01; $p = 0.070$) and American (RR = 0.70, 95% CI 0.50–1.00; $p = 0.051$) studies	[113]
	Meta-analysis of 19 studies (17 case-control and 2 cohort studies; 994,258 participants) Consumption of WGs or RGs: Low: <1/month Moderate: 1–2 times/week High: >3 times/week	WG consumption: 13% decrease in GC risk (OR = 0.87, 95% CI 0.79–0.95; $p = 0.003$) High consumption: 44% reduced risk ($OR_{\text{high vs. low}}$ = 0.56, 95% CI 0.45–0.69; $p < 0.001$) No significant correlation for moderate consumption RG consumption: 36% increase in GC risk (OR = 1.36, 95% CI 1.21–1.54; $p < 0.001$) 63% increased GC risk in high consumers (OR = 1.63, 95% CI 1.49–1.79; $p < 0.001$) 28% increased GC risk in moderate consumers (OR = 1.28, 95% CI 1.18–1.39; $p < 0.001$) 53% increased GC risk in rice consumers (OR = 1.53, 95% CI 1.31–1.79; $p < 0.001$) 28% increased GC risk in RG, not-rice consumers (OR = 1.28, 95% CI 1.11–1.49; $p = 001$) No correlation between small amounts of RG intake and GC risk	[159]
Whole grain fiber	Prospective 14-year cohort Iowa Women's Health Study (34,651 initially free-cancer women; 56 GC) Tertiles of WG fiber intake (g/day): T1: 0–1.49 T2: 1.50–3.98 T3: 3.99–35.75 Tertiles of RG fiber intake (g/day): T1: 0–1.37 T2: 1.37–2.35 T3: 2.35–16.93	WG fiber intake inversely related to GC risk ($HRR_{\text{T3 vs. T1}}$ = 0.53) No association for RG fiber intake	[95]

* Findings on WG intake per se, after adjusting for confounding factors (e.g., age, sex, education, smoking, dietary habits, alcohol, physical activity, etc.) through multivariate models. CI: Confidence Interval; GC: gastric cancer; HRR: Hazard Rate Ratio; OR: Odds Ratio; RG: refined grain; RR: Relative Risk; WG: whole grain.

To date, no conclusions on the role of fiber in WG/gastric cancer association can be drawn since available investigations are somehow misleading and difficult to interpret. Except for the cohort Iowa Women's Health Study (demonstrating strong protective effects of WG fiber against stomach cancer) [95], almost all studies consider only total dietary fiber intake and/or report no association at all for fiber from grains [55,161,162].

Among polyphenols, gallic acid has been shown to inhibit *H. pylori* proliferation, as well as invasion and metastasis of cancer cells [164]. Similarly, Ho and colleagues [165] demonstrated that gallic acid can in vitro reduce migration of human gastric carcinoma cells, through inhibition of RhoB expression and modulation of Akt signaling. Polyphenols also activate apoptosis: caffeic acid induces cell death by modulating cellular Ca^{2+} homeostasis [166], ferulic acid activates caspase-3 and caspase-9 [75], and apigenin modulates expression of pro- (Bax and caspase-3) and anti-apoptotic (Akt and Bad) proteins [167,168]. Lastly, polyphenols are able to modulate activity of specific miRNAs: luteolin inhibits Bcl-2 expression by upregulating miR-34a, while *p*-coumaric acid exerts antitumor effects by regulating hsa-miR-30a-5p, hsa-miR-125a-5p, and hsa-miR-7-5p [169–172].

7. Whole Grains in Esophageal Cancer

According to Globocan 2020, 508,585 cancer victims (5.3% of all cancer cases) were affected by esophageal cancer, the sixth cause of cancer deaths [84]. Incidence of esophageal cancer (3.1% of all cases) is higher in males than females; the highest mortality (78.2%) is registered in Asian continent [84].

Esophageal tumors are distinguished into esophageal squamous cell carcinomas, affecting upper layer cells lining esophagus, and adenocarcinomas, arising in glandular cells located between the esophagus–stomach junction [173,174]; esophageal squamous cell carcinomas are more frequent in developing countries, while esophageal adenocarcinomas predominate in eastern Asia and Africa [175]. Distinct risk factor profiles have been identified: tobacco smoking and alcohol abuse are main risk factors for esophageal squamous cell carcinoma, while obesity and gastro-esophageal reflux disease are key risk factors for adenocarcinoma. Specific dietary items and nutrients impact risk of both types of cancer [176]: for example, red, pork and processed meat, moldy food and pickled vegetable consumption are risks attributable to the entire population, while more varied diet, raw and cooked vegetables, vitamins, fiber, and carbohydrates are included among protective dietary factors [157,177–180].

Higher frequency of WG food consumption may be accounted among indicators of reduced risk of esophageal cancer (Table 4). In a small case-control study, for example, Levi and co-workers reported significant decrease in cancer risk in individuals consuming high amounts of WG foods (whole wheat bread and cereals), while cancer onset was directly related to consumption of RG items (white bread and biscuits, pizza, pasta, and rice) [91]. Decreased risk for high WG intake has been reported by retrospective and prospective studies, although with different ratios: for example, the above mentioned Italian case-control study from La Vecchia's group [108] reported 60% decreased risk for the highest WG intake, while the recent HELGA cohort study from Skeie and co-workers showed 35–45% reduction [181]. Noticeably, authors observed that such association varied with cereals and food products, with WG wheat and bread being associated with lower risk. This finding can be explained considering that cereals and cereal-based foods have different composition and concentration of nutrients and bioactive compounds, which cooperate to exert positive effects [182,183]. In this context, dietary fiber may play a crucial role, as inverse correlation exists between dietary fiber intake and risk of both Barrett's esophagus, an intermediate pre-neoplastic lesion, and esophageal cancer [180,184]. Potential mechanisms of protective action include modification of gastroesophageal reflux and/or weight control, neutralization of carcinogens contained in food, amelioration of cancer-associated esophageal dysbiosis, and direct action on cancer cells [180,184–188]. The prospective 14-year Iowa Women's Health Study, enrolling a cohort of 34,651 post-menopausal, initially cancer-free women, reported that malignancy incidence was inversely associated with WG intake, as well as with total fiber intake. In this context, some interesting data emerged: (i) none of inverse associations observed for fruit fiber, vegetable fiber, and total grain fiber was statistically significant; (ii) no protective effect was found for fiber from RGs (according to the evidence that milling process lowers content of fiber and bioactive compounds); (iii) the relationship with dietary fiber was driven by strong inverse association for WG fiber [95]. In the light of these findings, it should be advised to distinguish WGs or RGs as source of fiber, in order to avoid biased data [55,180].

Table 4. Some epidemiological studies on whole grains/whole grain fiber on esophageal cancer.

	Study Type and Design	Main Findings *	Refs
Whole grains	Swiss 7-year retrospective hospital-based case-control study (349 controls and 101 EC patients). Whole (whole wheat bread and cereals) and refined (white bread and biscuits, pizza, pasta and rice) grain foods Tertiles for WG intake (times/week): T1: <4 T2: 4–10 T3: >10 Tertiles for RG intake (times/week): T1: <9 T2: 9–17 T3: 17	EC risk inversely correlated to WG intakes ($OR_{T3\ vs.\ T1}$ = 0.30, CI 95% 0.1–0.6) and directly correlated to RG intakes ($OR_{T2\ vs.\ T1}$ = 2.6, CI 95% 1.1–6.2; $OR_{T3\ vs.\ T1}$ = 3.7, CI 95% 1.8–7.9)	[91]
	Italian 14-year hospital-based case-control studies (1983–1997) 10058 controls and 11.990 cancer patients (410 EC cases). Tertiles for WG food intake (day/week): T1: no or rare consumption T2: 1–3 T3: >3	WG consumption associated with reduced risk ($OR_{T3\ vs.\ T1}$ = 0.4, 95% CI 0.2–0.7 and $OR_{T2\ vs.\ T1}$ = 0.4, 95% CI 0.3–0.7)	[108]
	Meta-analysis of 34 studies of WG intake and risk of digestive tract cancer (EC: 4 case-control studies and 2 cohort studies (151,742 participants and 1223 EC cases))	WG consumption associated with reduced risk (RR = 0.54, 95% CI 0.44–0.67, $p < 0.001$) No statistically significant heterogeneity (I^2 = 27.7%, p = 0.217)	[113]
	Scandinavian 11-year prospective population-based case-control study (113,993 members from HELGA cohort including 56 EAC patients and 54 ESCC patients; 73.2% male and 33.8% women) Sex-specific tertiles of total WG intake (g/day): T1: F: 0–37.6; M: 0–37.8 T2: F: 37.7–60.5; M: 37.9–62.1 T3: F: 60.6–160.0; M: 62.2–160.0 Sex-specific tertiles of WG wheat intake (g/day): T1 F: 0–6.1, M: 0–2.5 71 T2 F: 6.2–32.1, M: 2.6–8.0 T3 F: 32.2–94.0, M: 8.1–65.7 Sex-specific tertiles of WG bread intake (g/day): T1: 0–59.6 for men; 0–65.8 for women T2: 62.0–125.0 for men; 68.2–113.2 for women T3: 129.6–520.0 for men; 113.8–520.0 for women	Inverse correlation between EC risk and total WGs ($HR_{T3\ vs.\ T1}$ = 0.55, 95% CI 0.31–0.97) and WG products ($HR_{T3\ vs.\ T1}$ = 0.51, 95% CI 0.30–0.88 per 25 g) Only wheat showed significant associations in adjusted models (adjusted $HR_{T3\ vs.\ T1}$ = 0.32, 95% CI 0.16–0.63) Only WG bread showed significant associations in adjusted model (adjusted HR $_{T3\ vs.\ T1}$ = 0.88, 95% CI 0.80–0.96 per 25 g WG bread) EAC: adjusted HR = 0.81, 95% CI 0.65–1.02 per 50 g WG products and HR = 0.85, 95% CI 0.66–1.09 per 20 g WGs ECCS: adjusted HR = 0.66, 95% CI 0.51–0.86 per 50 g WG products and adjusted HR = 0.75, 95% CI 0.56–1.00 per 20 g WGs	[181]

Table 4. *Cont.*

	Study Type and Design	Main Findings *	Refs
Whole grains/whole grain fiber	Prospective 14-year cohort Iowa Women's Health Study (34,651 initially free-cancer women; 21 EC and 56 GC) Tertiles of WG intake (servings/week): T1: 0–6.5 T2: 6.9–12.5 T3: 13.0–108.5 Tertiles of WG fiber intake (g/day): T1: 0–1.49 T2: 1.50–3.98 T3: 3.99–35.75	Inverse correlation between EC risk and WG ($HRR_{T3\ vs.\ T1}$ = 0.47) or WG fiber ($HRR_{T3\ vs.\ T1}$ = 0.35) intake	[95]

* Findings on WG intake per se, after adjusting for confounding factors (e.g., age, sex, education, smoking, dietary habits, alcohol, physical activity, etc.) through multivariate models. CI: Confidence Interval; EAC: esophageal adenocarcinoma; EC: esophageal cancer; ESCC: esophageal squamous cell carcinoma; HR: Hazard Rate; HRR: Hazard Rate Ratio; OR: Odds Ratio; RG: refined grain; RR: Relative Risk; WG: whole grain.

Additionally, polyphenols could be beneficial in esophageal cancer, thanks to their antioxidant activity, ability to improve esophageal reflux-related inflammation, and modulation of cell proliferation and survival [189,190]. Gallic acid, for example, induces cell death in human squamous esophagus carcinoma cells, much likely by activating both extrinsic and intrinsic apoptotic pathways, as well as by downregulating the Akt/mTOR survival signaling cascade [191]. Recently, protective roles of apigenin have been confirmed in esophageal tumors: in in vitro and in vivo experimental models, this flavonoid has been reported to (i) induce apoptosis of tumor cells, (ii) inhibit tumor-dependent angiogenesis, and (iii) attenuate inflammatory responses, by inhibiting gene expression of the pro-inflammatory cytokine interleukin-6, whose levels are elevated in tumor tissues [192,193].

8. Conclusions and Future Perspectives

Cancer onset, progression, and outcome are strictly dependent on interaction among genetic, metabolic, and environmental factors. Remarkably, besides some unhealthy habits (such as tobacco use, alcohol abuse, and sedentary lifestyle), consumption of harmful foodstuff and nutrients increases cancer risk; coherently, healthy dietary behaviors, which involve consumption of healthy foods (fruits, vegetables, cereals, legumes, fish, olive oil) and nutrients (antioxidants, phytochemicals, fiber, vitamins, mono- and poly-unsaturated fatty acids), are worldwide recognized as a valid strategy for primary cancer prevention. Scientific organizations of several countries encourage WG intake in maintaining health and reducing risk of chronic diseases, such as type 2 diabetes, cardiovascular disease, and cancer [119,194–197]. It has been estimated, in fact, that low WG intake resulted in almost 270,000 avoidable deaths and almost 4 million disability-adjusted life years in the European Union in 2015 [198].

Herein, we focused on inverse relationship between WGs, whose consumption is increasingly recommended, and gastrointestinal cancer onset and outcomes. What emerged is that WGs, unlike refined counterparts, consistently protect against gastrointestinal cancer, especially colorectal type; such differences can mainly be ascribed to reduction (or loss) of beneficial nutrients and phytochemicals during milling process. WG is indeed a complex food matrix containing different bioactive compounds, which synergistically act in chronic disease prevention. It is therefore difficult to identify which constituent is responsible for protection; for this reason, attention should be shifted not to single compounds, but instead to WG food matrix. For example, some WG positive effects essentially depends on fiber, but fiber varies from grain to grain and is present in other food items (vegetable, fruits, and legumes) that are equally consumed by high WG users. Thereby, although both fiber and WGs have healthy benefits, they are not interchangeable and consumers should pay particular attention to high-fiber products, sometimes containing bran or other added fiber without actually having much, if any, WG.

Noteworthily, WG consumers are more likely to consume less sugar, alcohol, fat, red meat, and indulgent foods, while consuming more fruits, vegetables, and fish; moreover, they have high education and socioeconomic status, as well as healthy lifestyle (physically active, no smoking). For this reason, most of studies investigated WG effects after accurate statistical adjustments for all of these possible confounders, thus removing potential bias from data and providing authentic and real relationship between WG intake and gastrointestinal cancer. Nonetheless, several elements point out that we are far from a solid, scientific-based knowledge for developing individualized WG-based regimens to prevent and manage cancer. WG consumption, indeed, reduces risk of digestive tract tumors with significant heterogeneity because of additional confounding factors, including differences in (i) type, duration, quality, and sample size of investigations; (ii) methods of collecting WG intake (food-frequency questionnaires vs. more quantitative questionnaires); (iii) type of WG foods; (iv) racial and ethnic demographic groups displaying different nutritional habits.

Despite all these limitations, beneficial effects of WGs cannot be denied, and therefore programs aimed at increasing WG consumption should be implemented through a broad partnership involving both public (Government authorities) and private (industries) stakeholders. Several barriers to WG consumption should, indeed, be removed by effective strategies: (i) univocal, quantitative, and international recommendations; (ii) nutritional education programs; (iii) improvement of sensory characteristics and increase of variety of WG foods (in order to satisfy different eating habits of consumers of all ages); (iv) better identification of WG-containing products; (v) reduction of purchase costs.

Author Contributions: Conceptualization, literature data search and drafting of manuscript V.T. and V.G.; manuscript correction and final approval M.V.C. and I.S. All authors have read and agreed to the published version of the manuscript.

Funding: This research received no external funding.

Conflicts of Interest: The authors declare no conflict of interest.

References

1. Ross, A.B.; van der Kamp, J.W.; King, R.; Lê, K.A.; Mejborn, H.; Seal, C.J.; Thielecke, F. Perspective: A definition for whole-grain food products—Recommendations from the Healthgrain Forum. *Adv. Nutr.* **2017**, *8*, 525–531. [CrossRef] [PubMed]
2. Ferruzzi, M.G.; Jonnalagadda, S.S.; Liu, S.; Marquart, L.; McKeown, N.; Reicks, M.; Riccardi, G.; Seal, C.; Slavin, J.; Thielecke, F.; et al. Developing a Standard Definition of Whole-Grain Foods for Dietary Recommendations: Summary Report of a Multidisciplinary Expert Roundtable Discussion. *Adv. Nutr.* **2014**, *5*, 164–176. [CrossRef] [PubMed]
3. Carcea, M. Nutritional Value of Grain-Based Foods. *Foods* **2020**, *9*, 504. [CrossRef] [PubMed]
4. Nugent, A.P.; Thielecke, F. Wholegrains and health: Many benefits but do contaminants pose any risk? *Nutr. Bull.* **2019**, *44*, 107–115. [CrossRef]
5. Cummings, J.H.; Engineer, A. *Nutr. Res. Rev.* **2018**, *31*, 1–15. [CrossRef] [PubMed]
6. Huang, T.; Xu, M.; Lee, A.; Cho, S.; Qi, L. Consumption of whole grains and cereal fiber and total and cause-specific mortality: Prospective analysis of 367,442 individuals. *BMC Med.* **2015**, *13*, 59. [CrossRef]
7. Mozaffarian, D.; Hao, T.; Rimm, E.B.; Willett, W.C.; Hu, F.B. Changes in diet and lifestyle and long-term weight gain in women and men. *N. Engl. J. Med.* **2011**, *364*, 2392–2404. [CrossRef]
8. Mellen, P.B.; Walsh, T.F.; Herrington, D.M. Whole grain intake and cardiovascular disease: A meta-analysis. *Nutr. Metab. Cardiovasc. Dis.* **2008**, *18*, 283–290. [CrossRef]
9. Larsson, S.C.; Giovannucci, E.; Bergkvist, L.; Wolk, A. Whole grain consumption and risk of colorectal cancer: A population-based cohort of 60,000 women. *Br. J. Cancer* **2005**, *92*, 1803–1807. [CrossRef]
10. Wu, H.; Flint, A.J.; Qi, Q.; Van Dam, R.M.; Sampson, L.A.; Rimm, E.B.; Holmes, M.D.; Willett, W.C.; Hu, F.B.; Sun, Q. Association between dietary whole grain intake and risk of mortality: Two large prospective studies in US Men and Women. *JAMA Intern. Med.* **2015**, *175*, 373–384. [CrossRef]
11. Zong, G.; Gao, A.; Hu, F.B.; Sun, Q. Whole grain intake and mortality from all causes, cardiovascular disease, and cancer. *Circulation* **2016**, *133*, 2370–2380. [CrossRef] [PubMed]

12. Giacco, R.; Costabile, G.; Fatati, G.; Frittitta, L.; Maiorino, M.I.; Marelli, G.; Parillo, M.; Pistis, D.; Tubili, C.; Vetrani, C.; et al. Effects of polyphenols on cardio-metabolic risk factors and risk of type 2 diabetes. A joint position statement of the Diabetes and Nutrition Study Group of the Italian Society of Diabetology (SID), the Italian Association of Dietetics and Clinical Nutrit. *Nutr. Metab. Cardiovasc. Dis.* **2020**, *30*, 355–367. [CrossRef] [PubMed]
13. Polito, R.; Costabile, G.; Nigro, E.; Giacco, R.; Vetrani, C.; Anniballi, G.; Luongo, D.; Riccardi, G.; Daniele, A.; Annuzzi, G. Nutritional factors influencing plasma adiponectin levels: Results from a randomised controlled study with whole-grain cereals. *Int. J. Food Sci. Nutr.* **2020**, *71*, 509–515. [CrossRef] [PubMed]
14. Vetrani, C.; Costabile, G.; Luongo, D.; Naviglio, D.; Rivellese, A.A.; Riccardi, G.; Giacco, R. Effects of whole-grain cereal foods on plasma short chain fatty acid concentrations in individuals with the metabolic syndrome. *Nutrition* **2016**, *32*, 217–221. [CrossRef]
15. US Department of Health and Human Services. US Department of Agriculture 2015–2020 Dietary Guidelines for Americans, 8th ed. Available online: https://health.gov/sites/default/files/2019-09/2015-2020_Dietary_Guidelines.pdf (accessed on 26 November 2020).
16. ALTOMKOST. Danish Veterinary and Food Administration The Official Dietary Guidelines. Available online: https://altomkost.dk/raad-og-anbefalinger/de-officielle-kostraad/ (accessed on 26 November 2020).
17. Helsedirektoratet. Norwegian Directorate of Health Recommendations about Diet, Nutrition and Physical Activity. Available online: https://www.helsedirektoratet.no/brosjyrer/helsedirektoratets-kostrad-brosjyre-og-plakat/Helsedirektoratets%20kostr%C3%A5d%20-%20engelsk.pdf/_/attachment/inline/80f68126-68af-4cec-b2aa-d04069d02471:dcb8efdbe6b6129470ec4969f6639be21a8afd82/Helsedirektoratets%20kostr%C3%A5d%20-%20engelsk.pdf (accessed on 26 November 2020).
18. Sweden Countries Swedish Food Agency Find Your Way to Eat Greener, not too much and Be Active. Available online: https://www.livsmedelsverket.se/globalassets/publikationsdatabas/rapporter/2015/rapp-hanteringsrapport-engelska-omslag--inlaga--bilagor-eng-version.pdf (accessed on 26 November 2020).
19. Division of Nutrition National Food Institute Technical University of Denmark. Wholegrain Intake of Danes 2011–2012. Available online: https://www.food.dtu.dk/english/-/media/Institutter/Foedevareinstituttet/Publikationer/Pub-2013/Rapport_Fuldkornsindtag_11-12_UK.ashx?la=da&hash=8B2A20C3ED33A0B8564E5403DFD8225CB25EE42D (accessed on 27 November 2020).
20. Albertson, A.M.; Reicks, M.; Joshi, N.; Gugger, C.K. Whole grain consumption trends and associations with body weight measures in the United States: Results from the cross sectional National Health and Nutrition Examination Survey 2001–2012. *Nutr. J.* **2016**, *15*, 8. [CrossRef]
21. O'Donovan, C.B.; Devlin, N.F.; Buffini, M.; Walton, J.; Flynn, A.; Gibney, M.J.; Nugent, A.P.; McNulty, B.A. Whole grain intakes in Irish adults: Findings from the National Adults Nutrition Survey (NANS). *Eur. J. Nutr.* **2019**, *58*, 541–550. [CrossRef]
22. Devlin, N.F.C.; McNulty, B.A.; Gibney, M.J.; Thielecke, F.; Smith, H.; Nugent, A.P. Whole grain intakes in the diets of Irish children and teenagers. *Br. J. Nutr.* **2013**, *110*, 354–362. [CrossRef]
23. Galea, L.M.; Beck, E.J.; Probst, Y.C.; Cashman, C.J. Whole grain intake of Australians estimated from a cross-sectional analysis of dietary intake data from the 2011–13 Australian Health Survey. *Public Health Nutr.* **2017**, *20*, 2166–2172. [CrossRef]
24. Mann, K.D.; Pearce, M.S.; McKevith, B.; Thielecke, F.; Seal, C.J. Low whole grain intake in the UK: Results from the National Diet and Nutrition Survey rolling programme 2008–11. *Br. J. Nutr.* **2015**, *113*, 1643–1651. [CrossRef]
25. Bellisle, F.; Hébel, P.; Colin, J.; Reyé, B.; Hopkins, S. Consumption of whole grains in French children, adolescents and adults. *Br. J. Nutr.* **2014**, *112*, 1674–1684. [CrossRef]
26. Sette, S.; D'Addezio, L.; Piccinelli, R.; Hopkins, S.; Le Donne, C.; Ferrari, M.; Mistura, L.; Turrini, A. Intakes of whole grain in an Italian sample of children, adolescents and adults. *Eur. J. Nutr.* **2017**, *56*, 521–533. [CrossRef]
27. Ruggiero, E.; Bonaccio, M.; Di Castelnuovo, A.; Bonanni, A.; Costanzo, S.; Persichillo, M.; Bracone, F.; Cerletti, C.; Donati, M.B.; de Gaetano, G.; et al. Consumption of whole grain food and its determinants in a general Italian population: Results from the INHES study. *Nutr. Metab. Cardiovasc. Dis.* **2019**, *29*, 611–620. [CrossRef] [PubMed]
28. Watson, R.R.; Preedy, V.; Zibadi, S. *Wheat and Rice in Disease Prevention and Health*; Elsevier Inc.: Amsterdam, The Netherlands, 2014; ISBN 9780124017160. [CrossRef]
29. van der Kamp, J.W.; Poutanen, K.; Seal, C.J.; Richardson, D.P. The Healthgrain definition of 'whole grain'. *Food Nutr. Res.* **2014**, *58*, 22100. [CrossRef] [PubMed]
30. Mathews, R.; Chu, Y.F. Global review of whole grain definitions and health claims. *Nutr. Rev.* **2020**, *78*, 98–106. [CrossRef]
31. Khan, K.; Shrewry, P.R. *Wheat: Chemistry and Technology*, 4th ed.; Elsevier Inc.: Amsterdam, The Netherlands, 2009; ISBN 9780128104545.
32. Carcea, M.; Turfani, V.; Narducci, V.; Melloni, S.; Galli, V.; Tullio, V. Stone Milling versus Roller Milling in Soft Wheat: Influence on Products Composition. *Foods* **2019**, *9*, 3. [CrossRef] [PubMed]
33. Apprich, S.; Tirpanalan, Ö.; Hell, J.; Reisinger, M.; Böhmdorfer, S.; Siebenhandl-Ehn, S.; Novalin, S.; Kneifel, W. Wheat bran-based biorefinery 2: Valorization of products. *LWT Food Sci. Technol.* **2014**, *56*, 222–231. [CrossRef]
34. Onipe, O.O.; Jideani, A.I.O.; Beswa, D. Composition and functionality of wheat bran and its application in some cereal food products. *Int. J. Food Sci. Technol.* **2015**, *50*, 2509–2518. [CrossRef]
35. Oghbaei, M.; Prakash, J. Effect of primary processing of cereals and legumes on its nutritional quality: A comprehensive review. *Cogent Food Agric.* **2016**, *2*, 1136015. [CrossRef]

36. Bailey, R.L.; Dodd, K.W.; Gahche, J.J.; Dwyer, J.T.; McDowell, M.A.; Yetley, E.A.; Sempos, C.A.; Burt, V.L.; Radimer, K.L.; Picciano, M.F. Total folate and folic acid intake from foods and dietary supplements in the United States: 2003–2006. *Am. J. Clin. Nutr.* **2010**, *91*, 231–237. [CrossRef]
37. Kristensen, M.; Jensen, M.G.; Riboldi, G.; Petronio, M.; Bügel, S.; Toubro, S.; Tetens, I.; Astrup, A. Wholegrain vs. refined wheat bread and pasta. Effect on postprandial glycemia, appetite, and subsequent ad libitum energy intake in young healthy adults. *Appetite* **2010**, *54*, 163–169. [CrossRef]
38. Cioffi, I.; Ibrugger, S.; Bache, J.; Thomassen, M.T.; Contaldo, F.; Pasanisi, F.; Kristensen, M. Effects on satiation, satiety and food intake of wholegrain and refined grain pasta. *Appetite* **2016**, *107*, 152–158. [CrossRef] [PubMed]
39. De Angelis, M.; Piccolo, M.; Vannini, L.; Siragusa, S.; de Giacomo, A.; Serrazzanetti, D.I.; Cristofori, F.; Guerzoni, M.E.; Gobbetti, M.; Francavilla, R. Fecal Microbiota and Metabolome of Children with Autism and Pervasive Developmental Disorder Not Otherwise Specified. *PLoS ONE* **2013**, *8*, e76993. [CrossRef] [PubMed]
40. Wu, G.D.; Chen, J.; Hoffmann, C.; Bittinger, K.; Chen, Y.-Y.; Keilbaugh, S.A.; Bewtra, M.; Knights, D.; Walters, W.A.; Knight, R.; et al. Linking Long-Term Dietary Patterns with. *Science* **2011**, *334*, 105–109. [CrossRef] [PubMed]
41. Tuohy, K.M.; Conterno, L.; Gasperotti, M.; Viola, R. Up-regulating the human intestinal microbiome using whole plant foods, polyphenols, and/or fiber. *J. Agric. Food Chem.* **2012**, *60*, 8776–8782. [CrossRef]
42. Claesson, M.J.; Jeffery, I.B.; Conde, S.; Power, S.E.; O'connor, E.M.; Cusack, S.; Harris, H.M.B.; Coakley, M.; Lakshminarayanan, B.; O'sullivan, O.; et al. Gut microbiota composition correlates with diet and health in the elderly. *Nature* **2012**, *488*, 178–184. [CrossRef] [PubMed]
43. De Filippis, F.; Pellegrini, N.; Vannini, L.; Jeffery, I.B.; La Storia, A.; Laghi, L.; Serrazanetti, I.D.; Di Cagno, R.; Ferrocino, I.; Lazzi, C.; et al. High-level adherence to a Mediterranean diet beneficially impacts the gut microbiota and associated metabolome. *Gut* **2016**, *65*, 1812–1821. [CrossRef]
44. D'Alessandro, A.; Lampignano, L.; De Pergola, G. Mediterranean Diet Pyramid: A Proposal for Italian People. A Systematic Review of Prospective Studies to Derive Serving Sizes. *Nutrients* **2019**, *11*, 1296. [CrossRef]
45. Vitiello, V.; Germani, A.; Capuzzo Dolcetta, E.; Donini, L.M.; del Balzo, V. The new modern mediterranean diet italian pyramid. *Ann. Ig* **2016**, *28*, 179–186. [CrossRef]
46. Bach-Faig, A.; Berry, E.M.; Lairon, D.; Reguant, J.; Trichopoulou, A.; Dernini, S.; Medina, F.X.; Battino, M.; Belahsen, R.; Miranda, G.; et al. Mediterranean diet pyramid today. Science and cultural updates. *Public Health Nutr.* **2011**, *14*, 2274–2284. [CrossRef]
47. Vanegas, S.M.; Meydani, M.; Barnett, J.B.; Goldin, B.; Kane, A.; Rasmussen, H.; Brown, C.; Vangay, P.; Knights, D.; Jonnalagadda, S.; et al. Substituting whole grains for refined grains in a 6-wk randomized trial has a modest effect on gut microbiota and immune and inflammatory markers of healthy adults. *Am. J. Clin. Nutr.* **2017**, *105*, 635–650. [CrossRef]
48. Saura-Calixto, F. Dietary fiber as a carrier of dietary antioxidants: An essential physiological function. *J. Agric. Food Chem.* **2011**, *59*, 43–49. [CrossRef]
49. Gianfredi, V.; Nucci, D.; Salvatori, T.; Dallagiacoma, G.; Fatigoni, C.; Moretti, M.; Realdon, S. Rectal Cancer: 20% Risk Reduction Thanks to Dietary Fibre Intake. Systematic Review and Meta-Analysis. *Nutrients* **2019**, *11*, 1579. [CrossRef]
50. AACC. The definition of dietary fiber. In *Cereal Foods World*; AACC (the American Association of Cereal Chemists): Eagan, MN, USA, 2001; Volume 46, pp. 112–129.
51. Jones, J.M. Dietary fiber methods in Codex Alimentarius: Current status and ongoing discussions. *Cereal Foods World* **2013**, *58*, 148–152. [CrossRef]
52. Caprita, R.; Caprita, A.; Cretescu, I. Effect of extraction conditions on the solubility of non-starch polysaccharides of wheat and barley. *J. Food Agric. Environ.* **2011**, *9*, 41–43.
53. Ain, H.B.U.; Saeed, F.; Ahmad, N.; Imran, A.; Niaz, B.; Afzaal, M.; Imran, M.; Tufail, T.; Javed, A. Functional and health-endorsing properties of wheat and barley cell wall's non-starch polysaccharides. *Int. J. Food Prop.* **2018**, *21*, 1463–1480. [CrossRef]
54. Chotinsky, D. The use of enzymes to improve utilization of nutrient in poultry diets. *Bulg. J. Agric. Sci.* **2015**, *21*, 429–435.
55. Terry, P.; Lagergren, J.; Ye, W.; Wolk, A.; Nyrén, O. Inverse association between intake of cereal fiber and risk of gastric cardia cancer. *Gastroenterology* **2001**, *120*, 387–391. [CrossRef]
56. Mirmiran, P.; Bahadoran, Z.; Moghadam, S.K.; Vakili, A.Z.; Azizi, F. A prospective study of different types of dietary fiber and risk of cardiovascular disease: Tehran lipid and glucose study. *Nutrients* **2016**, *8*, 686. [CrossRef]
57. CREA Tabelle di composizione degli alimenti. Available online: https://www.crea.gov.it/en/-/tabella-di-composizione-degli-alimenti (accessed on 12 November 2020).
58. USDA National Nutrient Database. Available online: https://data.nal.usda.gov/dataset/composition-foods-raw-processed-prepared-usda-national-nutrient-database-standard-reference-release-27 (accessed on 7 November 2020).
59. Carcea, M.; Narducci, V.; Turfani, V.; Giannini, V. Polyphenols in Raw and Cooked Cereals/Pseudocereals/Legume Pasta and Couscous. *Foods* **2017**, *6*, 80. [CrossRef]
60. Dykes, L.; Rooney, L.W. Phenolic Compounds in Cereal Grains and Their Health Benefits. *Cereal Foods World* **2007**, *52*, 105–111. [CrossRef]
61. Adom, K.K.; Liu, R.H. Antioxidant Activity of Grains. *J. Agric. Food Chem.* **2002**, *50*, 6182–6187. [CrossRef] [PubMed]
62. Oliver Chen, C.; Costa, S.M.; Carolo, K. Phenolic Acids. In *Whole Grains and Their Bioactives*; Wiley Online Library: Hoboken, NJ, USA, 2019; pp. 357–382.

63. Liu, Z.; Liu, Y.; Pu, Z.; Wang, J.; Zheng, Y.; Li, Y.; Wei, Y. Regulation, evolution, and functionality of flavonoids in cereal crops. *Biotechnol. Lett.* **2013**, *35*, 1765–1780. [CrossRef]
64. Quirós-Sauceda, A.E.; Palafox-Carlos, H.; Sáyago-Ayerdi, S.G.; Ayala-Zavala, J.F.; Bello-Perez, L.A.; Álvarez-Parrilla, E.; De La Rosa, L.A.; González-Córdova, A.F.; González-Aguilar, G.A. Dietary fiber and phenolic compounds as functional ingredients: Interaction and possible effect after ingestion. *Food Funct.* **2014**, *5*, 1063–1072. [CrossRef] [PubMed]
65. Sidhu, J.S.; Kabir, Y.; Huffman, F.G. Functional foods from cereal grains. *Int. J. Food Prop.* **2007**, *10*, 231–244. [CrossRef]
66. Kaur, P.; Purewal, S.S.; Sandhu, K.S.; Kaur, M.; Salar, R.K. Millets: A cereal grain with potent antioxidants and health benefits. *J. Food Meas. Charact.* **2019**, *13*, 793–806. [CrossRef]
67. Parada, J.; Aguilera, J.M. Food Microstructure Affects the Bioavailability of Several Nutrients. *J. Food Sci.* **2007**, *72*, R21–R32. [CrossRef]
68. Okarter, N.; Liu, C.S.; Sorrells, M.E.; Liu, R.H. Phytochemical content and antioxidant activity of six diverse varieties of whole wheat. *Food Chem.* **2010**, *119*, 249–257. [CrossRef]
69. Xu, Z.; Godber, J.S. Purification and identification of components of γ-oryzanol in rice bran oil. *J. Agric. Food Chem.* **1999**. [CrossRef]
70. Collins, F.W. Oat Phenolics: Avenanthramides, Novel Substituted N-Cinnamoylanthranilate Alkaloids from Oat Groats and Hulls. *J. Agric. Food Chem.* **1989**. [CrossRef]
71. McCarty, M.F.; Assanga, S.B.I. Ferulic acid may target MyD88-mediated pro-inflammatory signaling—Implications for the health protection afforded by whole grains, anthocyanins, and coffee. *Med. Hypotheses* **2018**, *118*, 114–120. [CrossRef]
72. Abdel-Aal, E.-S.M.; Young, J.C.; Rabalski, I. Anthocyanin Composition in Black, Blue, Pink, Purple, and Red Cereal Grains. *J. Agric. Food Chem.* **2006**, *54*, 4696–4704. [CrossRef] [PubMed]
73. Durazzo, A.; Zaccaria, M.; Polito, A.; Maiani, G.; Carcea, M. Lignan Content in Cereals, Buckwheat and Derived Foods. *Foods* **2013**, *2*, 53–63. [CrossRef] [PubMed]
74. Cao, X.; Cai, C.; Wang, Y.; Zheng, X. The inactivation kinetics of polyphenol oxidase and peroxidase in bayberry juice during thermal and ultrasound treatments. *Innov. Food Sci. Emerg. Technol.* **2018**, *45*, 169–178. [CrossRef]
75. Zhang, H.; Tsao, R. Dietary polyphenols, oxidative stress and antioxidant and anti-inflammatory effects. *Curr. Opin. Food Sci.* **2016**, *8*, 33–42. [CrossRef]
76. Gani, A.; Wani, S.M.; Masoodi, F.A.; Hameed, G. Whole-Grain Cereal Bioactive Compounds and Their Health Benefits: A Review. *J. Food Process. Technol.* **2012**, *3*, 146–156. [CrossRef]
77. Vitaglione, P.; Napolitano, A.; Fogliano, V. Cereal dietary fibre: A natural functional ingredient to deliver phenolic compounds into the gut. *Trends Food Sci. Technol.* **2008**, *19*, 451–463. [CrossRef]
78. Ribas-Agustí, A.; Seda, M.; Sarraga, C.; Montero, J.I.; Castellari, M.; Muñoz, P. Municipal solid waste composting: Application as a tomato fertilizer and its effect on crop yield, fruit quality and phenolic content. *Renew. Agric. Food Syst.* **2017**, *32*, 358–365. [CrossRef]
79. Vitaglione, P.; Mennella, I.; Ferracane, R.; Rivellese, A.A.; Giacco, R.; Ercolini, D.; Gibbons, S.M.; La Storia, A.; Gilbert, J.A.; Jonnalagadda, S.; et al. Whole-grain wheat consumption reduces inflammation in a randomized controlled trial on overweight and obese subjects with unhealthy dietary and lifestyle behaviors: Role of polyphenols bound to cereal dietary fiber. *Am. J. Clin. Nutr.* **2015**, *101*, 251–261. [CrossRef]
80. González-Aguilar, G.A.; Blancas-Benítez, F.J.; Sáyago-Ayerdi, S.G. Polyphenols associated with dietary fibers in plant foods: Molecular interactions and bioaccessibility. *Curr. Opin. Food Sci.* **2017**, *13*, 84–88. [CrossRef]
81. Bishehsari, F.; Engen, P.; Preite, N.; Tuncil, Y.; Naqib, A.; Shaikh, M.; Rossi, M.; Wilber, S.; Green, S.; Hamaker, B.; et al. Dietary Fiber Treatment Corrects the Composition of Gut Microbiota, Promotes SCFA Production, and Suppresses Colon Carcinogenesis. *Genes* **2018**, *9*, 102. [CrossRef]
82. Tang, Y.; Chen, Y.; Jiang, H.; Nie, D. The role of short-chain fatty acids in orchestrating two types of programmed cell death in colon cancer. *Autophagy* **2011**, *7*, 235–237. [CrossRef] [PubMed]
83. Lizarraga, D.; Vinardell, M.P.; Noé, V.; van Delft, J.H.; Alcarraz-Vizán, G.; van Breda, S.G.; Staal, Y.; Günther, U.L.; Reed, M.A.; Ciudad, C.J.; et al. A Lyophilized red grape pomace containing proanthocyanidin-rich dietary fiber induces genetic and metabolic alterations in colon mucosa of female C57Bl/6J mice. *J. Nutr.* **2011**, *141*, 1597–1604. [CrossRef] [PubMed]
84. Ferlay, J.; Ervik, M.; Lam, F.; Colombet, M.; Mery, L.; Piñeros, M.; Znaor, A.; Soerjomataram, I.B.F. *Global Cancer Observatory: Cancer Today*; International Agency for Research on Cancer: Lyon, France; Available online: https://gco.iarc.fr/today (accessed on 13 November 2020).
85. Nagtegaal, I.D.; Odze, R.D.; Klimstra, D.; Paradis, V.; Rugge, M.; Schirmacher, P.; Washington, K.M.; Carneiro, F.; Cree, I.A. The 2019 WHO classification of tumours of the digestive system. *Histopathology* **2020**, *76*, 182–188. [CrossRef]
86. Mereiter, S.; Balmaña, M.; Gomes, J.; Magalhães, A.; Reis, C.A. Glycomic approaches for the discovery of targets in gastrointestinal cancer. *Front. Oncol.* **2016**, *6*, 55. [CrossRef] [PubMed]
87. Arnold, M.; Abnet, C.C.; Neale, R.E.; Vignat, J.; Giovannucci, E.L.; McGlynn, K.A.; Bray, F. Global Burden of 5 Major Types of Gastrointestinal Cancer. *Gastroenterology* **2020**, *159*, 335–349.e15. [CrossRef]
88. Yang, W.; Ma, Y.; Liu, Y.; Smith-Warner, S.A.; Simon, T.G.; Chong, D.Q.; Qi, Q.; Meyerhardt, J.A.; Giovannucci, E.L.; Chan, A.T.; et al. Association of Intake of Whole Grains and Dietary Fiber with Risk of Hepatocellular Carcinoma in US Adults. *JAMA Oncol.* **2019**, *5*, 879–886. [CrossRef]

89. Lei, Q.; Zheng, H.; Bi, J.; Wang, X.; Jiang, T.; Gao, X.; Tian, F.; Xu, M.; Wu, C.; Zhang, L.; et al. Whole Grain Intake Reduces Pancreatic Cancer Risk. *Medicine* **2016**, *95*, e2747. [CrossRef]
90. Winn, D.M.; Ziegler, R.G.; Pickle, L.W.; Gridley, G.; Blot, W.J.; Hoover, R.N. Diet in the Etiology of Oral and Pharyngeal Cancer among Women from the Southern United States. *Cancer Res.* **1984**, *44*, 1216–1222.
91. Levi, F.; Pasche, C.; Lucchini, F.; Chatenoud, L.; Jacobs, D.; La Vecchia, C. Refined and whole grain cereals and the risk of oral, oesophageal and laryngeal cancer. *Eur. J. Clin. Nutr.* **2000**, *54*, 487–489. [CrossRef]
92. Chatenoud, L.; La Vecchia, C.; Franceschi, S.; Tavani, A.; Jacobs, D.R.; Parpinel, M.T.; Soler, M.; Negri, E. Refined-cereal intake and risk of selected cancers in Italy. *Am. J. Clin. Nutr.* **1999**, *70*, 1107–1110. [CrossRef]
93. Franceschi, S.; Favero, A.; Conti, E.; Talamini, R.; Volpe, R.; Negri, E.; Barzan, L.; Vecchia, C. La Food groups, oils and butter, and cancer of the oral cavity and pharynx. *Br. J. Cancer* **1999**, *80*, 614–620. [CrossRef] [PubMed]
94. Uzcudun, A.E.; Retolaza, I.R.; Fernández, P.B.; Sánchez Hernández, J.J.; Grande, A.G.; García, A.G.; Olivar, L.M.; de Diego Sastre, I.; Barón, M.G.; Bouzas, J.G. Nutrition and pharyngeal cancer: Results from a case-control study in Spain. *Head Neck* **2002**, *24*, 830–840. [CrossRef] [PubMed]
95. Kasum, C.M.; Jacobs, D.R.; Nicodemus, K.; Folsom, A.R. Dietary risk factors for upper aerodigestive tract cancers. *Int. J. Cancer* **2002**, *99*, 267–272. [CrossRef] [PubMed]
96. De Stefani, E.; Ronco, A.; Mendilaharsu, M.; Deneo-Pellegrini, H. Diet and risk of cancer of the upper aerodigestive tract—II. Nutrients. *Oral Oncol.* **1999**, *35*, 22–26. [CrossRef]
97. Toporcov, T.N.; Antunes, J.L.F.; Tavares, M.R. Fat food habitual intake and risk of oral cancer. *Oral Oncol.* **2004**, *40*, 925–931. [CrossRef]
98. Zheng, T.; Boyle, P.; Willett, W.C.; Hu, H.; Dan, J.; Evstifeeva, T.V.; Niu, S.; MacMahon, B. A case-control study of oral cancer in Beijing, People's Republic of China. Associations with nutrient intakes, foods and food groups. *Eur. J. Cancer. Part B Oral Oncol.* **1993**, *29*, 45–55. [CrossRef]
99. Sánchez, M.J.; Martínez, C.; Nieto, A.; Castellsagué, X.; Quintana, M.J.; Bosch, F.X.; Muñoz, N.; Herrero, R.; Franceschi, S. Oral and oropharyngeal cancer in Spain: Influence of dietary patterns. *Eur. J. Cancer Prev.* **2003**, *12*, 49–56. [CrossRef]
100. Franceschi, S.; Bidoli, E.; Barón, A.E.; Barra, S.; Talamini, R.; Serraino, D.; La Vecchia, C. Nutrition and cancer of the oral cavity and pharynx in north-east italy. *Int. J. Cancer* **1991**, *47*, 20–25. [CrossRef]
101. Nandakumar, A.; Thimmasetty, K.T.; Sreeramareddy, N.M.; Venugopal, T.C.; Rajanna; Vinutha, A.T.; Srinivas; Bhargava, M.K. A population-based case-control investigation on cancers of the oral cavity in Bangalore, India. *Br. J. Cancer* **1990**, *62*, 847–851. [CrossRef]
102. Garrote, L.F.; Herrero, R.; Reyes, R.M.O.; Vaccarella, S.; Anta, J.L.; Ferbeye, L.; Muoz, N.; Franceschi, S. Risk factors for cancer of the oral cavity and oro-pharynx in Cuba. *Br. J. Cancer* **2001**, *85*, 46–54. [CrossRef]
103. Schatzkin, A.; Park, Y.; Leitzmann, M.F.; Hollenbeck, A.R.; Cross, A.J. Prospective Study of Dietary Fiber, Whole Grain Foods, and Small Intestinal Cancer. *Gastroenterology* **2008**, *135*, 1163–1167. [CrossRef] [PubMed]
104. Siegel, R.L.; Fedewa, S.A.; Anderson, W.F.; Miller, K.D.; Ma, J.; Rosenberg, P.S.; Jemal, A. Colorectal Cancer Incidence Patterns in the United States, 1974–2013. *J. Natl. Cancer Inst.* **2017**, *109*, 27–32. [CrossRef] [PubMed]
105. Wang, L.; Lo, C.-H.; He, X.; Hang, D.; Wang, M.; Wu, K.; Chan, A.T.; Ogino, S.; Giovannucci, E.L.; Song, M. Risk Factor Profiles Differ for Cancers of Different Regions of the Colorectum. *Gastroenterology* **2020**, *159*, 241–256.e13. [CrossRef] [PubMed]
106. Lakatos, L.; Mester, G.; Erdelyi, Z.; David, G.; Pandur, T.; Balogh, M.; Fischer, S.; Vargha, P.; Lakatos, P.L. Risk factors for ulcerative colitis-associated colorectal cancer in a Hungarian cohort of patients with ulcerative colitis: Results of a population-based study. *Inflamm. Bowel Dis.* **2006**, *12*, 205–211. [CrossRef] [PubMed]
107. Center, M.M.; Jemal, A.; Ward, E. International trends in colorectal cancer incidence rates. *Cancer Epidemiol. Biomark. Prev.* **2009**, *18*, 1688–1694. [CrossRef] [PubMed]
108. La Vecchia, C.; Chatenoud, L.; Negri, E.; Franceschi, S. Session: Whole cereal grains, fibre and human cancer wholegrain cereals and cancer in Italy. *Proc. Nutr. Soc.* **2003**, *62*, 45–49. [CrossRef]
109. Um, C.Y.; Campbell, P.T.; Carter, B.; Wang, Y.; Gapstur, S.M.; McCullough, M.L. Association between grains, gluten and the risk of colorectal cancer in the Cancer Prevention Study-II Nutrition Cohort. *Eur. J. Nutr.* **2020**, *59*, 1739–1749. [CrossRef]
110. Vieira, A.R.; Abar, L.; Chan, D.S.M.; Vingeliene, S.; Polemiti, E.; Stevens, C.; Greenwood, D.; Norat, T. Foods and beverages and colorectal cancer risk: A systematic review and meta-analysis of cohort studies, an update of the evidence of the WCRF-AICR Continuous Update Project. *Ann. Oncol. Off. J. Eur. Soc. Med. Oncol.* **2017**, *28*, 1788–1802. [CrossRef]
111. Tieri, M.; Ghelfi, F.; Vitale, M.; Vetrani, C.; Marventano, S.; Lafranconi, A.; Godos, J.; Titta, L.; Gambera, A.; Alonzo, E.; et al. Whole grain consumption and human health: An umbrella review of observational studies. *Int. J. Food Sci. Nutr.* **2020**, *71*, 668–677. [CrossRef]
112. Schwingshackl, L.; Schwedhelm, C.; Hoffmann, G.; Knüppel, S.; Laure Preterre, A.; Iqbal, K.; Bechthold, A.; De Henauw, S.; Michels, N.; Devleesschauwer, B.; et al. Food groups and risk of colorectal cancer. *Int. J. Cancer* **2018**, *142*, 1748–1758. [CrossRef]
113. Zhang, X.-F.; Wang, X.-K.; Tang, Y.-J.; Guan, X.-X.; Guo, Y.; Fan, J.-M.; Cui, L.-L. Association of whole grains intake and the risk of digestive tract cancer: A systematic review and meta-analysis. *Nutr. J.* **2020**, *19*, 1–14. [CrossRef] [PubMed]
114. Sun, H.; Liu, Y.; Huang, H.; Li, D.; Zhao, Y. Diet quality score and survival rate in patients with colorectal cancer. *Asia Pac. J. Clin. Nutr.* **2019**, *28*, 601–606. [CrossRef] [PubMed]

115. Alegria-Lertxundi, I.; Aguirre, C.; Bujanda, L.; Fernández, F.J.; Polo, F.; Ordovás, J.M.; Etxezarraga, M.C.; Zabalza, I.; Larzabal, M.; Portillo, I.; et al. Food groups, diet quality and colorectal cancer risk in the Basque Country. *World J. Gastroenterol.* **2020**, *26*, 4108–4125. [CrossRef]
116. Hullings, A.G.; Sinha, R.; Liao, L.M.; Freedman, N.D.; Graubard, B.I.; Loftfield, E. Whole grain and dietary fiber intake and risk of colorectal cancer in the NIH-AARP Diet and Health Study cohort. *Am. J. Clin. Nutr.* **2020**, *112*, 603–612. [CrossRef] [PubMed]
117. Song, M.; Wu, K.; Meyerhardt, J.A.; Ogino, S.; Wang, M.; Fuchs, C.S.; Giovannucci, E.L.; Chan, A.T. Fiber Intake and Survival After Colorectal Cancer Diagnosis. *JAMA Oncol.* **2018**, *4*, 71–79. [CrossRef] [PubMed]
118. He, X.; Wu, K.; Zhang, X.; Nishihara, R.; Cao, Y.; Fuchs, C.S.; Giovannucci, E.L.; Ogino, S.; Chan, A.T.; Song, M. Dietary intake of fiber, whole grains and risk of colorectal cancer: An updated analysis according to food sources, tumor location and molecular subtypes in two large US cohorts. *Int. J. Cancer* **2019**, *145*, 3040–3051. [CrossRef] [PubMed]
119. WCRF/AICR. *Diet, Nutrition, Physical Activity and Colorectal Cancer*; Word Cancer Research Fund International: London, UK, 2018; pp. 1–111. ISBN 9781912259007.
120. Pan, P.; Yu, J.; Wang, L.-S. Colon Cancer. *Surg. Oncol. Clin. N. Am.* **2018**, *27*, 243–267. [CrossRef]
121. Nogacka, A.M.; Gómez-Martín, M.; Suárez, A.; González-Bernardo, O.; de los Reyes-Gavilán, C.G.; González, S. Xenobiotics Formed during Food Processing: Their Relation with the Intestinal Microbiota and Colorectal Cancer. *Int. J. Mol. Sci.* **2019**, *20*, 2051. [CrossRef]
122. Bingham, S.A.; Day, N.E.; Luben, R.; Ferrari, P.; Slimani, N.; Norat, T.; Clavel-Chapelon, F.; Kesse, E.; Nieters, A.; Boeing, H.; et al. Dietary fibre in food and protection against colorectal cancer in the European Prospective Investigation into Cancer and Nutrition (EPIC): An observational study. *Lancet* **2003**, *361*, 1496–1501. [CrossRef]
123. Bingham, S.A.; Day, N.E.; Luben, R. DEPARTMENT OF ERROR. *Lancet* **2003**, *362*, 1000. [CrossRef]
124. Kranz, S.; Dodd, K.W.; Juan, W.Y.; Johnson, L.A.K.; Jahns, L. Whole grains contribute only a small proportion of dietary fiber to the U.S. diet. *Nutrients* **2017**, *9*, 153. [CrossRef] [PubMed]
125. Janicke, B.; Hegardt, C.; Krogh, M.; Onning, G.; Åkesson, B.; Cirenajwis, H.M.; Oredsson, S.M. The antiproliferative effect of dietary fiber phenolic compounds ferulic acid and p-coumaric acid on the cell cycle of Caco-2 cells. *Nutr. Cancer* **2011**, *63*, 611–622. [CrossRef] [PubMed]
126. Hammond, W.A.; Swaika, A.; Mody, K. Pharmacologic resistance in colorectal cancer: A review. *Ther. Adv. Med. Oncol.* **2016**, *8*, 57–84. [CrossRef] [PubMed]
127. Pabla, B.; Bissonnette, M.; Konda, V.J. Colon cancer and the epidermal growth factor receptor: Current treatment paradigms, the importance of diet, and the role of chemoprevention. *World J. Clin. Oncol.* **2015**, *6*, 133–141. [CrossRef]
128. Roy, N.; Narayanankutty, A.; Nazeem, P.; Valsalan, R.; Babu, T.; Mathew, D. Plant phenolics ferulic acid and P-coumaric acid inhibit colorectal cancer cell proliferation through EGFR down- regulation. *Asian Pacific J. Cancer Prev.* **2016**, *17*, 4017–4021.
129. Qu, H.; Madl, R.L.; Takemoto, D.J.; Baybutt, R.C.; Wang, W. Lignans Are Involved in the Antitumor Activity of Wheat Bran in Colon Cancer SW480 Cells. *J. Nutr.* **2005**, *135*, 598–602. [CrossRef]
130. Ayella, A.; Lim, S.; Jiang, Y.; Iwamoto, T.; Lin, D.; Tomich, J.; Wang, W. Cytostatic inhibition of cancer cell growth by lignan secoisolariciresinol diglucoside. *Nutr. Res.* **2010**, *30*, 762–769. [CrossRef]
131. Shah, N.R.; Patel, B.M. Secoisolariciresinol diglucoside rich extract of L. usitatissimum prevents diabetic colon cancer through inhibition of CDK4. *Biomed. Pharmacother.* **2016**, *83*, 733–739. [CrossRef]
132. Yang, L.; Allred, K.F.; Dykes, L.; Allred, C.D.; Awika, J.M. Enhanced action of apigenin and naringenin combination on estrogen receptor activation in non-malignant colonocytes: Implications on sorghum-derived phytoestrogens. *Food Funct.* **2015**, *6*, 749–755. [CrossRef]
133. Turktekin, M.; Konac, E.; Onen, H.I.; Alp, E.; Yilmaz, A.; Menevse, S. Evaluation of the effects of the flavonoid apigenin on apoptotic pathway gene expression on the colon cancer cell line (HT29). *J. Med. Food* **2011**, *14*, 1107–1117. [CrossRef]
134. Takagaki, N.; Sowa, Y.; Oki, T.; Nakanishi, R.; Yogosawa, S.; Sakai, T. Apigenin induces cell cycle arrest and p21/WAF1 expression in a p53-independent pathway. *Int. J. Oncol.* **2005**, *26*, 185–189. [CrossRef] [PubMed]
135. Wang, W.; VanAlstyne, P.C.; Irons, K.A.; Chen, S.; Stewart, J.W.; Birt, D.F. Individual and interactive effects of apigenin analogs on G2/M cell-cycle arrest in human colon carcinoma cell lines. *Nutr. Cancer* **2004**, *48*, 106–114. [CrossRef] [PubMed]
136. Yao, Y.; Rao, C.; Zheng, G.; Wang, S. Luteolin suppresses colorectal cancer cell metastasis via regulation of the miR-384/pleiotrophin axis. *Oncol. Rep.* **2019**, *42*, 131–141. [CrossRef] [PubMed]
137. Hadi, L.A.; Di Vito, C.; Marfia, G.; Ferraretto, A.; Tringali, C.; Viani, P.; Riboni, L. Sphingosine kinase 2 and ceramide transport as key targets of the natural flavonoid luteolin to induce apoptosis in colon cancer cells. *PLoS ONE* **2015**, *10*, e0143384. [CrossRef]
138. Kang, K.A.; Piao, M.J.; Ryu, Y.S.; Hyun, Y.J.; Park, J.E.; Shilnikova, K.; Zhen, A.X.; Kang, H.K.; Koh, Y.S.; Jeong, Y.J.; et al. Luteolin induces apoptotic cell death via antioxidant activity in human colon cancer cells. *Int. J. Oncol.* **2017**, *51*, 1169–1178. [CrossRef] [PubMed]
139. Celesia, A.; Morana, O.; Fiore, T.; Pellerito, C.; D'Anneo, A.; Lauricella, M.; Carlisi, D.; De Blasio, A.; Calvaruso, G.; Giuliano, M.; et al. ROS-Dependent ER Stress and Autophagy Mediate the Anti-Tumor Effects of Tributyltin (IV) Ferulate in Colon Cancer Cells. *Int. J. Mol. Sci.* **2020**, *21*, 8135. [CrossRef] [PubMed]
140. Hassani, A.; Azarian, M.M.S.; Ibrahim, W.N.; Hussain, S.A. Preparation, characterization and therapeutic properties of gum arabic-stabilized gallic acid nanoparticles. *Sci. Rep.* **2020**, *10*, 1–18. [CrossRef]

141. Colquhoun, A.; Arnold, M.; Ferlay, J.; Goodman, K.J.; Forman, D.; Soerjomataram, I. Global patterns of cardia and non-cardia gastric cancer incidence in 2012. *Gut* **2015**, *64*, 1881–1888. [CrossRef]
142. Cook, M.B. Editorial: Non-Acid reflux: The missing link between gastric atrophy and esophageal squamous cell carcinoma. *Am. J. Gastroenterol.* **2011**, *106*, 1930–1932. [CrossRef]
143. Freedman, N.D.; Abnet, C.C.; Leitzmann, M.F.; Mouw, T.; Subar, A.F.; Hollenbeck, A.R.; Schatzkin, A. A prospective study of tobacco, alcohol, and the risk of esophageal and gastric cancer subtypes. *Am. J. Epidemiol.* **2007**, *165*, 1424–1433. [CrossRef]
144. Hoyo, C.; Cook, M.B.; Kamangar, F.; Freedman, N.D.; Whiteman, D.C.; Bernstein, L.; Brown, L.M.; Risch, H.A.; Ye, W.; Sharp, L.; et al. Body mass index in relation to oesophageal and oesophagogastric junction adenocarcinomas: A pooled analysis from the international BEACON consortium. *Int. J. Epidemiol.* **2012**, *41*, 1706–1718. [CrossRef] [PubMed]
145. Whiteman, D.C.; Sadeghi, S.; Pandeya, N.; Smithers, B.M.; Gotley, D.C.; Bain, C.J.; Webb, P.M.; Green, A.C. Combined effects of obesity, acid reflux and smoking on the risk of adenocarcinomas of the oesophagus. *Gut* **2008**, *57*, 173–180. [CrossRef] [PubMed]
146. Ye, W.; Chow, W.H.; Lagergren, J.; Yin, L.; Nyrén, O. Risk of adenocarcinomas of the esophagus and gastric cardia in patients with gastroesophageal reflux diseases and after antireflux surgery. *Gastroenterology* **2001**, *121*, 1286–1293. [CrossRef]
147. Binh, T.T.; Tuan, V.P.; Dung, H.D.Q.; Tung, P.H.; Tri, T.D.; Thuan, N.P.M.; Van Khien, V.; Hoan, P.Q.; Suzuki, R.; Uchida, T.; et al. Advanced non-cardia gastric cancer and Helicobacter pylori infection in Vietnam. *Gut Pathog.* **2017**, *9*, 46. [CrossRef] [PubMed]
148. Kamangar, F.; Dawsey, S.M.; Blaser, M.J.; Perez-Perez, G.I.; Pietinen, P.; Newschaffer, C.J.; Abnet, C.C.; Albanes, D.; Virtamo, J.; Taylor, P.R. Opposing risks of gastric cardia and noncardia gastric adenocarcinomas associated with Helicobacter pylori seropositivity. *J. Natl. Cancer Inst.* **2006**, *98*, 1445–1452. [CrossRef]
149. Wang, T.; Cai, H.; Sasazuki, S.; Tsugane, S.; Zheng, W.; Cho, E.R.; Jee, S.H.; Michel, A.; Pawlita, M.; Xiang, Y.B.; et al. Fruit and vegetable consumption, Helicobacter pylori antibodies, and gastric cancer risk: A pooled analysis of prospective studies in China, Japan, and Korea. *Int. J. Cancer* **2017**, *140*, 591–599. [CrossRef]
150. Chen, M.J.; Wu, D.C.; Lin, J.M.; Wu, M.T.; Sung, F.C. Etiologic factors of gastric cardiac adenocarcinoma among men in Taiwan. *World J. Gastroenterol.* **2009**, *15*, 5472–5480. [CrossRef]
151. Yamaji, Y.; Watabe, H.; Yoshida, H.; Kawabe, T.; Wada, R.; Mitsushima, T.; Omata, M. High-risk population for gastric cancer development based on serum pepsinogen status and lifestyle factors. *Helicobacter* **2009**, *14*, 81–86. [CrossRef]
152. Flores-Luna, L.; Bravo, M.M.; Kasamatsu, E.; Lazcano Ponce, E.C.; Martinez, T.; Torres, J.; Camorlinga-Ponce, M.; Kato, I. Risk factors for gastric precancerous and cancers lesions in Latin American counties with difference gastric cancer risk. *Cancer Epidemiol.* **2020**, *64*, 101630. [CrossRef]
153. Gaddy, J.A.; Radin, J.N.; Loh, J.T.; Zhang, F.; Kay Washington, M.; Peek, R.M.; Scott Algood, H.M.; Cover, T.L. High dietary salt intake exacerbates Helicobacter pylori-induced gastric carcinogenesis. *Infect. Immun.* **2013**, *81*, 2258–2267. [CrossRef]
154. Zaidi, S.F.; Ahmed, K.; Saeed, S.A.; Khan, U.; Sugiyama, T. Can Diet Modulate Helicobacter pylori-associated Gastric Pathogenesis? An Evidence-Based Analysis. *Nutr. Cancer* **2017**, *69*, 979–989. [CrossRef] [PubMed]
155. Rawla, P.; Barsouk, A. Epidemiology of gastric cancer: Global trends, risk factors and prevention. *Prz. Gastroenterol.* **2019**, *14*, 26. [CrossRef]
156. McCullough, M.L.; Robertson, A.S.; Jacobs, E.J.; Chao, A.; Calle, E.E.; Thun, M.J. A prospective study of diet and stomach cancer mortality in United States men and women. *Cancer Epidemiol. Biomark. Prev.* **2001**, *10*, 1201–1205.
157. Li, K.; Zhang, B. The association of dietary β-carotene and vitamin A intake on the risk of esophageal cancer: A meta-analysis. *Rev. Esp. Enferm. Dig.* **2020**, *112*. [CrossRef]
158. Xu, Y.; Yang, J.; Du, L.; Li, K.; Zhou, Y. Association of whole grain, refined grain, and cereal consumption with gastric cancer risk: A meta-analysis of observational studies. *Food Sci. Nutr.* **2019**, *7*, 256–265. [CrossRef] [PubMed]
159. Wang, T.; Zhan, R.; Lu, J.; Zhong, L.; Peng, X.; Wang, M.; Tang, S. Grain consumption and risk of gastric cancer: A meta-analysis. *Int. J. Food Sci. Nutr.* **2020**, *71*, 164–175. [CrossRef] [PubMed]
160. Boeing, H.; Frentzel-Beyme, R.; Berger, M.; Berndt, V.; Göres, W.; Körner, M.; Lohmeier, R.; Menarcher, A.; Männl, H.F.K.; Meinhardt, M.; et al. Case-control study on stomach cancer in Germany. *Int. J. Cancer* **1991**, *47*, 858–864. [CrossRef] [PubMed]
161. Mendez, M.A.; Pera, G.; Agudo, A.; Bas Bueno-de-Mesquita, H.; Palli, D.; Boeing, H.; Carneiro, F.; Berrino, F.; Sacerdote, C.; Tumino, R.; et al. Cereal fiber intake may reduce risk of gastric adenocarcinomas: The EPIC-EURGAST study. *Int. J. Cancer* **2007**, *121*, 1618–1623. [CrossRef]
162. Bravi, F.; Scotti, L.; Bosetti, C.; Bertuccio, P.; Negri, E.; La Vecchia, C. Dietary fiber and stomach cancer risk: A case–control study from Italy. *Cancer Causes Control* **2009**, *20*, 847–853. [CrossRef]
163. Gaesser, G.A. Whole Grains, Refined Grains, and Cancer Risk: A Systematic Review of Meta-Analyses of Observational Studies. *Nutrients* **2020**, *12*, 3756. [CrossRef]
164. Díaz-Gómez, R.; López-Solís, R.; Obreque-Slier, E.; Toledo-Araya, H. Comparative antibacterial effect of gallic acid and catechin against Helicobacter pylori. *LWT Food Sci. Technol.* **2013**, *54*, 331–335. [CrossRef]
165. Ho, H.H.; Chang, C.-S.; Ho, W.C.; Liao, S.Y.; Lin, W.L.; Wang, C.J. Gallic acid inhibits gastric cancer cells metastasis and invasive growth via increased expression of RhoB, downregulation of AKT/small GTPase signals and inhibition of NF-κB activity. *Toxicol. Appl. Pharmacol.* **2013**, *266*, 76–85. [CrossRef]
166. Chang, H.T.; Chen, I.L.; Chou, C.T.; Liang, W.Z.; Kuo, D.H.; Shieh, P.; Jan, C.R. Effect of caffeic acid on Ca2+ homeostasis and apoptosis in SCM1 human gastric cancer cells. *Arch. Toxicol.* **2013**, *87*, 2141–2150. [CrossRef]

167. Chen, J.; Chen, J.; Li, Z.; Liu, C.; Yin, L. The apoptotic effect of apigenin on human gastric carcinoma cells through mitochondrial signal pathway. *Tumor Biol.* **2014**, *35*, 7719–7726. [CrossRef] [PubMed]
168. Wu, K.; Yuan, L.H.; Xia, W. Inhibitory effects of apigenin on the growth of gastric carcinoma SGC-7901 cells. *World J. Gastroenterol.* **2005**, *11*, 4461–4464. [CrossRef] [PubMed]
169. Zhou, Y.; Ding, B.Z.; Lin, Y.P.; Wang, H.B. MiR-34a, as a suppressor, enhance the susceptibility of gastric cancer cell to luteolin by directly targeting HK1. *Gene* **2018**, *644*, 56–65. [CrossRef]
170. Hu, L.; Fan, Z.Y.; Wang, H.X.; Zhu, Z.L.; Cao, S.; Wu, X.Y.; Li, J.F.; Su, L.P.; Li, C.; Zhu, Z.G.; et al. Luteolin suppresses gastric cancer progression by reversing epithelial-mesenchymal transition via suppression of the Notch signaling pathway. *J. Transl. Med.* **2017**, *15*, 1–11. [CrossRef]
171. Jang, M.G.; Ko, H.C.; Kim, S.J. Effects of p-coumaric acid on microRNA expression profiles in SNU-16 human gastric cancer cells. *Genes Genom.* **2020**, *42*, 817–825. [CrossRef]
172. Wu, H.; Huang, M.; Liu, Y.; Shu, Y.; Liu, P. Luteolin Induces Apoptosis by Up-regulating miR-34a in Human Gastric Cancer Cells. *Technol. Cancer Res. Treat.* **2015**, *14*, 747–755. [CrossRef]
173. Pennathur, A.; Gibson, M.K.; Jobe, B.A.; Luketich, J.D. Oesophageal carcinoma. *Lancet* **2013**, *381*, 400–412. [CrossRef]
174. Chevallay, M.; Bollschweiler, E.; Chandramohan, S.M.; Schmidt, T.; Koch, O.; Demanzoni, G.; Mönig, S.; Allum, W. Cancer of the gastroesophageal junction: A diagnosis, classification, and management review. *Ann. N. Y. Acad. Sci.* **2018**, *1434*, 132–138. [CrossRef] [PubMed]
175. Malhotra, G.K.; Yanala, U.; Ravipati, A.; Follet, M.; Vijayakumar, M.; Are, C. Global trends in esophageal cancer. *J. Surg. Oncol.* **2017**, *115*, 564–579. [CrossRef] [PubMed]
176. Arnal, M.J.D.; Arenas, Á.F.; Arbeloa, Á.L. Esophageal cancer: Risk factors, screening and endoscopic treatment in Western and Eastern countries. *World J. Gastroenterol.* **2015**, *21*, 7933. [CrossRef] [PubMed]
177. Sun, L.-P.; Yan, L.-B.; Liu, Z.-Z.; Zhao, W.-J.; Zhang, C.-X.; Chen, Y.-M.; Lao, X.Q.; Liu, X. Dietary factors and risk of mortality among patients with esophageal cancer: A systematic review. *BMC Cancer* **2020**, *20*, 1–13. [CrossRef]
178. Xuan, F.; Li, W.; Guo, X.; Liu, C. Dietary carbohydrate intake and the risk of esophageal cancer: A meta-analysis. *Biosci. Rep.* **2020**, *40*. [CrossRef]
179. Levi, F.; Pasche, C.; Lucchini, F.; Bosetti, C.; Franceschi, S.; Monnier, P.; Vecchia La, C. Food groups and oesophageal cancer risk in Vaud, Switzerland. *Eur. J. Cancer Prev.* **2000**, *9*, 257–264. [CrossRef]
180. Kubo, A.; Block, G.; Quesenberry, C.P.; Buffler, P.; Corley, D.A. Effects of dietary fiber, fats, and meat intakes on the risk of barrett's esophagus. *Nutr. Cancer* **2009**, *61*, 607–616. [CrossRef]
181. Skeie, G.; Braaten, T.; Olsen, A.; Kyrø, C.; Tjønneland, A.; Landberg, R.; Nilsson, L.M.; Wennberg, M.; Overvad, K.; Åsli, L.A.; et al. Intake of whole grains and incidence of oesophageal cancer in the HELGA Cohort. *Eur. J. Epidemiol.* **2016**, *31*, 405–414. [CrossRef]
182. Slavin, J.L. Mechanisms for the Impact of Whole Grain Foods on Cancer Risk. *J. Am. Coll. Nutr.* **2000**, *19*, 300S–307S. [CrossRef]
183. Frølich, W.; Åman, P.; Tetens, I. Whole grain foods and health—A Scandinavian perspective. *Food Nutr. Res.* **2013**, *57*, 18503. [CrossRef]
184. Sun, L.; Zhang, Z.; Xu, J.; Xu, G.; Liu, X. Dietary fiber intake reduces risk for Barrett's esophagus and esophageal cancer. *Crit. Rev. Food Sci. Nutr.* **2017**, *57*, 2749–2757. [CrossRef] [PubMed]
185. Mayne, S.T.; Navarro, S.A. Diet, Obesity and Reflux in the Etiology of Adenocarcinomas of the Esophagus and Gastric Cardia in Humans. *J. Nutr.* **2002**, *132*, 3467S–3470S. [CrossRef] [PubMed]
186. Mulholland, H.G.; Cantwell, M.M.; Anderson, L.A.; Johnston, B.T.; Watson, R.G.P.; Murphy, S.J.; Ferguson, H.R.; McGuigan, J.; Reynolds, J.V.; Comber, H.; et al. Glycemic index, carbohydrate and fiber intakes and risk of reflux esophagitis, Barrett's esophagus, and esophageal adenocarcinoma. *Cancer Causes Control* **2009**, *20*, 279–288. [CrossRef]
187. McFadden, D.; Riggs, D.; Jackson, B.; Cunningham, C. Corn-derived carbohydrate inositol hexaphosphate inhibits Barrett's adenocarcinoma growth by pro-apoptotic mechanisms. *Oncol. Rep.* **2008**, *19*, 563–566. [CrossRef]
188. Nobel, Y.R.; Snider, E.J.; Compres, G.; Freedberg, D.E.; Khiabanian, H.; Lightdale, C.J.; Toussaint, N.C.; Abrams, J.A. Increasing Dietary Fiber Intake Is Associated with a Distinct Esophageal Microbiome. *Clin. Transl. Gastroenterol.* **2018**, *9*, e199. [CrossRef] [PubMed]
189. Kang, J.-W.; Lee, S.-M. Protective Effects of Chlorogenic Acid against Experimental Reflux Esophagitis in Rats. *Biomol. Ther.* **2014**, *22*, 420–425. [CrossRef] [PubMed]
190. Zhan, Y.; Li, R.; Feng, C.; Li, X.; Huang, S.; Wang, L.; Liu, Z.; Jiang, J.; Han, Y. Chlorogenic acid inhibits esophageal squamous cell carcinoma growth in vitro and in vivo by downregulating the expression of BMI1 and SOX2. *Biomed. Pharmacother.* **2020**, *121*, 109602. [CrossRef]
191. Faried, A.; Kurnia, D.; Faried, L.S.; Usman, N.; Miyazaki, T.; Kato, H.; Kuwano, H. Anticancer effects of gallic acid isolated from Indonesian herbal medicine, Phaleria macrocarpa (Scheff.) Boerl, on human cancer cell lines. *Int. J. Oncol.* **2007**, *30*, 605–613. [CrossRef]
192. Qiu, J.G.; Wang, L.; Liu, W.J.; Wang, J.F.; Zhao, E.J.; Zhou, F.M.; Ji, X.B.; Wang, L.H.; Xia, Z.K.; Wang, W.; et al. Apigenin inhibits IL-6 transcription and suppresses esophageal carcinogenesis. *Front. Pharmacol.* **2019**, *10*, 1002. [CrossRef]
193. Shioga, T.; Matsushima, S.; Yamada, E.; Uchiyama, T.; Noto, H.; Suzuki, D.; Nonaka, T.; Miyazawa, S.; Komatsu, T.; Yamamoto, Y.; et al. Esophageal carcinosarcoma that was diagnosed as a granulocyte-colony stimulating factor and interleukin-6-producing tumor with a tumor fever. *Intern. Med.* **2018**, *57*, 2819–2825. [CrossRef]

194. SACN. *Carbohydrates and Health*; The Stationery Office: London, UK, 2015; ISBN 9780117082847.
195. CREA. *Linee Guida per una sana Alimentazione*; CREA: Rome, Italy, 2018; pp. 1689–1699.
196. Joint WHO/FAO Expert Consultation. Diet, nutrition and the prevention of chronic diseases. In *World Health Organization Technical Report Series*; WHO: Geneva, Switzerland, 2003.
197. PHE. *Government Dietary Recommendations*; Public Health England: London, UK, 2016.
198. GBD 2017 Risk Factor Collaborators. Global, regional, and national comparative risk assessment of 84 behavioural, environmental and occupational, and metabolic risks or clusters of risks for 195 countries and territories, 1990–2017: A systematic analysis for the Global Burden of Disease Study 2017. *Lancet* **2018**, *392*, 1923–1994. [CrossRef]

nutrients

MDPI

Article

Preliminary Study on Pasta Samples Characterized in Antioxidant Compounds and Their Biological Activity on Kidney Cells

Federico Di Marco [1], Francesco Trevisani [1], Pamela Vignolini [2], Silvia Urciuoli [2], Andrea Salonia [1,3], Francesco Montorsi [1,3], Annalisa Romani [2], Riccardo Vago [1,3,*] and Arianna Bettiga [1,*]

[1] Division of Experimental Oncology, Urological Research Institute (URI), IRCCS San Raffaele Scientific Institute, 20132 Milan, Italy; federico.dimarco91@gmail.com (F.D.M.); trevisani.francesco@hsr.it (F.T.); salonia.andrea@hsr.it (A.S.); montorsi.francesco@hsr.it (F.M.)

[2] PHYTOLAB (Pharmaceutical, Cosmetic, Food Supplement, Technology and Analysis), DiSIA, University of Florence, 50019 Sesto Fiorentino, Italy; pamela.vignolini@unifi.it (P.V.); silvia.urciuoli@gmail.com (S.U.); annalisa.romani@unifi.it (A.R.)

[3] Department of Urology, Università Vita-Salute San Raffaele, 20132 Milan, Italy

* Correspondence: vago.riccardo@hsr.it (R.V.); bettiga.arianna@hsr.it (A.B.)

Citation: Di Marco, F.; Trevisani, F.; Vignolini, P.; Urciuoli, S.; Salonia, A.; Montorsi, F.; Romani, A.; Vago, R.; Bettiga, A. Preliminary Study on Pasta Samples Characterized in Antioxidant Compounds and Their Biological Activity on Kidney Cells. *Nutrients* **2021**, *13*, 1131. https://doi.org/10.3390/nu13041131

Academic Editor: Margarida Castell Escuer

Received: 9 March 2021
Accepted: 26 March 2021
Published: 30 March 2021

Publisher's Note: MDPI stays neutral with regard to jurisdictional claims in published maps and institutional affiliations.

Abstract: Pasta is one of the basic foods of the Mediterranean diet and for this reason it was chosen for this study to evaluate its antioxidant properties. Three types of pasta were selected: buckwheat, rye and egg pasta. Qualitative–quantitative characterization analyses were carried out by HPLC-DAD to identify antioxidant compounds. The data showed the presence of carotenoids such as lutein and polyphenols such as indoleacetic acid, (carotenoids from 0.08 to 0.16 mg/100 g, polyphenols from 3.7 to 7.4 mg/100 g). To assess the effect of the detected metabolites, in vitro experimentation was carried out on kidney cells models: HEK-293 and MDCK. Standards of β-carotene, indoleacetic acid and caffeic acid, hydroalcoholic and carotenoid-enriched extracts from samples of pasta were tested in presence of antioxidant agent to determine viability variations. β-carotene and indoleacetic acid standards exerted a protective effect on HEK-293 cells while no effect was detected on MDCK. The concentrations tested are likely in the range of those reached in body after the consumption of a standard pasta meal. Carotenoid-enriched extracts and hydroalcoholic extracts showed different effects, observing rescues for rye pasta hydroalcoholic extract and buckwheat pasta carotenoid-enriched extract, while egg pasta showed milder dose depending effects assuming pro-oxidant behavior at high concentrations. The preliminary results suggest behaviors to be traced back to the whole phytocomplexes respect to single molecules and need further investigations.

Keywords: pasta; Mediterranean diet; polyphenols; carotenoids; antioxidant compounds; kidney health; HPLC/DAD analyses

1. Introduction

Despite that there is no unique definition of the Mediterranean diet—due to the variations and adaptations in culinary traditions not only around the Mediterranean area but also around the world [1]—the Mediterranean diet may be thought of as having several components that meet important criteria for a healthy diet: low content of saturated fatty acids, high content of composite carbohydrates, dietary fiber and antioxidant molecules, and abundance in vegetables.

It is difficult to establish which foods, or active ingredients therein contained, of the Mediterranean diet are most responsible for the health benefits; accumulating data suggest that its main benefits lie in the combination of the complex and wide variety of different nutrients interacting synergistically and additively [2].

In recent years, the growing literature (observational studies and randomized control trials) has demonstrated the health benefits associated with the Mediterranean diet, reducing risk of developing multiple chronic diseases as cardiovascular diseases [3–5], diabetes [6–10], cancer [11,12], obesity [13–15] or cognitive health [16–19] and increasing life expectancy [20].

Improvements in blood pressure [21], lipid profile [22,23], insulin resistance [24], and protection against oxidative stress, inflammation, platelet aggregation and endothelial dysfunction [25–28] seem to be the mechanisms of action responsible for the beneficial effects on the general health. All these biological effects promoted by the adherence to the Mediterranean lifestyle could be relevant not only for the above-mentioned chronic diseases, but also for the prevention of renal decay in general population. Chronic kidney disease (CKD) represents a global public health problem, affecting over 750 million persons worldwide and acts often silent and unhindered thanks to its asymptomatic clinical presentation. Different lines of evidence have underlined the strong relationship between Mediterranean diet and renal preservation [29].

Considering all the above-mentioned effects as risk factors for the decline of the renal function, the kidney can be considered one of the primary targets.

A recent meta-analysis of cohort studies has shown that plant-based diets such as the Mediterranean diet or dietary approach to stop hypertension (DASH) diet was associated with a lower risk of incident CKD and albuminuria in the general population [30]. One explanation could be that fruit and vegetables contain bioactive compounds protecting against inflammation and endothelial dysfunction that promote dynamic changes of filtration fraction, resulting in a progressive reduction of the glomerular filtration rate, extracellular fluid volume expansion, abnormal ion balance, and renal hypoxia, ultimately leading to loss kidney function [31–33]. Moreover, the increase of vegetable, fruit, cereal and legumes consumption may have led to a decrease in animal protein consumption positively influencing acid-base balance and glomerular hemodynamics of the kidney, protecting it to glomerular sclerosis and loss of function [34–37].

Despite this, there are several studies with contrasting data for the impact of a high protein diet on renal function decline in the general population [38–41] growing evidence suggest that the protein source plays an important role to preservation of renal function, and that a shift from animal to plant source of protein might be beneficial [42,43].

Pasta is one of the most consumed foods in the world and is one of the staple foods of the Mediterranean diet. Traditionally, Italy is the main producer and leader of the pasta consumption, even if pasta is consumed worldwide for low cost, palatability, and the longer shelf life than other bakery products [44]. Pasta is a good source of carbohydrates and energy (100 g of cooked pasta contains about 31 g of carbohydrates and about 158 kcal) but it is typically low in lipids, proteins, phosphorus and potassium that are nutrient components restricted in the most renal diets for kidney disease (available online at https://fdc.nal.usda.gov, accessed on 1 December 2020). Naturally enriched wheat pastas, which represent the majority of commercially available pastas, also offer good levels of thiamin, riboflavin, niacin, folate, iron, and selenium, as well as polyphenols. Furthermore, pasta is always consumed in combination with other food items that could be considered as a stronger source of polyphenols, such as olive oil and vegetables. Pasta is usually produced with durum wheat semolina, due to its excellent rheological properties, the superior color of the pasta, the quality of cooking and the acceptance by the consumer, but lately pasta can also be produced starting from other cereals such as rice, buckwheat, rye, spelled.

In recent years, consumers have become more aware that a balanced diet, such as the Mediterranean one, can positively affect health, which is why the food industry tries to meet these demands by producing quality and functional foods and food ingredients.

Despite fortified foods have been produce for healthy purpose, people could be induced to assume them instead of a balanced diet. This is a so important topic to prompt the FDA to publish an ad hoc document entitled "FDA's fortification policy" to give an answer on one side to the food industry and on the other to the academic world [44].

Balanced, adequate and varied diet is an important step towards a happy and healthy lifestyle without the risk of overloading with some added component contained in fortified food. Precisely for this reason non-traditional cereals and pseudo-cereals have been rediscovered which have paved the way for their use as functional food ingredients as they possess some nutritional and functional qualities that are absent or lacking in traditional cereals. Buckwheat and rye are a rich source of phytochemicals such as polyphenols, compounds that are strongly correlated with antioxidant activities [45,46].

Launched in March 2005, the MOLI-SANI Project involved about 25,000 citizens residing in Molise (Italy), to evaluate environmental and genetic factors underlying cardiovascular diseases, tumors and neurodegenerative diseases. According to a study from the MOLI-SANI project, pasta intake is associated with a lower BMI and lower levels of central obesity. Many studies have confirmed that high calorie diets lead to weight gain and not a diet that includes carbohydrates. In fact, if the portion of pasta is correct and the sauce is not too caloric, a plate of pasta can have a caloric content that respects the daily energy requirement [47]. Pasta is held in high regard as its characteristics have been associated not only with body weight control, but also with several positive health properties [48]. A randomized, controlled study conducted on healthy subjects showed that rye-based products improve glycemic regulation, increase intestinal hormones involved in the regulation of appetite and metabolism. Rye reduces postprandial appetite levels by reducing the desire to eat at subsequent meals, increasing satiety and reducing hunger by increasing the satiety hormones GLP-1 (glucagon-like peptide-1) and peptide YY (PYY). A lower postprandial glucose response was observed in a group of people consuming rye-based foods compared to a group of subjects consuming refined wheat. This happens because rye is a low glycemic index cereal that can also be used in diabetic patients [49]. Rye can improve the plasma lipid profile, in fact the intake of whole rye is inversely associated with the concentration of LDL cholesterol, the LDL/HDL ratio and the concentration of triglycerides. This effect is due to the presence of fibers and β-glucans [50].

Pasta is well known as a source of carbohydrates, but it can also contain minor compounds, such as carotenoids [51], especially lutein, and other antioxidant species such as polyphenols. These bioactive molecules can be enriched in different types of pasta, depending on the raw materials. For instance, the carotenoids content of durum wheat is higher than that of bread wheat [52] and the level of antioxidants can be increased by incorporating bran fraction and entire kernel of durum wheat to pasta products [53]. Buckwheat is a substantial source of phenolic compounds, vitamins and essential amino acids [54], while the addition of egg in the preparation of the pasta conveys carotenoids and lutein. Carotenoids participate in prevention activity, especially in inflammation syndrome and their role is due to their antioxidant properties [55] against free radicals and singlet oxygen [56]. Pasta contains different types of antioxidant molecules and for the frequent consumers, large consumption can be considered an important source of such healthy compounds.

Here we show that different types of pasta namely buckwheat, rye and egg pasta, contain various amounts of antioxidant compounds, which have a role to improve renal cell viability at a steady state or upon the induction of oxidative stress.

2. Materials and Methods

2.1. Samples

Three samples of cooked pasta were analyzed and tested: one sample of wholemeal buckwheat pasta (1), one sample of wholemeal rye pasta (2) and one sample of egg pasta of stone-ground durum wheat flour (3).

2.2. Pasta Samples Preparation

Totals of 50 g of whole meal buckwheat pasta or whole meal rye pasta were cooked in 500 mL of water for 7 min; 50 g of egg pasta of stone-ground durum wheat flour were cooked in 500 mL of water for 1 min, as indicated by the manufacturer.

To obtain extracts enriched in carotenoids, 10 g of cooked pasta were dissolved in 100 mL acetone, cold sonicated for 30 min. The sample was centrifuged for 5 min at 5000 rpm, the supernatant has been dry evaporated with a rotary evaporator and the residue was dissolved in acetone or DMSO depending on the downstream analysis. To obtain extracts enriched in polyphenols, 10 g of each sample of cooked pasta were dissolved in 50 mL of 70:30 EtOH/H_2O at pH 3.2. The samples were shaken for 24 h, centrifuged for 5 min at 1400 rpm. The sample was centrifuged for 5 min at 5000 rpm and the supernatant was collected and analyzed in HPLC-DAD, while an aliquot was dry evaporated, and the residue dissolved in DMSO for cell treatment.

2.3. *HPLC-DA-MS Analysis*

Quali-quantitative analyses of carotenoids and polyphenols were carried out using an HP 1100 liquid chromatography equipped with a DAD detector and managed by an HP 9000 workstation (Agilent Technologies, Palo Alto, CA, USA) and linked to a mass spectrometer with an API/electrospray interface (Agilent Technologies). The mass spectrometer operating conditions were as follows: gas temperature, 350 °C; nitrogen flow rate, 11.0 L/min; nebulizer pressure, 40 psi; quadrupole temperature, 100 °C; and capillary voltage, 4000 V. The mass spectrometer was operated in positive and negative modes at 80–180 eV.

Compounds were separated using a 250 × 4.6 mm i.d, 5 μm LUNA C18 column (Phenomenex, Torrance, CA, USA). UV/Vis spectra were recorded in the 190–600 nm range and the chromatograms were acquired at 250, 280, 330, 350 and 450 nm. The samples were analyzed by gradient elution at a flow rate of 0.8 mL/min. The mobile phase for carotenoids was a multistep linear solvent gradient system (solvent A: acetone, solvent B: H_2O, pH 3.2 by formic acid), starting from 80% acetone up to 100% in 30 min. polyphenols were eluted using the following gradient: from 90% H_2O (adjusted to pH 3.2 by formic acid) to 100% CH_3CN in 40 min. All solvents used were of HPLC grade purity (BDH Laboratory Supplies, Poole, UK).

2.4. *Quantitative Analysis*

Quantification of individual polyphenolic compounds was directly performed by HPLC-DAD using a five-point regression curve ($R^2 \geq 0.998$) in the range of 0–30 μg on the basis of authentic standards. In particular, flavonols were determined at 350 nm using quercetin 3-O-glucoside as reference compound while caffeic acid derivatives were determined at 330 nm using clorogenic acid as reference compound and indoleacetic acid derivative at 280 nm using 3 indoleacetic acid (Sigma-Aldrich, St. Louis, MO, USA). Carotenoids were determined at 450 nm using β beta-carotene as reference compound (Extrasynthese, Lione, Francia). In all cases, actual concentrations of the derivatives were calculated after applying corrections for differences in molecular weight.

2.5. *Total Phenolic Content*

The total phenolic content was determined using the Folin–Ciocalteu method, described by Singleton et al. [57] and slightly modified according to Dewanto et al. [58]. To 125 μL of the suitably diluted sample extract, 0.5 mL of deionized water and 125 μL of the Folin–Ciocalteu reagent were added. The mixture was kept for 6 min and then 1.25 mL of a 7% aqueous Na_2CO_3 solution were added. The final volume was adjusted to 3 mL with water. After 90 min, the absorption was measured at 760 nm against water as a blank. The amount of total phenolics is expressed as gallic acid equivalents (GAE, mg gallic acid/100 g sample) through the calibration curve of gallic acid. The calibration curve ranged from 20 to 500 μg/mL ($R^2 = 0.9969$) [57,58].

2.6. Cell Culture

Human Embryonic Kidney 293 (HEK-293) cells and Madin–Darby canine kidney (MDCK) cells were cultured in DMEM supplemented with GlutaMAX, 10% (*v*/*v*), fetal bovine serum, 100 U/mL penicillin and100 μg/mL streptomycin at 37 °C with 5% CO_2.

2.7. MTT Assay

5000 cells/well MDCK and HEK-293 cells were seeded in 96 well plates and allowed to adhere for 24 h. For viability studies, cells were incubated for 24 h to 72 h with beta-carotene, indole-3-acetic acid, caffeic acid (in DMSO) and pasta extracts alone or in presence of magnesium monoperoxyphtalate (MMPP, Sigma-Aldrich, St. Louis, MO, USA). DMSO in the medium at equal concentrations to those used for the tested compounds was used for untreated samples. The tested concentration for the standards were ranging from 2 pM to 10 μM, while for MMPP from 0.001 to 0.05 mg/mL. Cell viability were measured 3-(4,5-dimethylthiazol-2-yl)-2,5-diphenyl tetrazolium bromide (MTT, Sigma-Aldrich, St. Louis, MO, USA). The plates were incubated at 37 °C for 1 h and then the Formazan produced by the MTT reduction was solubilized in DMSO. Absorbance was determined on a micro plate reader (Mithras LB 940-Berthold) at 570 nm. The percentage of cell viability was calculated using the ratio Abs_{TEST}/Abs_{CTRL}.

2.8. Statistical Analysis

Each experiment was performed in quadruplicate and repeated at least three times. Differences in viability percentages were assessed using paired Student's t test and two-way ANOVA followed by a post-hoc test with Holm's correction for multiple comparison over concentrations metabolites ($p < 0.05$). Results were expressed as mean ± standard error and the analysis were performed by R-Studio environment [59] for R version 3.6.3 [60] using the package "Tidyverse" [61].

3. Results and Discussion

3.1. Determination of the Chemical Quality of Pasta Samples

The nutritional information of a food allows making informed food and dietary choices. Pasta is traditionally prepared from semolina but even with other cereals, all contain starch as a principal constituent followed by protein, fat, vitamins, minerals and bioactive compounds [62]. In Table S1 are reported the nutritional data.

To compare the nutritional values of several types of pasta enriched in bioactive compounds due to the different source of cereals or to the supplementation of specific ingredients, we analyzed buckwheat pasta, rye pasta and egg pasta. In order to obtain extracts enriched in the various subclasses of compounds present in the pasta samples, both hydroalcoholic and acetone extractions were carried out and analyzed. Individual polyphenols were tentatively identified using data from HPLC-DAD-MS analysis by comparison and combination of their retention times and mass spectrometry and UV spectra (see Figures S1–S5 for same example of MS spectra) and comparing results with standards (kaempferol, quercetin, rutin and ferulic acid) and previous bibliographic works. In particular, in hydroalcoholic extracts, flavonoids, caffeic and indoleacetic acid derivatives were identified (Table 1): the indoleacetic acid derivative is the main compound as described before even for semolina. [63]. Considering relative compositions, buckwheat pasta showed higher levels of flavonoids, in particular quercetin derivatives are the main flavonols as previously reported in buckwheat [64,65], and negligible presence of caffeic acid derivatives; rye pasta showed higher levels of caffeic acid derivatives, in particular ferulic acid [66] and low content of flavonoids; egg pasta presented only low concentration of flavonoids (apigenin derivative [67]) and no presence of caffeic acid derivatives.

Table 1. Caffeic derivatives (caffeic der), flavonoids and indoleacetic acid derivative (IAA der) content in cooked pasta samples. Data are the mean of three determinations (standard deviation < 5%). The percentage of single classes of identified compounds is shown in brackets.

	IAA der mg/100 g (%)	Flavonoids mg/100 g (%)	Caffeic der mg/100 g (%)
buckwheat pasta (1)	2.7 ± 0.086 (37)	4.1 ± 0.114 (56)	0.6 ± 0.025 (7)
rye pasta (2)	1.6 ± 0.041 (44)	0.6 ± 0.019 (17)	1.5 ± 0.052 (39)
Egg pasta (3)	4.8 ± 0.110 (85)	0.8 ± 0.021 (15)	-

As an example, the chromatographic profiles of cooked pasta samples, recorded at 350 nm, are presented in Figure 1. The figure reveals the qualitative composition of the samples analyzed.

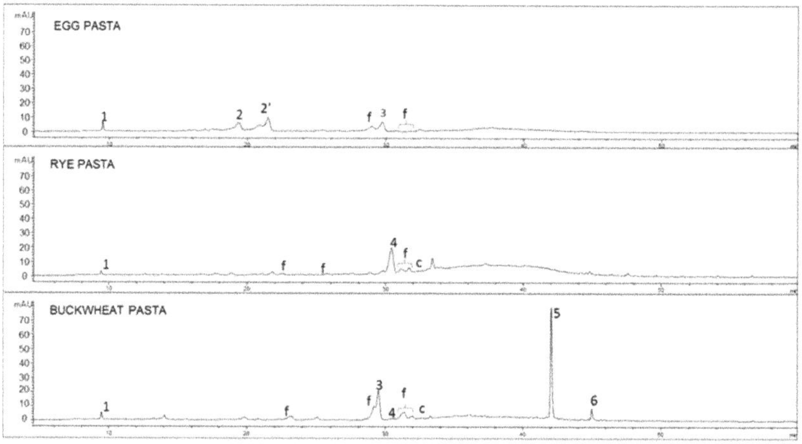

Figure 1. Chromatographic profile (Relative abundance as y axis vs. Retention time as x axis) acquired by HPLC-DAD (350 nm) of the hydroalcholic extracts of cooked buckwheat pasta, rye pasta, egg pasta. Identified compounds: 1 = indoleacetic acid derivative, 2,2′ = apigenin diglicosides, 3 = rutin, 4 = ferulic acid, 5 = quercetin, 6 = kaempferol, f = flavonoids.

Even carotenoids were pointed out in egg pasta and buckwheat pasta, and Figure 2 shows the HPLC-DAD chromatogram (450 nm) of the acetone extract of cooked egg pasta sample.

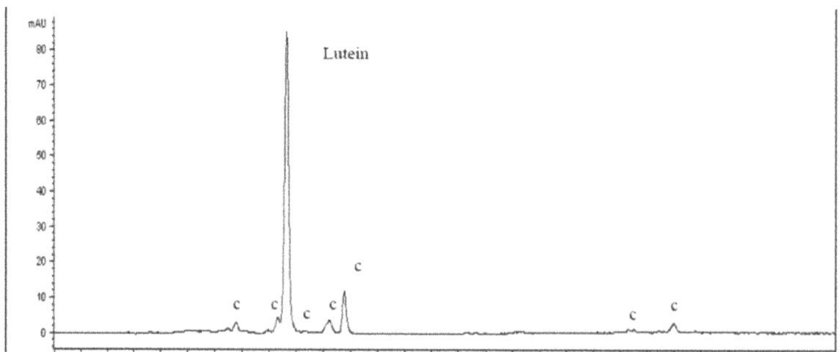

Figure 2. Chromatographic profile (Relative abundance as y axis vs. Retention time as x axis) at 450 nm of a carotenoid acetone fraction of egg cooked pasta extract. Identified compounds: Lutein, c = carotenoid derivatives.

Lutein was the main carotenoid identified in the acetone extracts, with similar relative content with respect to the overall carotenoids species, as shown in Table 2, in line with previous studies [68]. Carotenoids were not detected in rye pasta.

Table 2. Carotenoids and lutein content in cooked pasta samples. Data are the mean of three determinations (standard deviation < 5%).

	Carotenoids mg/100 g	Lutein mg/100 g
buckwheat pasta (1)	0.08 ± 0.002	0.06 ± 0.002
Egg pasta (3)	0.16 ± 0.006	0.11 ± 0.004

The presence of antioxidant secondary metabolites has been evaluated in the pasta samples as well (Table 3). The total antioxidant capacity of biocomponents, likely due to the activity of phenol/polyphenol was assessed and the amount of such molecules was deduced.

Table 3. Total phenolic content (GAE, mg gallic acid/100g, Folin–Ciocalteu method) in cooked pasta samples. Data are the mean of three determinations (standard deviation < 5%).

	GAE, mg Gallic Acid/100 g Pasta
buckwheat pasta (1)	33.23 ± 0.598
rye pasta (2)	21.03 ± 0.273
egg pasta (3)	11.55 ± 0.1848

Buckwheat pasta sample has higher values than other pasta's samples in terms of total phenolic content.

3.2. Effect of Indole-3-Acetic Acid, β-Carotene and Caffeic Acid on Cells Viability

Mediterranean diet has been recognized as a source of anti-oxidant molecules, which protect the organism against oxidative stress and support the healthy status [2]. Therefore, we wondered whether the biomolecules detected in our pasta samples can affect cell viability and at which extent. We used HEK-293 and MDCK healthy kidney cells to determine the activity of indoleacetic acid, β-carotene and caffeic acid standard compounds. The kidney is especially susceptible to oxidative stress and the accumulation of free radicals leads to renal impaired function [69]. Cells were incubated with scalar concentration of the standard compounds and the cell viability was measured at different time points (Figure 3 and Figure S6).

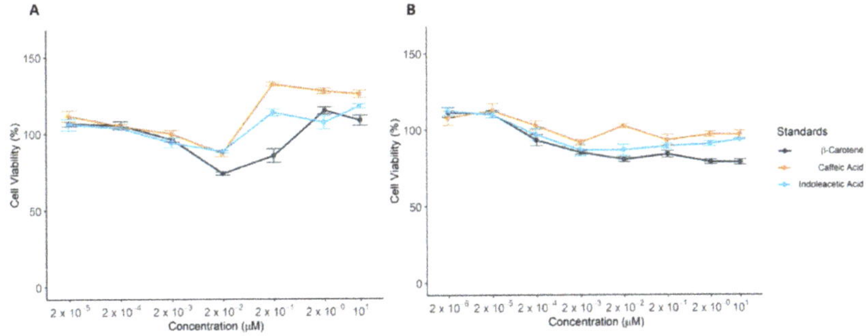

Figure 3. Detection cell viability of HEK-293 (**A**) and MDCK (**B**) incubated with increasing amount of indoleacetic acid, β-carotene and caffeic acid for 24 (**B**) or 72 h (**A**). Reported values correspond to mean of cell viability with standard error over three biological replicates. The percentage of cell viability was calculated using the ratio Abs_{TEST}/Abs_{CTRL}.

They were found not to exert any toxic effect on HEK-293 cells for all the tested concentrations at times of incubation. Only β-carotene reduced cell viability in a u-shape manner on HEK-293 cells and in a dose-dependent manner in MDCK cells (Figure 3). Remarkably, the concentrations likely reached in body after the consumption of a standard pasta meal (usually 80 gr) are in the range of the lowest tested (2×10^{-5}–2×10^{-4} μM, [70,71]), which provided a weak enhancement of the viability on both cell lines. On the other hand, at higher concentrations above 0.2 μM, which can be obtained only pharmacologically, caffeic acid showed a significant improvement of the cell viability, while indoleacetic acid and β-carotene showed comparable effects along the increase of concentration (2–20 μM) in HEK-293 cells along the time.

3.3. Effect of Indole-3-Acetic Acid, β-Carotene and Caffeic Acid on MMPP-Treated Cells

We next wondered if indoleacetic acid, β-carotene and caffeic acid can counteract a stress induced by the oxidizing agent MMPP. First, we tested the MMPP effect by exposing HEK-293 and MDCK cells to serial dilutions (0.005–0.05 mg/mL) to the drug and the cell viability was determined. A dose-dependent decrease following MMPP treatment was noticed and the IC50 was comparable in the two cell lines tested. After incubation with 20–25 μg/mL MMPP the cell viability was significantly decreased to around 50–60% (Figure 4) and the half maximal inhibitory concentration (IC_{50}) was calculated to be 26 μg/mL for HEK-293 and 22 μg/mL for MDCK.

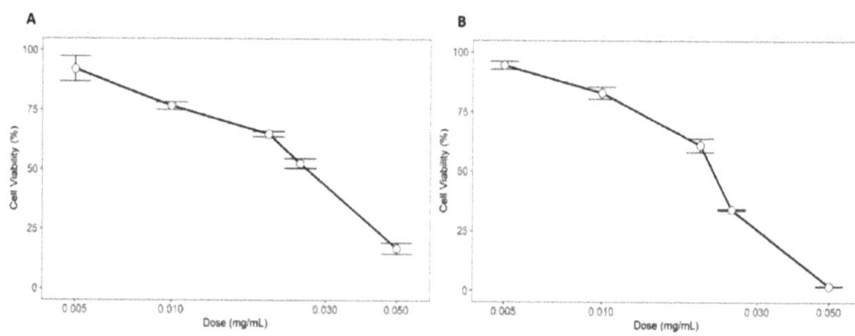

Figure 4. Effect of magnesium monoperoxyphtalate (MMPP) on cell viability of HEK-293 (**A**) and MDCK (**B**) after 24 (**B**) or 72 (**A**) hours incubation.

Then, we tested the protective activity of our anti-oxidant molecules on MMPP-treated cells, by supplementing the cell media with 2×10^{-5} nM indoleacetic, β-carotene or caffeic acid. A slightly positive effect on cell viability of both HEK-293 and MDCK was confirmed by employing the standards as such at the dose likely derived from a pasta portion (Figure 5).

As expected, MMPP reduced the call viability to 60–65%; the treatment with indoleacetic acid or β-carotene counteract the MMPP effect and rescued the cell viability by 30% ($p < 0.001$) and 27% ($p > 0.01$) respectively on HEK-293 cells, while caffeic acid did not exert any significant effect. In the presence of an oxidant agent, beta-carotene and indoleacetic acid showed a protective effect on the cells at a low concentration, whereas caffeic acid did not counteract oxidative stress at the concentration needed to enhance cell viability. Those positive effects are likely due to the antioxidant properties of indoleacetic acid and carotenoids that can significantly counterbalance the MMPP-mediated cytotoxicity at the dose used in this experimental condition. Nevertheless, no significant rescue was observed on MMPP-treated MDCK cells, suggesting that, even of standards alone displayed a significantly positive effect on cell viability, they are not sufficient to tackle the MMPP-induced oxidative stress. Embryonal HEK-293 cells seem to be more prompt to arrange a defense line against stress than adult MDCK, presumably due to additional cellular redox homeostasis systems.

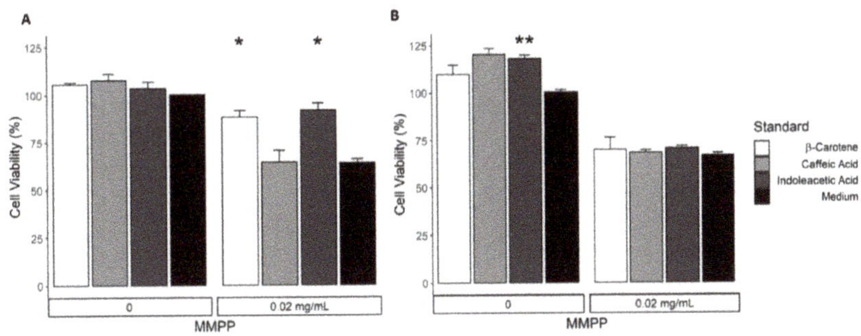

Figure 5. Effect of indoleacetic acid, β-carotene and caffeic acid on the cell viability of HEK-293 (**A**) and MDCK (**B**) in the absence or presence of MMPP after 24 (**B**) or 72 (**A**) hours incubation. Reported values correspond to mean of cell viability with standard error over three biological replicates. The percentage of cell viability was calculated using the ratio Abs_{TEST}/Abs_{CTRL}. * $p < 0.05$; ** $p < 0.01$.

3.4. Effect of the Pasta-Derived Carotenoid-Enriched Fraction on MMPP-Induced Cytotoxicity

To evaluate the effects of the carotenoid-enriched fractions derived from the acetone extraction of buckwheat pasta and egg pasta on renal cells in the presence of oxidative stress, HEK-293 cells were incubated with different concentrations of extracts ranking from 2×10^{-6} µM to 2×10^{-3} µM in presence of MMPP for 72 h (Figure 6 and Table S2). The concentration used refers to the of β-carotene concentration detected in the extract (Table 2). The lutein content ranged from 1.5×10^{-6} µM to 1.5×10^{-3} µM in Buckwheat pasta and from 1.34×10^{-6} µM to 1.34×10^{-3} µM in egg pasta.

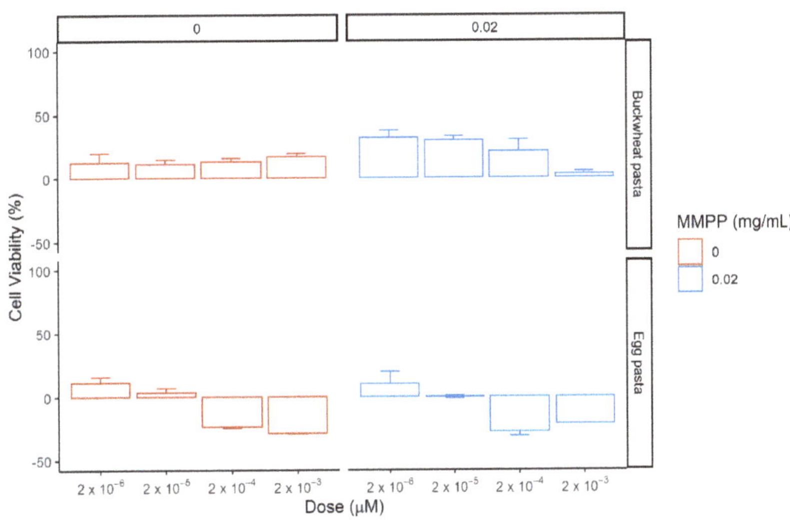

Figure 6. Effect of the pasta-derived carotenoid-enriched fractions on the cell viability of HEK-293 in the absence (left, red) or presence (right, blue) of MMPP after 72 h incubation. Reported values correspond to mean of the difference in cell viability between tested phytocomplexes and reference (medium in red, and MMPP in blue) with standard error over three biological replicates. The percentage of cell viability was calculated using the ratio Abs_{TEST}/Abs_{CTRL}.

In both acetone extracts, the presence of natural phytocomplexes, showed different effects with respect to the single standard, resulting more active at lower concentrations. As the concentration of extracts from egg pasta increases, a reduction of cell viability was observed. The lowest concentration of carotenoid extracts (2×10^{-6} μM) induced a modest positive effect alone or in the presence of MMPP on HEK-293 cells. Buckwheat pasta extract generated an enhancement of cell viability as the extract alone, but also in the presence of MMPP ranging from 32 ± 5% at lowest concentration to 21 ± 9% for 2×10^{-4} μM.

The activity of acetone extracts was assessed also on MDCK cells in presence or absence of MMPP and compared with that obtained in HEK-293 cells (Figure 7 and Table S2).

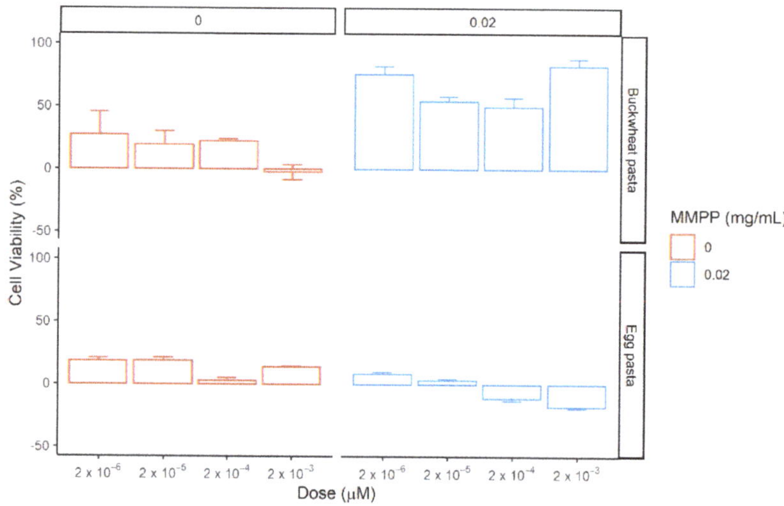

Figure 7. Effect of the pasta-derived carotenoid-enriched fractions on the cell viability of Madin–Darby canine kidney (MDCK) in the absence (left, red) or presence (right, blue) of MMPP after 24 h incubation. Reported values correspond to mean of the difference in cell viability between tested phytocomplexes and reference (medium in red, and MMPP in blue) with standard error over three biological replicates. The percentage of cell viability was calculated using the ratio Abs_{TEST}/Abs_{CTRL}.

Carotenoids-enriched extracts gave an overall pro-viability effect at all the tested concentrations for both types of pasta. With respect to HEK-293 buckwheat pasta extract gave a better rescue upon MMPP treatment for all the doses ranging from 50 ± 7% to 82 ± 5% compared to HEK-239 cells; extracts from egg pasta exerted a small rescue only at the lowest concentration of carotenoids 2×10^{-6} μM in the presence of MMPP, suggesting that the specific composition of extracts plays a key role in the response. Thus, buckwheat pasta showed a greater rescue capacity on the cell viability with respect to egg pasta and since the experiments were performed using the same amount of β-carotene, the difference could be due to the presence of minor compounds, such as flavonoids compounds, higher in the former.

Carotenoids species are involved in different chemical reactions exerting a dual nature of oxidant and anti-oxidant depending the biological context and their concentrations: this high reactivity is due to the presence of the conjugated double bounds [72]. Their antioxidant role was associated with the ability to delocalize the electrons given by oxidative species and radicals [73,74]. In the presence of high oxidative conditions their activity may be converted to pro-oxidant [75]. In our cases, this finding was observed in egg pasta for HEK-293 (both alone and in presence of MMPP) and MDCK cells in presence of MMPP, where different doses of acetone extract produced opposite effects.

3.5. Effect of Hydroalcoholic Extract on MMPP-Induced Cytotoxicity

Afterwards, we evaluated the effects of the hydroalcoholic extracts from of buckwheat pasta, rye pasta and egg pasta on renal cells by testing different concentrations ranking from

5×10^{-5} μM to 5×10^{-1} μM with or without MMPP (Figure 8 and Table S3). The concentration used refers to the concentration of total phenolic content detected in the extract (Table 3). According to the content of the pasta samples reported in Table 1, for Buckwheat pasta the concentration of Indoleacetic acid was ranking from 1.85×10^{-5} μM to 1.85×10^{-1} μM, for caffeic acid derivates was ranking from 3.5×10^{-6} μM to 3.5×10^{-2} μM and for flavonoids content was ranking from 2.8×10^{-5} μM to 2.8×10^{-1} μM; for rye pasta the concentration of Indoleacetic acid was ranking from 2.2×10^{-5} μM to 2.2×10^{-1} μM, for caffeic acid derivates from 1.95×10^{-5} μM to 1.95×10^{-1} μM and for flavonoids species from 8.5×10^{-6} μM to 8.5×10^{-2} μM; for egg pasta the concentration of Indoleacetic acid was ranking from 4.25×10^{-5} μM to 4.25×10^{-1} μM, for flavonoids species from 7.5×10^{-6} μM to 1.85×10^{-2} μM while no caffeic acid derivatives were present.

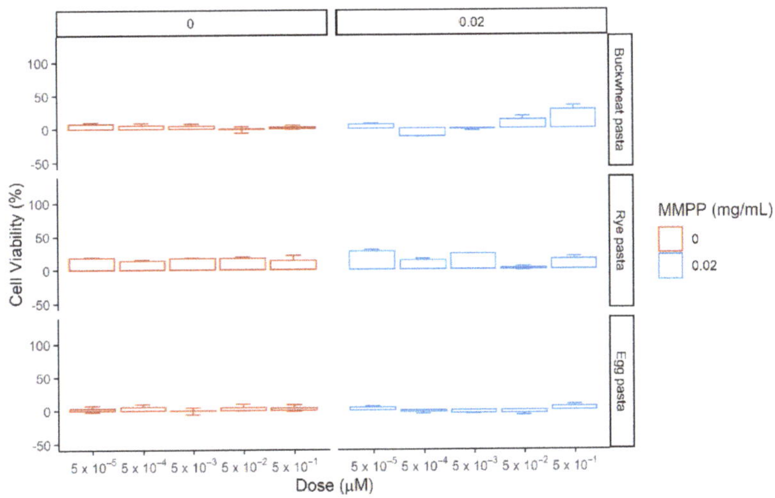

Figure 8. Effect of the pasta-derived hydroalcoholic extracts on the cell viability of HEK-293 in the absence (left, red) or presence (right, blue) of MMPP after 72 h incubation. Reported values correspond to mean of the difference in cell viability between tested phytocomplexes and reference (medium in red, and MMPP in blue) with standard error over three biological replicates. The percentage of cell viability was calculated using the ratio Abs_{TEST}/Abs_{CTRL}.

Hydroalcoholic extract from egg pasta did not exert any significant effect on HEK-293 viability independently on the presence of oxidative stress. Extract from buckwheat pasta showed an enhancement of cell viability upon MMPP treatment only at the highest concentration (28 ± 6%). Rye pasta extract alone showed a general enhancement of the cell viability for all the doses ranging from 14 ± 7% to 18 ± 1% as well as in the presence of MMPP from 27 ± 3% to 13 ± 3%. The phytocomplex presented in the extracts, resulted to be more active than single standards, exerting an effect on cell viability at lower concentrations compared to the latter. These results suggest that in the hydroalcoholic extracts, the major role in the rescue could be attributed to the presence of caffeic acid derivatives, observing higher rescue for rye pasta and lower/negligible effect for egg past, in which they are not detectable.

The hydroalcoholic extract were then tested on MDCK cells in the same conditions as before and the cell viability was assessed (Figure 9 and Table S4).

The hydroalcoholic extract from buckwheat and egg pasta displayed contrasting effects on cell viability: the first increases it, while the latter reduces it; no effect upon MMPP treatment was detected for both. Extract from rye pasta alone did not show any notable effect on cells viability, but it reveals a strong protective effect against MMPP-mediated oxidative stress at all the concentrations ranging from 50 ± 7% to 61 ± 21%. In standard conditions.

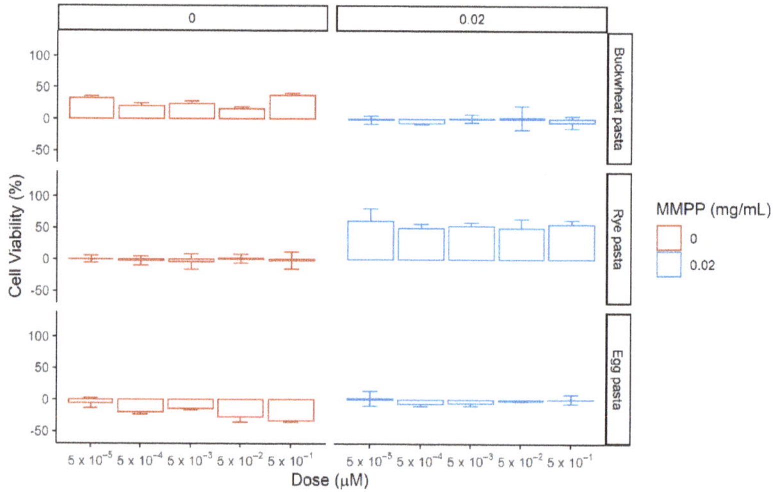

Figure 9. Effect of the pasta-derived hydroalcoholic extracts on the cell viability of MDCK in the absence (left, red) or presence (right, blue) of MMPP after 24 h incubation. Reported values correspond to mean of the difference in cell viability between tested phytocomplexes and reference (medium in red, and MMPP in blue) with standard error over three biological replicates. The percentage of cell viability was calculated using the ratio Abs_{TEST}/Abs_{CTRL}.

In the absence of MMPP, the high content of indoleacetic acid in the extract from egg pasta seems to exert a toxic effect on cell viability. Comparing the two cells models, we observed different behaviors with respect to the buckwheat and rye pasta in the absence of MMPP. In particular, HEK-293 showed an increase of viability with the latter, while MDCK with the former. The differences are related to the action of the phytocomplexes which are characterized by a higher content of flavonoids for the buckwheat pasta. The rescue on cell viability among samples, as observed for both cell lines, corresponds to the increase content of caffeic acid derivatives in rye pasta.

The study of the interaction between polyphenols containing phytocomplexes and renal cells, could help to investigate the known important effects on renal physiology exerted by those compounds [76], focusing on one of possible vehicle, pasta, typically associated with renal healthy diet. As a general rule, the metabolites of polyphenols have a plasma half-life of a few hours and are rapidly eliminated by the kidney and by the liver: the urinary and the biliary excretion route. The total amount of polyphenol metabolites excreted in urine is correlated not only with their maximum plasma concentration but also with their ability to be excreted by biliary rather than urinary tract. For example, urinary excretion of quercetin and its glycosides accounts for 0.3–1.4% of the ingested dose [77]; for caffeic acid and for ferulic acids accounts for 5.9% and 27% respectively. Therefore, pasta-derived polyphenols can have a positive impact on renal cells depending on their amount and metabolism. Further studies are needed to better define their contribution to kidney health.

4. Conclusions

The consumption of pasta is nowadays so diffused and accessible that this food could be one of the most common carriers for important metabolites, such as carotenoids and polyphenols that have been associated to general health and to protective mechanisms for kidney. In this study, we assessed the positive effects of bioactive compounds by comparing different type of pasta, highlighting how it represents an unconventional source of them. The comparison of the different cell responses between single molecules and extracts suggests considering the exerted activity as derived from the whole phytocomplex whit the presence of carotenoids such as lutein and polyphenols such as indoleacetic acid, respect to the individual species and the consequential attention to the raw materials. The scope of this study was (1) to strengthen the role of pasta as a food for the promotion of renal health in the healthy population and (2) for it to be a starting point for the improvement of the micronutrient profile in aproteic pastas. The results of this study suggest that the antioxidant species present in pasta play a major role in the protection of kidney cells from oxidative stress and reinforces the healthy role of quality organic pasta in the Mediterranean diet. High oxidative stress characterizes CKD and an anti-oxidant compounds can be beneficial to counteract it.

Even though the WHO and the FDA consider pasta as the most adequate food for fortification, in agreement with the very definition of fortified food (which is supplementing food with what is deficient in the population), we believe that pasta fortification is necessary only for special medical purpose, including protein-free pasta for CKD patients. For these patients, the micronutrient deficit in protein-free food and the necessary dietary restrictions to which they are subjected could generate nutritional deficiencies. In all the other cases, pasta fortification may not be necessary since pasta is naturally rich in complex carbohydrates and protein and low in fat and is therefore an already highly nutritious food by itself, especially if it is whole wheat based. Our study was meant to be an explorative preliminary study aimed at evaluating the possible beneficial effect that the polyphenolic components present in pasta could elicit on renal cells.

Supplementary Materials: The following are available online at https://www.mdpi.com/article/10.3390/nu13041131/s1, Table S1: Nutritional Table. Table S2—Detection of cell viability variation of HEK-293 and MDCK incubated with increasing amount of carotenoid-enriched fraction from Buckwheat pasta and Egg pasta on medium and MMPP for 72 (HEK-293) and 24 (MDCK) hours. * $p < 0.5$, ** $p < 0.01$, *** $p < 0.001$. Table S3—Detection of cell viability variation of HEK-293 incubated with increasing amount of hydroalcoholic extract from Buckwheat pasta, Rye pasta and Egg pasta on medium and MMPP for 72 h. * $p < 0.5$, ** $p < 0.01$, *** $p < 0.001$. Table S4—Detection of cell viability variation of MDCK incubated with increasing amount of hydroalcoholic extract from Buckwheat pasta, Rye pasta and Egg pasta on medium and MMPP for 24 h. * $p < 0.5$, ** $p < 0.01$, *** $p < 0.001$. Figure S1: MS spectra of Kaempferol in Buckwheat cooked pasta hydroalcoholic extract. Figure S2: MS spectra of quercetin in Buckwheat cooked pasta hydroalcoholic extract. Figure S3: MS spectra of rutin in cooked pasta hydroalcoholic extract. Figure S4: MS spectra of ferulic acid one in rye cooked pasta hydroalcoholic extract. Figure S5: MS spectra of apigenin diglicosides in egg cooked pasta hydroalcoholic extract. Figure S6: Detection cell viability of HEK-293 incubated with increasing amount of Indoleacetic acid, β-carotene and caffeic acid for 24 (A) and 48 (B) hours Reported values correspond to mean of cell viability with standard error over three biological replicates. The percentage of cell viability was calculated using the ratio Abs_{TEST}/Abs_{CTRL}.

Author Contributions: Conceptualization, A.R., R.V. and A.B.; methodology, F.D.M., P.V., A.R. and A.B.; software, F.D.M.; formal analysis, F.D.M.; investigation, R.V. and A.B.; resources, A.R., R.V. and A.B.; data curation, F.D.M.; writing—original draft preparation, F.D.M., S.U., R.V. and A.B.; writing—review and editing, F.T., S.U., P.V., A.S., F.M. and A.R.; visualization, F.D.M. and R.V.; supervision, A.R., R.V. and A.B.; project administration, A.R., R.V. and A.B.; funding acquisition, F.T., A.R., R.V. and A.B. All authors have read and agreed to the published version of the manuscript.

Funding: This work was supported by funds from the RE.ME.DIET. project.

Conflicts of Interest: The authors declare no conflict of interest.

References

1. Lăcătușu, C.M.; Grigorescu, E.D.; Floria, M.; Onofriescu, A.; Mihai, B.M. The mediterranean diet: From an environment-driven food culture to an emerging medical prescription. *Int. J. Environ. Res. Public Health* **2019**, *16*, 942. [CrossRef]
2. Jacobs, D.R.; Gross, M.D.; Tapsell, L.C. Food synergy: An operational concept for understanding nutrition. *Am. J. Clin. Nutr.* **2009**, *89*, 1543S. [CrossRef] [PubMed]

3. De Lorgeril, M.; Salen, P.; Martin, J.-L.; Monjaud, I.; Delaye, J.; Mamelle, N. Mediterranean Diet, Traditional Risk Factors, and the Rate of Cardiovascular Complications After Myocardial Infarction. *Circulation* **1999**, *99*, 779–785. [CrossRef]
4. Psaltopoulou, T.; Naska, A.; Orfanos, P.; Trichopoulos, D.; Mountokalakis, T.; Trichopoulou, A. Olive oil, the Mediterranean diet, and arterial blood pressure: The Greek European Prospective Investigation into Cancer and Nutrition (EPIC) study. *Am. J. Clin. Nutr.* **2004**, *80*, 1012–1018. [CrossRef]
5. Estruch, R.; Ros, E.; Salas-Salvadó, J.; Covas, M.-I.; Corella, D.; Arós, F.; Gómez-Gracia, E.; Ruiz-Gutiérrez, V.; Fiol, M.; Lapetra, J.; et al. Primary prevention of cardiovascular disease with a mediterranean diet. *Z. Gefassmedizin* **2013**, *10*, 28. [CrossRef] [PubMed]
6. Uusitupa, M.; Khan, T.A.; Viguiliouk, E.; Kahleova, H.; Rivellese, A.A.; Hermansen, K.; Pfeiffer, A.; Thanopoulou, A.; Salas-Salvadó, J.; Schwab, U.; et al. Prevention of type 2 diabetes by lifestyle changes: A systematic review and meta-analysis. *Nutrients* **2019**, *11*, 2611. [CrossRef] [PubMed]
7. Jannasch, F.; Kröger, J.; Schulze, M.B. Dietary patterns and Type 2 diabetes: A systematic literature review and meta-analysis of prospective studies. *J. Nutr.* **2017**, *147*, 1174–1182. [CrossRef]
8. Esposito, K.; Chiodini, P.; Maiorino, M.I.; Bellastella, G.; Panagiotakos, D.; Giugliano, D. Which diet for prevention of type 2 diabetes? A meta-analysis of prospective studies. *Endocrine* **2014**, *47*, 107–116. [CrossRef] [PubMed]
9. Huo, R.; Du, T.; Xu, Y.; Xu, W.; Chen, X.; Sun, K.; Yu, X. Effects of Mediterranean-style diet on glycemic control, weight loss and cardiovascular risk factors among type 2 diabetes individuals: A meta-analysis. *Eur. J. Clin. Nutr.* **2015**, *69*, 1200–1208. [CrossRef] [PubMed]
10. Martín-Peláez, S.; Fito, M.; Castaner, O. Mediterranean diet effects on type 2 diabetes prevention, disease progression, and related mechanisms. A review. *Nutrients* **2020**, *12*, 2236. [CrossRef] [PubMed]
11. Milajerdi, A.; Namazi, N.; Larijani, B.; Azadbakht, L. The Association of Dietary Quality Indices and Cancer Mortality: A Systematic Review and Meta-analysis of Cohort Studies. *Nutr. Cancer* **2018**, *70*, 1091–1105. [CrossRef] [PubMed]
12. Schwingshackl, L.; Schwedhelm, C.; Galbete, C.; Hoffmann, G. Adherence to mediterranean diet and risk of cancer: An updated systematic review and meta-analysis. *Nutrients* **2017**, *9*, 1063. [CrossRef]
13. Sayón-Orea, C.; Razquin, C.; Bulló, M.; Corella, D.; Fitó, M.; Romaguera, D.; Vioque, J.; Alonso-Gómez, Á.M.; Wärnberg, J.; Martínez, J.A.; et al. Effect of a Nutritional and Behavioral Intervention on Energy-Reduced Mediterranean Diet Adherence among Patients with Metabolic Syndrome: Interim Analysis of the PREDIMED-Plus Randomized Clinical Trial. *JAMA J. Am. Med. Assoc.* **2019**, *322*, 1486–1499. [CrossRef] [PubMed]
14. Bajerska, J.; Chmurzynska, A.; Muzsik, A.; Krzyżanowska, P.; Mądry, E.; Malinowska, A.M.; Walkowiak, J. Weight loss and metabolic health effects from energy-restricted Mediterranean and Central-European diets in postmenopausal women: A randomized controlled trial. *Sci. Rep.* **2018**, *8*, 11170. [CrossRef]
15. Estruch, R.; Martínez-González, M.A.; Corella, D.; Salas-Salvadó, J.; Fitó, M.; Chiva-Blanch, G.; Fiol, M.; Gómez-Gracia, E.; Arós, F.; Lapetra, J.; et al. Effect of a high-fat Mediterranean diet on bodyweight and waist circumference: A prespecified secondary outcomes analysis of the PREDIMED randomised controlled trial. *Lancet Diabetes Endocrinol.* **2019**, *7*, e6–e17. [CrossRef]
16. Molendijk, M.; Molero, P.; Ortuño Sánchez-Pedreño, F.; Van der Does, W.; Angel Martínez-González, M. Diet quality and depression risk: A systematic review and dose-response meta-analysis of prospective studies. *J. Affect. Disord.* **2018**, *226*, 346–354. [CrossRef]
17. Wu, L.; Sun, D. Adherence to Mediterranean diet and risk of developing cognitive disorders: An updated systematic review and meta-analysis of prospective cohort studies. *Sci. Rep.* **2017**, *7*, 1–9. [CrossRef]
18. Loughrey, D.G.; Lavecchia, S.; Brennan, S.; Lawlor, B.A.; Kelly, M.E. The impact of the mediterranean diet on the cognitive functioning of healthy older adults: A systematic review and meta-analysis. *Adv. Nutr.* **2017**, *8*, 571–586.
19. Arnold, S.E.; Arvanitakis, Z.; Macauley-Rambach, S.L.; Koenig, A.M.; Wang, H.Y.; Ahima, R.S.; Craft, S.; Gandy, S.; Buettner, C.; Stoeckel, L.E.; et al. Brain insulin resistance in type 2 diabetes and Alzheimer disease: Concepts and conundrums. *Nat. Rev. Neurol.* **2018**, *14*, 168–181. [CrossRef]
20. Trichopoulou, A.; Vasilopoulou, E. Mediterranean diet and longevity. *Br. J. Nutr.* **2020**, *84*, S205–S209. [CrossRef]
21. Toledo, E.; Hu, F.B.; Estruch, R.; Buil-Cosiales, P.; Corella, D.; Salas-Salvadó, J.; Covas, M.I.; Arós, F.; Gómez-Gracia, E.; Fiol, M.; et al. Effect of the Mediterranean diet on blood pressure in the PREDIMED trial: Results from a randomized controlled trial. *BMC Med.* **2013**, *11*, 207. [CrossRef] [PubMed]
22. Solá, R.; Fitó, M.; Estruch, R.; Salas-Salvadó, J.; Corella, D.; de La Torre, R.; Muñoz, M.A.; del Carmen López-Sabater, M.; Martínez-González, M.A.; Arós, F.; et al. Effect of a traditional Mediterranean diet on apolipoproteins B, A-I, and their ratio: A randomized, controlled trial. *Atherosclerosis* **2011**, *218*, 174–180. [CrossRef] [PubMed]
23. Hernáez, Á.; Castañer, O.; Elosua, R.; Pintó, X.; Estruch, R.; Salas-Salvadó, J.; Corella, D.; Arós, F.; Serra-Majem, L.; Fiol, M.; et al. Mediterranean Diet Improves High-Density Lipoprotein Function in High-Cardiovascular-Risk Individuals. *Circulation* **2017**, *135*, 633–643. [CrossRef]
24. Mirabelli, M.; Chiefari, E.; Arcidiacono, B.; Corigliano, D.M.; Brunetti, F.S.; Maggisano, V.; Russo, D.; Foti, D.P.; Brunetti, A. Mediterranean Diet Nutrients to Turn the Tide against Insulin Resistance and Related Diseases. *Nutrients* **2020**, *12*, 1066. [CrossRef]
25. Fitó, M.; Guxens, M.; Corella, D.; Sáez, G.; Estruch, R.; De La Torre, R.; Francés, F.; Cabezas, C.; López-Sabater, M.D.C.; Marrugat, J.; et al. Effect of a traditional Mediterranean diet on lipoprotein oxidation: A randomized controlled trial. *Arch. Intern. Med.* **2007**, *167*, 1195–1203. [CrossRef]

26. Calder, P.C.; Ahluwalia, N.; Brouns, F.; Buetler, T.; Clement, K.; Cunningham, K.; Esposito, K.; Jönsson, L.S.; Kolb, H.; Lansink, M.; et al. Dietary factors and low-grade inflammation in relation to overweight and obesity. *Br. J. Nutr.* **2011**, *106*, S1–S78. [CrossRef]
27. Torres-Peña, J.D.; Rangel-Zuñiga, O.A.; Alcala-Diaz, J.F.; Lopez-Miranda, J.; Delgado-Lista, J. Mediterranean Diet and Endothelial Function: A Review of its Effects at Different Vascular Bed Levels. *Nutrients* **2020**, *12*, 2212. [CrossRef]
28. Marin, C.; Ramirez, R.; Delgado-Lista, J.; Yubero-Serrano, E.M.; Perez-Martinez, P.; Carracedo, J.; Garcia-Rios, A.; Rodriguez, F.; Gutierrez-Mariscal, F.M.; Gomez, P.; et al. Mediterranean diet reduces endothelial damage and improves the regenerative capacity of endothelium. *Am. J. Clin. Nutr.* **2011**, *93*, 267–274. [CrossRef]
29. Huang, X.; Jiménez-Molén, J.J.; Lindholm, B.; Cederholm, T.; Ärnlöv, J.; Risérus, U.; Sjögren, P.; Carrero, J.J. Mediterranean diet, kidney function, and mortality in men with CKD. *Clin. J. Am. Soc. Nephrol.* **2013**, *8*, 1548–1555. [CrossRef] [PubMed]
30. Bach, K.E.; Kelly, J.T.; Campbell, K.L.; Palmer, S.C.; Khalesi, S.; Strippoli, G.F.M. Healthy dietary patterns and incidence of CKD: A meta-analysis of cohort studies. *Clin. J. Am. Soc. Nephrol.* **2019**, *14*, 1441–1449. [CrossRef] [PubMed]
31. Monsalve, B.; Concha-Meyer, A.; Palomo, I.; Fuentes, E. Mechanisms of endothelial protection by natural bioactive compounds from fruit and vegetables. *An. Acad. Bras. Cienc.* **2017**, *89*, 615–633. [CrossRef] [PubMed]
32. Zoccali, C. Endothelial dysfunction and the kidney: Emerging risk factors for renal insufficiency and cardiovascular outcomes in essential hypertension. *J. Am. Soc. Nephrol.* **2006**, *17*, 61–63. [CrossRef] [PubMed]
33. Galle, J.; Quaschning, T.; Seibold, S.; Wanner, C. Endothelial dysfunction and inflammation: What is the link? *Kidney Int. Suppl.* **2003**, *63*, 45–49. [CrossRef] [PubMed]
34. Kontessis, P.S.; Bossinakou, I.; Sarika, L.; Iliopoulou, E.; Papantoniou, A.; Trevisan, R.; Roussi, D.; Stipsanelli, K.; Grigorakis, S.; Souvatzoglou, A. Renal, metabolic, and hormonal responses to proteins of different origin in normotensive, nonproteinuric type I diabetic patients. *Diabetes Care* **1995**, *18*, 1233–1240. [CrossRef] [PubMed]
35. Metges, C.C.; Barth, C.A. Metabolic consequences of a high dietary-protein intake in adulthood: Assessment of the available evidence. *J. Nutr.* **2000**, *130*, 886–889. [CrossRef]
36. Epstein, F.H.; Brenner, B.M.; Meyer, T.W.; Hostetter, T.H. Dietary Protein Intake and the Progressive Nature of Kidney Disease. *N. Engl. J. Med.* **1982**, *307*, 652–659. [CrossRef]
37. Knight, E.L.; Stampfer, M.J.; Hankinson, S.E.; Spiegelman, D.; Curhan, G.C. The Impact of Protein Intake on Renal Function Decline in Women with Normal Renal Function or Mild Renal Insufficiency. *Ann. Intern. Med.* **2003**, *138*. [CrossRef] [PubMed]
38. Halbesma, N.; Bakker, S.J.L.; Jansen, D.F.; Stolk, R.P.; De Zeeuw, D.; De Jong, P.E.; Gansevoort, R.T. High protein intake associates with cardiovascular events but not with loss of renal function. *J. Am. Soc. Nephrol.* **2009**, *20*, 1797–1804. [CrossRef]
39. Cirillo, M.; Lombardi, C.; Chiricone, D.; De Santo, N.G.; Zanchetti, A.; Bilancio, G. Protein intake and kidney function in the middle-age population: Contrast between cross-sectional and longitudinal data. *Nephrol. Dial. Transplant.* **2014**, *29*, 1733–1740. [CrossRef]
40. Lew, Q.L.J.; Jafar, T.H.; Koh, H.W.L.; Jin, A.; Chow, K.Y.; Yuan, J.M.; Koh, W.P. Red meat intake and risk of ESRD. *J. Am. Soc. Nephrol.* **2017**, *28*, 304–312. [CrossRef]
41. Lin, J.; Fung, T.T.; Hu, F.B.; Curhan, G.C. Association of dietary patterns with albuminuria and kidney function decline in older white women: A subgroup analysis from the nurses health study. *Am. J. Kidney Dis.* **2011**, *57*, 245–254. [CrossRef]
42. Azadbakht, L.; Atabak, S.; Esmaillzadeh, A. Soy protein intake, cardiorenal indices, and C-reactive protein in type 2 diabetes with nephropathy. *Diabetes Care* **2008**, *31*, 648–654. [CrossRef]
43. Cai, Q.; Dekker, L.H.; Bakker, S.J.L.; de Borst, M.H.; Navis, G.J. Dietary Patterns Based on Estimated Glomerular Filtration Rate and Kidney Function Decline in the General Population: The Lifelines Cohort Study. *Nutrients* **2020**, *12*, 1099. [CrossRef] [PubMed]
44. Nilusha, R.A.T.; Jayasinghe, J.M.J.K.; Perera, O.D.A.N.; Perera, P.I.P. Development of pasta products with nonconventional ingredients and their effect on selected quality characteristics: A brief overview. *Int. J. Food Sci.* **2019**, *2019*, 6750726. [CrossRef] [PubMed]
45. Holasova, M.; Fiedlerova, V.; Smrcinova, H.; Orsak, M.; Lachman, J.; Vavreinova, S. Buckwheat—The source of antioxidant activity in functional foods. *Food Res. Int.* **2002**, *35*, 207–211. [CrossRef]
46. Velioglu, Y.S.; Mazza, G.; Gao, L.; Oomah, B.D. Antioxidant Activity and Total Phenolics in Selected Fruits, Vegetables, and Grain Products. *J. Agric. Food Chem.* **1998**, *46*, 4113–4117. [CrossRef]
47. Pounis, G.; Di Castelnuovo, A.; Costanzo, S.; Persichillo, M.; Bonaccio, M.; Bonanni, A.; Cerletti, C.; Donati, M.B.; De Gaetano, G.; Iacoviello, L. Association of pasta consumption with body mass index and waist-to-hip ratio: Results from Moli-sani and INHES studies. *Nutr. Diabetes* **2016**, *6*, e218. [CrossRef]
48. Huang, M.; Li, J.; Ha, M.A.; Riccardi, G.; Liu, S. A systematic review on the relations between pasta consumption and cardio-metabolic risk factors. *Nutr. Metab. Cardiovasc. Dis.* **2017**, *27*, 939–948. [CrossRef]
49. Sandberg, J.C.; Björck, I.M.E.; Nilsson, A.C. Rye-based evening meals favorably affected glucose regulation and appetite variables at the following breakfast; a randomized controlled study in healthy subjects. *PLoS ONE* **2016**, *11*, e0151985. [CrossRef]
50. Magnusdottir, O.K.; Landberg, R.; Gunnarsdottir, I.; Cloetens, L.; Åkesson, B.; Rosqvist, F.; Schwab, U.; Herzig, K.-H.; Hukkanen, J.; Savolainen, M.J.; et al. Whole Grain Rye Intake, Reflected by a Biomarker, Is Associated with Favorable Blood Lipid Outcomes in Subjects with the Metabolic Syndrome—A Randomized Study. *PLoS ONE* **2014**, *9*, e110827. [CrossRef]
51. Abdel-Aal, E.S.M.; Akhtar, H.; Zaheer, K.; Ali, R. Dietary sources of lutein and zeaxanthin carotenoids and their role in eye health. *Nutrients* **2013**, *5*, 1169–1185. [CrossRef] [PubMed]

52. Bustos, M.C.; Perez, G.T.; Leon, A.E. Structure and quality of pasta enriched with functional ingredients. *RSC Adv.* **2015**, *5*, 30780–30792. [CrossRef]
53. Ciccoritti, R.; Taddei, F.; Nicoletti, I.; Gazza, L.; Corradini, D.; D'Egidio, M.G.; Martini, D. Use of bran fractions and debranned kernels for the development of pasta with high nutritional and healthy potential. *Food Chem.* **2017**, *225*, 77–86. [CrossRef] [PubMed]
54. Kiprovski, B.; Mikulic-Petkovsek, M.; Slatnar, A.; Veberic, R.; Stampar, F.; Malencic, D.; Latkovic, D. Comparison of phenolic profiles and antioxidant properties of European Fagopyrum esculentum cultivars. *Food Chem.* **2015**, *185*, 41–47. [CrossRef]
55. Cervantes-Paz, B.; Victoria-Campos, C.I.; Ornelas-Paz, J. de J. Absorption of carotenoids and mechanisms involved in their health-related properties. *Subcell. Biochem.* **2016**, *79*, 415–454. [CrossRef] [PubMed]
56. Maiani, G.; Castón, M.J.P.; Catasta, G.; Toti, E.; Cambrodón, I.G.; Bysted, A.; Granado-Lorencio, F.; Olmedilla-Alonso, B.; Knuthsen, P.; Valoti, M.; et al. Carotenoids: Actual knowledge on food sources, intakes, stability and bioavailability and their protective role in humans. *Mol. Nutr. Food Res.* **2009**, *53*, 194–218. [CrossRef] [PubMed]
57. Singleton, V.L.; Orthofer, R.; Lamuela-Raventós, R.M. Analysis of total phenols and other oxidation substrates and antioxidants by means of folin-ciocalteu reagent. *Methods Enzymol.* **1999**, *299*, 152–178. [CrossRef]
58. Dewanto, V.; Xianzhong, W.; Adom, K.K.; Liu, R.H. Thermal processing enhances the nutritional value of tomatoes by increasing total antioxidant activity. *J. Agric. Food Chem.* **2002**, *50*, 3010–3014. [CrossRef]
59. R Core Team. *R: A Language and Environment for Statistical Computing*; R Foundation for Statistical Computing: Vienna, Austria, 2019.
60. R Studio Team. *RStudio: Integrated Development Environment for R*; RStudio, PBC: Boston, MA, USA, 2019.
61. Wickham, H.; Averick, M.; Bryan, J.; Chang, W.; McGowan, L.D.; François, R.; Grolemund, G.; Hayes, A.; Henry, L.; Hester, J.; et al. Welcome to the tidyverse. *J. Open Source Softw.* **2019**, *4*, 1686. [CrossRef]
62. Gupta, A.; Sharma, S.; Reddy Surasani, V.K. Quinoa protein isolate supplemented pasta: Nutritional, physical, textural and morphological characterization. *LWT* **2021**, *135*, 110045. [CrossRef]
63. Vignolini, P.; Urciuoli, S.; Heimler, D.; Romani, A. Carotenoids, Polyphenols and Antioxidant Activity Evaluation in Stone-Grinded Wheat Semolina. *J. Health Sci.* **2018**, *6*, 432–438. [CrossRef]
64. Salehi, A.; Fallah, S.; Kaul, H.P.; Zitterl-Eglseer, K. Antioxidant capacity and polyphenols in buckwheat seeds from fenugreek/buckwheat intercrops as influenced by fertilization. *J. Cereal Sci.* **2018**, *84*, 142–150. [CrossRef]
65. Francis Raguindin, P.; Adam Itodo, O.; Stoyanov, J.; Dejanovic, G.M.; Gamba, M.; Asllanaj, E.; Minder, B.; Bussler, W.; Metzger, B.; Muka, T.; et al. A systematic review of phytochemicals in oat and buckwheat. *Food Chem.* **2021**, *338*. [CrossRef] [PubMed]
66. Bondia-Pons, I.; Aura, A.M.; Vuorela, S.; Kolehmainen, M.; Mykkänen, H.; Poutanen, K. Rye phenolics in nutrition and health. *J. Cereal Sci.* **2009**, *49*, 323–336. [CrossRef]
67. Asenstorfer, R.E.; Wang, Y.; Mares, D.J. Chemical structure of flavonoid compounds in wheat (*Triticum aestivum* L.) flour that contribute to the yellow colour of Asian alkaline noodles. *J. Cereal Sci.* **2006**, *43*, 108–119. [CrossRef]
68. Mellado-Ortega, E.; Hornero-Méndez, D. Effect of long-term storage on the free and esterified carotenoids in durum wheat (Triticum turgidum conv. durum) and tritordeum (× Tritordeum Ascherson et Graebner) grains. *Food Res. Int.* **2017**, *99*, 877–890. [CrossRef]
69. Gyurászová, M.; Gurecká, R.; Bábíčková, J.; Tóthová, L'. Oxidative Stress in the Pathophysiology of Kidney Disease: Implications for Noninvasive Monitoring and Identification of Biomarkers. *Oxidative Med. Cell. Longev.* **2020**, *2020*, 5478708. [CrossRef]
70. Haskell, M.J. The challenge to reach nutritional adequacy for vitamin A: β-carotene bioavailability and conversion—Evidence in humans. *Am. J. Clin. Nutr.* **2012**, *96*, 1193S–1203S. [CrossRef]
71. Olthof, M.R.; Hollman, P.C.H.; Katan, M.B. Chlorogenic acid and caffeic acid are absorbed in humans. *J. Nutr.* **2001**, *131*, 66–71. [CrossRef]
72. Ribeiro, D.; Freitas, M.; Silva, A.M.S.; Carvalho, F.; Fernandes, E. Antioxidant and pro-oxidant activities of carotenoids and their oxidation products. *Food Chem. Toxicol.* **2018**, *120*, 681–699. [CrossRef]
73. Jomova, K.; Valko, M. Health protective effects of carotenoids and their interactions with other biological antioxidants. *Eur. J. Med. Chem.* **2013**, *70*, 102–110. [CrossRef]
74. Rao, A.V.; Rao, L.G. Carotenoids and human health. *Pharmacol. Res.* **2007**, *55*, 207–216. [CrossRef] [PubMed]
75. Kikugawa, K.; Hiramoto, K.; Tomiyama, S.; Asano, Y. β-Carotene effectively scavenges toxic nitrogen oxides: Nitrogen dioxide and peroxynitrous acid. *FEBS Lett.* **1997**, *404*, 175–178. [CrossRef]
76. Mafra, D.; Borges, N.A.; Lindholm, B.; Shiels, P.G.; Evenepoel, P.; Stenvinkel, P. Food as medicine: Targeting the uraemic phenotype in chronic kidney disease. *Nat. Rev. Nephrol.* **2020**, *17*, 153–171. [CrossRef] [PubMed]
77. Graefe, E.U.; Wittig, J.; Mueller, S.; Riethling, A.K.; Uehleke, B.; Drewelow, B.; Pforte, H.; Jacobasch, G.; Derendorf, H.; Veit, M. Pharmacokinetics and bioavailability of quercetin glycosides in humans. *J. Clin. Pharmacol.* **2001**, *41*, 492–499. [CrossRef] [PubMed]

nutrients

MDPI

Article

Adherence to Dietary Guidelines in Adults by Diabetes Status: Results From the 2012 Mexican National Health and Nutrition Survey

Nancy López-Olmedo [1], Satya Jonnalagadda [2], Ana Basto-Abreu [1], Alan Reyes-García [1], Carolyn J. Alish [2], Teresa Shamah-Levy [3] and Tonatiuh Barrientos-Gutierrez [1,*]

1 Center for Population Health Research, National Institute of Public Health, Cuernavaca, Morelos 62100, Mexico; nancy.lopez@insp.mx (N.L.-O.); ana.basto@insp.mx (A.B.-A.); rega.nutri@gmail.com (A.R.-G.)

2 Abbott Nutrition, Columbus, OH 43219, USA; satya.jonnalagadda@abbott.com (S.J.); calish@columbus.rr.com (C.J.A.)

3 Evaluation and Surveys Research Center, National Institute of Public Health, Cuernavaca, Morelos 62100, Mexico; tshamah@insp.mx

* Correspondence: tbarrientos@insp.mx; Tel.: +52-5487-1015

Received: 26 September 2020; Accepted: 23 October 2020; Published: 12 November 2020

Abstract: The aims of the present study were to compare the adherence to dietary guidelines and evaluate potential differences in nutrient profiles among adults by diabetes status. We used the Mexican Alternate Healthy Eating Index (MxAHEI) to evaluate adherence to dietary guidelines. We calculated the MxAHEI scores (total and by dietary component) with scales from 0 (non-adherence) to 100 (perfect adherence) based on a food frequency questionnaire. Mean daily intakes of macronutrients and micronutrients (g, mg, mcg/1000 kcal per day) were also estimated by diabetes status. Sex-specific, multivariable linear regression models were estimated to test whether MxAHEI scores as well as nutrient intakes were different by diabetes status. Mexican adults had low adherence to the dietary guidelines irrespective of their diabetes status (score < 50 points). Among men, the MxAHEI score was 2.6 points higher among those with diabetes than those without diabetes (46.9; 95% confidence intervals (CI): 44.6, 49.2 vs. 44.3; 95% CI: 44.2, 45.6, respectively). Among women, the total MxAHEI score was similar in individuals with diabetes compared to those without diabetes. Lower intakes of carbohydrates and added sugars and higher intakes of protein, calcium, and zinc were observed in individuals with diabetes. Our findings support the development of strategies focused on promoting dietary patterns that can help to prevent and control the disease.

Keywords: dietary patterns; diabetes; Mexican adults

---

1. Introduction

In 2016, type 2 diabetes was declared an epidemiological emergency in Mexico. Diabetes is the second cause of death in the country, where it accounts for 15.2% of all deaths annually and is the leading cause for years of life lost [1]. The prevalence of diagnosed diabetes in Mexico is 9.4% of all adults, representing 6.5 million people with diabetes. Complications of diabetes are frequent, 21.1% of people with type 2 diabetes mellitus experience microvascular complications (diabetic foot, retinopathy, or nephropathy), and 3.4% have macrovascular complications [2]. In 2011, diabetes costs in Mexico were estimated to be 7.7 billion US dollars each year [2], of which 44% were direct healthcare costs [3].

Nutrition therapy is a cornerstone of the management and prevention of diabetes. Recent updates to the American Diabetes Association (ADA) nutrition guidelines states that treatment and prevention of the disease should be focused on the overall diet more than a single nutrient or food group [4].

In individuals with diabetes, higher adherence to healthy dietary patterns have been associated with lower glycated hemoglobin levels and lower risk of diabetes complications [3]. However, adherence to healthy diets among people with diabetes can be challenging and remains a fundamental barrier to diabetes management worldwide [5,6].

Dietary indices have been recommended to evaluate adherence to dietary recommendations among individuals or populations. The Alternate Healthy Eating Index (AHEI) represents a measure of diet quality in relation to nutrition guidelines to manage and prevent chronic diseases such as diabetes because it was developed on evidence-based recommendations that specific foods and nutrients can have synergistic or antagonistic effects in the development or management of chronic diseases, particularly cardiometabolic diseases [2,7–9].

In Mexico, 22% of the population with diabetes reported modifying their diets as part of their diabetes management [2]. However, no study has evaluated the diet quality of adults with diabetes or examined whether diet quality differs based on diabetes status at the national level in Mexico. We aimed to compare the diet quality between adults with and without self-reported diabetes using food frequency data from the 2012 Mexican National Health and Nutrition Survey (ENSANUT), and the AHEI adapted to the Mexican context [10]. We also evaluated the intake of macronutrients and micronutrients by diabetes status as a secondary aim. Conducting this type of analysis is useful to understand the dietary behaviors of people with diabetes and could provide a platform to develop food and nutrition policies for this population.

2. Materials and Methods

2.1. Data Source and Population

The 2012 ENSANUT is a cross-sectional, multistage, stratified, and cluster-sampled survey representative of urban and rural areas, at the national, regional, and state levels in Mexico. The design and methods are described elsewhere [11]. Briefly, the 2012 ENSANUT was conducted between October 2011 and May 2012 and obtained information about sociodemographic, nutrition, and health characteristics from 96,031 people. Dietary information was obtained using a validated Semi-Quantitative Food Frequency Questionnaire (SFFQ) [12,13] applied to a random subsample of 11% of the sample. This study was conducted according to the guidelines in the Declaration of Helsinki, and all procedures involving human participants were approved by the Ethics Committee of the Mexican National Institute of Public Health (project number 1033, approval 1108). Written informed consent was obtained from all participants under study.

We included males and non-pregnant and non-lactating females >18 years (n = 7 512) and excluded individuals with incomplete health, anthropometric, or sociodemographic information (n = 400). We also excluded 56 individuals with extreme total energy intake (expressed as the ratio of total energy intake to estimated energy requirement in logarithmic scale +/−3 standard deviations (SD)), as previously described [14]. The analytical sample was composed of 2762 participants, who were stratified by sex and diabetes status.

2.2. Dietary Information

Trained interviewers collected dietary information using the 140-item Semi-Quantitative Food Frequency Questionnaire (SFFQ) adapted by the Mexican National Institute of Public Health. For each food, participants reported the frequency and the number of standard portions consumed over seven days prior to the interview. We then converted the reported intake into grams or milliliters per day. We considered edible portion and density factor to obtain net grams of solid food consumed and milliliters consumed from beverages, respectively. It is relevant to notice that the SFFQ has individual foods and mixtures (dishes with sets of default ingredients). For the present analysis, we disaggregated these mixtures into their component ingredients using standard recipes. Energy and nutrient values were estimated using a Food Composition Table compiled by the Mexican National Institute of Public

Health in Morelos, Mexico [15]. Added sugar values were estimated with methodology used by Sánchez-Pimienta et al. [16]. The portion sizes of each SFFQ item were estimated based on the Mexican Equivalent Food System [17].

2.3. Mexican Healthy Eating Index (MxAHEI)

We evaluated the adherence to dietary guidelines using the Mexican Alternate Healthy Eating Index (MxAHEI) [10], which is based on the original AHEI-2010 components and criteria for maximum and minimum scores [8,18]. Briefly, we identified portion intakes of 12 components: seven for which higher intakes are recommended (vegetables, whole fruit, fiber from cereals, legumes, nuts, and long chain omega-3 (n-3) fatty acids (eicosapentaenoic acid—EPA and docosahexaenoic acid—DHA), and polyunsaturated fats), and five that must be limited or avoided (sugar-sweetened beverages (SSBs), red and processed meats, sodium, trans fats, and alcohol). Each component was scored on a 0 to 5- or 9-point scale. The component scores were summed to obtain the total MxAHEI score, which can range from 0 (non-adherence) to 100 (perfect adherence). Details of the index components and the score standards are summarized in Table 1.

Table 1. Mexican Alternate Healthy Eating Index. Components and scoring criteria.

Food Component	MxAHEI		
	Maximum Score	**Criteria for Minimum Score (0)**	**Criteria for Maximum Score**
Higher Intake Recommended			
Vegetables	9	0 servings	≥5 servings
Whole fruit	9	0 servings	≥4 servings
Fiber from cereals	9	0 g	≥15 g
Legumes	5	0 servings	≥1 serving
Nuts	5	0 servings	≥1 serving
Long-chain (n-3) fatty acids (EPA + DHA)	9	0 mg	≥250 mg
Polyunsaturated fats [a]	9	≤2% of total energy intake	≥10% of total energy intake
Limited Intake Recommended			
SSBs	9	≥1 serving	0 servings
Red and processed meat	9	≥1.5 serving	0 servings
Sodium	9	>2 g	≤1.5 g
Trans fats	9	≥4% of total energy intake	≤0.5% of total energy intake
Alcohol			
Women	9	≥2.5 drinks	0.5–1.5 drinks
Men		≥3.5 drinks	0.5–2.0 drinks
Total score	100		

[a] Excluding long-chain (n-3) fatty acids. MxAHEI, Mexican Alternate Healthy Eating Index; DHA, docosahexaenoic acid; EPA, eicosapentaenoic acid; SSBs, sugar-sweetened beverages.

2.4. Health and Anthropometric Information

We determined self-reported diabetes status (yes/no) from responses to the question "Have you ever been told by a physician that you have diabetes or high sugar in your blood? Self-reported hypertension status was defined from responses to the question "Have you ever been told by a physician that you have high blood pressure or hypertension? Smoking status was self-reported and categorized as "current", for study participants who reported smoking at least 100 cigarettes during their lifetime and having smoked during the last 30 days; "former", for those who reported smoking at least 100 cigarettes during their lifetime and who did not currently smoke; and "never" for those who never smoked. We calculated the Metabolic Equivalents (MET) per week for each individual based on reported measures of minutes per week of walking, and moderate and vigorous physical activities. Trained personnel obtained anthropometric measurements using standard procedures [19]. Body weight was measured with participants wearing light clothing using digital scales (Seca © 872 digital floor scale, Gmbh & Co., Hamburg, Germany); stadiometers (Dyna-top ©, model E-1, Multimed, Monterrey, Mexico) were used to measure height. We calculated the body mass index (BMI) using the standard equation and categorized the results into normal, overweight, and obese based on World Health Organization classification [20].

2.5. Sociodemographic Variables

We defined rural areas as locations with < 2500 inhabitants and urban areas as locations with ≥ 2500 inhabitants, and we defined regions as North, Center, and South. A wealth index was constructed using principal components analysis that was applied to household characteristics and assets [21]. The index was then classified into three categories (low, medium, and high) using tertiles of the distribution of index as cut-off points. Education level was categorized as elementary, middle, high school, and college or more, according to the last grade of studies completed.

2.6. Statistical Analysis

We conducted a descriptive analysis by sex and diabetes status using means or percentages and 95% confidence intervals (CI). To test whether the MxAHEI diet scores (total and by dietary component) were different between individuals with and without diabetes, we performed a sex-specific linear regression models unadjusted and adjusted for age, hypertension status, BMI, smoking status, educational level, wealth index, rural/urban area and geographic region. We also estimated daily intakes of macronutrients, fiber and micronutrients (g, mg, or mcg per 1000 kcal per day). Similarly, we compared intakes of nutrients between adults with and without diabetes using sex-specific linear regression models unadjusted and adjusted for the same variables used to analyze the MxAHEI diet score.

Mean values and 95% confidence intervals were used to present the average dietary scores and nutrient intakes for men and women with and without diabetes. We used population weighted factors for all the statistical analyses and considered the survey's complex sampling design. Statistical tests were two-tailed and considered significant at $P < 0.05$. All analyses were carried out using Stata version 13 (StataCorp, Stata Statistical Software, Release 13, 2013. College Station, Texas, United Stated of America: StataCorp LP) [22].

3. Results

Participant characteristics are presented in Table 2. The 2762 individuals analyzed represent 62.1 million adults, of whom 41.3% were men; 7.2% of men and 9.8% of women reported being diagnosed with diabetes. Compared to participants without diabetes, participants with diabetes were older, less educated, and had lower physical activity. They also had a higher prevalence of hypertension and being overweight and had lower energy intake. In addition, among adults with diabetes, the time since diagnosis was longer among women than men. Moreover, regardless of the diabetes status, the mean of total MxAHEI score was less than half of the maximum score (100).

Among men, using the unadjusted models, the score of total MxAHEI and whole fruit and SSBs components were higher in those with diabetes than those without diabetes (Table S1). Similar results were observed in adjusted models (Table 3). The adherence to the total MxAHEI was higher in individuals with diabetes than those without diabetes (46.9; 95% CI: 44.6, 49.2 vs. 44.3; 95% CI: 44.2, 45.6, respectively). Specifically, participants who had diabetes had higher intakes of fruits than those without diabetes. Also, those with diabetes had lower intakes for SSBs (higher MxAHEI score) than men without diabetes. The intakes of the remaining components were similar between those with diabetes and those without diabetes. However, the intakes of most dietary components were slightly higher (higher MxAHEI score) among individuals with diabetes compared to those without diabetes. Inversely, the dietary components where the recommendation is to limit intake showed slightly lower intakes (higher MxAHEI score) among men with diabetes than those without diabetes, except for red and processed meats and sodium, where the intake observed was in the opposite direction.

Table 2. Sex-stratified characteristics of Mexican adults by diabetes status (n = 2762) [a].

Variables	Men				Women			
	Diabetes		Non-Diabetes		Diabetes		Non-Diabetes	
n	82		1060		159		1461	
N (millions)	2.5		26.3		3.7		29.5	
	Mean	SE	Mean	SE	Mean	SE	Mean	SE
Age (years)	61.3	3.9	41.8	0.7	56.9	1.9	41.1	0.6
Time since diagnosis (years)	8.9	0.9			11.1	1.4		
BMI (kg/m^2)	28.1	0.8	27.3	0.2	29.2	0.7	29.1	0.3
Total energy intake (kcal/day)	1936	92.7	2216	41.7	1695	82.7	1817	33.6
Total MxAHEI	48.0	1.2	44.4	0.3	46.6	0.6	45.1	0.4
Physical Activity (MET-min/week)	3273.4	855.8	7048.5	432.1	2847.7	635.4	4422.1	324.6
	%	SE	%	SE	%	SE	%	SE
Hypertension	53.6	9.0	9.5	1.4	45.0	5.6	17.1	1.5
BMI Categories								
Normal	23.0	7.0	37.0	2.3	23.5	5.8	27.8	1.8
Overweight	50.0	9.2	35.5	2.2	38.1	5.4	33.0	1.7
Obesity	27.0	6.9	28.0	2.2	38.4	5.5	39.2	1.9
Area								
Urban	83.6	4.3	75.4	1.8	85.4	3.5	75.0	1.5
Rural	16.4	4.3	24.6	1.8	14.6	3.5	25.0	1.5
Region								
North	16.2	4.4	19.8	1.5	23.4	4.3	19.4	1.3
Central	61.8	7.7	47.9	2.5	44.4	6.8	48.9	2.0
South	22.0	6.1	32.3	2.1	32.2	5.3	31.7	1.8
Tertiles of Wealth Index								
Low	17.3	5.9	27.3	1.9	29.6	5.5	25.2	1.6
Medium	17.5	4.6	31.5	2.1	31.1	4.7	33.2	2.2
High	65.2	7.6	41.2	2.1	39.3	6.4	41.6	2.2
Education level								
Elementary School or less	61.4	8.5	41.3	2.3	81.4	4.0	42.6	2.1
Middle School	13.1	4.5	32.5	2.7	12.3	3.1	25.9	1.7
High School	11.5	4.9	14.0	1.6	3.8	2.1	18.1	1.4
College or more	14.0	6.4	12.2	1.5	2.5	1.3	13.4	1.7
Smoking status								
Never	35.2	10.0	45.8	2.4	74.9	5.8	81.0	1.7
Former	39.1	10.7	28.0	2.6	12.6	3.2	10.0	1.3
Current	25.7	6.9	26.2	2.6	12.5	5.1	9.0	1.2

[a] Values are presented as means or percentages and SEs. Estimates were weighted to adjust for unequal probability of sampling and to be nationally representative. MET—Metabolic Equivalents; BMI—Body Mass Index. SE—standard error; MxAHEI—Mexican Alternate Healthy Eating Index.

Among women, the total MxAHEI score was similar in individuals with diabetes compared to those without diabetes (45.3; 95% CI: 43.9, 46.7 vs. 45.1; 95% CI: 44.5, 45.7, respectively). There were significant differences in omega-3 and SSBs scores. Women with diabetes had lower intakes for omega-3 (lower MxAHEI score); and SSBs (higher MxAHEI score) compared to those without diabetes in both, unadjusted and adjusted models (Tables S1 and S2, respectively). The intakes of most dietary components were similar across diabetes categories. Nevertheless, the intakes of most dietary healthy components were lower (lower MxAHEI score) among individuals with diabetes than among those who did not have diabetes.

Table S3 and Table 4 present sex-specific unadjusted and adjusted proportional macronutrient intakes by diabetes status. Among men with diabetes, we observed higher intakes of protein and total, monounsaturated and saturated fatty acids, and lower intakes of carbohydrates and added sugars compared to men without diabetes. Although not significant, the intakes of the remaining nutrients were higher among individuals with diabetes versus those without diabetes. The intake of total and monounsaturated fatty acids was not statistically different between men with and without diabetes in adjusted models.

Table 3. Sex-stratified adjusted Mexican Alternate Healthy Eating Index Score and components in Mexican adults by diabetes status (n = 2762) [a].

Variables		Men						Women					
		Diabetes		Non-Diabetes				Diabetes		Non-Diabetes			
N		82		1060				159		1461			
	Maximum Points	Mean	95% CI	Mean	95% CI	Diff Diab vs Non-Diab	*p*-Value	Mean	95% CI	Mean	95% CI	Diff Diab vs. Non-Diab	*p*-Value
Total Scores	100	46.9	44.6, 49.2	44.3	44.2, 45.6	2.6	0.018	45.3	43.9, 46.7	45.1	44.5, 45.7	0.2	0.807
Component Scores													
Higher intake recommended													
Vegetables	9	6.6	5.6, 7.5	6.0	5.7, 6.3	0.6	0.318	5.9	5.3, 6.5	6.3	6.0, 6.5	−0.4	0.245
Whole fruit	9	3.9	3.1, 4.6	2.9	2.7, 3.2	1.0	0.015	2.9	2.4, 3.3	3.3	3.1, 3.5	−0.4	0.081
Fiber from cereals	9	2.2	1.7, 2.8	2.2	2.0, 2.3	0.0	0.763	1.9	1.5, 2.3	2.1	1.9, 2.2	−0.2	0.335
Nuts	5	0.3	0.1, 0.5	0.3	0.2, 0.3	0.0	0.684	0.2	0.1, 0.2	0.2	0.1, 0.2	0.0	0.780
Legumes	5	3.6	3.1, 4.1	3.5	3.3, 3.7	0.1	0.696	3.0	2.5, 3.5	3.1	2.9, 3.3	−0.1	0.761
EPA + DHA	9	3.2	2.5, 3.9	2.9	2.7, 3.1	0.3	0.347	1.8	1.4, 2.2	2.6	2.4, 2.7	−0.8	0.001
Polyunsaturated Fats	9	6.0	5.3, 6.6	5.7	5.5, 5.9	0.3	0.443	5.2	4.7, 5.8	5.7	5.5, 5.9	−0.5	0.138
Limited intake recommended													
SSBs	9	3.9	2.6, 5.1	2.9	2.2, 3.8	1.0	0.010	5.7	4.9, 6.5	4.3	4.0, 4.6	1.4	0.001
Red/processed meat	9	3.9	3.2, 4.7	4.1	3.8, 4.4	−0.2	0.758	4.8	4.1, 5.5	4.8	4.5, 5.1	0.0	0.962
Sodium	9	2.7	1.6, 3.8	3.4	3.0, 3.8	−0.7	0.245	4.9	3.9, 5.9	3.9	3.6, 4.3	1.0	0.056
Trans Fats	9	9.0	9.0, 9.0	9.0	9.0, 9.0	0.0	0.471	9.0	9.0, 9.0	9.0	9.0, 9.0	0.0	0.995
Alcohol	9	1.60	0.8, 2.5	1.44	1.2, 1.7	0.2	0.713	0.04	−0.04, 0.12	0.01	0.001, 0.03	0.03	0.557

[a] Sex-specific multivariable linear regression models were used to predict dietary scores according to diabetes status and adjusting for age, hypertension status, body mass index, smoking status, educational level, wealth index, rural/urban area, and geographic region. Estimates were weighted to adjust for unequal probability of sampling and to be nationally representative. CI—confidence intervals; DHA—docosahexaenoic acid; EPA—eicosapentaenoic acid—SSBs, sugar-sweetened beverages.

Table 4. Sex-stratified adjusted intakes of macronutrients in Mexican adults by diabetes status (n = 2762) [a].

Variables	Men						Women					
	Diabetes		Non-Diabetes				Diabetes		Non-Diabetes			
N	82		1060				159		1461			
	Mean	95% CI	Mean	95% CI	Diff Diab vs Non-Diab	*p*-Value	Mean	95% CI	Mean	95% CI	Diff Diab vs. Non-Diab	*p*-Value
Macronutrients (Intake Per Day)												
Total protein (g/1000 kcal)	33.6	31.9, 35.2	31.0	30.4, 31.6	2.6	0.010	33.2	31.7, 34.6	31.7	31.1, 32.3	1.5	0.067
Total carbohydrates (g/1000 kcal)	137.1	130.9, 143.2	144.2	142.0, 146.4	−7.1	0.038	145.8	139.6, 152.0	148.9	147.1, 150.8	−3.1	0.354
Fiber (g/1000 kcal)	14.0	13.0, 14.9	13.0	12.7, 13.3	1.0	0.055	14.2	13.3, 15.1	14.0	13.7, 14.3	0.2	0.699
Total sugars (g/1000 kcal)	52.3	46.9, 57.7	56.0	54.1, 57.9	−3.7	0.187	55.1	50.3, 59.9	60.8	58.8, 62.9	−5.7	0.041
Added sugars (g/1000 kcal)	24.9	20.6, 29.2	34.7	32.8, 36.6	−9.8	0.000	26.8	22.0, 31.6	35.0	32.9, 37.1	−8.2	0.003
Total fatty acids (g/1000 kcal)	34.2	32.2, 36.3	32.1	31.5, 32.8	2.1	0.059	32.9	30.9, 34.9	32.8	32.1, 33.4	0.1	0.938
Saturated fatty acids (g/1000 kcal)	12.9	11.7, 14.1	11.5	11.2, 11.8	1.4	0.025	12.8	11.6, 13.9	12.1	11.9, 12.4	0.7	0.322
Monounsaturated fatty acids (g/1000 kcal)	12.0	11.0, 13.0	11.1	10.8, 11.3	0.9	0.082	10.5	9.6, 11.4	11.0	10.8, 11.3	−0.5	0.252
Polyunsaturated fatty acids (g/1000 kcal)	8.0	7.3, 8.6	7.7	7.5, 7.9	0.3	0.473	7.3	6.7, 7.8	7.8	7.6, 8.1	−0.5	0.062
Trans fat (g/1000 kcal)	0.3	0.2, 0.3	0.2	0.2, 0.2	0.1	0.206	0.3	0.2, 0.3	0.3	0.3, 0.3	0.0	0.556
Omega-3 (g/1000 kcal)	0.05	0.04, 0.07	0.05	0.04, 0.05	0.00	0.487	0.04	0.03, 0.06	0.05	0.04, 0.05	0.01	0.687

[a] Sex-specific multivariable linear regression models were used to predict macronutrient intakes according to diabetes status and adjusting for age, hypertension status, body mass index, smoking status, educational level, wealth index, rural/urban area, and geographic region. Estimates were weighted to adjust for unequal probability of sampling and to be nationally representative. CI—confidence intervals.

Women with diabetes had lower intakes of total and added sugars than women without diabetes. On the other hand, total protein intake was higher in women with diabetes than women without diabetes. While not significant, the intakes of carbohydrates, monounsaturated, and polyunsaturated fatty acids were lower in women with diabetes than those without diabetes. Conversely, the intakes of fiber, total and saturated fatty acids, and omega-3 were higher among women with diabetes versus those without diabetes. The difference in total protein intake between women with and without diabetes was no longer statistically significant in the adjusted models.

Table S3 and Table 5 show sex-specific unadjusted and adjusted proportional mean daily micronutrient intakes by diabetes status. Men with diabetes had higher daily intakes of vitamins A, D, B6, and B12 and folate, calcium, magnesium, zinc, and potassium than did men who did not have diabetes. While not significant, similar trends were observed for intakes of the remaining micronutrients. In the adjusted model, the difference in the intakes of vitamins A, D, B6, and B12 between men with and without diabetes was no longer significant.

Table 5. Sex-stratified adjusted intakes of micronutrients in Mexican adults by diabetes status (n = 2762) [a].

Variables	Men						Women					
	Diabetes		No Diabetes				Diabetes		No Diabetes			
N	82		1060				159		1461			
	Mean	95% CI	Mean	95% CI	Diff Diab vs Non-diab	p-Value	Mean	95% CI	Mean	95% CI	Diff Diab vs. Non-Diab	p-Value
Micronutrients (Intake Per Day)												
Vitamin A (RAE) (μg/1000 kcal)	336.5	295.9, 377.0	314.6	298.2, 331.0	21.9	0.345	428.4	373.8, 483.1	404.2	386.6, 421.8	24.2	0.390
Vitamin D * (μg/1000 kcal)	2.3	1.8, 2.8	1.8	1.7, 1.9	0.5	0.065	2.2	1.8, 2.5	2.0	1.9, 2.1	0.2	0.274
Vitamin E (mg/1000 kcal)	3.5	3.1, 3.9	3.1	3.0, 3.2	0.4	0.072	3.2	3.0, 3.4	3.4	3.3, 3.5	−0.2	0.087
Folate (μg/1000 kcal)	164.4	151.9, 177.0	149.7	144.6, 154.9	14.7	0.043	179.1	163.0, 195.3	177.7	172.4, 182.9	1.4	0.854
Vitamin C (mg/1000 kcal)	91.5	74.2, 108.8	79.6	74.0, 85.3	11.9	0.209	95.6	81.7, 109.6	102.5	95.9, 109.0	−6.9	0.370
Vitamin B-6 (mg/1000 kcal)	0.9	0.9, 1.0	0.9	0.9, 0.9	0.0	0.244	1.4	0.6, 2.1	1.0	0.9, 1.1	0.4	0.321
Vitamin B-12 (μg/1000 kcal)	2.3	1.9, 2.6	2.1	2.0, 2.2	0.2	0.314	2.4	2.1, 2.7	2.2	2.1, 2.3	0.2	0.254
Calcium (mg/1000 kcal)	494.9	450.2, 539.7	424.6	413.7, 435.4	70.3	0.005	549.7	508.5, 591.0	469.5	458.0, 481.0	80.2	0.000
Magnesium (mg/1000 kcal)	207.0	191.6, 222.3	188.1	184.1, 192.1	18.9	0.022	215.3	200.2, 230.4	197.1	193.4, 200.8	18.2	0.022
Zinc (mg/1000 kcal)	5.3	5.0, 5.6	5.0	4.9, 5.0	0.3	0.037	5.5	5.3, 5.7	5.1	5.0, 5.2	0.4	0.000
Potassium (mg/1000 kcal)	1420.1	1320.8, 1519.4	1223.7	1190.0, 1257.4	196.4	0.001	1428.1	1331.5, 1524.8	1370.1	1340.9, 1399.3	58.0	0.252
Sodium (mg/1000 kcal)	1155.0	1060.5, 1249.5	1081.6	1048.7, 1114.4	73.4	0.164	1103.2	1010.3, 1196.1	1134.2	1107.3, 1161.1	−31.0	0.541

[a] Sex-specific multivariable linear regression models were used to predict macronutrient intakes according to diabetes status and adjusting for age, hypertension status, body mass index, smoking status, educational level, wealth index, rural/urban area, and geographic region. Estimates were weighted to adjust for unequal probability of sampling and to be nationally representative. * p-value < 0.05, CI—confidence intervals; RAE—retinol activity equivalents.

Among women, the average intakes of calcium, magnesium, zinc, and potassium were higher among participants who had diabetes than among those who did not have diabetes. For the remaining micronutrients, although the results were not significant, the intakes were higher among women with diabetes versus women without diabetes, except for vitamin E, C, and sodium. The potassium intake was not different between women with and without diabetes in adjusted models.

4. Discussion

We compared the diet quality of self-reported declared Mexican adults with and without diabetes. Overall, the adherence to the MxAHEI was poor among Mexican adults, with a score below 50 out of 100, regardless of the diabetes status. We found a higher adherence to the MxAHEI among men with diabetes than among men who did not have diabetes (score of 46.9 vs. 44.3, respectively). Specifically, adherence to the recommended intakes for fruits and SSBs was greater among men with diabetes versus those without diabetes. Among women, the total MxAHEI score was similar in individuals with diabetes compared to those without diabetes. Specifically, women with diabetes had lower intakes of omega-3 and SSBs than did those who did not have diabetes. We also found higher intakes of protein, saturated fatty acids, folate, calcium, magnesium, zinc, and potassium, and lower intakes of carbohydrates and added sugars among men with diabetes compared to those without diabetes. Finally, among women, the intakes of total and added sugars were lower, while calcium and magnesium were higher among individuals with diabetes versus those without diabetes.

In our study, we found that overall adherence to the MxAHEI was poor in both men and women, a finding consistent with previous studies that reported low adherence to dietary recommendations in the general population, characterized by low consumption of fruits and vegetables and high consumption of meat [23–25]. In addition, our results are in line with the findings of a cross-sectional study conducted in Spain, where it was found that adherence to the AHEI in adults was low, regardless of diabetes status [26].

The difference in the total MxAHEI score among men by diabetes status can be attributed to different intakes of some dietary components, particularly a higher intake of fruits and a lower intake of SSBs among individuals with diabetes than among those without diabetes. Similar to men, the consumption of SSBs was significantly lower among women with diabetes compared to their counterparts who did not have the disease. However, adherence to recommended consumption was slightly lower in women with diabetes than women who did not have diabetes in most dietary components, particularly in omega-3 fatty acids, which may explain our finding that there was no significant differences in the total MxAHEI among women who had diabetes and those who did not.

These findings are consistent with other population-based studies that analyzed dietary behaviors by diabetes status. Fitzgerald et al. and Nöthlings et al., found lower intakes in SSBs and slight increases in intakes of food groups considered healthy among individuals with diabetes than in those without diabetes [27,28]. The better adherence to recommendations of fruits and SSBs among individuals with diabetes was an expected result in our study. As part of medical nutrition therapy, individuals with diabetes are encouraged to increase their consumption of fruits, vegetables, and legumes while reducing the consumption of foods containing refined carbohydrates and saturated fats [29]. In our study, both men and women with diabetes had a better score for SSBs, however, their consumption was far below the perfect adherence (zero consumption of SSBs). The SSBs score in men and women was 3.9 and 5.7, respectively, from a maximum of 9.

We also observed that the overall adherence to dietary recommendations was higher in men with diabetes versus women with diabetes. Despite changes in gender roles over the last decades in Mexican society, a considerable gender gap remains in the time spent on unpaid work, such as household chores (6.4 h for Mexican women, 2.3 h for Mexican men) [30]. It has also been reported that women, rather than men, spend more time as caregivers. [31]. We can postulate that women who spend more time engaged in unpaid work and in caregiver roles may perceive they have limited ability to engage in self-care behaviors to manage their diabetes or other chronic diseases. These gender inequalities might decrease the ability for women to take care of their own, and this may hamper their lifestyle changes, including dietary behaviors, when diagnosed with diseases. Also, it has been observed that women may not be willing to change their lifestyles to not disrupt the family [32]. Future studies should examine dietary change dynamics after chronic disease diagnosis with a gender perspective, as social roles could play an important role in dietary modifications. For instance, studies could compare the similarity of diets in members of the same household, to assess if men have more dissimilar diets when diagnosed with diabetes than women under the same disease conditions.

Our findings of reduced intakes of carbohydrates and added sugars among adults with diabetes are consistent with previous studies in adults from several European countries and among women in the United States [33–35]. These results are expected given the link between dietary sugar and serum glucose or "blood sugar" levels [36]. Reducing the consumption of foods with a high glycemic index is a key nutritional recommendation of many international diabetes associations [37,38], including those in Mexico [39]. Part of this recommendation involves evenly distributing carbohydrate intake across meals and snacks throughout the day. Because carbohydrates have the greatest impact on blood glucose levels in diabetes, individuals may opt to limit foods containing carbohydrates as a way to manage their diabetes. In our study, the findings for added sugar and SSBs were in the same direction, which was foreseeable given that SSBs are the greatest contributor to added sugar intake in Mexico [16].

We observed that protein intake was higher in men and women with diabetes than those who did not have diabetes, in line with findings among adults from European countries [33–35]. However, we also found that the intake of total fat, and specifically the intakes of saturated and monounsaturated fatty acids, were higher among men with diabetes than in men without diabetes. Only the study by Virtanen et al. found a higher intake of total fatty acids among men with diabetes from Finland and the Netherlands [33]. Given that the reduction of one macronutrient is substituted with others [40], we hypothesize that among individuals in our study, the reduction in carbohydrates was replaced by additional proteins and fatty acids.

We observed differences in select micronutrients based on diabetes status. In general, the intakes of calcium, magnesium and zinc were higher in adults with diabetes than in those without diabetes. However, the trend in these micronutrients are not generalized to the consumption of the components of the MxAHEI, particularly in vegetables, legumes, nuts, and red meat. The inconsistency in findings between nutrient intakes and dietary components can be partly explained by the truncated nature of the index score. In other words, the probable variability in food intake among individuals with the lowest or highest scores was no longer considered when the score is assigned. Furthermore, the MxAHEI does not consider dairy, which contributes to the intake of several micronutrients, particularly calcium. We also observed that potassium and folate intakes were higher among men with diabetes than in men without diabetes. This is similar to a previous study carried out in European populations with and without diabetes, where similar trends were observed [33]. These results are linked with the higher consumption of fruits among men with diabetes because dietary potassium and folate can be found in a wide variety of fruits [41]. Lower intakes of folate and potassium among individuals with diabetes, particularly among women, raises the issue of improving intakes in this population in view that these micronutrients play a significant role in insulin synthesis and sensitivity [42–44].

Our analysis has several limitations that are important to acknowledge. First, even though the SFFQ provides a good estimation of usual intake and therefore of diet quality, it is not the optimal instrument for estimating absolute nutrient intakes. We addressed this limitation by reporting proportional intakes. Second, there is a potential loss of accuracy to detect differences between micronutrient intake and MxAHEI components due to the truncated nature of the latter. However, the MxAHEI is an adapted version of the AHEI-2010, which is the only diet quality index validated against different chronic diseases, including diabetes [8,18,45–47]. Specifically, the MxAHEI and AHEI-2010, unlike other dietary indices, include the SSBs component, which largely accounted for the differences found by diabetes status, as also observed in a previous study [9]. Third, we cannot rule out the possibility of social desirability bias [48], which may impact the reported dietary intake data of adults who have diabetes, relative to those who do not have diabetes. People who have diabetes are more likely to receive dietary counseling and education about diet than are individuals who do not have diabetes. Fourth, participants were not asked to respond whether they were diagnosed with type 1 or type 2 diabetes. However, since type 1 diabetes is usually diagnosed in children and adolescents [49] and is a less common type of diabetes [50], a low proportion of Mexican adults likely have type 1 diabetes. In our representative sample, 2.7% was diagnosed with diabetes during childhood or adolescence. Therefore, it is more likely that our results characterize adults with type 2 diabetes. Moreover, diet recommendations are similar between people with type 1 and type 2 diabetes [49]. Fifth, the duration or time since diagnosis is another factor that can affect the adherence to dietary recommendations. However, we could not conduct analyses among men and women with diabetes by time since diagnosis because the sample size is limited. Future studies with larger samples will help to understand the role of time since diagnosis on dietary recommendations adherence. Finally, we did not include physical activity in the regression models due to the poor validity of the International Physical Activity Questionnaire short form for assessing physical activity among Mexican adults [51]. Despite these limitations, the results presented provide an overview of the adherence to dietary guidelines in adults and how this adherence may be different by diabetes status in a representative sample of Mexican men and women. This offers information to understand the disparities in diet and health in Mexico.

5. Conclusions

In conclusion, we observed higher adherence to dietary recommendations among Mexican adults with diabetes than adults who did not have diabetes, especially among men. These findings suggest that Mexican adults with diabetes are adhering to the dietary guidelines in some extent. However, we emphasize that MxAHEI scores, total and by component, were well below the recommendations. Therefore, it is necessary to reinforce the implementation and evaluation of programs that promote healthy dietary patterns in the Mexican population. It is also important to consider in such strategies the potential gender inequalities for reaching the dietary recommendations.

Supplementary Materials: The following are available online at http://www.mdpi.com/2072-6643/12/11/3464/s1, Table S1. Sex-stratified unadjusted Mexican Alternate Healthy Eating Index Score and components in Mexican adults by diabetes status. Table S2. Sex-stratified unadjusted intakes of micronutrients in Mexican adults by diabetes status. Table S3. Sex-stratified unadjusted intakes of macronutrients in Mexican adults by diabetes status.

Author Contributions: N.L.-O. and T.B.-G. were responsible for study design, results interpretation, and drafting and editing of the manuscript; A.B.-A. was responsible for results interpretation and writing the manuscript; A.R.-G. was responsible of performing statistical analysis, results interpretation, and writing the manuscript; N.L.-O., A.B.-A., A.R.-G., T.S.-L., and T.B.-G. contributed to study design; S.J. and C.J.A. were responsible for results interpretation and editing the manuscript, and all authors contributed to critical revision of the manuscript. T.B.-G. has final responsibility of this study. All authors have read and agreed to the published version of the manuscript.

Funding: Funding for conducting this research was provided by Abbott Nutrition, Columbus, Ohio, United States of America.

Conflicts of Interest: The authors declare no conflict of interest.

References

1. Institute for Health Metrics and Evaluation (IHME). *Mexico Profile*; IHME, University of Washington: Seattle, WA, USA, 2018; Available online: http://www.healthdata.org/mexico (accessed on 13 June 2019).
2. Rojas-Martínez, R.; Basto-Abreu, A.; Aguilar-Salinas, C.A.; Zárate-Rojas, E.; Villalpando, S.; Barrientos-Gutiérrez, T. Prevalence of previously diagnosed diabetes mellitus in Mexico. *Salud Publica Mex.* **2018**, *60*, 224–232.
3. Arredondo, A.; Reyes, G. Health Disparities from Economic Burden of Diabetes in Middle-income Countries: Evidence from México. *PLoS ONE* **2013**, *8*, e68443. [CrossRef]
4. Evert, A.B.; Dennison, M.; Gardner, C.D.; Garvey, W.T.; Lau, K.H.K.; MacLeod, J.; Mitri, J.; Pereira, R.F.; Rawlings, K.; Robinson, S.; et al. Nutrition Therapy for Adults With Diabetes or Prediabetes: A Consensus Report. *Diabetes Care* **2019**, *42*, 731–754. [CrossRef] [PubMed]
5. Ahola, A.J.; Groop, P.-H. Barriers to self-management of diabetes. *Diabet. Med.* **2013**, *30*, 413–420. [CrossRef]
6. Han, C.Y.; Chan, C.G.B.; Lim, S.L.; Zheng, X.; Woon, Z.W.; Chan, Y.T.; Bhaskaran, K.; Tan, K.F.; Mangaikarasu, K.; Chong, M.F.-F. Diabetes-related nutrition knowledge and dietary adherence in patients with Type 2 diabetes mellitus: A mixed-methods exploratory study. *Proc. Singap. Heal.* **2020**, *29*, 81–90. [CrossRef]
7. Kennedy, E.T.; Ohls, J.; Carlson, S.; Fleming, K. The healthy eating index: Design and applications. *J. Am. Diet. Assoc.* **1995**, *95*, 1103–1108. [CrossRef]
8. Chiuve, S.E.; Fung, T.T.; Rimm, E.B.; Hu, F.B.; McCullough, M.L.; Wang, M.; Stampfer, M.J.; Willett, W.C. Alternative Dietary Indices Both Strongly Predict Risk of Chronic Disease. *J. Nutr.* **2012**, *142*, 1009–1018. [CrossRef] [PubMed]
9. Al-Ibrahim, A.A.; Jackson, R.T. Healthy eating index versus alternate healthy index in relation to diabetes status and health markers in U.S. adults: NHANES 2007–2010. *Nutr. J.* **2019**, *18*, 26. [CrossRef] [PubMed]
10. López-Olmedo, N.; Popkin, B.M.; A Mendez, M.; Smith, T.L. The association of overall diet quality with BMI and waist circumference by education level in Mexican men and women. *Public Health Nutr.* **2019**, *22*, 2777–2792. [CrossRef] [PubMed]
11. Gutiérrez, J.P.; Rivera-Dommarco, J.; Shamah-Levy, T.; Villalpando-Hernández, S.; Franco, A.; Cuevas-Nasu, L.; Gutiérrez, J.P.; Rivera-Dommarco, J.Á. *Encuesta Nacional de Salud y Nutrición 2012*; Resultados Nacionales; Instituto Nacional de Salud Pública (MX): Cuernavaca, Mexico, 2012.

12. Ramírez-Silva, I.; Jiménez-Aguilar, A.; Valenzuela-Bravo, D.; Martínez-Tapia, B.; Rodríguez-Ramírez, S.; Gaona-Pineda, E.B.; Angulo-Estrada, S.; Shamah-Levy, T. Metodología para la estimación de información dietética del cuestionario semicuantitativo de frecuencia de consumo de alimentos de la Encuesta Nacional de Salud y Nutrición 2012. *Salud Publica Mex.* **2016**, *58*, 629–638. [PubMed]
13. Denova-Gutiérrez, E.; Ramírez-Silva, I.; Rodríguez-Ramírez, S.; Jiménez-Aguilar, A.; Shamah-Levy, T.; A Rivera-Dommarco, J. Validity of a food frequency questionnaire to assess food intake in Mexican adolescent and adult population. *Salud Publica Mex.* **2016**, *58*, 617–628. [CrossRef] [PubMed]
14. López-Olmedo, N.; Carriquiry, A.L.; Rodríguez-Ramírez, S.; Ramírez-Silva, I.; Espinosa, J.; Hernandez, L.; Campirano, F.; Martínez-Tapia, B.; A Rivera, J. Usual Intake of Added Sugars and Saturated Fats Is High while Dietary Fiber Is Low in the Mexican Population. *J. Nutr.* **2016**, *146*, 1856S–1865S. [CrossRef]
15. Instituto Nacional de Salud Publica. *Base de Datos de Valor Nutritivo de los Alimentos*; [National Institute of Public Health. Food Composition Table.]; Compilación del Instituto Nacional de Salud Pública Cuernavaca (Mexico); National Institute of Public Health: Cuernavaca, Mexico, 2012.
16. Sánchez-Pimienta, T.G.; Batis, C.; Lutter, C.K.; A Rivera, J. Sugar-Sweetened Beverages Are the Main Sources of Added Sugar Intake in the Mexican Population. *J. Nutr.* **2016**, *146*, 1888S–1896S. [CrossRef]
17. Pérez, L.A.; Palacios, G.B.; Castro, B.A.; Flores, G.I. *Sistema Mexicano de Alimentos Equivalentes*, 4th ed.; Fomento de Nutrición y Salud A.C.: Mexico City, Mexico, 2014.
18. McCullough, M.L.; Feskanich, D.; Stampfer, M.J.; Giovannucci, E.L.; Rimm, E.B.; Hu, F.B.; Spiegelman, D.; Hunter, D.J.; A Colditz, G.; Willett, W.C. Diet quality and major chronic disease risk in men and women: Moving toward improved dietary guidance. *Am. J. Clin. Nutr.* **2002**, *76*, 1261–1271. [CrossRef]
19. Habitch, J.P. Estandarización de Métodos Epidemiológicos Cuantitativos Sobre el Terreno. *Boletín de la Oficina Sanitaria Panamericana* **1974**, *76*, 375–385.
20. World Health Organization. *Physical Status: The Use and Interpretation of Anthropometry*; Technical Report Series 894; WHO: Geneva, Switzerland, 1995.
21. Gutiérrez, J.P. Clasificación socioeconómica de los hogares en la ENSANUT 2012. *Salud Publica Mex.* **2013**, *55*, S341–S346. [CrossRef]
22. StataCorp. *Stata Statistical Software: Release 13*; StataCorp LP: College Station, TX, USA, 2013.
23. Moreira, P.R.S.; Rocha, N.P.; Milagres, L.C.; De Novaes, J.F. Análise crítica da qualidade da dieta da população brasileira segundo o Índice de Alimentação Saudável: Uma revisão sistemática. *Ciência & Saúde Coletiva* **2015**, *20*, 3907–3923. [CrossRef]
24. Ayele, A.A.; Emiru, Y.K.; Tiruneh, S.A.; Ayele, B.A.; Gebremariam, A.D.; Tegegn, H.G. Level of adherence to dietary recommendations and barriers among type 2 diabetic patients: A cross-sectional study in an Ethiopian hospital. *Clin. Diabetes Endocrinol.* **2018**, *4*, 1–7. [CrossRef] [PubMed]
25. Rivellese, A.A.; Boemi, M.; Cavalot, F.; Costagliola, L.; De Feo, P.; Miccoli, R.; Patti, L.; Trovati, M.; Vaccaro, O.; Zavaroni, I.; et al. Dietary habits in type II diabetes mellitus: How is adherence to dietary recommendations? *Eur. J. Clin. Nutr.* **2008**, *62*, 660–664. [CrossRef] [PubMed]
26. Alcubierre, N.; Granado-Casas, M.; Real, J.; Perpiñán, H.; Rubinat, E.; Falguera, M.; Castelblanco, E.; Franch-Nadal, J.; Mauricio, D. Spanish People with Type 2 Diabetes Show an Improved Adherence to the Mediterranean Diet. *Nutrients* **2020**, *12*, 560. [CrossRef]
27. Fitzgerald, N.; Damio, G.; Segura-Pérez, S.; Pérez-Escamilla, R. Nutrition Knowledge, Food Label Use, and Food Intake Patterns among Latinas with and without Type 2 Diabetes. *J. Am. Diet. Assoc.* **2008**, *108*, 960–967. [CrossRef] [PubMed]
28. Nöthlings, U.; Boeing, H.; Maskarinec, G.; Hutri-Kähönen, N.; Teucher, B.; Kaaks, R.; Tjønneland, A.; Halkjaer, J.; Dethlefsen, C.; Overvad, K.; et al. Food intake of individuals with and without diabetes across different countries and ethnic groups. *Eur. J. Clin. Nutr.* **2011**, *65*, 635–641. [CrossRef]
29. Secretaría de Salud. NOM-015-SSA2-1994, Para la Prevención, Tratamiento y Control de la Diabetes. *Diario Oficial de la Federación 2000.* Available online: http://www.salud.gob.mx/unidades/cdi/nom/m015ssa24.html (accessed on 22 March 2020).
30. Organization for Economic Co-operation and Development (OECD). Gender Equality. *Balancing Paid Work, Unpaid Work and Leisure. Paris, France.* 2018. Available online: https://www.oecd.org/gender/balancing-paid-work-unpaid-work-and-leisure.htm# (accessed on 20 June 2020).
31. Swinkels, J.; Tilburg, T.V.; Verbakel, E.; van Broese Groenou, M. Explaining the Gender Gap in the Caregiving Burden of Partner Caregivers. *J. Gerontol. B Psychol. Sci. Soc. Sci.* **2019**, *74*, 309–317. [PubMed]

32. A Siddiqui, M.; Khan, M.F.; E Carline, T. Gender Differences in Living with Diabetes Mellitus. *Mater. Sociomed.* **2013**, *25*, 140–142. [CrossRef]
33. Virtanen, S.M.; Feskens, E.; Räsänen, L.; Fidanza, F.; Tuomilehto, J.; Giampaoli, S.; Nissinen, A.; Kromhout, D. Comparison of diets of diabetic and non-diabetic elderly men in Finland, The Netherlands and Italy. *Eur. J. Clin. Nutr.* **2000**, *54*, 181–186. [CrossRef]
34. Shimakawa, T.; Herrera-Acena, M.G.; Colditz, G.A.; E Manson, J.; Stampfer, M.J.; Willett, W.C.; Stamper, M.J. Comparison of Diets of Diabetic and Nondiabetic Women. *Diabetes Care* **1993**, *16*, 1356–1362. [CrossRef] [PubMed]
35. Gauthier-Chelle, K.; Mennen, L.; Arnault, N.; Rigalleau, V.; Hercberg, S.; Gin, H. Comparison of the diet of self-declared diabetics with non-diabetic patients in the SU.VI.MAX study: Did the diabetics modify their nutritional behavior? *Diabetes Metab.* **2004**, *30*, 535–542. [CrossRef]
36. Clay, D. Public Perceptions of Sugar and Health: Implications for Consumption. In Proceedings of the Sugar Cuba Conference; FAO: La Habana, Cuba, 1999. Available online: http://www.fao.org/3/x4988e/x4988e11.htm (accessed on 24 May 2019).
37. American Diabetes Association. 4. Lifestyle management: Standards of Medical Care in Diabetes 2018. *Diabetes Care* **2018**, *41* (Suppl. S1), S38–S50. [CrossRef]
38. World Health Organization. *Diet, Nutrition and the Prevention of Chronic Diseases*; Report of a Joint WHO/FAO Expert Consultation, WHO Technical Report Series No. 916; WHO: Geneva, Switzerland, 2003.
39. Instituto Mexicano del Seguro Social. *Tratamiento de la Diabetes Mellitus tipo 2 en el primer nivel de Atención*; Instituto Mexicano del Seguro Social: Mexico City, Mexico, 2014; Available online: http://www.cenetec.salud.gob.mx/interior/catalogoMaestroGPC.html (accessed on 2 July 2019).
40. Willett, W. *Nutritional Epidemiology*, 2nd ed.; Oxford University Press: New York, NY, USA, 1998.
41. National Institutes of Health. Office of Dietary Supplements. In *Folate. Fact Sheet for Health Professionals*; Office of Dietary Supplements. Available online: https://ods.od.nih.gov/factsheets/Folate-HealthProfessional/ (accessed on 11 March 2020).
42. Valdes-Ramos, R.; Laura, G.-L.A.; Elina, M.-C.B.; Donají, B.-A.A. Vitamins and Type 2 Diabetes Mellitus. *Endocrine Metab. Immune Disord.Drug Targets* **2015**, *15*, 54–63. [CrossRef]
43. Martini, L.A.; Catania, A.S.; Ferreira, S.R.G. Role of vitamins and minerals in prevention and management of type 2 diabetes mellitus. *Nutr. Rev.* **2010**, *68*, 341–354. [CrossRef]
44. Durlach, J.; Collery, P. Magnesium and potassium in diabetes and carbohydrate metabolism. Review of the present status and recent results. *Magnesium* **1984**, *3*, 315–323.
45. Jannasch, F.; Kröger, J.; Boeing, H. Dietary Patterns and Type 2 Diabetes: A Systematic Literature Review and Meta-Analysis of Prospective Studies. *J. Nutr.* **2017**, *147*, 1174–1182. [CrossRef] [PubMed]
46. Chen, G.-C.; Koh, W.-P.; Neelakantan, N.; Yuan, J.-M.; Qin, L.; Van Dam, R.M. Diet Quality Indices and Risk of Type 2 Diabetes Mellitus: The Singapore Chinese Health Study. *Am. J. Epidemiol.* **2018**, *187*, 2651–2661. [CrossRef] [PubMed]
47. McCullough, M.L.; Willett, W.C. Evaluating adherence to recommended diets in adults: The Alternate Healthy Eating Index. *Public Health Nutr.* **2006**, *9*, 152–157. [CrossRef] [PubMed]
48. Kipnis, V.; Midthune, D.; Freedman, L.; Bingham, S.; E Day, N.; Riboli, E.; Ferrari, P.; Carroll, R.J. Bias in dietary-report instruments and its implications for nutritional epidemiology. *Public Health Nutr.* **2002**, *5*, 915–923. [CrossRef] [PubMed]
49. International Diabetes Federation. About Diabetes. Type 1 Diabetes. 2020. Available online: https://www.idf.org/aboutdiabetes/type-1-diabetes.html (accessed on 9 January 2020).

50. International Diabetes Federation. *IDF Diabetes Atlas*, 9th ed.; International Diabetes Federation: Brussels, Belgium, 2019.
51. Medina, C.; Barquera, S.; Janssen, I.I. Validity and reliability of the International Physical Activity Questionnaire among adults in Mexico. *Rev. Panam. Salud Publica.* **2013**, *34*, 21. [PubMed]

Publisher's Note: MDPI stays neutral with regard to jurisdictional claims in published maps and institutional affiliations.

nutrients

Article

Mediterranean Personalized Diet Combined with Physical Activity Therapy for the Prevention of Cardiovascular Diseases in Italian Women

Laura Di Renzo [1,2,*,†], Giulia Cinelli [3,4,†], Maria Dri [5], Paola Gualtieri [1], Alda Attinà [3], Claudia Leggeri [3], Giuseppe Cenname [6], Ernesto Esposito [7], Alberto Pujia [2], Gaetano Chiricolo [2], Chiara Salimei [8] and Antonino De Lorenzo [1]

1. Section of Clinical Nutrition and Nutrigenomic, Department of Biomedicine and Prevention, University of Rome Tor Vergata, 00133 Rome, Italy; paola.gualtieri@uniroma2.it (P.G.); delorenzo@uniroma2.it (A.D.L.)
2. Department of Biomedicine and Prevention, University of Rome Tor Vergata, 00133 Rome, Italy; albpujia@gmail.com (A.P.); nucciochiricolo@gmail.com (G.C.)
3. School of Specialization in Food Sciences, University of Rome Tor Vergata, 00133 Rome, Italy; giuliacinelli88@gmail.com (G.C.); alda.attina@gmail.com (A.A.); claudialeggeri@gmail.com (C.L.)
4. Predictive and Preventive Medicine Research Unit, Bambino Gesù Children's Hospital, IRCCS, 00165 Rome, Italy
5. Department of Surgical Sciences, School of Applied Medical-Surgical Sciences, University of Rome Tor Vergata, 00133 Rome, Italy; marystenr@hotmail.it
6. Comando Generale Arma Carabinieri, Direzione di Sanità, 00197 Rome, Italy; giuseppe.cenname@carabinieri.it
7. Department of Human Policies (General Directorate) of Basilicata Region, 85100 Potenza, Italy; ernesto.esposit1962@gmail.com
8. Department of Systems Medicine, University of Rome Tor Vergata, 00133 Rome, Italy; chiara.salimei@gmail.com

* Correspondence: laura.di.renzo@uniroma2.it; Tel.: +390-672-596-855

† These authors contributed equally to this work.

Received: 20 October 2020; Accepted: 9 November 2020; Published: 11 November 2020

Abstract: Cardiovascular diseases (CVDs) and inflammatory risk indexes are used to calculate the exposure to morbidity. Most of them are suggested by the American College of Cardiology/American Heart Association to predict the risk of CVDs diagnosis in primary prevention, instead of treating the ongoing pathology. Prevention starts from habit changes with the prescription of diet and physical activity (PA). The aim of the study is to investigate the effectiveness of a personalized Mediterranean Diet (MD) and a PA intervention, on the risk indexes Atherogenic Index of Plasma (AIP), Lipid Accumulation Product (LAP) and Fatty Liver Index (FLI) in a population of women at risk of CVDs with different pathological conditions. After treatment, patients achieved the best results in body composition (BC) and laboratory tests. The BC analysis showed a significant reduction of total body Fat Mass (FM). CVDs risk indexes significantly decreased, except for Neutrophil/Lymphocyte (NLR) and Platelet/Lymphocyte Ratios (PLR). The reduction of the CVDs indexes associated with lipid profile was linked to both weight and FM decrease. AIP and LAP were significantly reduced when losing fat mass and body weight, respectively. A personalized MD therapy plus a PA program led to body weight loss, BC remodelling and risk indexes reduction.

Keywords: cardiovascular disease; Mediterranean diet; non-communicable disease; obesity; physical activity

1. Introduction

The World Health Organization (WHO) declares obesity as a risk factor for non-communicable diseases (NCD) [1,2], consisting of a multifactorial pathology and chronic low-grade inflammatory disease [3]. Obesity is a complex, heterogeneous and multifactorial morbid condition, to which both environmental and genetic factors contribute. It occurs due to an imbalance between caloric intake and energy expenditure, with consequent accumulation of an excess of adipose tissue, such as to increase the risk of morbidity and mortality of the individual himself [4]. As adiposopathy consisting not so much in weight gain rather in excess fat mass [5], the Body Mass Index (BMI) cannot define obesity [6,7], and it is not sufficient to evaluate the risk of developing a chronic disease. Therefore, obesity must be evaluated with tools that measure the amount of fat mass and consider the relationship between environment-genes and metabolic diseases [8].

Obesity may worsen metabolic disease in adverse endocrine and immune responses triggered by body fat dysfunction [9]. Among the most known pathologies caused by fat mass accumulation, there is dyslipidemia and cardiovascular diseases (CVDs) [10].

According to the most recent guidelines on the management of obesity, prevention must start from therapeutic lifestyle changes, which include the prescription of personalized diet therapy following the evaluation of the nutritional status, the prescription of a physical activity (PA) program and behavioural therapy to maintain and strengthen patient's adherence to the treatment [11].

A key goal of cardiovascular disease prevention efforts is to predict events over medium to long periods of time [12]. Valid predictive indexes of cardiovascular and Myocardial Infarction (MI) risk, atherosclerosis and coronary heart disease have been determined: the total cholesterol/high-density lipoprotein cholesterol (cHDL) ratio [13], the triglycerides (TG)/cHDL ratio [14]; the low-density lipoprotein cholesterol (cLDL)/cHDL ratio [15]; the Atherogenic Index of Plasma (AIP) [16]; the Lipid Accumulation Product (LAP) [17,18]; Fatty Liver Index (FLI) [19]. Among different inflammatory markers, such as C-Reactive Protein (CRP) and Erythrocyte Sedimentation Rate (ESR) [20], the evaluation of Neutrophil/Lymphocyte Ratio (NLR) and the Platelet/Lymphocyte Ratio (PLR) may have a high value in predicting the prognosis of different CVDs including MI, acute coronary syndrome, Heart Failure (HF) and Atherosclerosis [21,22]. Moreover, to predict the risk and recommend management strategies for those at risk of Atherosclerotic Cardiovascular Diseases (ASCVD), a specific Risk Algorithm was standardized (ASCVD risk Algorithm) [23,24]. In addition, the BARD score (BMI, Alanine aminotransferase (ALT)/Aspartate aminotransferase (AST) ratio (AAR), Diabetes Mellitus (DM)), one of the most used algorithms for fibrosis evaluation, could be useful to predict a CVDs risk linked to the cHDL value in patients with hepatic fibrosis [25,26].

The guidelines of the American College of Cardiology/American Heart Association (ACC/AHA) for the prevention of ASCVD focus on the evaluation of cardiovascular risk [23,24], the implementation of a healthy lifestyle, the management of overweight and obesity and the treatment of high blood pressure and hypercholesterolemia.

Di Renzo et al. highlighted the personalization of the Mediterranean Diet (MD) plays a key role in the prevention and treatment of Non-Communicable Chronic Diseases (NCCDs) [11]. Moreover, numerous studies evaluated the role of MD on different parameters and confirmed that adherence to the MD was associated with a reduced incidence, prevalence and mortality from coronary heart disease, as well as with other CVDs and a reduced all-cause mortality [27]. As reported by Martini et al., the positive action of MD on the cardiometabolic risk was due to the capability of decreasing the risk of diabetes and metabolic-related conditions [28]. Study on gene–diet interaction showed that MD represented a good nutritional treatment to reduce the body fat mass [29]. Moreover, several parameters, i.e., BMI, waist circumference, waist-to-hip (WHR) ratio, liver enzymes, serum glucose and CVDs risk indexes, were impaired by MD and a more active lifestyle [30].

For a therapeutic change in lifestyle, in addition to a healthy diet, an increase in motor activity is also necessary, from walking to endurance activity. Regular PA is indicated to either prevent or delay the onset of CVDs [31]. Different types of PA, increased walking, and to a lesser extent, exercise intensity, were independently associated with a lower risk of CVDs.

Considering this scientific evidence, the main aim of this prospective observational study was to evaluate, for the first time, the effectiveness of a combined MD therapy intervention and a PA plan, on the adipocyte risk indexes AIP, LAP, BARD score and FLI, in a population of women at risk of CVDs and with different pathological conditions.

The secondary purpose was to evaluate the effects on cardiovascular and inflammatory risk indexes (ASCVD, total cholesterol/cHDL, TG/cHDL, cLDL/cHDL, NLR and PLR), the changes of the body composition, evaluated through anthropometry, Dual-energy X-ray absorptiometry (DXA) and Bioelectrical Impedance Analysis (BIA).

2. Materials and Methods

2.1. Subjects and Study Design

The prospective observational study in a single cohort of adult women, at the Section of Clinical Nutrition and Nutrigenomics, Department of Biomedicine and Prevention of the University of Rome Tor Vergata (Rome, Italy) was conducted between May 2018 and July 2019. We consecutively enrolled all the women, who voluntarily came up at the Section of Clinical Nutrition and Nutrigenomics, Department of Biomedicine and Prevention of the University of Rome Tor Vergata, for nutritional-medical check-up.

To be eligible, each woman had to be Caucasian, Italian, older than 18 years old and had to be affected by at least one pathological condition in absence of drug treatment (obesity, pre-diabetes, diabetes, metabolic syndrome, osteopenia, arterial hypertension, dyslipidemia, ischemic heart disease, hepatosteatosis, hyperuricemia, Obstructive Sleep Apnea Syndrome (OSAS), chronic kidney disease). Patients with a diagnosis of cancer, hepatitis, viral infections, underweight or in therapy with antioxidant supplements were excluded.

All the patients had a specialist nutritional consultation for the evaluation of their nutritional status at baseline (T0), and after 6 months (T1). The nutritional evaluation consisted of a medical examination, anthropometric measurements, laboratory tests, the determination of body composition analysis carried out using both BIA (BIA 101S, Akern/RJL Systems, Pontassieve, Florence, Italy) and DXA (I-DXA, GE Medical Systems, Milwaukee, WI, USA). All the measurements were performed after a 12 h overnight fast.

For what concerns weight, the population was divided according to a loss of the initial body weight equal to or greater than 10% and lower than 10%. The choice of the cut-off of 10% was taken in relation to the well-known data of the scientific literature according to which a 10% reduction of the initial body weight is sufficient to reduce the risk of complications and mortality from NCCDs [32]. Conversely, concerning the FM loss from the DXA analysis, the cut-off of 15% was chosen considering the sample's median fat loss at T1.

A personalized dietetic plan and a scheduled physical activity were prescribed for each patient and the follow-up consultations were carried out after 6 months. To ensure adherence to the diet and PA plan, patients were monitored during the 6 months by monthly telephone interviews or e-mails.

2.2. Body Composition Assessmen

2.2.1. Anthropometric Assessment

After a 12-h overnight fast, all subjects underwent a body composition assessment. Anthropometric parameters were measured for all participants according to standard methods: body weight, height, hip and waist circumferences. Candidates were instructed to take off their clothes and shoes before performing all the measurements.

Body weight and height were evaluated using a scale and a stadiometer (Invernizzi, Rome, Italy), while the subject was standing and wearing underwear. Data were collected to the nearest 0.1 kg and 0.1 cm, respectively. BMI was calculated according to the following formula:

$$\text{BMI} = \text{body weight/height}^2\ (\text{kg/m}^2)$$

Body circumferences (neck, abdomen, waist and hip) were measured with flexible and non-extensible metric tape [33], according to the International Society for the Advancement of Kinanthropometry protocol and National Institute of Health Guidelines [34].

The WHR was calculated as a predictor of MI risk [35].

2.2.2. Bioelectrical Impedance Analysis (BIA)

BIA allowed for measuring the impedance of the human body to the passage of an alternating current, with constant intensity (400 μA) and frequency (50 KHz), according to the amount of water and electrolytes in the body. It generates Resistance (Rz) and Reactance (Xc) values for each patient that are processed through the BIA Akern software with validated algorithms, providing graphical and numerical data on the analysis of body composition [36].

BIA was used to evaluate resistance (Rz), reactance (Xc), hydration, total body water (TBW), extracellular water (ECW), intracellular water (ICW) and phase angle (PA).

2.2.3. Dual-Energy X-ray Absorptiometry (DXA)

DXA was performed to assess total body FM (percentage and kg), total body lean mass (LM) (kg), and total body bone mass (TBBone) [7]. Before each session, the standard DXA instrument quality control and calibration measures were conducted. They laid supine on the DXA, without moving for 20 min when the DXA scan recorded their results. The coefficient of variation (CV% = 100 × SD/mean) intra and inter subjects ranged from 1% to 5%. The coefficient of variation for bone measurements was less than 1%; CVs on this instrument for five subjects scanned six times over a nine-month period were 2.2% for FM and 1.1% for lean body mass (LM). The radiation dose of the procedure was 0.01 mSv.

FM% was calculated as FM (kg) divided by the total mass of all tissues, including the LM and TBBone, as the following [37]:

$$\text{FM\%} = (\text{FM}/(\text{FM} + \text{LM} + \text{TBBone})) \times 100.$$

In order to define muscle mass status the Appendicular Skeletal Muscle Mass Index (ASMMI) was calculated using the following formula [29,38]:

$$\text{Legs Muscle Mass (kg)} + \text{Arms Muscle Mass (kg)/Height}^2\ (\text{m}^2).$$

Subjects with the percentage of total body fat mass (FM) <30% were considered normal weight, otherwise they were considered pre-obese/obese [39].

2.3. *Laboratory Tests, Cardiovascular and Inflammatory Risk Indexes*

The evaluated parameters were blood count, glycemia, insulinemia, hepatic transaminases, γ-glutamyl transferase, creatinine, lipidemic profile (total cholesterol, cLDL, cHDL, TG), C-reactive protein (CRP) and Erythrocyte Sedimentation Rate (ESR).

Blood tests were prescribed by clinicians of the Section of Clinical Nutrition and Nutrigenomics, Department of Biomedicine and Prevention of the University of Rome Tor Vergata, and subjects were asked to perform them at the same accredited laboratory of Tor Vergata Hospital (Rome, Italy). To assess the cardiovascular and inflammatory risk, the following indexes and ratios were calculated.

Currently, the ACC/AHA guidelines, updated in June 2019, define severe hypercholesterolemia when the plasma Low-Density Lipoprotein cholesterol (cLDL) value is greater or equal to 190 mg/dL or 4.9 mmol/L. In this condition, the recommendation is to undergo high-intensity statin therapy immediately without calculating 10-year ASCVD risk [24].

To assess the cardiovascular and inflammatory risk, the following indexes and ratios were calculated:

(1) Total cholesterol/cHDL ratio was calculated according to the formula [13,40].

$$\text{Total cholesterol/cHDL ratio} = \text{Total cholesterol (mg/dL)/cHDL (mg/dL)}, \quad (1)$$

with normal values <3.

(2) Lipoproteins cholesterol (cLDL/cHDL) ratio was calculated according to the formula [15]:

$$\text{cLDL/cHDL ratio} = \text{cLDL (mg/dL)/cHDL (mg/dL)}, \quad (2)$$

with normal values <2.

(3) Triglycerides (TG) /cHDL ratio was calculated according to the formula [14]:

$$\text{TG/cHDL ratio} = \text{TG (mg/dL)/cHDL (mg/dL)}, \quad (3)$$

with normal values <1.

(4) AIP was calculated according to the formula [16,41]:

$$\text{AIP} = \log\text{(TG/cHDL)} \quad (4)$$

(5) The Fatty Liver Index (FLI) was calculated according to the formula [42]:

$$\text{FLI} = (\text{e}0.953 \times \text{loge(triglycerides)} + 0.139 \times \text{BMI} + 0.718 \times \text{loge(ggt)} + 0.053 \times \text{waistcircumference} - 15.745)/(1 + \text{e}0.953 \times \text{loge(triglycerides)} + 0.139 \times \text{BMI} + 0.718 \times \text{loge(ggt)} + 0.053 \times \text{waistcircumference} - 15.745) \times 100, \quad (5)$$

with absence of steatosis for values <30.

(6) ASCVD Risk Algorithm was calculated using the ACC/AHA calculator [23].

(7) LAP was calculated according to the formula [17]:

$$\text{LAP} = (\text{WC-65}) \times \text{TG for men and (WC-58)} \times \text{TG for women.} \quad (6)$$

(8) BARD score consists of the weighted sum of three variables: body mass index ≥28 represents 1 point, the Aspartate Aminotransferase (AST)/Alanine Aminotransferase (ALT) ratio ≥0.8 represents 2 points, and diabetes mellitus represents 1 point. A score of 2–4 had an odds ratio of 17 (confidence interval: 9.2–31.9) to determine advanced fibrosis and a negative predictive value of 96% [43].

(9) NLR is easily calculated by dividing the absolute neutrophil count by the absolute lymphocyte count from a complete blood count with differential [22]. The values for Low risk <1.6, Medium risk 1.6–2.4 and High risk >2.4.

(10) PLR is calculated by dividing the platelet count by the lymphocytes [44]. The cut off is <150.

(11) CRP and ESR were used to evaluate inflammatory risk [45].

2.4. Personalized Diet Therapy

At baseline, subjects were instructed to record weight and/or measures of foods and beverages consumed for a 3-day food intake assessment [46]. The estimated intake of macronutrients was calculated by using Dietosystem dietary software (DS Medica S.r.l., Milan, Italy).

Considering the results of the food intake, food tastes and calculating the daily energy expenditure, a personalized MD was prescribed to each patient. The diet was personalized, considering a caloric restriction of 20% compared to the daily energy expenditure for overweight and obese patients, while an isocaloric diet was prescribed to normal weight ones. The daily energy expenditure was calculated estimating Basal Metabolism according to the De Lorenzo Formula [47].

Daily macronutrient intake was distributed as follows: 55% of total kcal/day of carbohydrates, 20% of total kcal/day of protein (>50% of vegetable derivation), <25% of total kcal/day of lipids (on total daily energy intake: saturated fat <10%, 6–10% polyunsaturated fatty acids (PUFA), n-6/n-3 PUFA ratio of 3:1, 15% of monounsaturated fatty acids (MUFA); <1% trans-fatty acids) and 25 g of fiber.

The MD was balanced starting from the daily protein amount, with a protein intake of 2 g/Kg of total LM, according to Colica et al. [48,49]. The diet therapy prescribed was formulated according to the MD model based on whole grains, fresh fruits and vegetables (5 portions per day), legumes, nuts and daily use of olive oil. Plant proteins and fresh fish were preferred to red meat. Processed foods were excluded from the diet. A small amount (125 g) of red wine was allowed once a day. The adherence to the diet therapy was assessed remotely through a complete dietetic anamnesis, including the 24 h recall method and a specific food and beverages consumption investigation.

2.5. Planned Physical Activity

The levels of physical activity (PA) in diverse domains, such as working activity, leisure time activity and sedentary activities and participation in organized sport, were evaluated for each patient, requesting them to answer the question of the Finnish Diabetes Risk Score (FINDRISC) [16]: "Do you exercise during your free time and/or for work for at least 30 min almost every day?" All patients were recommended to engage in physical activity for 150 min per week of moderate-intensity aerobic activity (50–70% Heart Rate max) and/or 90 min per week of high-intensity activity (>70% HR max) distributed in at least three days a week, with no more than 2 consecutive days between bouts of aerobic activity, according to the American College of Sports Medicine and the American Diabetes Association [50]. PA was considered moderate when the time/week spent was on 60 min. To estimate sedentary PA, we considered the h/day spent on sedentary behaviours. To estimate vigorous PA, we considered time/week spent on 20 min of high-intensity activity PA. The level of physical activity was monitored and self-reported by patients during the monthly interviews.

2.6. Statistical Analysis

Statistical analysis was performed using IBM SPSS Statistics V15.0 (SPSS, Chicago, IL, USA). The Kolmogorov–Smirnov test was performed to evaluate variables in distribution. Continuous variables are represented as mean and standard deviation. The paired t-test and the Wilcoxon signed-rank test were performed to compare normal and skewed continuous variables, respectively, between pre- (T0) and post- (T1) treatment. The McNemar test was performed to compare dichotomous data between T0 and T1. Pearson's correlation analysis was carried out to evaluate a possible linear correlation between the risk indexes compared to the weight loss and the loss of body fat mass. Results were significant for p-value < 0.05.

2.7. Ethics Approval

All participants enrolled in the study approved their participation after studying and signing the informed consent, carried out in accordance with the Helsinki Declaration of 1975 as revised in 1983. The study protocol was approved by the ethical committee of the Calabria Region Center Area Section (Register Protocol No. 146 17/05/2018).

3. Results

3.1. Subjects

A total of 71 patients were invited to participate to the study. Of those, 11 declined the invitation, while 8 were not eligible because already in pharmacological treatment. Finally, 52 of them met the criteria for eligibility (age 47.3 ± 12.5 years) and completed the study after 6 months (T1). At baseline, the 94% of the studied population was obese, and comorbidities of obesity were highlighted: mainly prediabetes, metabolic syndrome, dyslipidemia and arterial hypertension. In total, 54% had MS, 31% pre-diabetes, 13% T2DM, 37% arterial hypertension, 69% dyslipidemia, 4% ischemic heart disease, 29% hepatosteatosis, 8% hyperuricemia, 15% OSAS, 73% hypovitaminosis D, 6% chronic kidney disease and 38% osteopenia.

Considering the subjects on the basis of the number of comorbidities, the prevalence of multimorbidity of the sample was also assessed on the basis of the number of pathologies found at baseline for each patient: 15% of the population had 2 comorbidities, 25% had 3, 10 % had 4, 17% had 5 and 6 and 4% had 7.

During the study, the patients showed a good compliance to the diet and PA prescription, according to the self-report data collected through the monthly interviews.

3.2. Anthropometry and Body Composition

Table 1 summarizes the anthropometric and body composition's characteristics of the population both at T0 and T1. At T1, after the nutritional and physical activity intervention, weight, BMI, as well as all the circumferences, significantly decreased ($p < 0.0001$). The TBW was significantly reduced, in particular for what concerned the extracellular compartment.

Table 1. Anthropometric and body composition's characteristics at T0 and T1.

	T0	T1	*p*-Value *
Weight	88.4 ± 24.9	79.7 ± 18.7	<0.0001
BMI	32.3 ± 8.0	29.2 ± 6.0	<0.0001
Neck circumference	38.8 ± 4.4	37.1 ± 3.9	<0.0001
Waist circumference	98.6 ± 18.0	90.7 ± 13.6	<0.0001
Abdomen circumference	109.3 ± 19.6	100.0 ± 13.2	<0.0001
Hip circumference	113.3 ± 14.6	107.3 ± 11.1	<0.0001
WHR	0.869 ± 0.098	0.846 ± 0.093	0.0002
Rz	470 ± 79	468 ± 76	0.85
Xc	50 ± 10	52 ± 10	0.21
PA°	6.17 ± 1.11	6.38 ± 1.07	0.1
TBW (L)	42.6 ± 9.3	41.5 ± 8.6	0.004
TBW (%)	49.3 ± 7.3	52.4 ± 6.7	<0.0001
ECW (L)	19.2 ± 4.1	18.3 ± 3.7	0.003
ECW (%)	45.4 ± 5.0	44.5 ± 4.6	0.11
ICW (L)	23.4 ± 6.1	23.3 ± 5.8	0.8
ICW (%)	54.6 ± 5.0	55.5 ± 4.6	0.10
FM (kg)	36.75 ± 17.16	29.80 ± 12.24	<0.0001
FM (%)	40.3 ± 9.1	36.5 ± 8.8	<0.0001
LM (kg)	48.56 ± 10.71	47.42 ± 10.14	<0.0001
ASMMI	8.32 ± 1.82	8.06 ± 1.71	0.003

ASMMI, appendicular skeletal mass index; BMI, body mass index; ECW, Extra-Cellular Water; FM, fat mass; ICW, intracellular index; LM, lean mass; PA, Phase Angle; Rz, Resistance; TBW, Total Body Water; WHR, waist-to-hip ratio; Xc, Reactance. * The paired *t*-test and the Wilcoxon signed-rank test were performed in order to compare normal and skewed continuous variables, respectively, between pre- (T0) and post- (T1) treatment. Statistical significance for $p < 0.05$.

Patients after diet and physical activity therapies achieved the best results in terms of body composition and laboratory parameters. Bone mineral density remained almost unchanged in this period and did not undergo statistically significant changes.

Moreover, the DXA analysis showed a statistically significant reduction of the total body FM (kg and %). Bone Mineral Density (BMD) has been maintained from T0 to T1 (data not shown). Finally, the loss of FM is associated with a slight decrease in LM ($p < 0.0001$), according to the reduction of ECW measured by BIA.

When dividing the population into two groups according to the percentage of weight loss (less than 10% or equal/greater than 10%) LM resulted to be significantly reduced in subjects with a weight loss ≥10%, corresponding to a mean value of 2.4 kg ($p < 0.0001$).

Conversely, when considering the population according to an FM loss greater than or equal to 15% or lower than 15% of the initial FM, we found LM to be significantly decreased in both groups at T1 (subjects ≥15% FM loss: −1.2 kg of LM, $p < 0.005$; subjects <15% of FM loss: −1.1 kg of LM, $p < 0.0004$).

3.3. Cardiovascular Risk Indexes

At T0 most of the study population showed altered risk indexes, especially the ones associated with plasma lipoproteins. In particular, 98%, 92% and 81% of them had an impaired TG/cHDL, Total cholesterol/cHDL and cLDL/cHDL ratio, respectively; while 46%, 35% and 32% of them showed altered AIP, PLR and ASCVD risk, respectively. At T1, all risk indexes showed a significant statistical reduction, except for NLR, PLR and BARD (Table 2).

Table 2. Modification of risk markers after nutritional intervention and prescription of physical activity.

Cardiovascular Risk Indexes	T0	T1	*p* Value *
ASCVD risk	6.27 ± 7.21	2.84 ± 3.44	0.0001
NLR	1.73 ± 0.74	1.84 ± 0.57	0.31
PLR	116.24 ± 37.58	125.78 ± 65.29	0.29
Total cholesterol/cHDL	4.64 ± 1.38	3.50 ± 0.87	<0.0001
cLDL/cHDL	2.95 ± 1.09	2.12 ± 0.69	<0.0001
TG/cHDL	3.38 ± 2.53	1.89 ± 1.07	<0.0001
AIP	0.06 ± 0.27	−0.14 ± 0.22	<0.0001
FLI	59.74 ± 32.26	42.06 ± 30.73	<0.0001
LAP	58.66 ± 42.79	34.72 ± 23.43	<0.0001
BARD	3.42 ± 0.11	2.89 ± 0.23	0.06

ASCVD, Atherosclerotic Cardiovascular Diseases, AIP, Atherogenic Index of Plasma; BARD, (BMI, Alanine aminotransferase (ALT)/Aspartate aminotransferase (AST) ratio (AAR), Diabetes Mellitus (DM), cLDL/cHDL, low-density lipoprotein cholesterol-to-high density lipoprotein cholesterol ratio; FLI, Fatty Liver Index; LAP, Lipid Accumulation Product; NLR, neutrophils-to-lymphocytes ratio; PLR, platelets-to-lymphocytes ratio; Total cholesterol/cHDL, total cholesterol-to-high density lipoprotein cholesterol ratio; TG/cHDL, triglycerides-to-high density lipoprotein cholesterol ratio. * The paired t-test was performed to compare continuous variables between pre- (T0) and post- (T1) treatment. Statistical significance for $p < 0.05$.

The ASCVD risk reduction was positively correlated to the decrease of both weight and FM ($p < 0.0001$, $p = 0.004$), regardless of the extent of the loss. When considering a reduction of body weight ≥ 10% and FM ≥ 15%, a switch from a higher to a lower risk class was observed ($p < 0.0001$ and $p = 0.001$, respectively). The FLI index significantly decreased with body weight and FM loss, regardless of the extent of the loss. Moreover, when considering a reduction of body weight ≥ 10% and FM ≥ 15 a switch from a higher to a lower risk class was observed ($p < 0.0001$). The AIP was significantly reduced with the loss of FM ($p = 0.001$), and more markedly with a loss greater than 15% ($p < 0.0001$). Conversely, a significant reduction of the AIP with a body weight loss of≥ 10% ($p = 0.1$) was not observed. Finally, the LAP index significantly decreased with the reduction of body weight ($p < 0.0001$). The cardiovascular risk indexes associated with the lipid profile were significantly reduced with both the decrease of body weight and FM, regardless of the amount of decrease. A greater reduction was observed with a loss of FM ≥ 15% ($p < 0.0001$) (Figure 1).

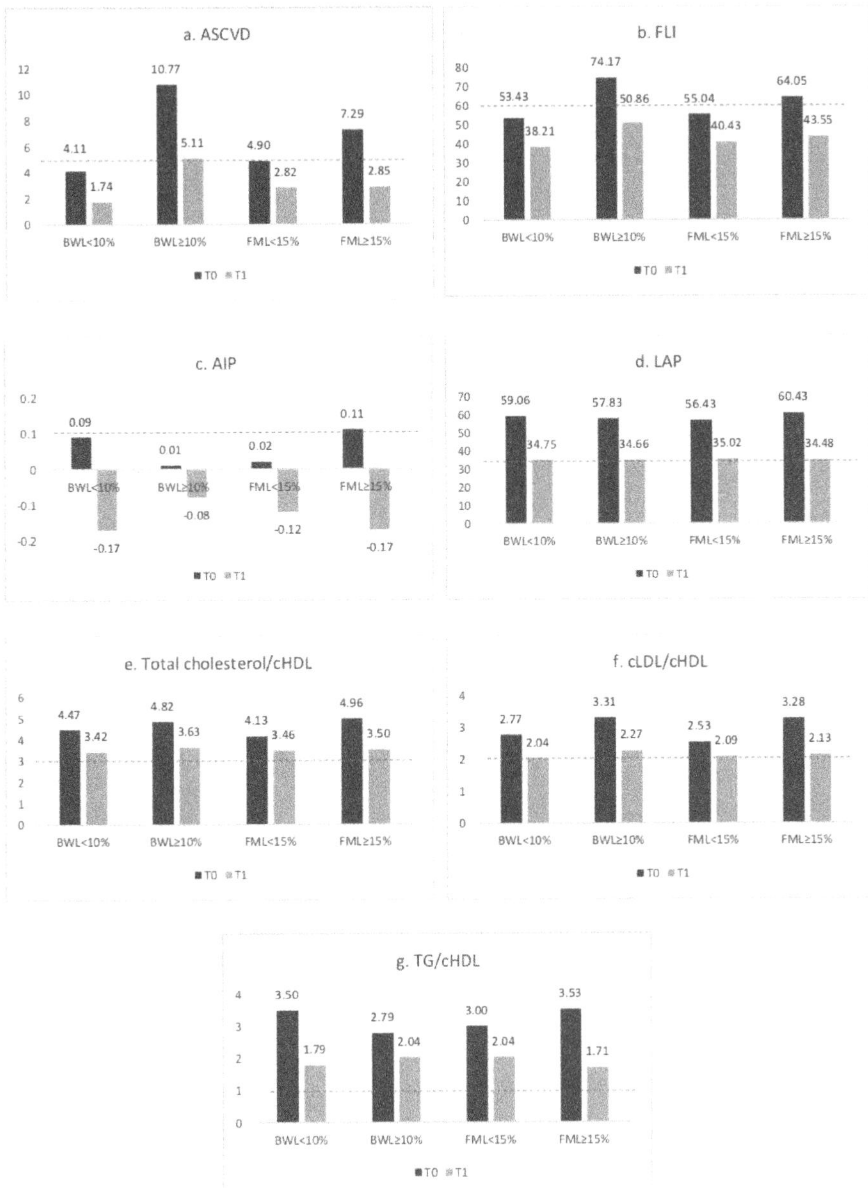

Figure 1. Modification of Cardiovascular Diseases and lipid profile-related risk indexes before (T0) and after the intervention (T1) according to Body Weight Loss (BWL) <10% and ≥10% and Fat Mass Loss (FML) <15% or ≥15%. (**a**) ASCVD, Atherosclerotic Cardiovascular Diseases; (**b**) FLI, Fatty Liver Index; (**c**) AIP, Atherogenic Index of Plasma; (**d**) LAP, Lipid Accumulation Product; (**e**) Total cholesterol/cHDL, total cholesterol-to-high density lipoprotein cholesterol ratio (**f**) cLDL/cHDL: low-density lipoprotein cholesterol-to-high density lipoprotein cholesterol ratio; (**g**) TG/cHDL, triglycerides-to-high density lipoprotein cholesterol ratio. The dotted lines refer to the cut-off for each index.

4. Discussion

In recent literature, the CVDs risk reversibility has been widely demonstrated. Around 80% of cardiovascular events can be avoided by adopting a correct lifestyle consisting of eating healthy and staying physically active [51]. Nevertheless, the improvement of strategies aimed at controlling risk factors is still unsatisfactory today.

In the United States, in the context of modifiable risk factors, high blood pressure is the primary cause of atherosclerotic cardiovascular diseases [52].

Consequently, in all adults with high blood pressure, a non-pharmacological intervention is recommended in the first instance, while in hypertensive patients already in drug therapy, a pressure target of 130/80 mmHg must be pursued [53].

Therefore, it underlines the necessity of a multidisciplinary approach. It would be aimed at controlling risk factors through the implementation of preventive strategies, the importance of sharing decisions with patient and paying a deep look to the social determinants of health (obstacles to care, educational level, economic difficulties, environmental and socio-economic factors) [11].

The most recent epidemiological studies on nutrition have shown how dietary models instead of isolated nutrients can be a more accurate tool for studying eating habits and preparing a therapeutic plan for the prevention and treatment of NCD [54]. Healthy, personalized nutrition can reduce cardiovascular risk factors, such as obesity, diabetes, dyslipidemia and high blood pressure and therefore plays a crucial role in preventing the recurrence of chronic ischemic heart diseases [55]. Among dietary models, MD was associated with a lower risk of CVDs incidence and mortality, including coronary heart disease and MI [56].

Moreover, it has been shown that engaging in regular physical activity reduces the risk of developing and dying from cardiovascular outcomes [57,58].

However, to date, there are few clinical studies aimed at understanding the independent and even synergistic impacts of the effect of a dietary plan adhering to MD with physical inactivity [58].

Therefore, we set up a dietary and PA plan, aimed at improving the risk indexes of adiposity, inflammatory and cardio-vascular diseases, as well as weight loss of $< 10\%$ or $\geq 10\%$ of the initial body weight and reduction of fat mass.

We analysed the statistical relationship between the improvement of risk indexes at baseline (T0) and after 6 months of therapy (T1). The reduction in the evaluated indexes was statistically associated with bodyweight regardless of the extent of the loss. The reduction of most of the evaluated indexes reflects a positive health change linked to the amelioration of laboratory tests from T0 to T1.

Considering the ASCVD risk index, which is a predictor for forecasting the potential impact of different interventions on patient risk [59], we observed an effective 10-year risk reduction for cardiovascular atherosclerotic disease when patients lost at least 10% of body weight or at least 15% of FM. In this context, the ASCVD risk reduction was strongly significant, leading to a transition from the higher to a lower risk category. Currently, the ACC/AHA guidelines, updated in June 2019, define severe hypercholesterolemia when the plasma cLDL value is greater or equal to 190 mg/dL or 4.9 mmol/L. In this condition, the recommendation is to undergo high-intensity statin therapy immediately without calculating 10-year ASCVD risk [14].

In our study, the dietary treatment with MD combined with a physical activity plan, resulted in a significant reduction of all blood parameters of the lipid profile, even in the absence of drug therapy.

In our study, we observed a statistically significant reduction of plasma lipoproteins, TG and total cholesterol, especially in subjects with reduced FM by at least 15%.

Total cholesterol/cHDL and cLDL/cHDL ratios are risk indicators with greater predictive value than isolated parameters used independently, particularly cLDL. The total cholesterol/cHDL ratio, known as the atherogenic or Castelli risk index [13], and the cLDL/cHDL ratio are two important components and indicators of cardiovascular risk, the predictive value of which is greater than the isolated parameters [11]. TG/cHDL ratio is a deputy marker of LDL particle size (small and dense) to observe the link with insulin resistance and thyroid co-morbidity [12].The reduction of most of the evaluated indexes reflects a positive health change linked to the amelioration of laboratory tests from T0 to T1.

The ratios between plasma lipoproteins showed a statistically significant reduction, thanks to the proper nutritional intervention, and were marked in subjects with reduced FM by at least 15%.

The same trend was also observed for the LAP, AIP and FLI index. Both LAP and FLI reflected the metabolic health of the patient. It is known that LAP is strongly correlated with visceral fat [60] and it is associated with metabolic syndrome [61], type-2 diabetes mellitus [17], hypertension [62] and CVDs [63]. FLI particularly is a simple and accurate predictor of hepatic steatosis in the general population [64]. Furthermore, AIP is a predictive indicator for the coronary artery disease in postmenopausal women [65] and it is a strong marker to predict the risk of atherosclerosis and coronary heart disease [40]. For what concerns the inflammation milieu, both NLR and PLR have the potential to be inflammatory markers that reflect the activity of many inflammatory diseases [66]. Obese subjects undergo a chronic inflammatory condition due to the adipose tissue dysfunction of immune-related activities, involving a transient infiltration of neutrophils within the abdominal fat and their binding to adipocytes. NLR and PLR are considered cost-effective markers for the detection of subclinical inflammation [67]. Nevertheless, the result was not statistically significant for the two indexes. This latter probably considering that dietary intervention should be followed for longer in order to show a higher effect on neutrophils, platelets, lymphocytes and liver transaminases.

The BARD score was established specifically for assessing and predicting Non-alcoholic Fatty Liver Disease (NAFLD). It is widely used to predict liver fibrosis in NAFLD patients and requires simple clinical data [68]. A BARD score of 2–4 was associated with an OR for advanced fibrosis of 17 (confidence interval 9.2 to 31.9) and a negative predictive value of 96% [26]. It was developed to exclude the presence of advanced fibrosis in patients with NAFLD [27]. However, in our study, the values of BARD score were in the moderate risk cut-off, and no significant changes were observed at T1 ($p = 0.06$).

After MD and PA therapy, we observed an improvement of the main anthropometric and body composition parameters. The LM remains stable in subjects who reduce FM <15%. In patients who lost more than 15% of FM, the reduction of LM observed could be linked to extracellular liquids, occurring mostly at the beginning of weight loss, and therefore also could be related to fat loss.

The proof of the above conditions can be demonstrated with the PA. Indeed, PA is a deep marker for sarcopenia, fragility and risk of mortality in obese people [69]. In our study, the PA reveals a slight improvement, even if it cannot be considered as significant data, due to the restricted period of intervention. Significant maintenance of LM after nutritional and physical interventions is undoubtedly a clinically important factor due to the preservation of muscle mass achieved to the dietetic nutritional composition in terms of proteins and amino acids. Future investigations should focus on ASMMI as a marker for the prevention of sarcopenia obesity [38]. It would be important to consider this effect when a nutritional and physical intervention is prescribed to a patient.

Finally, we can assume that the combination of diet therapy and PA lead to the best results in terms of body composition and laboratory tests.

The strength of the present study is that data are observed on a population with concomitant risk factors, and that it is the first one monitoring AIP, LAP and FLI during a combination between MD personalized diet therapy and PA program. Conversely, the main limitations are the small sample size, the lack of a longer follow-up and the self-report data about diet and PA adherence.

5. Conclusions

In conclusion, our findings suggest that a combination of both personalized MD diet therapy and PA program lead to body weight loss, body composition remodelling and risk indexes reduction. The evaluation of specific pathologies risk indexes, such as adiposity (AIP, LAP, FLI) and CVDs risk indexes, may represent a predictor factor to guide clinicians in the nutritional status assessment and improve the personalization of diet and PA prescription. As the main limitations of the study are the small sample size and the lack of a longer follow-up, additional investigations are needed in this field. Furthermore, future clinical studies should apply innovative and objective physical activity measurement methods, during the time of examination.

Author Contributions: Conceptualization, L.D.R.; Formal analysis, L.D.R.; Investigation, M.D., P.G., A.A., C.L., G.C. (Giuseppe Cenname), E.E., A.P., G.C. (Gaetano Chiricolo) and C.S.; Methodology, L.D.R.; Project administration, A.D.L.; Supervision, A.D.L.; Writing—original draft, L.D.R. and G.C. (Giulia Cinelli); Writing—review and editing, G.C. (Giulia Cinelli), P.G., A.A. and C.L. All authors have read and agreed to the published version of the manuscript.

Funding: No financial or otherwise are declared by the authors.

Acknowledgments: The authors thanks to all the subjects who participated volunteered in the clinical trial. The authors thank Dott. Fulvia Mariotti for the editorial and English language revisions and Dott. Anna Anzidei for supporting the study for the clinical practice.

Conflicts of Interest: No conflicts of interest are declared by the authors.

References

1. WHO|Diet. Nutrition and the Prevention of Chronic Diseases. Available online: https://www.who.int/dietphysicalactivity/publications/trs916/en/ (accessed on 4 November 2020).
2. *Shaping the Future;* The world health report; Weltgesundheitsorganisation: Geneva, Switzerland, 2003; ISBN 978-92-4-156243-0.
3. De Lorenzo, A.; Gratteri, S.; Gualtieri, P.; Cammarano, A.; Bertucci, P.; Di Renzo, L. Why primary obesity is a disease? *J. Transl. Med.* **2019**, *17*, 169. [CrossRef] [PubMed]
4. WHO. Obesity: Preventing and Managing the Global Epidemic. Available online: http://www.who.int/entity/nutrition/publications/obesity/WHO_TRS_894/en/index.html (accessed on 8 September 2020).
5. Bays, H. Adiposopathy, metabolic syndrome, quantum physics, general relativity, chaos and the Theory of Everything. *Expert Rev. Cardiovasc. Ther.* **2005**, *3*, 393–404. [CrossRef] [PubMed]
6. Di Angelantonio, E.; Bhupathiraju, S.; Wormser, D.; Gao, P.; Kaptoge, S.; Berrington de Gonzalez, A.; Cairns, B.; Huxley, R.; Jackson, C.; Joshy, G.; et al. Body-mass index and all-cause mortality: Individual-participant-data meta-analysis of 239 prospective studies in four continents. *Lancet* **2016**, *388*, 776–786. [CrossRef]
7. De Lorenzo, A.; Bianchi, A.; Maroni, P.; Iannarelli, A.; Di Daniele, N.; Iacopino, L.; Di Renzo, L. Adiposity rather than BMI determines metabolic risk. *Int. J. Cardiol.* **2013**, *166*, 111–117. [CrossRef] [PubMed]
8. Bays, H.E. Adiposopathy is "sick fat" a cardiovascular disease? *J. Am. Coll. Cardiol.* **2011**, *57*, 2461–2473. [CrossRef] [PubMed]
9. Bays, H.E.; González-Campoy, J.M.; Bray, G.A.; Kitabchi, A.E.; Bergman, D.A.; Schorr, A.B.; Rodbard, H.W.; Henry, R.R. Pathogenic potential of adipose tissue and metabolic consequences of adipocyte hypertrophy and increased visceral adiposity. *Expert Rev. Cardiovasc. Ther.* **2008**, *6*, 343–368. [CrossRef] [PubMed]
10. Zhao, D.; Guallar, E.; Ouyang, P.; Subramanya, V.; Vaidya, D.; Ndumele, C.E.; Lima, J.A.; Allison, M.A.; Shah, S.J.; Bertoni, A.G.; et al. Endogenous Sex Hormones and Incident Cardiovascular Disease in Post-Menopausal Women. *J. Am. Coll. Cardiol.* **2018**, *71*, 2555–2566. [CrossRef] [PubMed]
11. Di Renzo, L.; Gualtieri, P.; Romano, L.; Marrone, G.; Noce, A.; Pujia, A.; Perrone, M.A.; Aiello, V.; Colica, C.; De Lorenzo, A. Role of Personalized Nutrition in Chronic-Degenerative Diseases. *Nutrients* **2019**, *11*, 1707. [CrossRef] [PubMed]
12. Ambale-Venkatesh, B.; Yang, X.; Wu, C.O.; Liu, K.; Hundley, W.G.; McClelland, R.; Gomes, A.S.; Folsom, A.R.; Shea, S.; Guallar, E.; et al. Cardiovascular Event Prediction by Machine Learning: The Multi-Ethnic Study of Atherosclerosis. *Circ. Res.* **2017**, *121*, 1092–1101. [CrossRef]
13. Castelli, W.P.; Abbott, R.D.; McNamara, P.M. Summary estimates of cholesterol used to predict coronary heart disease. *Circulation* **1983**, *67*, 730–734. [CrossRef]
14. Jayanthi, R.; Srinivasan, A.R.; Hanifah, M.; Maran, A.L. Associations among Insulin Resistance, Triacylglycerol/High Density Lipoprotein (TAG/HDL ratio) and Thyroid hormone levels-A study on Type 2 diabetes mellitus in obese and overweight subjects. *Diabetes Metab. Syndr.* **2017**, *11* (Suppl. 1), S121–S126. [CrossRef] [PubMed]

15. Millán, J.; Pintó, X.; Muñoz, A.; Zúñiga, M.; Rubiés-Prat, J.; Pallardo, L.F.; Masana, L.; Mangas, A.; Hernández-Mijares, A.; González-Santos, P.; et al. Lipoprotein ratios: Physiological significance and clinical usefulness in cardiovascular prevention. *Vasc. Health Risk Manag.* **2009**, *5*, 757–765. [PubMed]
16. Dobiásová, M.; Frohlich, J. The plasma parameter log (TG/HDL-C) as an atherogenic index: Correlation with lipoprotein particle size and esterification rate in apoB-lipoprotein-depleted plasma (FER(HDL)). *Clin. Biochem.* **2001**, *34*, 583–588. [CrossRef]
17. Kahn, H.S. The "lipid accumulation product" performs better than the body mass index for recognizing cardiovascular risk: A population-based comparison. *BMC Cardiovasc. Disord.* **2005**, *5*, 26. [CrossRef] [PubMed]
18. Rotter, I.; Rył, A.; Szylińska, A.; Pawlukowska, W.; Lubkowska, A.; Laszczyńska, M. Lipid Accumulation Product (LAP) as an Index of Metabolic and Hormonal Disorders in Aging Men. *Exp. Clin. Endocrinol. Diabetes* **2017**, *125*, 176–182. [CrossRef]
19. Khang, A.R.; Lee, H.W.; Yi, D.; Kang, Y.H.; Son, S.M. The fatty liver index, a simple and useful predictor of metabolic syndrome: Analysis of the Korea National Health and Nutrition Examination Survey 2010–2011. *Diabetes Metab. Syndr. Obes* **2019**, *12*, 181–190. [CrossRef] [PubMed]
20. Yildirim, I.; Hur, E.; Kokturk, F. Inflammatory Markers: C-Reactive Protein, Erythrocyte Sedimentation Rate, and Leukocyte Count in Vitamin D Deficient Patients with and without Chronic Kidney Disease. *Int. J. Endocrinol.* **2013**, *2013*, 802165. [CrossRef]
21. Haybar, H.; Pezeshki, S.M.S.; Saki, N. Evaluation of complete blood count parameters in cardiovascular diseases: An early indicator of prognosis? *Exp. Mol. Pathol.* **2019**, *110*, 104267. [CrossRef]
22. Bazzi, W.M.; Dejbakhsh, S.Z.; Bernstein, M.; Russo, P. Neutrophil-lymphocyte ratio in small renal masses. *Int. Sch. Res. Not.* **2014**, *2014*, 759253. [CrossRef]
23. Arnett Donna, K.; Blumenthal Roger, S.; Albert Michelle, A.; Buroker Andrew, B.; Goldberger Zachary, D.; Hahn Ellen, J.; Himmelfarb Cheryl, D.; Khera, A.; Lloyd-Jones, D.; McEvoy, J.W.; et al. 2019 ACC/AHA Guideline on the Primary Prevention of Cardiovascular Disease: Executive Summary: A Report of the American College of Cardiology/American Heart Association Task Force on Clinical Practice Guidelines. *Circulation* **2019**, *140*, e563–e595. [CrossRef]
24. Grundy, S.M.; Stone, N.J.; Bailey, A.L.; Beam, C.; Birtcher, K.K.; Blumenthal, R.S.; Braun, L.T.; de Ferranti, S.; Faiella-Tommasino, J.; Forman, D.E.; et al. 2018 AHA/ACC/AACVPR/AAPA/ABC/ACPM/ADA/AGS/APhA/ASPC/NLA/PCNA Guideline on the Management of Blood Cholesterol: A Report of the American College of Cardiology/American Heart Association Task Force on Clinical Practice Guidelines. *J. Am. Coll. Cardiol.* **2019**, *73*, e285–e350. [CrossRef] [PubMed]
25. Cichoż-Lach, H.; Celiński, K.; Prozorow-Król, B.; Swatek, J.; Słomka, M.; Lach, T. The BARD score and the NAFLD fibrosis score in the assessment of advanced liver fibrosis in nonalcoholic fatty liver disease. *Med. Sci. Monit.* **2012**, *18*, CR735–CR740. [CrossRef] [PubMed]
26. Klisic, A.; Abenavoli, L.; Fagoonee, S.; Kavaric, N.; Kocic, G.; Ninić, A. Older age and HDL-cholesterol as independent predictors of liver fibrosis assessed by BARD score. *Minerva Med.* **2019**, *110*, 191–198. [CrossRef] [PubMed]
27. Bray, G.A.; Heisel, W.E.; Afshin, A.; Jensen, M.D.; Dietz, W.H.; Long, M.; Kushner, R.F.; Daniels, S.R.; Wadden, T.A.; Tsai, A.G.; et al. The Science of Obesity Management: An Endocrine Society Scientific Statement. *Endocr. Rev.* **2018**, *39*, 79–132. [CrossRef]
28. Martini, D. Health Benefits of Mediterranean Diet. *Nutrients* **2019**, *11*, 1802. [CrossRef]
29. Di Renzo, L.; Cioccoloni, G.; Falco, S.; Abenavoli, L.; Moia, A.; Sinibaldi Salimei, P.; De Lorenzo, A. Influence of FTO rs9939609 and Mediterranean diet on body composition and weight loss: A randomized clinical trial. *J. Transl. Med.* **2018**, *16*, 308. [CrossRef]
30. Gelli, C.; Tarocchi, M.; Abenavoli, L.; Di Renzo, L.; Galli, A.; De Lorenzo, A. Effect of a counseling-supported treatment with the Mediterranean diet and physical activity on the severity of the non-alcoholic fatty liver disease. *World J. Gastroenterol.* **2017**, *23*, 3150–3162. [CrossRef]
31. Soares-Miranda, L.; Siscovick, D.S.; Psaty, B.M.; Longstreth, W.T.; Mozaffarian, D. Physical Activity and Risk of Coronary Heart Disease and Stroke in Older Adults: The Cardiovascular Health Study. *Circulation* **2016**, *133*, 147–155. [CrossRef]

32. Pasanisi, F.; Contaldo, F.; de Simone, G.; Mancini, M. Benefits of sustained moderate weight loss in obesity. *Nutr. Metab. Cardiovasc. Dis.* **2001**, *11*, 401–406.
33. De Lorenzo, A.; Siclari, M.; Gratteri, S.; Romano, L.; Gualtieri, P.; Marchetti, M.; Merra, G.; Colica, C. Developing and cross-validation of new equations to estimate fat mass in Italian population. *Eur. Rev. Med. Pharmacol. Sci.* **2019**, *23*, 2513–2524. [CrossRef]
34. Lohman, T.G.; Roche, A.F.; Martorell, R. (Eds.) *Anthropometric Standardization Reference Manual;* Human Kinetics Books: Champaign, IL, USA, 1988; ISBN 978-0-87322-121-4.
35. Cao, Q.; Yu, S.; Xiong, W.; Li, Y.; Li, H.; Li, J.; Li, F. Waist-hip ratio as a predictor of myocardial infarction risk: A systematic review and meta-analysis. *Medicine* **2018**, *97*, e11639. [CrossRef]
36. Colica, C.; Di Renzo, L.; Trombetta, D.; Smeriglio, A.; Bernardini, S.; Cioccoloni, G.; Costa de Miranda, R.; Gualtieri, P.; Sinibaldi Salimei, P.; De Lorenzo, A. Antioxidant Effects of a Hydroxytyrosol-Based Pharmaceutical Formulation on Body Composition, Metabolic State, and Gene Expression: A Randomized Double-Blinded, Placebo-Controlled Crossover Trial. *Oxidative Med. Cell Longev.* **2017**, *2017*, 2473495. [CrossRef] [PubMed]
37. Encyclopedia of Human Nutrition—3rd Edition. Available online: https://www.elsevier.com/books/encyclopedia-of-human-nutrition/unknown/978-0-12-375083-9 (accessed on 9 September 2020).
38. Di Renzo, L.; Sarlo, F.; Petramala, L.; Iacopino, L.; Monteleone, G.; Colica, C.; De Lorenzo, A. Association between -308 G/A TNF-α polymorphism and appendicular skeletal muscle mass index as a marker of sarcopenia in normal weight obese syndrome. *Dis. Markers* **2013**, *35*, 615–623. [CrossRef] [PubMed]
39. De Lorenzo, A.; Soldati, L.; Sarlo, F.; Calvani, M.; Di Lorenzo, N.; Di Renzo, L. New obesity classification criteria as a tool for bariatric surgery indication. *World J. Gastroenterol.* **2016**, *22*, 681–703. [CrossRef] [PubMed]
40. Niroumand, S.; Khajedaluee, M.; Khadem-Rezaiyan, M.; Abrishami, M.; Juya, M.; Khodaee, G.; Dadgarmoghaddam, M. Atherogenic Index of Plasma (AIP): A marker of cardiovascular disease. *Med. J. Islam. Repub. Iran* **2015**, *29*, 240.
41. Kanthe, P.S.; Patil, B.S.; Bagali, S.; Deshpande, A.; Shaikh, B.; Aithala, M. Atherogenic Index as a Predictor of Cardiovascular Risk among Women with Different Grades of Obesity. *Public Health* **2012**, *4*, 8.
42. Kantartzis, K.; Rettig, I.; Staiger, H.; Machann, J.; Schick, F.; Scheja, L.; Gastaldelli, A.; Bugianesi, E.; Peter, A.; Schulze, M.B.; et al. An extended fatty liver index to predict non-alcoholic fatty liver disease. *Diabetes Metab.* **2017**, *43*, 229–239. [CrossRef]
43. Stål, P. Liver fibrosis in non-alcoholic fatty liver disease—Diagnostic challenge with prognostic significance. *World J. Gastroenterol.* **2015**, *21*, 11077–11087. [CrossRef]
44. Erdal, E.; İnanir, M. Platelet-to-lymphocyte ratio (PLR) and Plateletcrit (PCT) in young patients with morbid obesity. *Rev. Assoc. Med. Bras.* **2019**, *65*, 1182–1187. [CrossRef]
45. Lapić, I.; Padoan, A.; Bozzato, D.; Plebani, M. Erythrocyte Sedimentation Rate and C-Reactive Protein in Acute Inflammation. *Am. J. Clin. Pathol.* **2020**, *153*, 14–29. [CrossRef]
46. Block, G. Human dietary assessment: Methods and issues. *Prev. Med.* **1989**, *18*, 653–660. [CrossRef]
47. De Lorenzo, A.; Di Renzo, L.; Morini, P.; de Miranda, R.C.; Romano, L.; Colica, C. New equations to estimate resting energy expenditure in obese adults from body composition. *Acta Diabetol.* **2018**, *55*, 59–66. [CrossRef] [PubMed]
48. Colica, C.; Avolio, E.; Bollero, P.; Costa de Miranda, R.; Ferraro, S.; Sinibaldi Salimei, P.; De Lorenzo, A.; Di Renzo, L. Evidences of a New Psychobiotic Formulation on Body Composition and Anxiety. *Mediat. Inflamm.* **2017**, *2017*, 5650627. [CrossRef] [PubMed]
49. Colica, C.; Merra, G.; Gasbarrini, A.; De Lorenzo, A.; Cioccoloni, G.; Gualtieri, P.; Perrone, M.A.; Bernardini, S.; Bernardo, V.; Di Renzo, L.; et al. Efficacy and safety of very-low-calorie ketogenic diet: A double blind randomized crossover study. *Eur. Rev. Med. Pharmacol. Sci.* **2017**, *21*, 2274–2289. [PubMed]
50. Colberg, S.R.; Sigal, R.J.; Fernhall, B.; Regensteiner, J.G.; Blissmer, B.J.; Rubin, R.R.; Chasan-Taber, L.; Albright, A.L.; Braun, B.; American College of Sports Medicine; et al. Exercise and type 2 diabetes: The American College of Sports Medicine and the American Diabetes Association: Joint position statement. *Diabetes Care* **2010**, *33*, e147–e167. [CrossRef] [PubMed]

51. Wald, N.J.; Law, M.R. A strategy to reduce cardiovascular disease by more than 80%. *BMJ* **2003**, *326*, 1419. [CrossRef]
52. Garies, S.; Hao, S.; McBrien, K.; Williamson, T.; Peng, M.; Khan, N.A.; Padwal, R.S.; Quan, H.; Leung, A.A. Prevalence of Hypertension, Treatment, and Blood Pressure Targets in Canada Associated With the 2017 American College of Cardiology and American Heart Association Blood Pressure Guidelines. *JAMA Netw. Open* **2019**, *2*. [CrossRef]
53. Arnett, D.K.; Blumenthal, R.S.; Albert, M.A.; Buroker, A.B.; Goldberger, Z.D.; Hahn, E.J.; Himmelfarb, C.D.; Khera, A.; Lloyd-Jones, D.; McEvoy, J.W.; et al. 2019 ACC/AHA Guideline on the Primary Prevention of Cardiovascular Disease: Executive Summary: A Report of the American College of Cardiology/American Heart Association Task Force on Clinical Practice Guidelines. *J. Am. Coll. Cardiol.* **2019**, *74*, 1376–1414. [CrossRef]
54. Kalea, A.Z.; Drosatos, K.; Buxton, J.L. Nutriepigenetics and cardiovascular disease. *Curr. Opin. Clin. Nutr. Metab. Care* **2018**, *21*, 252–259. [CrossRef]
55. Mozaffarian, D. Dietary and Policy Priorities for Cardiovascular Disease, Diabetes, and Obesity: A Comprehensive Review. *Circulation* **2016**, *133*, 187–225. [CrossRef]
56. Martínez-González, M.A.; Gea, A.; Ruiz-Canela, M. The Mediterranean Diet and Cardiovascular Health. *Circ. Res.* **2019**, *124*, 779–798. [CrossRef] [PubMed]
57. Brooks, J.; Ahmad, I.; Easton, G. Promoting physical activity: The general practice agenda. *Br. J. Gen. Prac.* **2016**, *66*, 454–455. [CrossRef] [PubMed]
58. Lacombe, J.; Armstrong, M.E.G.; Wright, F.L.; Foster, C. The impact of physical activity and an additional behavioural risk factor on cardiovascular disease, cancer and all-cause mortality: A systematic review. *BMC Public Health* **2019**, *19*. [CrossRef] [PubMed]
59. Lloyd-Jones, D.M.; Huffman, M.D.; Karmali, K.N.; Sanghavi, D.M.; Wright, J.S.; Pelser, C.; Gulati, M.; Masoudi, F.A.; Goff, D.C. Estimating Longitudinal Risks and Benefits From Cardiovascular Preventive Therapies Among Medicare Patients: The Million Hearts Longitudinal ASCVD Risk Assessment Tool: A Special Report From the American Heart Association and American College of Cardiology. *J. Am. Coll. Cardiol.* **2017**, *69*, 1617–1636. [CrossRef]
60. Roriz, A.K.C.; Passos, L.C.S.; de Oliveira, C.C.; Eickemberg, M.; Moreira, P.d.A.; Sampaio, L.R. Evaluation of the accuracy of anthropometric clinical indicators of visceral fat in adults and elderly. *PLoS ONE* **2014**, *9*, e103499. [CrossRef]
61. Taverna, M.J.; Martínez-Larrad, M.T.; Frechtel, G.D.; Serrano-Ríos, M. Lipid accumulation product: A powerful marker of metabolic syndrome in healthy population. *Eur. J. Endocrinol.* **2011**, *164*, 559–567. [CrossRef]
62. Gao, X.; Wang, G.; Wang, A.; Xu, T.; Tong, W.; Zhang, Y. Comparison of lipid accumulation product with body mass index as an indicator of hypertension risk among Mongolians in China. *Obes. Res. Clin. Pract.* **2013**, *7*, e308–e314. [CrossRef]
63. Hosseinpanah, F.; Barzin, M.; Mirbolouk, M.; Abtahi, H.; Cheraghi, L.; Azizi, F. Lipid accumulation product and incident cardiovascular events in a normal weight population: Tehran Lipid and Glucose Study. *Eur. J. Prev. Cardiol.* **2016**, *23*, 187–193. [CrossRef]
64. Bedogni, G.; Bellentani, S.; Miglioli, L.; Masutti, F.; Passalacqua, M.; Castiglione, A.; Tiribelli, C. The Fatty Liver Index: A simple and accurate predictor of hepatic steatosis in the general population. *BMC Gastroenterol.* **2006**, *6*, 33. [CrossRef]
65. Atherogenic Index of Plasma (AIP): A Novel Predictive Indicator for the Coronary Artery Disease in Postmenopausal Women | Lipids in Health and Disease | Full Text. Available online: https://lipidworld.biomedcentral.com/articles/10.1186/s12944-018-0828-z (accessed on 5 November 2020).
66. Al-Osami, M.H.; Awadh, N.I.; Khalid, K.B.; Awadh, A.I. Neutrophil/lymphocyte and platelet/lymphocyte ratios as potential markers of disease activity in patients with Ankylosing spondylitis: A case-control study. *Adv. Rheumatol.* **2020**, *60*, 13. [CrossRef]
67. Early Inflammatory Status Related to Pediatric Obesity—Abstract—Europe PMC. Available online: https://europepmc.org/article/pmc/6591428 (accessed on 5 November 2020).

68. Harrison, S.A.; Oliver, D.; Arnold, H.L.; Gogia, S.; Neuschwander-Tetri, B.A. Development and validation of a simple NAFLD clinical scoring system for identifying patients without advanced disease. *Gut* **2008**, *57*, 1441–1447. [CrossRef] [PubMed]
69. Toselli, S.; Badicu, G.; Bragonzoni, L.; Spiga, F.; Mazzuca, P.; Campa, F. Comparison of the Effect of Different Resistance Training Frequencies on Phase Angle and Handgrip Strength in Obese Women: A Randomized Controlled Trial. *Int. J. Environ. Res. Public Health* **2020**, *17*, 1163. [CrossRef] [PubMed]

nutrients

MDPI

Review

Potential Cardiovascular and Metabolic Beneficial Effects of ω-3 PUFA in Male Obesity Secondary Hypogonadism Syndrome

Annalisa Noce [1,*], Giulia Marrone [1,2,*], Francesca Di Daniele [1,2], Manuela Di Lauro [1], Anna Pietroboni Zaitseva [1], Georgia Wilson Jones [1], Antonino De Lorenzo [3] and Nicola Di Daniele [1]

1 UOC of Internal Medicine-Center of Hypertension and Nephrology Unit, Department of Systems Medicine, University of Rome Tor Vergata, 00133 Rome, Italy; francesca.didaniele@gmail.com (F.D.D.); dilauromanuela@gmail.com (M.D.L.); annapietroboni@icloud.com (A.P.Z.); georgia.wilson.jones@gmail.com (G.W.J.); didaniele@med.uniroma2.it (N.D.D.)

2 PhD School of Applied Medical, Surgical Sciences, University of Rome Tor Vergata, 00133 Rome, Italy

3 Section of Clinical Nutrition and Nutrigenomic, Department of Biomedicine and Prevention, University of Rome Tor Vergata, 00133 Rome, Italy; delorenzo@uniroma2.it

* Correspondence: annalisa.noce@uniroma2.it (A.N.); giul.marr@gmail.com (G.M.); Tel.: +39-06-2090-2188 (A.N. & G.M.); Fax: +39-06-2090-2096 (A.N. & G.M.)

Received: 17 July 2020; Accepted: 17 August 2020; Published: 20 August 2020

Abstract: Long-chain ω-3 polyunsaturated fatty acids (PUFAs) are fundamental biocomponents of lipids and cell membranes. They are involved in the maintenance of cellular homeostasis and they are able to exert anti-inflammatory and cardioprotective actions. Thanks to their potential beneficial effects on the cardiovascular system, metabolic axis and body composition, we have examined their action in subjects affected by male obesity secondary hypogonadism (MOSH) syndrome. MOSH syndrome is characterized by the presence of obesity associated with the alteration of sexual and metabolic functions. Therefore, this review article aims to analyze scientific literature regarding the possible benefits of ω-3 PUFA administration in subjects affected by MOSH syndrome. We conclude that there are strong evidences supporting ω-3 PUFA administration and/or supplementation for the treatment and management of MOSH patients.

Keywords: male obesity secondary hypogonadism (MOSH) syndrome; ω-3 PUFA; adipose tissue; body weight; testosterone

---

1. Introduction

It is truly fascinating to study how lifestyle modification can alter the course of a disease by modifying genetic expression and protein synthesis patterns. Thanks to modern epigenetics, researchers have found that changes in daily habits coupled with healthy nutrition can literally modulate our gene expression, in order to achieve better metabolic profiles and decrease the risk of developing an array of diseases [1]. Exploring the properties of natural compounds such as ω-3 polyunsaturated fatty acids (PUFAs) and how they can be optimally integrated in the diet is of paramount importance. Obesity represents a major public health burden and it can be defined as a pathological increase in weight and therefore in body mass index (BMI).

PUFA ω-3 would seem to exert a cardioprotective role as they improve heart rate variability, a non-invasive marker of cardiac autonomic system function, with a subsequent reduction in the risk of sudden cardiac death and arrhythmias [2]. A further beneficial effect induced by PUFAs is linked to their anti-inflammatory capacity [3] and their ability to modulate the inflammatory response. Moreover, their effects in terms of improving body composition have also been recently demonstrated [4].

Obesity is defined as a condition characterized by a pathological increase in weight and therefore in body mass index (BMI). Its interpretation is based on weight status groupings, calculated by weight in kg divided by the square of the height in meters. A BMI exceeding 30 kg/m^2 is indicative of obesity, as BMI rises, its values can be further subdivided into different classes correlating with different degrees of severity and cardiovascular disease (CVD) risk (class I between 30 and 34.9 kg/m^2, class II between 35 and 39.9 kg/m^2 and class III ≥40kg/m^2) [5]. A BMI greater than 40 kg/m^2 is defined as extreme, severe or morbid, whilst having a BMI between 25 and 30 kg/m^2 is described as being in a state termed pre-obesity [6,7]. A sedentary lifestyle coupled with unhealthy eating habits, characterized by the excessive consumption of high energy foods, are the root of the growing prevalence of obesity worldwide. The mechanisms which have led to such a dramatic increase in the incidence and prevalence of obesity are complex and are intertwined with environmental and societal trends [8]. It is not uncommon nowadays to see the term obesity flanked by the term epidemic or even pandemic. This is due to the sheer statistics regarding obesity, which estimate that in 2016 there were 1.9 billion overweight adults worldwide [9,10]. Obesity is the pathophysiological state determined by weight and adipose excess, which is characterized by the alteration of body composition starting from peripheral tissues such as adipose tissue, liver and muscles [11]. These alterations lead to an increased risk of the onset of arterial hypertension, CVDs and other chronic non-communicable degenerative diseases (CNCDs), such as type 2 diabetes mellitus (T2DM), male obesity secondary hypogonadism (MOSH), respiratory diseases, cancer, chronic kidney disease and psychopathological alterations that negatively impact on both quality of life and longevity [12–15].

In obese men, MOSH syndrome leads to a plethora of symptoms such as impaired fertility and sexual function, deficient bone mineralization, altered fat metabolism and body composition and the deterioration of muscle mass [16]. Epidemiological data obtained by population studies state that the prevalence of MOSH syndrome is above 45–57.5% of male obese subjects and it correlates with high-rate morbidity and mortality [17,18].

In this review article we analyzed the possible beneficial effects of ω-3 PUFA on clinical signs and symptoms of MOSH syndrome.

2. Methods

Current literature investigating the possible positive impact of ω-3 PUFA consumption on MOSH syndrome is analyzed and contextualized in this review. Specifically, research has been conducted on Medline (Pubmed) and Scopus. Such a review article analyzes studies (both in vivo and in vitro studies) published up to June 2020.

3. Structure, Metabolic Pathways and Dietary Sources of PUFA

Fatty acids (FAs) are fundamental biocomponents of lipids and cell membranes. They are made up by a hydrocarbon backbone and a carboxylic head group. FAs are classified according to the length of the hydrocarbon backbone (generally 12 to 24 carbon atoms long), and according to the presence and the number of double bonds. We can distinguish between saturated fatty acids (SFAs), which are characterized by the absence of double bonds, monounsaturated fatty acids (MUFAs), which only have one double bond and PUFAs, in which more than one double bond may be found. FAs can be further classified according to the position of the first double bond compared to carbon ω (the furthest carbon from the carboxylic group), forming two classes: ω-3 and ω-6 PUFA [19,20].

The human body can produce almost all fatty acids, except α-linolenic acid (ALA, C18:3 ω-3) and linolenic acid (LA, C18:2 ω-6) which are precursors of ω-3 and ω-6 PUFAs. These are termed "essential fatty acids" because they can only be obtained through diet [21]. Through endogenous conversion (elongation and desaturation) the organism is capable of synthesizing longer-chain counterparts such as eicosapentaenoic acid (EPA) and docosahexaenoic acid (DHA) in the ω-3 family, and γ-linolenic acid (GLA), dihomo-γ-linolenic acid (DGLA) and arachidonic acid (AA) in the ω-6 family [20,22].

Long-chain ω-3 PUFAs and long-chain ω-6 PUFAs are precursors of molecules with important biological activity called eicosanoids such as prostaglandins (PG), thromboxanes (Tx), leukotrienes (LTS), lipoxins (LXS) and resolvins. Depending on which precursor family they belong to, PUFAs can perform different biological functions. In fact, while ω-3 PUFAs carry out an anti-inflammatory function, ω-6 PUFAs elicit a proinflammatory function.

The ω-3 and ω-6 long-chain PUFAs compete to bind enzymes such as cyclooxygenase, lipoxygenase and epoxygenases, which are responsible for the release of inflammatory mediators. Thus, the equilibrium between ω-3 and ω-6 PUFA intracellular concentrations is fundamental for the maintenance of cellular homeostasis and cardiovascular (CV) protection [20,23,24]. In order for them to perform their correct biological actions, it is necessary to have a balanced PUFA intake. Recent studies suggest that an ideal ratio between ω-6/ω-3 is between 1:1 and 1:5, whilst the actual intake ratio in Western countries is of 15:1–16.7:1. Therefore, it appears necessary to maintain an adequate and balanced intake of ω-6/ω-3 in order to prevent CVD onset [25–27].

Regarding main food sources, PUFAs are present as precursors (ALA and LA) in plant-based products and as derivatives (EPA, DHA, AA) in meat (Tables 1 and 2). Fish is the main source of long-chain ω-3 PUFAs, including EPA, DHA and docosapentaenoic acid (DPA), while ALA is a plant and ω-3 PUFAs are mainly found in seeds and nuts and their oils. Plant sources of ω-3 PUFAs cannot currently be considered as a replacement for seafood-derived ω-3 PUFAs. This suggests that ω-3 PUFAs, derived from different sources, have their own specific effects. Therefore, it appears necessary to have a varied and balanced diet [20,27].

Table 1. Main dietary sources of ω-3 fatty acids.

Ω-3 Series	Foods
ALA	Green leafy vegetables, nuts, soybeans, flaxseed
EPA	Fish oil(herring, salmon, sardine, cod liver), Fish(caviar black and red, herring, salmon)
DHA	Fish oil, herring, caviar black and red, salmon

Abbreviations: ALA, α-linoleic acid; DHA, docosahexaenoic acid; EPA, eicosapentaenoic acid.

Table 2. Main dietary sources of ω-6 fatty acids.

Ω-6 Series	Foods
LA	Olive oil, seed oil
GLA	Borage oil, black currant oil
DGLA	Human milk
AA	Meat, dairy, shelfish, human milk

Abbreviations: AA, arachidonic acid; LA, linoleic acid; DGLA, dihomo-γ-linolenic acid; GLA, γ-linolenic acid.

ω-3 PUFAs are also known as "vitamin F", not only are they needed for basic cellular functions such as cell signaling, membrane fluidity and structural integrity, but also for nervous system regulation [28,29]. They have a role in regulating blood pressure, clotting, glucose metabolism and inflammation [28]. Moreover, they have been related to be preventative in the occurrence of CVevents and to slow down the progression of CVDs. These concepts will be further explored in the following section [30].

4. Male Obesity Secondary Hypogonadism (MOSH) Syndrome Definition

MOSH syndrome is a clinical condition found in obese middle-aged men and epidemiological reports assert that in the last 10 years its prevalence has enhanced, even if it is currently an underestimated and underdiagnosed condition [31].

In MOSH syndrome, obesity corroborates hypogonadism to give rise to reduced levels of testosterone (T). This reduction is due to the alteration of metabolic patterns such as lipid metabolism, chronic inflammation and insulin resistance (Figure 1) [32].

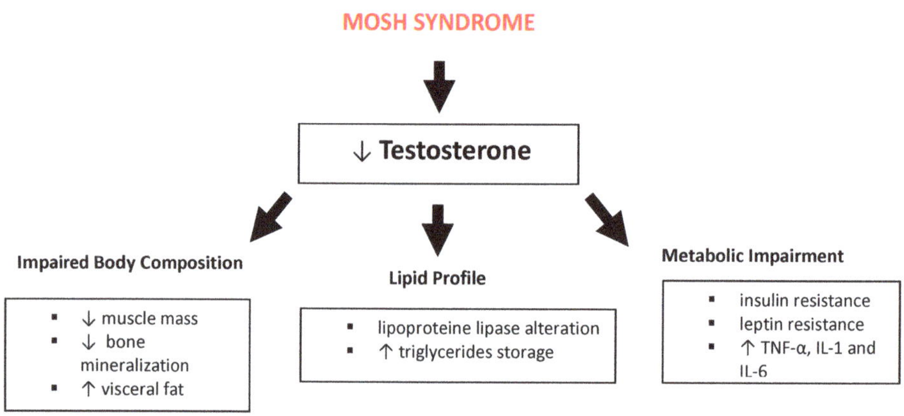

Figure 1. Impact of male obesity secondary hypogonadism (MOSH) syndrome on body composition, lipid profile and metabolic pathways. Abbreviations: TNF-α, tumor necrosis factor-α; IL, interleukin; ↑: increase; ↓: decrease.

The pathophysiological mechanisms linking obesity with hypogonadism are complex and multifactorial [32]. Obese male subjects show a significant reduction of T levels caused by an increase of aromatase enzymes levels, released by the adipose tissue and enhanced by estrogen hormones [33], coupled with the negative feedback produced by the estrogen on the hypothalamic-pituitary axis, this is another factor decreasing the circulating T levels. Such pattern affects the lipid profile through the alteration of lipoprotein lipase presence on adipocytes and increase triglycerides (TG) storage, leading to an increase in visceral adipose deposition and total body fat. These alterations are considered particularly harmful and are highly associated with CV disease risk [33]. Moreover, these lipid profile alterations create a sort of self-perpetuating cycle between obesity and hypogonadism.

The hypertrophy of adipose tissue, characteristic of obese subjects, leads to the lowering of T levels. Metabolic impairment caused by body fat enhancement is responsible for insulin and leptin resistance, and for the increase of pro-inflammatory cytokines (such as Tumor Necrosis Factor-α - TNF-α, interleukins 1 and 6 - IL-1,IL-6) which influence hypothalamic function, in particular decreasing kisspeptin signaling [34]. Such a decrease entails the reduction of gonadotropin-releasing hormone (GnRH), which in turn decreases luteinizing hormone (LH) and follicle-stimulating hormone (FSH) secretion by anterior pituitary gonadotrophs, resulting in a T reduction and in the alteration of fertility [35].

Subjects affected by MOSH are often characterized by reduced osseous mineral density, which can be explained by the T deficiency that is strongly involved in the modulation of bone mineralization, as T is implicated in the regulation of the proliferation and differentiation of osteoblasts [16].

T induces skeletal muscle hypertrophy through numerous mechanisms including its effects in modulating pluripotent mesenchymal cell engagement. Studies have shown that elevated T levels are associated with an increase in the size of motor neurons [36]. Therefore, in subjects with MOSH, the reduction of T levels can lead to a reduction in muscle mass.

MOSH syndrome is potentially reversible. Its treatment, in addition to exogenous T administration, includes lifestyle changes such as diet therapy and physical activity aimed to reduce obesity [16,37,38].

4.1. Role of PUFA in Cardiovascular Disease

In the last few years, the role of ω-3 PUFAs has been widely debated within the scientific and medical communities in virtue of the possible role they may play in contrasting CV diseases (Table 3).

On the one hand observational studies reported an inverse association between CV diseases and dietary intake or plasma concentrations of ω-3 PUFAs (primarily EPA and DHA), suggesting that their supplementation might exert cardio protective effects, on the other hand successive clinical trials and metanalyses have speculated the absence of true benefits induced by ω-3 PUFA consumption on the CV system [39–42]. This discrepancy may be justified by the multiple variables that influence CV diseases which may lead to contrasting results. These variables render CV diseases quite heterogeneous, resulting in different responses to ω-3 PUFA treatment. We must take into consideration that this kind of treatment does not carry out the action of a "pharmaceutical" drug, but rather acts by producing a modulatory effect on the subject's metabolism which can be more or less susceptible to a response, depending not only on the degree and type of pathological involvement but also on the subject's genetic susceptibility. This renders the task even more articulated, particularly as an individual's genetic susceptibility is determined by the genotype and by environmental and epigenetic changes. Even if the debate on ω-3 PUFAs is currently unresolved, it is worth underlining that their consumption has never been associated with deleterious effects on health and therefore their use can either induce positive CV effects, or in the worst case scenario, can induce a neutral effect [43]. For such reason, the following section will comment on the possible beneficial health effects induced by PUFA consumption in subjects with an elevated CV risk and in patients affected by MOSH syndrome. The cardioprotective role of ω-3 PUFAs was hypothesized for the first time in the 1950s in the Eskimo population, which presented elevated levels of plasma cholesterol but an exiguous CV mortality rate [44]. Successively, such observation was also made in the Japanese and Icelandic populations, in which there was an evidently low mortality from CV pathologies compared to Western populations [45,46]. This cardio protective effect was attributed to eating habits, in particular to elevated fish consumption. Further epidemiological studies confirmed this correlation and described the cardioprotective effects induced by ω-3 PUFA consumption [47]. In light of the data published by two large clinical randomized trials, the American Heart Association (AHA) in 2002 suggested the consumption of 1g/day EPA+DHA in patients with coronary artery disease in virtue of their cardioprotective potential [48–50]. Successively, the Gruppo Italiano per lo Studio della Streptochinasi nell'Infarto (GISSI) [50,51] and Diet And Reinfarction Trial (DART) [48] studies have demonstrated a reduction in CV risk following treatment with ω-3 PUFAs, representing the milestones of clinical recommendations for ω-3 PUFA treatment in cardiopathic subjects since it was observed that the benefits outweighed any possible side effect related to their consumption [52,53]. The main cardio protective effects induced by ω-3 PUFA consumption are achieved through actions such as the reduction of plasma TG and of chronic low-grade inflammatory status, an improvement of endothelial function, cardiac functional remodeling and of cardiac contractility [51,54,55]. An in vitro study conducted in bovine aortic endothelial cells demonstrated that treatment with adiponectin is able to increase nitric oxide (NO) production by 3-fold in endothelial cells. This action is due to the phosphorylation of endothelial-nitric oxide synthase (e-NOS) by phosphatidylinositol 3-kinase-dependent pathways [56]. In 2002, the AHA affirmed that a dose between 2 and 4 g/day of ω-3 PUFA was able to treat hypertriglyceridemia [57]. In the wake of this finding, one of the principal studies, aimed at underlining an improvement in plasma TG, was conducted by Harris et al. [58] These authors observed a dose-dependent plasma TG reduction after ω-3 PUFA administration, especially in subjects who presented basal TG levels >500 mg/dL [58]. This was confirmed in subsequent clinical trials performed on subjects presenting very high triglyceride (VHT) levels (TG > 500 mg/dL) and high triglyceride (HT) levels (TG between 200 and 499 mg/dL). Results showed a 30% reduction in plasma TG in the VHT group and a reduction between 20 and 30% in the HT group following the consumption of 4 g/day of ω-3, confirming that the reduction in percentage of TG correlated with their plasma levels before treatment [59–61]. ω-3 PUFAs are able to contrast chronic inflammation via the reduction of macrophage-monocyte adhesion, caused by oxidized low-density lipoprotein (LDL) to the endothelial lining of the coronary vessels. This effect is coupled with the increased expression of e-NOS induced by DHA, with a consequent increase in NO release and therefore, vasodilation [62]. DHA is also able to modulate endothelial function by inducing the transcription of the gene coding for the proinflammatory cytokine TNF-α, and the inhibition of the pathway generated by nuclear factor kappa-light-chain-enhancer of activated B cells (NF-κB), which causes a reduction in vascular cell adhesion molecule-1 (VCAM-1) [63]. Therefore, the actions carried out by DHA at the endothelial level suggest its vasoprotective role.

Moreover, ω-3 PUFAs induce the suppression of thromboxane A_2 (a factor responsible for platelet aggregation, vasoconstriction and fibrinogen reduction) synthesis, and favor the synthesis of thromboxane A_3 [64–66]. In this context, animal models highlighted that EPA consumption also plays a role in stabilizing the atheromatous plaque [67].

EPA and DHA inhibit a series of processes linked to inflammation, such as leukocyte chemotaxis, adhesion interactions between leukocytes and the endothelium, eicosanoid production and T cell reactivity [68]. Finally, an increase in EPA and DHA availability modifies the equilibrium between ω-3 and ω-6 PUFAs, favoring anti-inflammatory eicosanoid synthesis [69].

ω-3 PUFA consumption is associated with a better vascular function, playing a protective role in atherosclerosis, in which endothelial dysfunction is at the basis of the pathogenic process [62,70]. ω-3 PUFAs improve arterial wall rigidity [71] and it was observed that their supplementation induces a reduction in endothelial damage biomarkers such as E-selectin [72].

ω-3 PUFA supplementation was also associated with the reduction of heart rate at rest [73,74], the reduction of systolic and diastolic blood [75,76], and the increase in early and late left ventricular ejection fraction [77].

ω-3 PUFA treatment can lead to a reduction in hospitalization and CV mortality incidence [52]. Finally, the study OMEGA-REMODEL has demonstrated a reduction in cardiac remodeling and fibrosis markers in patients with acute myocardial infarction (AMI), following a supplementation of ω-3 PUFAs (4 g/day) in the diet [78]. It is hypothesized that this beneficial effect is correlated with the reduction of macrophage activation and with the inhibition of galectin-3 (Gal-3), a factor which reflects cardiac function impairment and remodeling [79]. In an elderly population in which subjects had recently undergone an AMI, there were significant inverse correlations between ω-3 PUFA content in serum phospholipids and serum levels of Gal-3, confirming the beneficial effects of ω-3 PUFAs on cardiac remodeling [79].

Table 3. Studies on polyunsaturated fatty acids (PUFAs) and cardiovascular disease.

Type of the Study	Reference	Year	Type of Intervention	Primary Outcome	Conclusions
Animal	Wang, T.M. et al. [63]	2011	40 ApoE (-/-) knockout mice randomized into 5 groups: 1 control group fed normal chow diet, 4 groups fed chow diet supplemented with 200 mg/kg/day of (i) DHA, (ii) EPA, (iii) LA or (iv) AA, for 10 weeks.	DHA supplementation reduced the expression of VCAM-1 in a dose-dependent manner in TNF-α -activated aortic endothelial cells.	DHA supplementation acts at endothelial level by exercising a vasoprotective role.
	Matsumoto, M. et al. [67]	2008	12 ApoE-deficient mice were fed with Western diet and randomized into two groups: (i) 5% EPA supplementation (ii) without EPA supplementation, for 13 weeks.	The EPA supplementation group showed a reduction in the development of atherosclerotic lesions. Lesions presented a great amount collagen and smooth muscle cells and lower amount of macrophages.	EPA has anti-inflammatory and stabilizing effects on the atherosclerotic plaque.
Human	Rhee J.J. et al. [42]	2017	Prospective cohort study on a total of 39,876 women aged ≥45 years without CV diseases subjected to questionnaires on food frequency.	During the follow-up (1993–2014) period there were no associations between the consumption of fish and ω-3 PUFA and CV disease.	The consumption of ω-3 PUFA does not reduce CV risk.
	GISSI-Prevenzione Investigators [51]	1999	Controlled study conducted on 11,324 patients who survived myocardial infarction. They were randomized into 4 groups taking: (i) ω-3 PUFA supplements (1 g per day), (ii) vitamin E (300 mg per day), (iii) both or (iv) neither, for 3–5 years.	The groups with ω-3 PUFA supplementation and with ω-3 PUFA + vitamin E supplementation, presented a reduced risk of death due to CV causes, equally.	Supplementation with ω-3 PUFA reduces the risk of CV mortality.
	Bays, H.E. et al. [59]	2011	Double blind controlled study conducted on 229 subjects with highly elevated blood levels of triglycerides (≥500 mg/dL), which were randomized into 3 groups: (i) 4 g/day supplementation of EPA ethyl ester (ii) 2 g/day supplementation of EPA ethyl ester or (iii) placebo, for 12 weeks.	The supplementation of 4 g/day EPA ethyl ester reduced triglyceride levels by 33.1% whereas 2 g/day supplementation led to 19.2% reduction, both compared to placebo.	ω-3 PUFA supplementation can be useful to counteract hypertriglyceridemia.
	Kastelein, J.J. et al. [60]	2014	Double-blind controlled study conducted on subjects with highly elevated blood levels of triglycerides (≥500 mg/dL) which were randomized into 4 groups: (i) control group (4 g/day of olive oil), (ii) 2g/day of ω-3 PUFA, (iii) 3 g/day of ω-3 PUFA, (iv) 4 g/day of ω-3 PUFA, for 12 weeks in combination with a nutrition education program.	25.9%, 25.5% and 30.9% reduction in blood triglycerides with supplementation of 2, 3 and 4 g/day of ω-3 PUFA respectively, compared to placebo group.	ω-3 PUFA supplementation can be used in lowering hypertriglyceridemia.

Table 3. *Cont.*

Type of the Study	Reference	Year	Type of Intervention	Primary Outcome	Conclusions
Human	Maki, K.C. et al. [61]	2013	647 patients with triglyceride values ≥200 mg/dL and <500 mg/dL were randomized to 6 weeks of treatment with control capsules (4 g/d of olive oil), 2 g/day ω-3 PUFA + 2 g/day olive oil or 4 g/d of ω-3 PUFA.	14.6% and 20.6% reduction in triglyceride levels and 3.9% and 6.9% reduction in non-HDL cholesterol levels with supplementation of 2 g/day and 4 g/d respectively ω-3 PUFA.	ω-3 PUFA supplementation can be useful in the control of dyslipidemias.
	Casanova, M.A. et al. [71]	2017	29 adults with hypertriglyceridemia were divided into: high risk CV patients and low risk CV patients, randomized to take ω -3 PUFA (1800 mg/day) or ciprofibrate (100 mg/day) for 12 weeks.	In high-risk patients, pulse wave velocity was reduced and flow-mediated dilation was increased by ω-3 PUFA.	ω -3 PUFA improved arterial stiffness and endothelial function.
	Huang, F. et al. [72]	2019	69 normal weight adolescents and 70 obese adolescents with hypertriglyceridemia were treated with a lifestyle intervention and randomized for ω -3 PUFA or placebo supplementation for 12 weeks.	After 12 weeks of omega-3 supplementation, adolescents showed a significant reduction in triglycerides, leptin, RBP4, ADMA and E-selectin compared to the placebo group and compared to lifestyle intervention alone.	ω-3 PUFA supplementation associated with a healthy lifestyle improves dyslipidemia, insulin resistance and endothelial dysfunction.
	Rantanen, J.M. et al. [73]	2018	Randomized, double-blind, controlled trial of 112 chronic dialysis patients from Denmark randomized for daily supplementation of 2 g marine ω-3 PUFA or control group, for three months.	In the group with daily supplementation with ω-3 PUFA there was a reduction in heart rate of 2.5 beats per minute, evaluated by 48-h Holter monitoring.	ω-3 PUFA could contribute to vagal modulation that could be protective against malignant ventricular arrhythmias.
	Sagara, M. et al. [75]	2011	38 men with arterial hypertension and/or hypercholesterolemia were randomized for a five-week dietary supplement with 2 g of DHA vs active placebo (1 g of olive oil).	Significant reduction in systolic and diastolic blood pressure, heart rate and HDL-C increase in the group with supplementation of DHA.	DHA supplementation can reduce coronary heart disease risk factors.
	Lee, J.B. et al. [76]	2019	Randomized double-blind study of 86 healthy young men and women to evaluate the effects of oral supplementation of 3 g/day of (i) EPA, (ii) DHA or (iii) olive oil for 12 weeks.	Reduction of systolic and diastolic BP at rest after DHA and olive oil supplementation compared to EPA; DHA supplementation enhances peripheral vasoconstrictor outflow.	DHA supplementation may represent a valid support for patients with high chronic BP.

Table 3. *Cont.*

Type of the Study	Reference	Year	Type of Intervention	Primary Outcome	Conclusions
Human	Ghio, S. et al. [77]	2010	608 patients with chronic symptomatic heart failure were randomized to take (i) ω-3 PUFA or (ii) placebo supplementation. Echocardiography was performed at baseline and after 1, 2 and 3 years.	Left ventricular ejection fraction increase after ω-3 PUFA supplementation of 8.1% at 1 year, 11.1% at 2 years and 11.5% at 3 years.	ω-3 PUFA supplementation can significantly improve left ventricular ejection fraction in patients with symptomatic heart failure.
	Tavazzi, L. et al. [52]	2008	Controlled, double-blind study in patients with chronic heart failure randomized for the supplementation of 1 g/day of (i) ω-3 PUFA or (ii) placebo, followed for a median of 3.9 years.	57% of patients in the ω-3 PUFA supplement group, compared to 59% in the placebo group, died or were hospitalized for CV reasons.	ω-3 PUFA supplementation may provide a small benefit in terms of mortality and hospitalization for CV reasons in heart failure patients.
	Heydari, B. et al. [78]	2016	Controlled, double-blind study on 358 patients after acute MI, randomized for (i) ω-3 PUFA supplementation (4 g/die) or (ii) placebo and underwent baseline assessment by CMR 4-28 days after MI, with 6 months follow-up.	After 6 months of PUFA treatment, the follow-up CMR revealed a significant reduction in left ventricular end-systolic volume indexed and myocardial extra-cellular volume fraction and ST2, fibrosis marker, compared to placebo.	PUFA have an important effect on phenotypes of myocardial tissue after MI.
	Laake, L. et al. [79]	2017	Evaluation of the relationship between serum level of marine PUFA ω-3 and ω-6 and biomarkers of fibrosis and cardiac remodeling (ST2 and Galectin-3) in 299 elderly patients 2–8 weeks after acute MI.	Galectin-3 levels were inversely related to EPA and DHA and positively related to the ω -6/ω -3 ratio.	ω-3 PUFA display positive effect on cardiac remodeling in acute MI elderly patients.

Abbreviations: AA, arachidonic acid; ADMA, asymmetric dimethylarginine; ApoE, apolipoprotein E; BP, blood pressure; CMR, Cardiac Magnetic Resonance; CV, Cardiovascular; DHA, docosahexaenoic acid; EPA, eicosapentaenoic acid; HDL, high-density lipoprotein; LA, linoleic acid; MI, myocardial infarction; PUFA, polyunsaturated fatty acids; RBP4, retinol binding protein 4, TNF-α, tumor necrosis factor-α.;VCAM, vascular cell adhesion molecule.

4.2. Impact of PUFA Consumption on Body Weight

The Mediterranean diet is known to provide a balanced supply of PUFAs [14]. In vivo studies have demonstrated that the consumption ω-3 FA is correlated with the improvement of body composition. Specifically, it is observed that there is a reduction in adipose tissue thanks to the interactions with metabolic pathways, including the glucose one [80]. A meta-analysis conducted in 2014 [4] has explored the relationship between the consumption of long-chain ω-3 PUFAs and body composition in Caucasian subjects (Table 4). The study examined 934 subjects who were getting long-chain ω-3 PUFAs from fish or from supplements. The authors have found statistically significant variations comparing results obtained between the study group and healthy subjects. The examined parameters were: body weight; BMI; fat mass (FM) %; and waist circumference (WC). Moreover, the authors have also investigated the possible gender effect tied to the consumption of long-chain ω-3 PUFAs, highlighting that in male subjects the WC diminished significantly more than in females.

There is considerable evidence showing that, at the cellular level, PUFAs are potent transcription regulators of genes involved in lipid metabolism. In fact, PUFAs have an important role in the inhibition of genes involved in lipogenesis, and in the promotion of genes involved in lipid oxidation [81]. Other than being prone to rapid oxidation and peroxidation, PUFAs are able to favor the synthesis of proteins involved in detoxification processes that counteract oxidative stress [82]. A study by Di Nunzio et al. [83] has shed light on the antioxidant and pro-oxidant properties of different PUFAs. The authors have demonstrated that only DHA is able to diminish susceptibility to hydrogen peroxide, which stimulates the transcription and the activation of the peroxisome proliferator-activated receptor α (PPARα). PPARα is able to favor the activity of antioxidant enzymes, such as Catalase- CAT and superoxide dismutase- SOD [83]. Therefore, the consumption of PUFAs, specifically DHA, allows an adequate antioxidant protection at the cellular level if the ω-3/ω-6 at 1:5 ratio is followed [84].

The enhancement in lipid oxidation, and the increased use of lipids as an energy source, can translate into a reduction in FM. In fact, some studies demonstrate that increased PUFA intake is associated with substantial FM loss, especially in the abdominal region [85].

Couet et al. have examined a population of lean and healthy individuals who were administered 6 g/day of visible fat for 3 weeks followed by a wash-out period lasting 10–12 weeks, followed in turn by the administration of 6 g/day of fish oil for 3 weeks. The authors have reported a statistically significant reduction in FM, whilst body weight was maintained [86].

A study by Huang et al. [87] has examined the possible genetic–dietary interactions in a population of 24,357 subjects. The authors have analyzed all known 77 single-nucleotide polymorphisms (SNPs) correlated with BMI. The data showed that consumption of fish-derived long-chain ω-3 are able to modulate gene expression related to weight gain and BMI modifications. In fact, long-chain ω-3 PUFAs were able to modify the genetic associations that determine adipose tissue accumulation in various body regions [88]. Therefore, the consumption of long-chain ω-3 PUFAs plays an important role in phenotype manifestation, modulating the expression of weight regulatory genes.

The notion that adipose tissue is simply an inert tissue that stores fat has become obsolete. On the contrary, it is now recognized as a metabolically active endocrine organ, which has the capacity to synthesize biological mediators called adipocytokines, which regulate the body's metabolic status and influence homeostasis [89]. Adipose tissue is not solely comprised of adipocytes, but also of blood vessels and stroma, which contain the precursor cells. It is also useful to distinguish white adipose tissue (WAT) from brown adipose tissue (BAT). WAT is made up of unilocular adipocytes and is better suited for storage, while BAT adipocytes are multilocular, contain copious amounts of mitochondria and are involved in thermogenesis [90]. Diet-induced thermogenesis is a metabolic process linked with energy expenditure following the ingestion of various macronutrients (such as carbohydrates, proteins, fats and alcohol). A study by Casas-Agustench et al. has examined a population of 29 healthy males and compares the thermogenic effects induced by three isocaloric meals: the first contained high levels of PUFAs from walnuts, the second contained high levels of MUFAs from olive oil and the third contained high levels of fat from dairy products. Thermogenesis induced 5 hours after the first meal was 28% greater than the one induced by the third meal. Therefore, the quality of fats can influence the thermogenic response, even if the properties which influence lipid substrate oxidation are still not known [91].

A further feature of obese subjects is the low-grade chronic inflammatory state. The postulation that obesity is inherently linked to the latter goes hand in hand with the notion that adipose tissue in an obese individual undergoes compelling alterations in both composition and function, a process named "adipose tissue remodeling" [92]. The inflammatory status is characterized by pro-inflammatory molecules such as TNF-α, interleukin (IL)-1β, IL-6, IL-8, transforming growth factor-β, nerve growth factor and acute phase response molecules such as plasminogen activator inhibitor-1, haptoglobin; serum amyloid A, has been recognized as a driver of metabolic disease in obese subjects [93]. Therefore, a reduction of the low-grade chronic inflammatory status, consequent to a decrease of body weight, would lead to an improvement in the clinical conditions of MOSH syndrome.

A study by Lund et al. [94], other than attaining positive results regarding BMI, WC and hip circumference (HC) reduction following PUFA consumption, has highlighted an inverse correlation between ALA ω-3 consumption and levels of macrophage inflammatory protein (MIP)-1α. The latter is a chemokine which is overexpressed in obese subjects who present abdominal visceral fat accumulation. Therefore, PUFAs are able to act beneficially on MIP-1α levels, and therefore on central adiposity.

Table 4. Studies on impact of PUFA consumption on body weight.

Type of the Study	Reference	Year	Type of Intervention	Primary Outcome	Conclusions
In vitro	Di Nunzio, M. et al. [83]	2011	Supplementation of HepG2 cells with different PUFAs produced various effects on cytotoxicity, oxidation and on antioxidant defenses.	Supplementation with ARA highlighted the induction of oxidative damage, on the contrary, DHA supplementation induced an enhancement in antioxidant defenses.	Each PUFA seems to exert certain actions, on the basis of chemical structure.
Human	Summers, L.K. et al. [85]	2002	17 subjects (6 with T2DM, 6 non-obese and 5 obese without T2DM) were randomized in a crossover study to follow two 5-week periods (one period with a diet rich in saturated fatty acids and one period with a diet rich in polyunsaturated fatty acids).	Insulin sensitivity and plasma LDL cholesterol concentrations ameliorated in subjects that followed a diet rich in PUFA compared with the subjects that followed a diet rich in saturated fatty acids. The authors observed a reduction in abdominal subcutaneous fat.	These dietary patterns suggest an improvement in insulin sensitivity, reducing the risk of developing T2DM.
	Couet, C. et al. [86]	1997	Six volunteers were fed with a control diet (52% carbohydrates, 16% protein, 32% fat; no FO) *ad libitum* for 3 weeks and, 10–12 weeks later, 6 g of fats, of the same diet, were replaced with 6 g/d of FO for a further 3 weeks.	After the dietetic treatment with FO, there was observed a decrease in body fat mass and basal respiratory quotient and an increase of basal lipid oxidation.	Dietary FO causes a reduction of body fat mass and induces lipid oxidation in healthy adults.
	Huang, T. et al. [87]	2019	The authors tested interactions of ω-3 PUFA habitual consumption and overall genetic susceptibility on long-term weight change.	Food-sourced ω-3 PUFA assumption showed substantial interactions with GRS on long-term changes in body weight.	High intake of ω-3 PUFAs is related to an attenuation of genetic association with long-term weight gain.
	Vaughan, L.K. et al. [88]	2015	The authors evaluated body composition, plasma adipokines and ghrelin in 982 subjects. They investigated gene–diet interactions.	The authors observed a linkage for all obesity-related traits. They identified new regions of interest for adiponectin and body circumferences. They reported that ω-3 PUFAs are able to modify the link with obesity-related traits.	These authors speculated on the interaction between gene-obesity tract and the pathophysiology obesity.
	Casas-Agustench, P. et al. [91]	2009	29 healthy men were randomized in a crossover trial. The authors compared the thermogenic effects of 3 isocaloric sources: (i) high in polyunsaturated fatty acids from walnuts, (ii) high in monounsaturated fatty acids from olive oil and (iii) high in saturated fatty acids from fat-rich dairy products.	Five hours postprandial thermogenesis was greater after the high-polyunsaturated meal (i), and after the high-monounsaturated meal (ii) compared to the high-in-saturated meal (iii).	The thermogenic response was influenced by the fat quality, although the action on substrate oxidation or satiety was unknown.
	Lund, A.S. et al. [94]	2013	1212 healthy individuals were enrolled and the authors collected information on nutritional habits associated with different measures of body fat, and inflammatory biomarkers.	Absolute ω-3 PUFA intake presents inverse correlation with anthropometric measures of body fat and among ω-3 PUFA derivatives. In particular, only ALA was inversely associated with body fat measures.	Intake of ω-3 PUFA, in particular ALA, is positively associated with body fat.

Abbreviations: ALA, α-linolenic acid; ARA, arachidonic acid;; DHA, docosahexaenoic acid; FO, fish oil; GRS, genetic risk score; LDL, low-density lipoproteins; PUFA, polyunsaturated fatty acid;T2DM, type 2 diabetes mellitus.

4.3. PUFA and Metabolic Axis

Long-chain ω-3 PUFAs are able to regulate numerous metabolic mechanisms apt to contrast weight gain. They enable better control of the hunger and satiety mechanism and allow better perfusion of metabolically active tissues (such as skeletal muscle) through the modulation of gene expression. They also induce fatty acid oxidation and can cause an increase in energy expenditure associated with a reduction in fat deposits [95].

Several studies suggest that long-chain ω-3 PUFAs can suppress appetite and regulate thermogenesis by inducing an increase in blood concentration of adipocyte hormones such as leptin and adiponectin [80,96,97] (Table 5). Leptin was the first hormone to be recognized for having a regulatory action at the hypothalamic level [98]. Its principal function is to control food-intake, undertaking an anorexigenic effect, however, it can also regulate energy expenditure and body weight [99]. Leptin acts upon the metabolism and food consumption, reducing appetite and increasing energy expenditure [100]. The expression and release of this hormone are positively correlated with the amount of fat mass and adipocyte dimension, and they are stimulated by hormones such as cortisol and insulin [101].

Different studies report that the reduction in leptin plasma concentration represents a short-term adaptation to the mechanism of hunger or fasting and therefore, in response to diet-induced weight loss, the levels of leptin decrease significantly [102,103]. In normal weight subjects, leptin is released into circulation and acts through hypothalamic and extra-hypothalamic brain receptors (arcuate nucleus and dorsomedial hypothalamus, respectively), inhibiting hunger and increasing thermogenesis following food intake. Moreover, in non-obese subjects, leptin acts through hypothalamic receptors, inhibiting the hunger mechanism and increasing thermogenesis during the fasting period. Decreased leptin levels provoke a reduction in central sympathetic nervous outflow and mobilize stored adipose tissue through glucocorticoid stimulation [104]. Whereas in obese subjects, even if plasma leptin concentration seems to be increased, it does not decrease food consumption and increase energy expenditure. Such a phenomenon suggests that obese subjects become leptin-resistant as reported by different authors since the 1990s [98,104]. The "leptin resistance hypothesis" was demonstrated by Enriori et al. and observed an attenuation of the phosphorylation of signal transducer and activator of transcription 3 (STAT3) in obese mice, which is a crucial factor for the action of leptin on the hypothalamic arcuate nucleus [105].

Hyperleptinemia is also associated with an increased production and release into the bloodstream of pro-inflammatory cytokines (such as TNF-α, C- reactive protein- CRP, etc.) [106,107] and to an increase of platelet aggregation and thrombosis [108]. Thus, the persistent condition of hyperleptinemia could play an unfavorable role in different organs and systems such as the CV system.

A study by Pérez-Matute et al. [109] investigated the potential anti-obesogenic and insulin-sensitizing properties associated with long-chain ω-3 PUFA consumption in an animal model, this was done by feeding the animals two different dietary regimens for the duration of 5 weeks. The control group was administered a standard laboratory diet, whilst the study group was administered a fat-rich hyperenergetic diet. These groups were further divided into two subgroups, differentiated by whether or not they were administered EPA. Results showed that EPA consumption during a fat-rich hyperenergetic diet is able to restrain weight gain and consequently leads to an increase in fat mass. This effect could be correlated with an increase in leptin levels, which causes reduced hunger. Another finding shows that the group consuming a fat-rich hyperenergetic diet and EPA supplementation showed significant weight loss, greater than the standard laboratory diet and EPA supplementation group. It can be speculated that the metabolic effects related to a fat-rich hyperenergetic diet could be correlated to its bromatological composition.

Adiponectin is a protein which regulates the endocrine functions of adipocytes, which perform autocrine and paracrine functions. Adiponectin seems to improve lipid storage, contrasting ectopic deposition of lipids [110] favoring healthy adipose tissue composition. Moreover, it can regulate energy homeostasis by modulating lipid and the glucose metabolism as well as fatty acid oxidation. A study highlighted that adiponectin is able to ameliorate insulin sensitivity in the liver and in skeletal muscles, regulating healthy adipose tissue expansion [111,112]. A study was conducted by Dimiter [113] to investigate the relationship between ω-3 PUFA consumption and circulating adiponectin levels on 35 subjects with metabolic syndrome. The subjects were subdivided into two groups: one was treated with ω-3 PUFA supplements and the control group was given a placebo for a period of three months. The results showed that the treated group demonstrated a statistically significant increase in plasma adiponectin and high-density lipoprotein (HDL) cholesterol, with a concomitant decrease in TGs, Homeostatic model assessment - insulin resistance (HOMA-IR) and CRP. These findings highlighted that supplementation with ω-3 PUFAs can contribute to a bettering of the clinical profile of metabolic syndrome patients by reducing inflammation, improving dyslipidemia and endocrine function through adiponectin-dependent mechanisms.

Long-chain ω-3 PUFAs can alter gene expression in skeletal muscle, suppressing catabolic pathways and upregulating anabolic ones. These mechanisms attenuate muscular mass loss while maintaining muscular functionality and metabolic rate [95]. The restriction of energetic intake results in efficacious fat mass reduction; however, it can often cause the loss of fat-free mass and skeletal muscle. This may negatively impact on physical performance and cause a reduction in metabolic rate by reducing lipid oxidation capacity [114]. The principal pathway responsible for muscle catabolism during energetic intake restriction is the ubiquitin-proteasome pathway [115]. EPA is able to inhibit the activity of such a pathway during periods of severe energy intake restriction. In this context, long-chain ω-3 PUFAs can augment the activation of the Protein kinase B (Akt)—Mammalian target of rapamycin (mTOR)—the Ribosomal protein S6 kinase beta-1 (S6K1) anabolic pathway in skeletal muscle-promoting anti-catabolites and anabolites [116]. In a study by Howe et al. [117], long-chain ω-3 PUFAs were able to attenuate muscle mass loss during an energy restriction diet. Moreover, an improvement of lean mass and energy balance was observed [95,118]. Successively, the same authors have observed an increase in lean mass percentage, suggesting a direct relationship between the consumption of ω-3 PUFAs and lean mass improvement [117].

Table 5. Studies on PUFAs and metabolic axis.

Type of the Study	Reference	Year	Type of Intervention	Primary Outcome	Conclusions
In vitro	Andrade-Vieira, R. et al. [116]	2013	In vitro tests on MCF7 and HeLaS cell lines.	ω-3 PUFA increases the activation of the Akt-mTOR-S6K1 anabolic pathway.	ω-3 PUFA supplementation can improve cellular metabolism by the promotion of anticatabolite and anabolites production.
Human	Perez-Matute, P. et al. [109]	2007	Male Wistar rats were fed a high-fat diet for 5 weeks: (i) with oral administration of EPA (1 g/kg) or (ii) without EPA administration.	The increase in body weight and FM was lower in the group treated with EPA. Moreover, EPA administration induced a decrease in food intake and an increase in leptin production and was able to prevent the increase in TNFα.	EPA supplementation can increase the feeling of satiety and reduce the inflammatory state induced by a high-fat diet.
	Takahashi, Y. et al. [80]	2000	4 groups of rats were fed for 21 days as follows: (i) low-fat diet with 20 g of safflower oil/kg; (ii) high-fat diet (200 g/kg) rich in ω-6 with safflower oil; (iii) high-fat diet (200 g/kg) rich in ω-3 with perilla; (iv) high-fat diet (200 g/kg) rich in ω-3 with fish oil.	The high-fat diets rich in ω-3, compared to a low-fat diet, did not increase the WAT mass, but increased BAT. Moreover, the diets rich in ω-3, reducing the expression of GLUT-4 mRNA in WAT.	In rats, the gene expression of GLUT-4 mRNA in adipose tissue by ω-3, prevents body fat accumulation and regulates glucose metabolism.
	Whitehouse, A.S. et al. [114]	2001	Evaluation of the effect of EPA administration on soleus muscle proteolysis during acute fasting in rats compared to control group (olive oil).	Significant reduction of soleus muscle proteolysis in an EPA-treated group and attenuation of the proteasome "chymotryptic-like" enzyme activity.	EPA is able to inhibit protein proteolysis during acute starvation.
	Madsen, E.L. et al. [103]	2009	Treatment of 68 subjects in abdominal obesity with a low-calorie diet (600–800 kcal/die) for 8 weeks followed by 36 months of randomized treatment with (i) orlistat or (ii) placebo, in association with the lifestyle intervention.	The decrease in body weight is associated with a significant reduction of IL-18, MMP-9 and leptin levels.	Long-term weight loss reduces non-traditional CV risk factors.
	Parra, D. et al. [97]	2008	Appetite monitoring in 233 volunteers during the last 2 weeks of an 8-week low-calorie diet, associated with: (i) low ω-3 intake (<260 mg/die) or (ii) high ω-3 intake (>1300 mg/die).	The evaluation of VAS reveals lower hunger in the high-ω-3 group after dinner after 120 min.	ω-3 intake modulates postprandial satiety in obese and overweight subjects during the weight loss.
	Shamsuzzaman, A.S. et al. [107]	2004	Association study between plasma leptin and CRP in 100 healthy subjects.	Strong positive and significant association between leptin and CRP, even after adjustment for age, BMI, waist-hip ratio, smoking and alcohol consumption.	The study confirms a strong correlation between metabolic and inflammatory mechanisms.
	Dimiter, V. [113]	2007	35 overweight and obese adults with metabolic syndrome were randomized into 2 groups: (i) treated with ω-3 and (ii) treated with placebo, for 3 months on a normal diet, without lifestyle changes.	Increased plasma concentrations of HDL-C and adiponectin and decreased plasma concentrations of triglycerides, HOMA-IR and CRP after ω-3 treatment.	ω-3 PUFA supplementation can improve the inflammatory status and lipid profile through adiponectin-dependent mechanisms in patients with metabolic syndrome.

Abbreviations: Akt- mTOR, protein kinase B- mammalian target of rapamycin; BAT, brown adipose tissue; BMI, body mass index;CRP, C-reactive protein; CV, cardiovascular; EPA, Eicosapentaenoic acid; FM, fat mass; GLUT-4, glucose transporter type-4; HDL-C, high density lipoprotein-cholesterol; HOMA-ir, homeostatic model assessment for insulin resistance; IL-18, interleukin-18; MMP-9, matrix metalloproteinase-9; PUFA, polyunsaturated fatty acids; TNFα, tumor necrosis factor-α; VAS, visual analogue scale WAT, white adipose tissue.

5. Summary and Future Perspectives

In conclusion, there seems to be evidence that ω-3 PUFA consumption may be clinically beneficial in the treatment and clinical management of MOSH patients.

The ability of ω-3 PUFAs to act on some pathological aspects of MOSH, such as obesity, inflammation, metabolic and cardiovascular disorders, coupled with their optimal safety profile, leads us to the postulation that ω-3 PUFA assumption could be a valuable tool in ameliorating the clinical manifestations of MOSH syndrome.

In pursuance to assess this conclusion in a definitive manner and in order to define the most advantageous dosage, randomized clinical trials on a large population sample are required. Moreover, it would be interesting and useful to conduct experimental studies exploring the possible effects of ω-3 PUFA consumption on hormone profile, on the sexual sphere (T concentrations) and on body composition.

Author Contributions: Conceptualization, A.N. and N.D.D.; writing—original draft preparation, A.N., G.M., F.D.D., M.D.L., A.P.Z. and G.W.J.; writing—review and editing, N.D.D. and A.D.L.; supervision, N.D.D. and A.N. All authors have read and agreed to the published version of the manuscript.

Funding: This research received no external funding.

Conflicts of Interest: The authors declare no conflict of interest.

Abbreviation List

AA	Arachidonic acid
AHA	American Heart Association
Akt	Protein kinase B
ALA	α-linoleic acid
AMI	Myocardial infarction
BAT	Brown adipose tissue
BMI	Body mass index
CAD	Caspase-activated DNase
CAT	Catalase
CNCD	Chronic non-communicable degenerative disease
CRP	C-reactive protein
CV	Cardiovascular
CVD	Cardiovascular disease
DART	Diet And Reinfarction Trial
DGLA	Dihomo-γ-linoleic acid
DHA	Docosahexaenoic acid
DPA	Docosapentaenoic acid
eNOS	Endothelial nitric oxide synthase
EPA	Eicosapentaenoic acid
FAs	Fatty acids
FM	Fat mass
FSH	Follicle-stimulating hormone
Gal-3	Galectin-3
GISSI	Gruppo Italiano per lo Studio della Streptochinasi nell'Infarto
GLA	γ-linoleic acid
GnRH	Gonadotropin-Releasing Hormone
HC	Hip circumference
HDL	High-density lipoprotein
HOMA-IR	Homeostatic model assessment-insulin resistance
HT	High triglyceride
IL	Interleukin
LA	Linoleic acid
LDL	Low-density lipoprotein

LH	Luteinizing hormone
LTS	Leucotriens
LXS	Lipoxines
MIP	Macrophage inflammatory protein
MOSH	Male obesity secondary hypogonadism
mTOR	Mammalian target of rapamycin
MUFAs	Monounsaturated fatty acids
NF-κB	Nuclear factor kappa-light-chain-enhancer of activated B cells
PG	Prostaglandins
PPARα	Peroxisome proliferator- activated receptor α
PUFA	Polyunsaturated fatty acid
S6K1	Ribosomal protein S6 kinase beta-1
SFAs	Saturated fatty acids
SNPs	Single-nucleotide polymorphism
SOD	Superoxide dismutase
STAT3	Signal transducer and activator of transcription 3
T	Testosterone
T2DM	Type 2 diabetes mellitus
TG	Triglycerides
TNF-α	Tumor necrosis factor- α
Tx	Thromboxanes
VCAM-1	Vascular cell adhesion molecule-1
VHT	Very high triglyceride
WAT	White adipose tissue
WC	Waist circumference
WHO	World Health Organization

References

1. Barnes, S. Nutritional genomics, polyphenols, diets, and their impact on dietetics. *J. Am. Diet. Assoc.* **2008**, *108*, 1888–1895. [CrossRef] [PubMed]
2. Christensen, J.H. Omega-3 polyunsaturated Fatty acids and heart rate variability. *Front. Physiol.* **2011**, *2*, 84. [CrossRef] [PubMed]
3. Raphael, W.; Sordillo, L.M. Dietary polyunsaturated fatty acids and inflammation: The role of phospholipid biosynthesis. *Int. J. Mol. Sci.* **2013**, *14*, 21167–21188. [CrossRef] [PubMed]
4. Bender, N.; Portmann, M.; Heg, Z.; Hofmann, K.; Zwahlen, M.; Egger, M. Fish or n3-PUFA intake and body composition: A systematic review and meta-analysis. *Obes. Rev.* **2014**, *15*, 657–665. [CrossRef]
5. Ofei, F. Obesity—A preventable disease. *Ghana. Med. J.* **2005**, *39*, 98–101.
6. de Mello, A.H.; Uberti, M.F.; de Farias, B.X.; de Souza, N.A.R.; Rezin, G.T. n-3 PUFA and obesity: From peripheral tissues to the central nervous system. *Br. J. Nutr.* **2018**, *119*, 1312–1323. [CrossRef]
7. Obesity, W. Obesity Classification. Available online: https://www.worldobesity.org/about/about-obesity/obesity-classification (accessed on 1 June 2020).
8. Cohen, D.A. Obesity and the built environment: Changes in environmental cues cause energy imbalances. *Int. J. Obes.* **2008**, *32* (Suppl. 7), S137–S142. [CrossRef]
9. Obesity, W. Prevalence of Obesity. Available online: https://www.worldobesity.org/about/about-obesity/prevalence-of-obesity (accessed on 3 April 2020).
10. Chooi, Y.C.; Ding, C.; Magkos, F. The epidemiology of obesity. *Metabolism* **2019**, *92*, 6–10. [CrossRef]
11. Knight, J.A. Diseases and disorders associated with excess body weight. *Ann. Clin. Lab. Sci.* **2011**, *41*, 107–121.
12. Kyrou, I.; Randeva, H.S.; Tsigos, C.; Kaltsas, G.; Weickert, M.O. Clinical Problems Caused by Obesity. In *Endotext*; Feingold, K.R., Anawalt, B., Boyce, A., Chrousos, G., Dungan, K., Grossman, A., Hershman, J.M., Kaltsas, G., Koch, C., Kopp, P., et al., Eds.; Endotext.com: South Dartmouth, MA, USA, 2000.
13. World Health Organization. Cardiovascular Diseases. Available online: https://www.who.int/health-topics/cardiovascular-diseases#tab=tab_1 (accessed on 1 June 2020).

14. Di Daniele, N.; Noce, A.; Vidiri, M.F.; Moriconi, E.; Marrone, G.; Annicchiarico-Petruzzelli, M.; D'Urso, G.; Tesauro, M.; Rovella, V.; De Lorenzo, A. Impact of Mediterranean diet on metabolic syndrome, cancer and longevity. *Oncotarget* **2017**, *8*, 8947–8979. [CrossRef]
15. Canale, M.P.; Manca di Villahermosa, S.; Martino, G.; Rovella, V.; Noce, A.; De Lorenzo, A.; Di Daniele, N. Obesity-related metabolic syndrome: Mechanisms of sympathetic overactivity. *Int. J. Endocrinol.* **2013**, *2013*, 865965. [CrossRef] [PubMed]
16. De Lorenzo, A.; Noce, A.; Moriconi, E.; Rampello, T.; Marrone, G.; Di Daniele, N.; Rovella, V. MOSH Syndrome (Male Obesity Secondary Hypogonadism): Clinical Assessment and Possible Therapeutic Approaches. *Nutrients* **2018**, *10*, 474. [CrossRef] [PubMed]
17. Calderon, B.; Gomez-Martin, J.M.; Vega-Pinero, B.; Martin-Hidalgo, A.; Galindo, J.; Luque-Ramirez, M.; Escobar-Morreale, H.F.; Botella-Carretero, J.I. Prevalence of male secondary hypogonadism in moderate to severe obesity and its relationship with insulin resistance and excess body weight. *Andrology* **2016**, *4*, 62–67. [CrossRef]
18. Anderson, J.L.; May, H.T.; Lappe, D.L.; Bair, T.; Le, V.; Carlquist, J.F.; Muhlestein, J.B. Impact of Testosterone Replacement Therapy on Myocardial Infarction, Stroke, and Death in Men With Low Testosterone Concentrations in an Integrated Health Care System. *Am. J. Cardiol.* **2016**, *117*, 794–799. [CrossRef] [PubMed]
19. Wiktorowska-Owczarek, A.; Berezinska, M.; Nowak, J.Z. PUFAs: Structures, Metabolism and Functions. *Adv. Clin. Exp. Med.* **2015**, *24*, 931–941. [CrossRef] [PubMed]
20. Saini, R.K.; Keum, Y.S. Omega-3 and omega-6 polyunsaturated fatty acids: Dietary sources, metabolism, and significance—A review. *Life Sci.* **2018**, *203*, 255–267. [CrossRef]
21. Das, U.N. Essential Fatty acids—A review. *Curr. Pharm. Biotechnol.* **2006**, *7*, 467–482. [CrossRef]
22. Tvrzicka, E.; Kremmyda, L.S.; Stankova, B.; Zak, A. Fatty acids as biocompounds: Their role in human metabolism, health and disease–a review. Part 1: Classification, dietary sources and biological functions. *Biomed. Pap. Med. Fac. Univ. Palacky Olomouc. Czech. Repub.* **2011**, *155*, 117–130. [CrossRef]
23. Adkins, Y.; Kelley, D.S. Mechanisms underlying the cardioprotective effects of omega-3 polyunsaturated fatty acids. *J. Nutr. Biochem.* **2010**, *21*, 781–792. [CrossRef]
24. Dessi, M.; Noce, A.; Bertucci, P.; Noce, G.; Rizza, S.; De Stefano, A.; Manca di Villahermosa, S.; Bernardini, S.; De Lorenzo, A.; Di Daniele, N. Plasma and erythrocyte membrane phospholipids and fatty acids in Italian general population and hemodialysis patients. *Lipids Health Dis.* **2014**, *13*, 54. [CrossRef]
25. Yang, L.G.; Song, Z.X.; Yin, H.; Wang, Y.Y.; Shu, G.F.; Lu, H.X.; Wang, S.K.; Sun, G.J. Low n-6/n-3 PUFA Ratio Improves Lipid Metabolism, Inflammation, Oxidative Stress and Endothelial Function in Rats Using Plant Oils as n-3 Fatty Acid Source. *Lipids* **2016**, *51*, 49–59. [CrossRef] [PubMed]
26. Duan, Y.; Li, F.; Li, L.; Fan, J.; Sun, X.; Yin, Y. n-6:n-3 PUFA ratio is involved in regulating lipid metabolism and inflammation in pigs. *Br. J. Nutr.* **2014**, *111*, 445–451. [CrossRef] [PubMed]
27. Simopoulos, A.P. Evolutionary aspects of diet, the omega-6/omega-3 ratio and genetic variation: Nutritional implications for chronic diseases. *Biomed. Pharmacother.* **2006**, *60*, 502–507. [CrossRef] [PubMed]
28. Gammone, M.A.; Riccioni, G.; Parrinello, G.; D'Orazio, N. Omega-3 Polyunsaturated Fatty Acids: Benefits and Endpoints in Sport. *Nutrients* **2018**, *11*, 46. [CrossRef] [PubMed]
29. Merida-Ortega, A.; Rothenberg, S.J.; Torres-Sanchez, L.; Schnaas, L.; Hernandez-Alcaraz, C.; Cebrian, M.E.; Garcia-Hernandez, R.M.; Ogaz-Gonzalez, R.; Lopez-Carrillo, L. Polyunsaturated fatty acids and child neurodevelopment among a population exposed to DDT: A cohort study. *Environ. Health* **2019**, *18*, 17. [CrossRef]
30. Liu, W.; Xie, X.; Liu, M.; Zhang, J.; Liang, W.; Chen, X. Serum omega-3 Polyunsaturated Fatty Acids and Potential Influence Factors in Elderly Patients with Multiple Cardiovascular Risk Factors. *Sci. Rep.* **2018**, *8*, 1102. [CrossRef]
31. Gore, A.C.; Chappell, V.A.; Fenton, S.E.; Flaws, J.A.; Nadal, A.; Prins, G.S.; Toppari, J.; Zoeller, R.T. EDC-2: The Endocrine Society's Second Scientific Statement on Endocrine-Disrupting Chemicals. *Endocr. Rev.* **2015**, *36*, E1–E150. [CrossRef]
32. Fernandez, C.J.; Chacko, E.C.; Pappachan, J.M. Male Obesity-related Secondary Hypogonadism—Pathophysiology, Clinical Implications and Management. *Eur. Endocrinol.* **2019**, *15*, 83–90. [CrossRef]
33. Kelly, D.M.; Jones, T.H. Testosterone: A metabolic hormone in health and disease. *J. Endocrinol.* **2013**, *217*, R25–R45. [CrossRef]

34. Clarke, H.; Dhillo, W.S.; Jayasena, C.N. Comprehensive Review on Kisspeptin and Its Role in Reproductive Disorders. *Endocrinol. Metab.* **2015**, *30*, 124–141. [CrossRef]
35. Roseweir, A.K.; Millar, R.P. The role of kisspeptin in the control of gonadotrophin secretion. *Hum. Reprod. Update* **2009**, *15*, 203–212. [CrossRef] [PubMed]
36. Herbst, K.L.; Bhasin, S. Testosterone action on skeletal muscle. *Curr. Opin. Clin. Nutr. Metab. Care* **2004**, *7*, 271–277. [CrossRef] [PubMed]
37. Giagulli, V.A.; Castellana, M.; Murro, I.; Pelusi, C.; Guastamacchia, E.; Triggiani, V.; De Pergola, G. The Role of Diet and Weight Loss in Improving Secondary Hypogonadism in Men with Obesity with or without Type 2 Diabetes Mellitus. *Nutrients* **2019**, *11*, 2975. [CrossRef] [PubMed]
38. Di Renzo, L.; Gualtieri, P.; Romano, L.; Marrone, G.; Noce, A.; Pujia, A.; Perrone, M.A.; Aiello, V.; Colica, C.; De Lorenzo, A. Role of Personalized Nutrition in Chronic-Degenerative Diseases. *Nutrients* **2019**, *11*, 1707. [CrossRef] [PubMed]
39. Denke, M.A. Diet, lifestyle, and nonstatin trials: Review of time to benefit. *Am. J. Cardiol.* **2005**, *96*, 3F–10F. [CrossRef] [PubMed]
40. Abdelhamid, A.S.; Martin, N.; Bridges, C.; Brainard, J.S.; Wang, X.; Brown, T.J.; Hanson, S.; Jimoh, O.F.; Ajabnoor, S.M.; Deane, K.H.; et al. Polyunsaturated fatty acids for the primary and secondary prevention of cardiovascular disease. *Cochrane Database Syst. Rev.* **2018**, *7*, CD012345. [CrossRef] [PubMed]
41. Chowdhury, R.; Warnakula, S.; Kunutsor, S.; Crowe, F.; Ward, H.A.; Johnson, L.; Franco, O.H.; Butterworth, A.S.; Forouhi, N.G.; Thompson, S.G.; et al. Association of dietary, circulating, and supplement fatty acids with coronary risk: A systematic review and meta-analysis. *Ann. Intern. Med.* **2014**, *160*, 398–406. [CrossRef]
42. Rhee, J.J.; Kim, E.; Buring, J.E.; Kurth, T. Fish Consumption, Omega-3 Fatty Acids, and Risk of Cardiovascular Disease. *Am. J. Prev. Med.* **2017**, *52*, 10–19. [CrossRef]
43. Siscovick, D.S.; Barringer, T.A.; Fretts, A.M.; Wu, J.H.; Lichtenstein, A.H.; Costello, R.B.; Kris-Etherton, P.M.; Jacobson, T.A.; Engler, M.B.; Alger, H.M.; et al. Omega-3 Polyunsaturated Fatty Acid (Fish Oil) Supplementation and the Prevention of Clinical Cardiovascular Disease: A Science Advisory From the American Heart Association. *Circulation* **2017**, *135*, e867–e884. [CrossRef]
44. Wilber, C.G.; Levine, V.E. Fat metabolism in Alaskan eskimos. *Exp. Med. Surg.* **1950**, *8*, 422–425.
45. Kagawa, Y.; Nishizawa, M.; Suzuki, M.; Miyatake, T.; Hamamoto, T.; Goto, K.; Motonaga, E.; Izumikawa, H.; Hirata, H.; Ebihara, A. Eicosapolyenoic acids of serum lipids of Japanese islanders with low incidence of cardiovascular diseases. *J. Nutr. Sci. Vitaminol.* **1982**, *28*, 441–453. [CrossRef] [PubMed]
46. Bang, H.O.; Dyerberg, J.; Sinclair, H.M. The composition of the Eskimo food in north western Greenland. *Am. J. Clin. Nutr.* **1980**, *33*, 2657–2661. [CrossRef] [PubMed]
47. Peter, S.; Chopra, S.; Jacob, J.J. A fish a day, keeps the cardiologist away!—A review of the effect of omega-3 fatty acids in the cardiovascular system. *Indian. J. Endocrinol. Metab.* **2013**, *17*, 422–429. [CrossRef] [PubMed]
48. Burr, M.L.; Fehily, A.M.; Gilbert, J.F.; Rogers, S.; Holliday, R.M.; Sweetnam, P.M.; Elwood, P.C.; Deadman, N.M. Effects of changes in fat, fish, and fibre intakes on death and myocardial reinfarction: Diet and reinfarction trial (DART). *Lancet* **1989**, *2*, 757–761. [CrossRef]
49. Kris-Etherton, P.M.; Harris, W.S.; Appel, L.J.; American Heart Association; Nutrition, C. Fish consumption, fish oil, omega-3 fatty acids, and cardiovascular disease. *Circulation* **2002**, *106*, 2747–2757. [CrossRef]
50. Marchioli, R.; Schweiger, C.; Tavazzi, L.; Valagussa, F. Efficacy of n-3 polyunsaturated fatty acids after myocardial infarction: Results of GISSI-Prevenzione trial. Gruppo Italiano per lo Studio della Sopravvivenza nell'Infarto Miocardico. *Lipids* **2001**, *36* (Suppl. 1), S119–S126. [CrossRef]
51. GISSI-Prevenzione Investigators. Dietary supplementation with n-3 polyunsaturated fatty acids and vitamin E after myocardial infarction: Results of the GISSI-Prevenzione trial. Gruppo Italiano per lo Studio della Sopravvivenza nell'Infarto miocardico. *Lancet* **1999**, *354*, 447–455. [CrossRef]
52. Tavazzi, L.; Maggioni, A.P.; Marchioli, R.; Barlera, S.; Franzosi, M.G.; Latini, R.; Lucci, D.; Nicolosi, G.L.; Porcu, M.; Tognoni, G.; et al. Effect of n-3 polyunsaturated fatty acids in patients with chronic heart failure (the GISSI-HF trial): A randomised, double-blind, placebo-controlled trial. *Lancet* **2008**, *372*, 1223–1230. [CrossRef]
53. Ness, A.R.; Hughes, J.; Elwood, P.C.; Whitley, E.; Smith, G.D.; Burr, M.L. The long-term effect of dietary advice in men with coronary disease: Follow-up of the Diet and Reinfarction trial (DART). *Eur. J. Clin. Nutr.* **2002**, *56*, 512–518. [CrossRef]

54. Zehr, K.R.; Walker, M.K. Omega-3 polyunsaturated fatty acids improve endothelial function in humans at risk for atherosclerosis: A review. *Prostaglandins Lipid. Mediat.* **2018**, *134*, 131–140. [CrossRef]
55. Marion-Letellier, R.; Savoye, G.; Ghosh, S. Polyunsaturated fatty acids and inflammation. *IUBMB Life* **2015**, *67*, 659–667. [CrossRef] [PubMed]
56. Chen, H.; Montagnani, M.; Funahashi, T.; Shimomura, I.; Quon, M.J. Adiponectin stimulates production of nitric oxide in vascular endothelial cells. *J. Biol. Chem.* **2003**, *278*, 45021–45026. [CrossRef] [PubMed]
57. Miller, M.; Stone, N.J.; Ballantyne, C.; Bittner, V.; Criqui, M.H.; Ginsberg, H.N.; Goldberg, A.C.; Howard, W.J.; Jacobson, M.S.; Kris-Etherton, P.M.; et al. Triglycerides and cardiovascular disease: A scientific statement from the American Heart Association. *Circulation* **2011**, *123*, 2292–2333. [CrossRef] [PubMed]
58. Harris, W.S.; Bulchandani, D. Why do omega-3 fatty acids lower serum triglycerides? *Curr. Opin. Lipidol.* **2006**, *17*, 387–393. [CrossRef] [PubMed]
59. Bays, H.E.; Ballantyne, C.M.; Kastelein, J.J.; Isaacsohn, J.L.; Braeckman, R.A.; Soni, P.N. Eicosapentaenoic acid ethyl ester (AMR101) therapy in patients with very high triglyceride levels (from the Multi-center, plAcebo-controlled, Randomized, double-blINd, 12-week study with an open-label Extension [MARINE] trial). *Am. J. Cardiol.* **2011**, *108*, 682–690. [CrossRef] [PubMed]
60. Kastelein, J.J.; Maki, K.C.; Susekov, A.; Ezhov, M.; Nordestgaard, B.G.; Machielse, B.N.; Kling, D.; Davidson, M.H. Omega-3 free fatty acids for the treatment of severe hypertriglyceridemia: The EpanoVa fOr Lowering Very high triglyceridEs (EVOLVE) trial. *J. Clin. Lipidol.* **2014**, *8*, 94–106. [CrossRef]
61. Maki, K.C.; Orloff, D.G.; Nicholls, S.J.; Dunbar, R.L.; Roth, E.M.; Curcio, D.; Johnson, J.; Kling, D.; Davidson, M.H. A highly bioavailable omega-3 free fatty acid formulation improves the cardiovascular risk profile in high-risk, statin-treated patients with residual hypertriglyceridemia (the ESPRIT trial). *Clin. Ther.* **2013**, *35*, 1400–1411.e3. [CrossRef]
62. Balakumar, P.; Taneja, G. Fish oil and vascular endothelial protection: Bench to bedside. *Free Radic. Biol. Med.* **2012**, *53*, 271–279. [CrossRef]
63. Wang, T.M.; Chen, C.J.; Lee, T.S.; Chao, H.Y.; Wu, W.H.; Hsieh, S.C.; Sheu, H.H.; Chiang, A.N. Docosahexaenoic acid attenuates VCAM-1 expression and NF-kappaB activation in TNF-alpha-treated human aortic endothelial cells. *J. Nutr. Biochem.* **2011**, *22*, 187–194. [CrossRef]
64. Dyerberg, J.; Bang, H.O.; Stoffersen, E.; Moncada, S.; Vane, J.R. Eicosapentaenoic acid and prevention of thrombosis and atherosclerosis? *Lancet* **1978**, *2*, 117–119. [CrossRef]
65. Haglund, O.; Mehta, J.L.; Saldeen, T. Effects of fish oil on some parameters of fibrinolysis and lipoprotein(a) in healthy subjects. *Am. J. Cardiol.* **1994**, *74*, 189–192. [CrossRef]
66. DeFilippis, A.P.; Rai, S.N.; Cambon, A.; Miles, R.J.; Jaffe, A.S.; Moser, A.B.; Jones, R.O.; Bolli, R.; Schulman, S.P. Fatty acids and TxA(2) generation, in the absence of platelet-COX-1 activity. *Nutr. Metab. Cardiovasc. Dis.* **2014**, *24*, 428–433. [CrossRef]
67. Matsumoto, M.; Sata, M.; Fukuda, D.; Tanaka, K.; Soma, M.; Hirata, Y.; Nagai, R. Orally administered eicosapentaenoic acid reduces and stabilizes atherosclerotic lesions in ApoE-deficient mice. *Atherosclerosis* **2008**, *197*, 524–533. [CrossRef]
68. Calder, P.C. Omega-3 polyunsaturated fatty acids and inflammatory processes: Nutrition or pharmacology? *Br. J. Clin. Pharmacol.* **2013**, *75*, 645–662. [CrossRef]
69. Calder, P.C. n-3 polyunsaturated fatty acids, inflammation, and inflammatory diseases. *Am. J. Clin. Nutr.* **2006**, *83*, 1505S–1519S. [CrossRef]
70. Egert, S.; Stehle, P. Impact of n-3 fatty acids on endothelial function: Results from human interventions studies. *Curr. Opin. Clin. Nutr. Metab. Care* **2011**, *14*, 121–131. [CrossRef] [PubMed]
71. Casanova, M.A.; Medeiros, F.; Trindade, M.; Cohen, C.; Oigman, W.; Neves, M.F. Omega-3 fatty acids supplementation improves endothelial function and arterial stiffness in hypertensive patients with hypertriglyceridemia and high cardiovascular risk. *J. Am. Soc. Hypertens.* **2017**, *11*, 10–19. [CrossRef] [PubMed]
72. Huang, F.; Del-Rio-Navarro, B.E.; Leija-Martinez, J.; Torres-Alcantara, S.; Ruiz-Bedolla, E.; Hernandez-Cadena, L.; Barraza-Villarreal, A.; Romero-Nava, R.; Sanchez-Munoz, F.; Villafana, S.; et al. Effect of omega-3 fatty acids supplementation combined with lifestyle intervention on adipokines and biomarkers of endothelial dysfunction in obese adolescents with hypertriglyceridemia. *J. Nutr. Biochem.* **2019**, *64*, 162–169. [CrossRef]

73. Rantanen, J.M.; Riahi, S.; Johansen, M.B.; Schmidt, E.B.; Christensen, J.H. Effects of Marine n-3 Polyunsaturated Fatty Acids on Heart Rate Variability and Heart Rate in Patients on Chronic Dialysis: A Randomized Controlled Trial. *Nutrients* **2018**, *10*, 1313. [CrossRef]
74. Hidayat, K.; Yang, J.; Zhang, Z.; Chen, G.C.; Qin, L.Q.; Eggersdorfer, M.; Zhang, W. Effect of omega-3 long-chain polyunsaturated fatty acid supplementation on heart rate: A meta-analysis of randomized controlled trials. *Eur. J. Clin. Nutr.* **2018**, *72*, 805–817. [CrossRef]
75. Sagara, M.; Njelekela, M.; Teramoto, T.; Taguchi, T.; Mori, M.; Armitage, L.; Birt, N.; Birt, C.; Yamori, Y. Effects of docosahexaenoic Acid supplementation on blood pressure, heart rate, and serum lipids in Scottish men with hypertension and hypercholesterolemia. *Int. J. Hypertens.* **2011**, *2011*, 809198. [CrossRef] [PubMed]
76. Lee, J.B.; Notay, K.; Klingel, S.L.; Chabowski, A.; Mutch, D.M.; Millar, P.J. Docosahexaenoic acid reduces resting blood pressure but increases muscle sympathetic outflow compared with eicosapentaenoic acid in healthy men and women. *Am. J. Physiol. Heart Circ. Physiol.* **2019**, *316*, H873–H881. [CrossRef] [PubMed]
77. Ghio, S.; Scelsi, L.; Latini, R.; Masson, S.; Eleuteri, E.; Palvarini, M.; Vriz, O.; Pasotti, M.; Gorini, M.; Marchioli, R.; et al. Effects of n-3 polyunsaturated fatty acids and of rosuvastatin on left ventricular function in chronic heart failure: A substudy of GISSI-HF trial. *Eur. J. Heart Fail* **2010**, *12*, 1345–1353. [CrossRef] [PubMed]
78. Heydari, B.; Abdullah, S.; Pottala, J.V.; Shah, R.; Abbasi, S.; Mandry, D.; Francis, S.A.; Lumish, H.; Ghoshhajra, B.B.; Hoffmann, U.; et al. Effect of Omega-3 Acid Ethyl Esters on Left Ventricular Remodeling After Acute Myocardial Infarction: The OMEGA-REMODEL Randomized Clinical Trial. *Circulation* **2016**, *134*, 378–391. [CrossRef]
79. Laake, K.; Seljeflot, I.; Schmidt, E.B.; Myhre, P.; Tveit, A.; Norseth, J.; Arnesen, H.; Solheim, S. Galectin-3, a marker of cardiac remodeling, is inversely related to serum levels of marine omega-3 fatty acids. A cross-sectional study. *JRSM Cardiovasc. Dis.* **2017**, *6*, 2048004017729984. [CrossRef]
80. Takahashi, Y.; Ide, T. Dietary n-3 fatty acids affect mRNA level of brown adipose tissue uncoupling protein 1, and white adipose tissue leptin and glucose transporter 4 in the rat. *Br. J. Nutr.* **2000**, *84*, 175–184. [CrossRef]
81. Clarke, S.D. Polyunsaturated fatty acid regulation of gene transcription: A mechanism to improve energy balance and insulin resistance. *Br. J. Nutr.* **2000**, *83* (Suppl. 1), S59–S66. [CrossRef]
82. Ma, Q. Role of nrf2 in oxidative stress and toxicity. *Annu. Rev. Pharmacol. Toxicol.* **2013**, *53*, 401–426. [CrossRef]
83. Di Nunzio, M.; Valli, V.; Bordoni, A. Pro- and anti-oxidant effects of polyunsaturated fatty acid supplementation in HepG2 cells. *Prostaglandins Leukot. Essent. Fatty Acids* **2011**, *85*, 121–127. [CrossRef]
84. CREA. Linee Guida per Una Sana Alimentazione. Available online: https://www.crea.gov.it/web/alimenti-e-nutrizione/-/linee-guida-per-una-sana-alimentazione-2018 (accessed on 15 June 2020).
85. Summers, L.K.; Fielding, B.A.; Bradshaw, H.A.; Ilic, V.; Beysen, C.; Clark, M.L.; Moore, N.R.; Frayn, K.N. Substituting dietary saturated fat with polyunsaturated fat changes abdominal fat distribution and improves insulin sensitivity. *Diabetologia* **2002**, *45*, 369–377. [CrossRef]
86. Couet, C.; Delarue, J.; Ritz, P.; Antoine, J.M.; Lamisse, F. Effect of dietary fish oil on body fat mass and basal fat oxidation in healthy adults. *Int. J. Obes. Relat. Metab. Disord* **1997**, *21*, 637–643. [CrossRef] [PubMed]
87. Huang, T.; Wang, T.; Heianza, Y.; Zheng, Y.; Sun, D.; Kang, J.H.; Pasquale, L.R.; Rimm, E.B.; Manson, J.E.; Hu, F.B.; et al. Habitual consumption of long-chain n-3 PUFAs and fish attenuates genetically associated long-term weight gain. *Am. J. Clin. Nutr.* **2019**, *109*, 665–673. [CrossRef] [PubMed]
88. Vaughan, L.K.; Wiener, H.W.; Aslibekyan, S.; Allison, D.B.; Havel, P.J.; Stanhope, K.L.; O'Brien, D.M.; Hopkins, S.E.; Lemas, D.J.; Boyer, B.B.; et al. Linkage and association analysis of obesity traits reveals novel loci and interactions with dietary n-3 fatty acids in an Alaska Native (Yup'ik) population. *Metabolism* **2015**, *64*, 689–697. [CrossRef]
89. Coelho, M.; Oliveira, T.; Fernandes, R. Biochemistry of adipose tissue: An endocrine organ. *Arch. Med. Sci.* **2013**, *9*, 191–200. [CrossRef] [PubMed]
90. Zhu, Q.; Glazier, B.J.; Hinkel, B.C.; Cao, J.; Liu, L.; Liang, C.; Shi, H. Neuroendocrine Regulation of Energy Metabolism Involving Different Types of Adipose Tissues. *Int. J. Mol. Sci.* **2019**, *20*, 2707. [CrossRef] [PubMed]
91. Casas-Agustench, P.; Lopez-Uriarte, P.; Bullo, M.; Ros, E.; Gomez-Flores, A.; Salas-Salvado, J. Acute effects of three high-fat meals with different fat saturations on energy expenditure, substrate oxidation and satiety. *Clin. Nutr.* **2009**, *28*, 39–45. [CrossRef]

92. Itoh, M.; Suganami, T.; Hachiya, R.; Ogawa, Y. Adipose tissue remodeling as homeostatic inflammation. *Int. J. Inflam.* **2011**, *2011*, 720926. [CrossRef]
93. Fuentes, E.; Fuentes, F.; Vilahur, G.; Badimon, L.; Palomo, I. Mechanisms of chronic state of inflammation as mediators that link obese adipose tissue and metabolic syndrome. *Mediators. Inflamm.* **2013**, *2013*, 136584. [CrossRef]
94. Lund, A.S.; Hasselbalch, A.L.; Gamborg, M.; Skogstrand, K.; Hougaard, D.M.; Heitmann, B.L.; Kyvik, K.O.; Sorensen, T.I.; Jess, T. N-3 polyunsaturated fatty acids, body fat and inflammation. *Obes. Facts* **2013**, *6*, 369–379. [CrossRef]
95. Howe, P.; Buckley, J. Metabolic health benefits of long-chain omega-3 polyunsaturated fatty acids. *Mil. Med.* **2014**, *179*, 138–143. [CrossRef]
96. Gray, B.; Steyn, F.; Davies, P.S.; Vitetta, L. Omega-3 fatty acids: A review of the effects on adiponectin and leptin and potential implications for obesity management. *Eur. J. Clin. Nutr.* **2013**, *67*, 1234–1242. [CrossRef] [PubMed]
97. Parra, D.; Ramel, A.; Bandarra, N.; Kiely, M.; Martinez, J.A.; Thorsdottir, I. A diet rich in long chain omega-3 fatty acids modulates satiety in overweight and obese volunteers during weight loss. *Appetite* **2008**, *51*, 676–680. [CrossRef] [PubMed]
98. Friedman, J.M.; Halaas, J.L. Leptin and the regulation of body weight in mammals. *Nature* **1998**, *395*, 763–770. [CrossRef]
99. Halaas, J.L.; Gajiwala, K.S.; Maffei, M.; Cohen, S.L.; Chait, B.T.; Rabinowitz, D.; Lallone, R.L.; Burley, S.K.; Friedman, J.M. Weight-reducing effects of the plasma protein encoded by the obese gene. *Science* **1995**, *269*, 543–546. [CrossRef]
100. Auwerx, J.; Staels, B. Leptin. *Lancet* **1998**, *351*, 737–742. [CrossRef]
101. Houseknecht, K.L.; Baile, C.A.; Matteri, R.L.; Spurlock, M.E. The biology of leptin: A review. *J. Anim. Sci.* **1998**, *76*, 1405–1420. [CrossRef]
102. Cameron, A.J.; Welborn, T.A.; Zimmet, P.Z.; Dunstan, D.W.; Owen, N.; Salmon, J.; Dalton, M.; Jolley, D.; Shaw, J.E. Overweight and obesity in Australia: The 1999-2000 Australian Diabetes, Obesity and Lifestyle Study (AusDiab). *Med. J. Aust.* **2003**, *178*, 427–432. [CrossRef]
103. Madsen, E.L.; Bruun, J.M.; Skogstrand, K.; Hougaard, D.M.; Christiansen, T.; Richelsen, B. Long-term weight loss decreases the nontraditional cardiovascular risk factors interleukin-18 and matrix metalloproteinase-9 in obese subjects. *Metabolism* **2009**, *58*, 946–953. [CrossRef]
104. Legradi, G.; Emerson, C.H.; Ahima, R.S.; Flier, J.S.; Lechan, R.M. Leptin prevents fasting-induced suppression of prothyrotropin-releasing hormone messenger ribonucleic acid in neurons of the hypothalamic paraventricular nucleus. *Endocrinology* **1997**, *138*, 2569–2576. [CrossRef]
105. Enriori, P.J.; Sinnayah, P.; Simonds, S.E.; Garcia Rudaz, C.; Cowley, M.A. Leptin action in the dorsomedial hypothalamus increases sympathetic tone to brown adipose tissue in spite of systemic leptin resistance. *J. Neurosci.* **2011**, *31*, 12189–12197. [CrossRef]
106. Finck, B.N.; Johnson, R.W. Tumor necrosis factor-alpha regulates secretion of the adipocyte-derived cytokine, leptin. *Microsc. Res. Tech.* **2000**, *50*, 209–215. [CrossRef]
107. Shamsuzzaman, A.S.; Winnicki, M.; Wolk, R.; Svatikova, A.; Phillips, B.G.; Davison, D.E.; Berger, P.B.; Somers, V.K. Independent association between plasma leptin and C-reactive protein in healthy humans. *Circulation* **2004**, *109*, 2181–2185. [CrossRef] [PubMed]
108. Corsonello, A.; Perticone, F.; Malara, A.; De Domenico, D.; Loddo, S.; Buemi, M.; Ientile, R.; Corica, F. Leptin-dependent platelet aggregation in healthy, overweight and obese subjects. *Int. J. Obes. Relat. Metab. Disord.* **2003**, *27*, 566–573. [CrossRef] [PubMed]
109. Perez-Matute, P.; Perez-Echarri, N.; Martinez, J.A.; Marti, A.; Moreno-Aliaga, M.J. Eicosapentaenoic acid actions on adiposity and insulin resistance in control and high-fat-fed rats: Role of apoptosis, adiponectin and tumour necrosis factor-alpha. *Br. J. Nutr.* **2007**, *97*, 389–398. [CrossRef] [PubMed]
110. Xu, A.; Wang, Y.; Keshaw, H.; Xu, L.Y.; Lam, K.S.; Cooper, G.J. The fat-derived hormone adiponectin alleviates alcoholic and nonalcoholic fatty liver diseases in mice. *J. Clin. Invest.* **2003**, *112*, 91–100. [CrossRef]
111. Berg, A.H.; Combs, T.P.; Du, X.; Brownlee, M.; Scherer, P.E. The adipocyte-secreted protein Acrp30 enhances hepatic insulin action. *Nat. Med.* **2001**, *7*, 947–953. [CrossRef]
112. Dadson, K.; Liu, Y.; Sweeney, G. Adiponectin action: A combination of endocrine and autocrine/paracrine effects. *Front. Endocrinol.* **2011**, *2*, 62. [CrossRef]

113. Dimiter, D. Effect of omega-3 fatty acids on plasma adiponectin levels in Metabolic syndrome subjects. *Inter. J. Obes.* **2007**, *31*.
114. Whitehouse, A.S.; Tisdale, M.J. Downregulation of ubiquitin-dependent proteolysis by eicosapentaenoic acid in acute starvation. *Biochem. Biophys. Res. Commun.* **2001**, *285*, 598–602. [CrossRef]
115. Noce, A.; Marrone, G.; Rovella, V.; Cusijmano, A.; Di Daniele, N. Beneficial effects of physical activity on uremic sarcopenia. *Med. Dello Sport* **2018**. [CrossRef]
116. Andrade-Vieira, R.; Han, J.H.; Marignani, P.A. Omega-3 polyunsaturated fatty acid promotes the inhibition of glycolytic enzymes and mTOR signaling by regulating the tumor suppressor LKB1. *Cancer Biol. Ther.* **2013**, *14*, 1050–1058. [CrossRef] [PubMed]
117. Howe, P.; Coates, A.; Murphy, K.; Pettman, T.; Milte, C.; Buckley, J. Higher erythrocyte LCn-3 PUFA content is associated with a healthier body composition. *Australas. Med. J.* **2014**, *1*, 60.
118. Wing, S.S.; Goldberg, A.L. Glucocorticoids activate the ATP-ubiquitin-dependent proteolytic system in skeletal muscle during fasting. *Am. J. Physiol.* **1993**, *264*, E668–E676. [CrossRef] [PubMed]

nutrients

MDPI

Article

Do Diet and Lifestyles Play a Role in the Pathogenesis of NMSCs?

Nevena Skroza [1], Ilaria Proietti [1,*], Anna Marchesiello [1], Salvatore Volpe [1], Veronica Balduzzi [1], Nicoletta Bernardini [1], Patrizia Maddalena [1], Alessandra Mambrin [1], Simone Michelini [1], Ersilia Tolino [1], Giuseppe La Torre [2] and Concetta Potenza [1]

1 Dermatology Unit "Daniele Innocenzi", Department of Medical-Surgical Sciences and Bio-Technologies, Sapienza University of Rome, Fiorini Hospital, Polo Pontino, 04019 Terracina, Italy; nevena.skroza@uniroma1.it (N.S.); anna.marchesiello90@gmail.com (A.M.); sal.volpe@hotmail.it (S.V.); veronica.balduzzi@gmail.com (V.B.); nicoletta.bernardini@libero.it (N.B.); patrizia.maddalena@gmail.com (P.M.); mambrinalessandra@gmail.com (A.M.); simo.mik@hotmail.it (S.M.); ersiliatolino@gmail.com (E.T.); concetta.potenza@uniroma1.it (C.P.)

2 Department of Public Health and Infectious Diseases, Sapienza University of Rome, 00161 Rome, Italy; giuseppe.latorre@uniroma1.it

* Correspondence: proiettilaria@gmail.com; Tel.: +39-333-468-4342

Received: 30 September 2020; Accepted: 10 November 2020; Published: 11 November 2020

Abstract: Background and Aims: Literature highlights the role of risk factors like age, body mass index (BMI), tobacco smoking, alcohol intake and diet in the pathogenesis of several cancer types but little is known for non-melanoma skin cancers (NMSC). The aim of this epidemiological study was to evaluate the correlation between modifiable risk factors (BMI, metabolic panel, diet, lifestyle, medical history) and not modifiable risk factors (gender, age) and NMSC development. Methods: From February 2018 to September 2019, 162 patients affected by NMSC were compared to a group of 167 controls. A univariate and multivariate analysis was conducted to elaborate the data collected through face-to-face interviews. Results: While our evidence did not always reach statistical significance, NMSC study group patients exhibited high rates of analyzed risk factors (male gender aging over 55 years, high BMI, reduced physical activity) compared to the control group. Conclusions: Our study indicates that practicing more than 30 min of physical activity daily could be a protective factor against the NMSC onset. Other risk factors were not correlated with NMSC, but more evidence is needed to establish a possible link.

Keywords: NMSC; lifestyle; smoking; alcohol; diet

1. Introduction

The incidence of non-melanoma skin cancer (NMSC), including predominantly basal cell carcinoma (BCC) and squamous cell carcinoma (SCC), has risen significantly among white populations, sparking an increasing scientific interest in these types of tumors [1,2].

Genetic factors and behavioral changes are known to promote skin cancer.

However, high frequency of UV radiation (UVR) exposure remains the predominant causative agent, altering cell genetic and immunological profile, leading to DNA damage, oxidative stress, free radical production and photo aging [3].

An additional well-established risk factor for NMSC includes Fitzpatrick I-II skin type.

Modifiable lifestyle risk factors, such as diet, exposure to exogenous hormones, alcohol intake, and tobacco smoking, generally noted to be linked to carcinogenesis, have not definitively been confirmed as risk factors for NMSC.

Recently, extensive research has highlighted the role of correct eating habits in controlling skin carcinogenesis and tumor growth in other models; nevertheless, a comprehensive immunological evaluation implicated in this relationship is still pending [4].

Moreover, obesity has increased alarmingly in the Western world during recent decades, becoming a risk factor for several cancers [5].

However, evidence for an association between obesity and malignant melanoma (MM) and non-melanoma skin cancer (NMSC) has not been fully established yet.

Currently, results from a large meta-analysis showed that males with high body mass index (BMI) had a high risk for MM, whereas this association was not found among females [6].

Fewer studies have addressed the associations between BMI and risk for NMSC but the results are conflicting [7].

In order to investigate how much lifestyle affects the risk of developing NMSC, we conducted an epidemiological study evaluating the metabolic syndrome indices, eating behavior, smoking habits, alcohol consumption and physical activity in NMSC patients in correlation with a control study group.

2. Materials & Methods

From February 2018 to September 2019, a community-based case-control was carried out enrolling 162 cases of NMSC outpatients of a dermatological ambulatory service and 167 clinically healthy NMSC-free subjects, as controls.

Data collection was carried out through structured face-to-face interviews.

Primarily, anamnestic data were collected regarding age, gender, work activity, health status and medications.

The second part of the interview concerned the dietary habits.

Normal eating habits of participants were assessed using the validated food frequency questionnaire (FFQ), which included 12 items corresponding to the 12 elemental characteristics of the Mediterranean diet: carbohydrates, vegetables, fruit, milk, extra virgin olive oil, white meat, red meat, sausages, fish, eggs, legumes and sweets [8].

For each participant, a score was constructed by adding the scores obtained for the 12 groups of foods. Adherence to the Mediterranean diet was assessed by a score created by Trichopoulou et al. [9].

Finally, participants were asked about their smoking habits and daily physical activity (for more or less than 30 min).

Furthermore, BMI was calculated for each patient complete with a metabolic panel (glycemia, cholesterolemia and triglyceridemia).

3. Statistical Analysis

The statistical analysis was carried out using SPSS (IBM Corp. Released 2017. IBM SPSS Statistics for Windows, Version 25.0. Armonk, NY, USA: IBM Corp), release 25.0.

The univariate analysis was conducted using the chi-square test in order to compare the cases and controls for different dichotomous variables.

The multivariate analysis was performed using a logistic model. The results are presented as odd ratio (OR) (95% CI). Two types of model were created, the first model including all the variables (full model), and the second one using a stepwise approach (backward elimination).

The significance level was set at p value < 0.05.

Since the variables used for the multivariate analysis were not continuous, no multicollinearity was present.

No missing data were present.

Composite scores based on the items were constructed.

4. Results

The survey included 326 patients: 162 (96 males and 66 females) affected by NMSC (BCC, actinic keratosis (AK), SCC), with a mean age of 68 years (range 36–95); the control group consisted of 164 subjects not suffering from skin cancers (82 males and 82 females), with a mean age of 57 years (range 32–88) (Tables 1 and 2). Male gender and age < 55 years were found to be associated with a higher risk of developing NMSC (p value > 0.05). BMI was used to categorize each person as underweight, normal weight, overweight or obese: the NMSC group was 1% underweight, 31% normal weight, 44% overweight, 20% first grade obesity, 4% second grade obesity and 0% third grade obesity; the control group was 1% underweight, 52% normal weight, 29% overweight, 15% first grade obesity, 2% second grade obesity and 1% third grade obesity. The condition of obesity or overweight was significantly higher in the NMSC group (p value < 0.001). There were no statistically significant differences between the two groups in the frequency and the amount of consumption of the aliments examined. Nevertheless, cold meat (p value = 0.055) and sugary drinks (p value = 0.076) were found to be widely consumed in the NMSC group. As far as concerns the Mediterranean diet score, levels of the score equal to or over 10 were associated with a lower risk of NMSC (p = 0.056).

Concerning the metabolic panel assessment (glycemia, cholesterolemia and triglyceridemia), blood tests were over the limits in 76 NMSC patients (46.9%) and in 74 (44.3%) control study group subjects. This difference was not statistically significant (p value = 0.636), although considerably high levels of glycemia were observed in the NMSC group. In the NMSC group 83 patients (51.2%) declared themselves to have never smoked and 79 patients (48.8%) declared themselves to be smokers or former smokers. In the control group, on the other hand, 96 patients (58.5%) reported never having smoked and 68 patients (41.5%) claimed to be smokers or former smokers. These data were not statistically significant (p value = 0.185). Differences in the frequency of alcohol consumption were minimal and not statistically significant. Among NMSC patients, only 33 (20.4%) claimed to practice physical activity for more than 30 min daily, as opposed to 60 patients (36.6%) from the control group.

Table 1. Univariate analysis table.

	Cases	Controls	p Value
Age			
<55 years	16 (9.9)	0 (0)	<0.001
≥55 years	146 (90.1)	167 (100)	
Sex			
Females	65 (40.1)	94 (56.3)	0.03
Males	97 (59.9)	73 (43.7)	
Tobacco smoking			
Non-smokers	83 (51.2)	99 (59.3)	0.142
Smokers/ex-smokers	79 (48.8)	68 (40.7)	
Metabolic panel			
Normal	86 (53.1)	93 (55.7)	0.636
Not normal	76 (46.9)	74 (44.3)	
Body mass index			
Normal	52 (32.1)	89 (53.3)	<0.001
Overweight/obesity	110 (67.9)	78 (46.7)	
Physical activity			
<30 min daily	129 (79.6)	107 (64.1)	0.002
≥ 30min daily	33 (20.4)	60 (35.9)	
Mediterranean diet score ≥ 10			
<10 (reference)	157 (96.9)	151 (92.1)	0.056
≥10	5 (3.1)	13 (7.9)	

Table 2. Multivariate analysis table.

	Full Model		Backward Elimination Model	
	OR	**CI 95%**	**OR**	**CI 95%**
Sex				
Females	0.59	0.36–0.96	0.59	0.37–0.94
Males (reference)	1		1	
Tobacco smoking				
Non-smokers (reference)	1			
Smokers/ex-smokers	1.01	0.62–1.63		
Metabolic panel				
Normal (reference)	1			
Not normal	0.80	0.50–1.28		
Body mass index				
Normal (reference)	1		1	
Overweight/obesity	1.94	1.20–3.15	1.90	1.18–3.05
Physical activity				
<30 min daily (reference)	1		1	
≥30 min daily	0.50	0.29–0.75	0.52	0.31–0.86
Mediterranean diet score > 10				
<10 (reference)	1	0.13–1.14	1	0.12–1.05
≥10	0.38		0.36	

5. Discussion

Skin cancer represents one of the most common worldwide malignancies and its incidence exhibits no signs of plateauing [10].

Whereas different types of cancers are causally related to a variety of endogenous and exogenous modifiable risk factors, the concept that lifestyle behavior may play a role in modulating cancer incidence has become largely investigated, recently, upon indirect epidemiologic evidences.

However, few studies have found correlations of dietary fat intake with NMSC incidence [11].

Commonly, dietary history and lifestyle habits questionnaires and surveys are widely used procedures for collecting epidemiological data about this relation. Nevertheless, with regard to NMSC, only a few epidemiological investigations found no association between its incidence and dietary intake [12,13].

Through the interviews performed, we observed a suggestive pattern of elevated risk of NMSC in male subjects, aged over 55 years. This trend reflects the recent incidence data [14].

No statistically significant correlation has been observed between the foods investigated, neither those typical of the Western diet nor those characteristic of the Mediterranean diet.

Despite the Mediterranean diet showing a favorable impact on some chronic inflammatory skin conditions such as acne, rosacea, hidradenitis suppurativa, psoriasis as well as certain malignancies, presumably due to its anti-inflammatory and antioxidant effects, more high-quality research is needed to verify a presumable correlation with NMSC [15–17].

Physical activity influences the risk of several types of malignancies. There is much evidence sustaining the role of regular physical activity in reducing risk of colorectal cancers, breast cancers, endometrial cancers, testicular cancers and minorly, lung and pancreas tumors [18–20].

Suggested biological pathways by which physical activity may influence cancer risk include changes in hormones, growth factors, inflammatory cytokines, and immune function.

Regarding skin cancer, physical activity explicates its beneficial effects by improving immune function, increasing detoxification and DNA repair, thus reducing sun exposure-induced DNA damage [21].

Nevertheless, some authors indicate a potential positive association between outdoor physical activity and keratinocyte cancer, due to high UV radiation exposure levels while performing outdoor sports [21].

Other studies also highlight existing sex differences in the association between recreational physical activity and the development of SCC due to more outdoor exercise, less clothing, and less use of sunscreens in men than women [22].

Our data suggest a possible protective effect of physical activity, for at least 30 min daily, on NMSC onset. In fact, among 162 NMSC patients, only 33 (20.4%) claimed to practice physical activity for more than 30 min daily, as opposed to 60 subjects (35.9%) of the control group.

The limit of this observation is attributable to the fact that these data were not adjusted for outdoor sun exposure or any other risk factor pertaining to sun exposure.

Although not statistically significant, higher BMI was observed in NMSC patients than in the control group. In addition, the obesity condition was significantly increased in the NMSC study population (p value <0.0001).

The BMI of adults and the global obesity prevalence is trending higher. Numerous epidemiologic studies reported, constantly, that severe and morbid obesity are associated with elevated risks of adverse health outcomes and highlighted a high BMI as a potential risk factor for many types of malignancies [23].

Recent evidence suggests that adipocytes could play an important role in the proliferation of cancers, although the mechanism underlying promotion of carcinogenesis is not fully established [24].

Specific to NMSC, the association with high BMI is controversial, since various studies reported an inverse relationship, while others have not found significant results [25].

Regarding skin cancer, the conflicting association with being overweight or obese could have both behavioral and biological explanations: on one side, presumably, overweight subjects are less likely to be exposed to UV light in public settings, the primary risk factor for skin cancer [26,27].

On the other side, a potential mechanism is due to the alterations in hormones and growth factors induced by high caloric intake, that lead either to height or to an increased number of cells that could potentially have an abnormal proliferation.

In support of this evidence, studies conducted in obese leptin-deficient mice skin tissue revealed stronger inflammatory response to UV radiation and greater oxidative stress resulting in altered cellular signaling [28].

Therefore, our observation of a suggestive positive association between high BMI and NMSC development could sustain this position.

6. Limitations of the Study

Regarding the role of physical activity as a presumable protective factor against NMSC development, the limit of our observation is attributable to the fact that our data were not adjusted for outdoor sun exposure or any other risk factor pertaining to sun exposure.

7. Conclusions

In these past years, NMSCs have acquired a clinical relevance that parallels their steadily increasing incidence.

Moreover, many patients often exhibit two or more types of NMSC simultaneously.

In fact, the main limitation of our study is the lack of stratification regarding the three main epithelial tumors (AK, BCC, SCC), as they often coexist in the same subject.

As in all malignancies, for NMSC it is advisable to investigate possible modifiable risk factors in order to act promptly on their prevention.

Our study has revealed the existence of some risk factors such as high BMI, male gender aging over 55 years and lack of physical activity, as potentially related to a high risk of developing NMSC.

Although the correlation of these factors and the development of NMSC is still controversial, additional studies are necessary to confirm or disprove our hypotheses.

Author Contributions: Conceptualization, C.P., N.S. and I.P.; methodology, C.P. and N.S.; software, G.L.T.; validation, N.S. and I.P.; data curation, N.B. and A.M. (Alessandra Mambrin); writing—original draft preparation, A.M. (Anna Marchesiello) and S.V.; writing—review and editing, V.B. and S.M.; visualization, E.T. and P.M.; supervision, N.S. and I.P. All authors have read and agreed to the published version of the manuscript.

Funding: This research received no external funding.

Conflicts of Interest: The authors declare no conflict of interest.

References

1. Han, S.H.; Kim, S.H.; Kim, C.K.; Jo, D.I. Multiple nonmelanocytic skin cancers in multiple regions. *Arch. Craniofac. Surg.* **2020**, *21*, 188–192. [CrossRef] [PubMed]
2. Kune, G.A.; Bannerman, S.; Field, B.; Watson, L.F.; Cleland, H.; Merenstein, D.; Vitetta, L. Diet, alcohol, smoking, serum beta-carotene, and vitamin A in male nonmelanocytic skin cancer patients and controls. *Nutr. Cancer* **1992**, *18*, 237–244. [CrossRef] [PubMed]
3. Corona, R.; Dogliotti, E.; D'Errico, M.; Sera, F.; Iavarone, I.; Baliva, G.; Chinni, L.M.; Gobello, T.; Mazzanti, C.; Puddu, P.; et al. Risk factors for basal cell carcinoma in a Mediterranean population: Role of recreational sun exposure early in life. *Arch. Dermatol.* **2001**, *137*, 1162–1166. [CrossRef] [PubMed]
4. Zheng, Y.Y.; Viswanathan, B.; Kesarwani, P.; Mehrotra, S. Dietary agents in cancer prevention: An immunological perspective. *Photochem. Photobiol.* **2012**, *88*, 1083–1098. [CrossRef]
5. Nathan, A. Berger Young Adult Cancer: Influence of the Obesity Pandemic Obesity (Silver Spring) Author manuscript; available in PMC 1 April 2019. *Obesity (Silver Spring)* **2018**, *26*, 641–650. [CrossRef]
6. Sergentanis, T.N.; Antoniadis, A.G.; Gogas, H.J.; Antonopoulos, C.N.; Adami, H.O.; Ekbom, A.; Petridou, E.T. Obesity and risk of malignant melanoma: A meta-analysis of cohort and case-control studies. *Eur. J. Cancer* **2013**, *49*, 642–657. [CrossRef]
7. Præstegaard, C.; Kjær, S.K.; Christensen, J.; Tjønneland, A.; Halkjær, J. Obesity and Risks for Malignant Melanoma and Non-Melanoma Skin Cancer: Results from a Large Danish Prospective Cohort Study. *J. Investig. Dermatol.* **2015**, *135*, 901–904. [CrossRef]
8. Martin-Moreno, J.M.; Boyle, P.; Gorgojo, L.; Maisonneuve, P.; Fernandez-Rodriguez, J.C.; Salvini, S.; Willett, W.C. Development and validation of a food frequency questionnaire in Spain. *Int. J. Epidemiol.* **1993**, *22*, 512–519. [CrossRef]
9. Trichopoulou, A.; Kouris-Blazos, A.; Wahlqvist, M.L.; Gnardellis, C.; Lagiou, P.; Polychronopoulos, E.; Vassilakou, T.; Lipworth, L.; Trichopoulos, D. Diet and overall survival in elderly people. *BMJ* **1995**, *311*, 1457–1460. [CrossRef]
10. Gordon, R. Skin cancer: An overview of epidemiology and risk factors. *Semin. Oncol. Nurs.* **2013**, *29*, 160–169. [CrossRef]
11. Black, H.S. Can diet prevent nonmelanoma skin cancer progression? *Expert. Rev. Anticancer Ther.* **2005**, *5*, 801–808. [CrossRef] [PubMed]
12. Graham, S. Results of case-controlled studies of diet and cancer in Buffalo. *N. Y. Cancer Res.* **1983**, *43*, 2409–2413.
13. Hunter, D.J.; Colditz, G.A.; Stampfer, M.J.; Rosner, B.; Willett, W.C.; SpeizerDiet, F.E. Risk of basal cell carcinoma of the skin in a prospective cohort of women. *Ann. Epidemiol.* **1992**, *2*, 231–239. [CrossRef]
14. Leiter, U.; Keim, U.; Eigentler, T.; Katalinic, A.; Holleczek, B.; Martus, P.; Garbe, C. Incidence, Mortality, and Trends of Nonmelanoma Skin Cancer in Germany. *J. Investig. Dermatol.* **2017**, *137*, 1860–1867. [CrossRef] [PubMed]
15. Barrea, L.; Fabbrocini, G.; Annunziata, G.; Muscogiuri, G.; Donnarumma, M.; Marasca, C.; Colao, A.; Savastano, S. Role of Nutrition and Adherence to the Mediterranean Diet in the Multidisciplinary Approach of Hidradenitis Suppurativa: Evaluation of Nutritional Status and Its Association with Severity of Disease. *Nutrients* **2018**, *11*, 57. [CrossRef] [PubMed]

16. Skroza, N.; Tolino, E.; Semyonov, L.; Proietti, I.; Bernardini, N.; Nicolucci, F.; La Torre, G. Mediterranean diet and familial dysmetabolism as factors influencing the development of acne. *Scand. J. Public Health* **2012**, *40*, 466–474. [CrossRef] [PubMed]
17. Sánchez-Sánchez, M.L.; García-Vigara, A.; Hidalgo-Mora, J.J.; García-Pérez, M.Á.; Tarín, J.; Cano, A. Mediterranean diet and health: A systematic review of epidemiological studies and intervention trials. *Maturitas* **2020**, *136*, 25–37. [CrossRef]
18. Oruç, Z.; Kaplan, M.A. Effect of exercise on colorectal cancer prevention and treatment. *World J. Gastrointest. Oncol.* **2019**, *11*, 348. [CrossRef]
19. Lahart, I.M.; Metsios, G.S.; Nevill, A.M.; Carmichael, A.R. Physical activity for women with breast cancer after adjuvant therapy. *Cochrane Database Syst. Rev.* **2018**, *1*, CD011292. [CrossRef]
20. Huang, S.; Signal, V.; Sarfati, D.; Shaw, C.; Stanley, J.; McGlynn, K.; Gurney, J. Physical activity and risk of testicular cancer: A systematic review. *BMC Cancer* **2018**, *18*, 189. [CrossRef]
21. Lahmann, P.H.; Russell, A.; Adèle, C. Green Prospective study of physical activity and risk of squamous cell carcinoma of the skin. *BMC Cancer* **2011**, *11*, 516. [CrossRef] [PubMed]
22. Schnohr, P.; Grønbaek, M.; Petersen, L.; Hein, H.O.; Sørensen, T.I. Physical activity in leisure-time and risk of cancer: 14-year follow-up of 28,000 Danish men and women. *Scand. J. Public Health* **2005**, *33*, 244–249. [CrossRef] [PubMed]
23. Bosy-Westphal, A.; Reinecke, U.; Schlörke, T.; Illner, K.; Kutzner, D.; Heller, M.; Müller, M.J. Effect of organ and tissue masses on resting energy expenditure in underweight, normal weight and obese adults. *Int. J. Obes. Relat. Metab. Disord.* **2004**, *28*, 72–79. [CrossRef] [PubMed]
24. Dirat, B.; Bochet, L.; Dabek, M.; Daviaud, D.; Dauvillier, S.; Majed, B.; Wang, Y.Y.; Meulle, A.; Salles, B.; Le Gonidec, S.; et al. Cancer-associated adipocytes exhibit an activated phenotype and contribute to breast cancer invasion. *Cancer Res.* **2011**, *71*, 2455–2465. [CrossRef] [PubMed]
25. Zhang, Y.; Cartmel, B.; Choy, C.C.; Molinaro, A.M.; Leffell, D.J.; Bale, A.E.; Mayne, S.T.; Ferrucci, L.M. Body mass index, height and early-onset basal cell carcinoma in a case-control study. *Cancer Epidemiol.* **2017**, *46*, 66–72. [CrossRef]
26. Pothiawala, S.; Qureshi, A.A.; Li, Y.; Han, J. Obesity and the incidence of skin cancer in US Caucasians. *Cancer Causes Control.* **2012**, *23*, 717–726. [CrossRef]
27. Gerstenblith, M.R.; Rajaraman, P.; Khaykin, E.; Doody, M.M.; Alexander, B.H.; Linet, M.S.; Freedman, D.M. Basal cell carcinoma and anthropometric factors in the U.S. radiologic technologists cohort study. *Int. J. Cancer* **2012**, *131*, E149-55. [CrossRef]
28. Sharma, S.D.; Katiyar, S.K. Leptin deficiency-induced obesity exacerbates ultraviolet B radiation-induced cyclooxygenase-2 expression and cell survival signals in ultraviolet B-irradiated mouse skin. *Toxicol. Appl. Pharmacol.* **2010**, *244*, 328–335. [CrossRef]

nutrients

MDPI

Article

Association between Nutrient-Based Dietary Patterns and Bladder Cancer in Italy

Valeria Edefonti [1], Carlo La Vecchia [1], Matteo Di Maso [1], Anna Crispo [2], Jerry Polesel [3], Massimo Libra [4], Maria Parpinel [5], Diego Serraino [3], Monica Ferraroni [1,†] and Francesca Bravi [1,*,†]

1 Department of Clinical Sciences and Community Health, Università degli Studi di Milano, via A. Vanzetti, 5, 20133 Milan, Italy; valeria.edefonti@unimi.it (V.E.); carlo.lavecchia@unimi.it (C.L.V.); matteo.dimaso@unimi.it (M.D.M.); monica.ferraroni@unimi.it (M.F.)

2 Epidemiology and Biostatistics Unit, Istituto Nazionale Tumori—IRCCS "Fondazione G. Pascale", via M. Semmola, 80131 Naples, Italy; a.crispo@istitutotumori.na.it

3 Cancer Epidemiology Unit, Centro di Riferimento Oncologico di Aviano (CRO) IRCCS, via F. Gallini, 2, 33080 Aviano, Italy; polesel@cro.it (J.P.); serrainod@cro.it (D.S.)

4 Laboratory of Translational Oncology & Functional Genomics, Department of Biomedical and Biotechnological Sciences, Università di Catania, Via Santa Sofia, 97, 95123 Catania, Italy; m.libra@unict.it

5 Department of Medicine, University of Udine, via Colugna, 50, 33100 Udine, Italy; maria.parpinel@uniud.it

* Correspondence: francesca.bravi@unimi.it; Tel.: +39-02-5032074; Fax: +39-02-50320866

† M.F. and F.B. equally contributed to the study.

Received: 14 April 2020; Accepted: 25 May 2020; Published: 28 May 2020

Abstract: Limited knowledge is available on dietary patterns and bladder cancer risk. We analyzed data from an Italian case-control study carried out between 2003 and 2014, including 690 incident bladder cancer cases and 665 hospital-controls. We derived nutrient-based dietary patterns applying principal component factor analysis on 28 selected nutrients. We categorized factor scores according to quartiles, and estimated the odds ratios (ORs) and the corresponding 95% confidence intervals (CIs) through logistic regression models, adjusted for major confounding factors. We identified four dietary patterns named "Animal products", "Vitamins and fiber", "Starch-rich", and "Animal unsaturated fatty acids". We found an inverse association between the "Vitamins and fiber" pattern and bladder cancer (OR = 0.70, 95% CI: 0.48–0.99, IV versus I quartile category). Inverse relationships of borderline significance were also found for the "Animal products" and the "Animal unsaturated fatty acids" dietary patterns. No significant association was evident for the "Starch-rich" pattern. The current study allowed us to identify major dietary patterns in this Italian population. Our study confirms available evidence and shows that scoring high on a fruit-and-vegetables pattern provides beneficial effects on bladder cancer risk.

Keywords: bladder cancer; case-control study; dietary patterns; diet; factor analysis

---

1. Introduction

Bladder cancer accounted for about 550,000 new cases worldwide in 2018 and ranked 15th among the causes of cancer mortality [1]. Tobacco smoking is the major recognized risk factor for this neoplasm, accounting for 28% of male cases and 18% of female cases in Europe, and 21% and 16%, respectively, in the USA [2]. Other well-known risk factors are past occupational exposures to aromatic amines, high level of arsenic in drinking water, and *Schistosoma haematobium* and other urinary tract infections, whereas the role of other lifestyle factors needs to be clarified [3–5]. Diet has been involved in the development of bladder cancer because the metabolites of the ingested foods have direct contact with bladder mucosa [6]. However, the role of specific foods and nutrients is still unclear, and the World

Cancer Research Fund updated report indicates that only limited/suggestive evidence has been reached for a favorable role of vegetables and fruit consumption and tea drinking, while limited/non-conclusive evidence was available for the other dietary items [7].

Most studies have investigated single foods or nutrients, while only a few studies have used *a priori* [8–11] or *a posteriori* dietary patterns [12,13] to describe dietary habits as a complex behavior and to assess their relationship with bladder cancer risk. Previous analyses of selected aspects of diet in the present Italian population showed an unfavorable role of meat consumption (particularly stewed and roasted meat) [14], carbohydrates and glycaemic load [15], and an inverse relationship with vegetables, milk/yogurt [14], and flavones and isoflavones [16]. Considering dietary habits as a whole in the same population, bladder cancer was positively associated with a pro-inflammatory diet [17] and inversely related to a Mediterranean diet [18].

A national food consumption survey conducted in Italy in early 2000 showed that cereals represented the primary source of energy (providing ~38% of energy), followed by oils and fats (~17%, mainly from oils), and milk products (~13%); meats and fish provided a limited contribution (~9% and 2% of energy, respectively) [19]. Cereals were the primary source of fiber, providing a fiber intake comparable to the one given by the sum of fruit and vegetables [19]. However, the mean daily intake of vegetables, legumes, and fruit was below the recommendations provided by the World Cancer Research Fund [20].

To further investigate the role of overall dietary habits on bladder cancer risk, we derived *a posteriori* dietary patterns—using an exploratory principal component factor analysis—in an Italian case-control study.

2. Materials and Methods

2.1. Design and Participants

A hospital-based case-control study was carried out between 2003 and 2014 in four Italian areas —i.e., Milan, Pordenone, Naples, and Catania. The study design and inclusion criteria have been described in detail elsewhere [21]. Briefly, cases were 690 patients (including 595 men and 95 women) younger than 85 (age range: 25–84, median: 67) years with incident urothelial carcinoma of the bladder, recruited in major teaching and general hospitals of the study areas, and with no previous history of other neoplasms. Almost all cancer cases (n = 645, 93%) were confirmed by histology or cytology. Cases were classified on the basis of the 2016 World Health Organization (WHO) grading system [22]: 268 cases (38.8%) were non-muscle invasive (i.e., TNM pTis/pTa), 192 (27.8%) were pT1, 159 (23.0%) were muscle-invasive (i.e., pT2–pT4); as to the grading, 307 cases (44.5%) were moderately or well-differentiated (i.e., G1–G2), and 312 (45.2%) were undifferentiated or poorly differentiated (i.e., G3–G4). Controls were 665 patients (including 561 men and 104 women), selected among subjects admitted to the same hospital networks of cases for acute non-neoplastic diseases unrelated to alcohol drinking or tobacco smoking, and not associated with long-term dietary modifications. The median age of controls was 66, range 27–84 years. Among controls, 39.9% were admitted to hospital for acute surgery, 28.9% for traumas, 22.1% for nontraumatic orthopedic conditions, and 9.8% for other miscellaneous diseases. Less than 5% of cases and controls approached did not participate. The study protocol was submitted to the Board of Ethics of the participating hospitals and received the approval required at the time of data collection. The Ethics Committee of the National Cancer Institute "Centro di Riferimento Oncologico, IRCCS", Aviano, updated the study approval on 14 December 2012 with the protocol number IRB-15-2012. The Ethics Committee of the Hospital "Niguarda Ca' Granda", Milan, provided the study approval on 23 March 2012, with register number 99_03/2012.

Details on selected characteristics of the study participants are provided in Table S1.

2.2. Data Collection

Participants' data were collected during their hospital stay by centrally trained interviewers through a structured questionnaire assessing information on sociodemographic characteristics, anthropometric factors, tobacco smoking, a personal medical history of selected diseases, and occupational exposures. Dietary habits in the two years before cancer diagnosis (or before interview for controls) were also assessed with a food frequency questionnaire (FFQ), including 80 foods and recipes, and a list of different beverages. The FFQ included the following sections: (I) milk and sweeteners, (II) bread, cereals, and first courses, (III) second courses, (IV) side dishes (i.e., raw and cooked vegetables), (V) fruit, and (VI) sweets and desserts. Additional sections concerned lifetime consumption of beverages, including (a) alcoholic beverages, (b) hot beverages, (c) soft drinks, and (d) tap and bottled water [14,23]. Participants were asked to recall their usual frequency of consumption per week for each dietary item. Occasional consumption (defined as frequency >1 per month and <1 per week) was coded as 0.5 per week. For each participant, we estimated total energy, selected nutrient intakes, and the amount of condiments (e.g., butter and different types of oil) used in the recipes, through an Italian food composition database [24,25]. The validity and reproducibility of the FFQ were tested using a 7-day dietary record repeated twice [26,27].

2.3. Statistical Analysis

2.3.1. Factorability of the Original Matrix

The analyses were carried out on a comprehensive list of 28 macronutrients, micronutrients, and minerals. We examined the potential relationships among nutrients to avoid over-representing any single nutrient or specific profiles of consumption, thus resulting in artificially higher correlation coefficients. We evaluated the factorability of the correlation matrix of the original nutrients (including both cases and controls) by visual examination of the matrix and through statistical procedures, namely Bartlett's test of sphericity, overall (Kaiser–Meyer–Olkin) and individual measures of sampling adequacy [28]. Since we obtained satisfactory results (see Table 1), we used an exploratory principal component factor analysis to derive the *a posteriori* dietary patterns.

Table 1. Factorability of the correlation matrix of the original nutrients: Bartlett's test of sphericity and measures of sampling adequacy.

Bartlett's Test of Sphericity: p-Value < 0.001	
Overall Measure of Sampling Adequacy (Kaiser–Meyer–Olkin statistic) [1]: 0.86	
Individual Measures of Sampling Adequacy:	
0.60–0.69	retinol, linoleic acid
0.70–0.79	total fiber, starch, vitamin E, monounsaturated fatty acids, vitamin D
0.80–0.89	lycopene, vegetable protein, other polyunsaturated fatty acids, riboflavin, animal protein, saturated fatty acids, sodium, calcium, iron, vitamin C, potassium, folate
≥0.90	phosphorus, niacin, zinc, thiamin, cholesterol, soluble carbohydrates, linolenic acid, vitamin B6, beta-carotene equivalents

[1] Overall and individual measures of sampling adequacy range between 0 and 1, with values > 0.60 indicating a satisfactory size.

2.3.2. Dietary Pattern Identification

We carried out a principal component factor analysis [29] on the correlation matrix of the 28 selected nutrients (including cases and controls together) to describe the variance–covariance structure of these variables in terms of a smaller number of underlying unobservable and randomly varying factors, which can be interpreted as dietary patterns. We selected the number of factors to retain, taking into account the following criteria: factor eigenvalue >1, scree-plot visual inspection, and factor interpretability [29]. We applied a varimax rotation to obtain a simpler and more interpretable loading structure. Each factor's interpretation and labeling were based on nutrients with rotated factor loadings

≥0.63 in absolute value. We set this cut-off because it implies a minimum contribution of any factor to any nutrient's total variance of approximately 40% (i.e., 0.63^2) [30]. Nutrients with such factor loadings are called "dominant nutrients" hereafter. Factor scores, as continuous measures, were calculated for each participant and each pattern, and quantify the degree of adherence of each subject's diet to the identified pattern. Factor scores were computed using the weighted least squares method [31].

2.3.3. Reproducibility, Reliability, and Validity of Dietary Patterns

To evaluate the internal reproducibility of the identified dietary patterns, we performed additional analyses using a different procedure for estimating factor scores (namely, multiple regression method), and different estimation methods (namely, principal axis factor analysis with generalized least squares estimation method, and maximum likelihood factor analysis, after logarithmic transformation of the original nutrients) [29]. Moreover, following a split-half approach, we split the original dataset into 2 randomly selected subsamples (with cases and controls equally distributed), and carried out the principal component factor analysis procedure separately in each subsample; this procedure was repeated several times using different starting seeds for the random assignment, in order to verify the stability of the identified patterns [31]. As a sensitivity analysis, the principal component factor analysis procedure was also carried out among controls only.

Since all these checks were satisfactory, we performed all the subsequent analyses on factor scores obtained from the main analysis based on principal component factor analysis performed on cases and controls together, with varimax rotation and weighted least squares method.

To evaluate factor reliability and refine the identified dietary patterns, we calculated the standardized Cronbach's coefficient alpha for those nutrients with a factor loading ≥0.40 in absolute value on any factor [32]. For each factor, we computed an overall coefficient alpha and several nutrient-specific coefficient alphas when-item-deleted, which assessed the importance of each nutrient within the corresponding pattern.

To further describe and interpret the identified dietary patterns, we calculated the Spearman rank correlation coefficients between the continuous factor scores derived from principal component factor analysis and the weekly intake of selected food groups and condiments, defined on the same data and derived from the original 80 food items [31].

2.3.4. Risk Estimates

For each dietary pattern, we categorized participants into 4 groups according to the quartiles of the distribution of factor scores among cases and controls combined. We estimated the odds ratios (ORs) and the corresponding 95% confidence intervals (CIs) for quartile categories of factor scores compared to the reference category (i.e., the lowest quartile category for patterns characterized by positive factor loadings, and the highest quartile for patterns characterized by negative factors loadings), using unconditional multiple logistic regression models. We fitted both separate models for each factor and a composite model, including all the factors simultaneously. We included in each model the following potential confounding variables: age (<55, 55–59, 60–64, 65–69, 70–74, 75–79, ≥80 years), sex, center of recruitment, education (<7, 7–11, ≥12 years), cigarette smoking status and intensity (never, former, current <15, current 15–25, current ≥25 cigarettes/day), alcohol drinking intensity (<1, 1–<2, 2–<4, ≥4 drinks/day), history of occupational exposure (yes vs. no) in selected sectors relevant for bladder cancer risk, history of diabetes, history of cystitis, family history of bladder cancer, year of interview, and body mass index (BMI, <20, 20–24, 25–29, ≥30 kg/m^2). The tests for linear trends were computed for all these models, scoring the quartile categories as numbers from 1 to 4. Moreover, we fitted a composite model including the factor scores of the identified dietary patterns in continuum (with measurement unit equal to 1 standard deviation), which estimates the mean variation in bladder cancer risk per 1 standard deviation increment of factor scores, accounting for the aforementioned confounding variables.

Calculations were performed using the open-source statistical computing environment R (Ihaka and Gentleman, 1996; R Core Team, 2019), with its libraries psych [33] and GPArotation [34].

3. Results

Visual inspection of the correlation matrix of the original nutrient variables (among cases and controls) indicated that it was adequate to carry out a factor analysis. Table 1 shows the results from statistical procedures for checking matrix factorability. In particular, the Bartlett's test of sphericity was statistically significant (p-value < 0.001), allowing to reject the null hypothesis that the correlation matrix is the identity matrix. The overall measure of sampling adequacy (Keiser–Meyer–Olkin statistic) was 0.86, thus indicating that the sample size was adequate compared to the number of nutrients included in the analysis. In addition, the individual measures of sampling adequacy were satisfactory for all considered variables, since 9 nutrients had measures ≥0.90, 12 had measures between 0.80 and 0.89, 5 had measures between 0.70 and 0.79, and only 2 nutrients had measures between 0.60 and 0.69.

Table 2 gives the factor loading matrix of the four selected dietary patterns and the corresponding communalities. The retained dietary patterns explained about 78% of the variance of the original nutrient variables. All the examined nutrients had one or more factor loadings ≥0.30, thus indicating that all the selected nutrients were relevant in this analysis. The greater (in absolute value) was the loading of a given nutrient to a factor, the higher was the contribution of that nutrient to the factor. The first dietary pattern was named "Animal products" and was characterized by high positive factor loadings on calcium, saturated fatty acids, riboflavin, animal protein, cholesterol, phosphorus, and zinc. The second one, labeled "Vitamins and fiber", was characterized by high negative factor loadings on vitamin C, total fiber, beta-carotene equivalents, vitamin E, potassium, and total folate. The third dietary pattern, named "Starch-rich", had high negative factor loadings on starch, vegetable protein, and sodium. The fourth one was labeled as "Animal unsaturated fatty acids" and showed high positive factor loadings on other polyunsaturated fatty acids and vitamin D.

Table 2. Factor loading matrix [1], communalities, and explained variances for the four major dietary patterns identified by principal component factor analysis.

Nutrient	Dietary Patterns				Communalities
	Animal Products	Vitamins and Fiber	Starch-Rich	Animal Unsaturated Fatty Acids	
Animal protein	**0.79**	−0.23	−0.20	0.43	0.90
Vegetable protein	0.13	−0.45	**−0.85**	0.18	0.97
Cholesterol	**0.78**	−0.12	−0.22	0.42	0.84
Saturated fatty acids	**0.82**	−0.26	−0.25	0.16	0.82
Monounsaturated fatty acids	0.48	−0.57	−0.24	0.33	0.72
Linoleic acid	0.46	−0.27	−0.25	0.45	0.54
Linolenic acid	0.60	−0.29	−0.23	0.31	0.59
Other polyunsaturated fatty acids	0.28	−0.19	−0.12	**0.87**	0.89
Soluble carbohydrates	0.43	−0.61	−0.16	–	0.59
Starch	0.13	−0.20	**−0.93**	0.14	0.94
Sodium	0.42	–	**−0.82**	–	0.87
Calcium	**0.83**	−0.30	−0.17	–	0.82
Potassium	0.48	**−0.72**	−0.30	0.25	0.90
Phosphorus	**0.77**	−0.39	−0.37	0.22	0.93
Iron	0.45	−0.49	−0.36	0.35	0.69
Zinc	**0.66**	−0.39	−0.43	0.42	0.94
Thiamin (vitamin B1)	0.56	−0.56	−0.41	0.25	0.85
Riboflavin (vitamin B2)	**0.82**	−0.40	−0.18	0.12	0.88
Vitamin B6	0.53	−0.61	0.29	0.39	0.89
Total folate	0.48	**−0.68**	−0.34	0.22	0.85
Niacin	0.40	−0.47	−0.33	0.58	0.83
Vitamin C	0.20	**−0.88**	–	–	0.82
Retinol	0.47	–	–	0.13	0.25

Table 2. *Cont.*

Nutrient	Dietary Patterns				Communalities
	Animal Products	Vitamins and Fiber	Starch-Rich	Animal Unsaturated Fatty Acids	
Beta-carotene equivalents	0.17	**−0.80**	–	0.23	0.73
Lycopene	–	−0.45	−0.22	0.37	0.40
Vitamin D	0.17	−0.18	–	**0.82**	0.74
Vitamin E	0.35	**−0.73**	−0.19	0.41	0.86
Total fiber (Englyst)	0.10	**−0.81**	−0.33	0.13	0.80
Proportion of explained variance (%)	26.59	24.31	13.88	13.31	
Cumulative explained variance (%)	26.59	50.90	64.78	78.09	

[1] Estimates from a principal component factor analysis carried out on 28 nutrients, as measured among cases and controls together. For each factor, loadings greater or equal to 0.63 (in absolute value) indicated important or "dominant nutrients" in the current paper and were shown in bold typeface; loadings smaller than 0.1 in absolute value were suppressed.

The factor loading matrix derived from the sensitivity analysis on controls only is presented in Table S2. Visual inspection of the two-factor loading matrices provided reassuring results, with all the dominant nutrients highlighted in Table 2 being confirmed in Table S2. In addition, the single component and overall explained variances were comparable across the two solutions.

Standardized Cronbach's coefficient alphas were high (at least 0.93) for all the factors, and most standardized coefficient alphas when-item-deleted were smaller than the corresponding coefficient alpha for the same factor, which supported the presence of internal consistency of the nutrients on each identified factor (data not shown).

Table 3 gives the Spearman correlation coefficients between the identified dietary patterns and selected food groups and condiments. The "Animal products" pattern score was positively correlated with the consumption of cheese, milk, liver, red meat, desserts, eggs, bread, and processed meat. The "Vitamins and fiber" pattern score had positive correlation coefficients with citrus fruit, other fruits, olive oil, fruiting vegetables, leafy vegetables, root vegetables, other vegetables, and pasta and rice. The "Starch-rich" pattern showed a week positive correlation with pasta and rice. The "Animal unsaturated fatty acids" pattern was positively correlated with the consumption of fish, unspecified seed oils, and red meat.

Table 3. Spearman rank correlation coefficients [1] between continuous factor scores derived from principal component factor analysis on nutrient intakes and weekly number of portions for selected food groups and condiments derived on the same data.

Food Group	Animal Products	Vitamins and Fiber	Starch-Rich	Animal Unsaturated Fatty Acids
Milk	**0.45**	0.15	–	−0.18
Coffee	–	–	–	–
Tea and decaffeinated coffee	–	–	–	–
Bread	**0.35**	–	0.17	0.19
Pasta and rice	0.18	**0.32**	**0.29**	0.20
Soup	–	0.10	0.13	–
Eggs	**0.35**	–	0.13	0.18
White meat	0.18	0.13	0.10	0.14
Red meat	**0.38**	–	0.18	**0.41**
Liver	**0.40**	−0.03	−0.08	0.14
Processed meat	**0.32**	−0.02	0.19	0.13
Fish	–	–	–	**0.64**
Cheese	**0.63**	–	0.18	–
Potatoes	0.17	0.13	0.12	0.12
Pulses	–	0.24	0.16	–
Leafy vegetables	0.13	**0.40**	–	–
Fruiting vegetables	–	**0.45**	–	0.11
Root vegetables	0.06	**0.39**	–	0.10
Cruciferous vegetables	–	0.24	–	0.10

Table 3. *Cont.*

Food Group	Animal Products	Vitamins and Fiber	Starch-Rich	Animal Unsaturated Fatty Acids
Other vegetables	0.23	**0.39**	–	0.10
Citrus fruit	–	**0.50**	–	–
Other fruits	–	**0.63**	–	–
Soft drinks and fruit juices	0.15	–	–	–
Desserts	**0.37**	–	0.17	–
Sugar and candies	0.24	0.22	–	–
Butter and margarine	0.24	–	–	–
Specified seed oils	–	–	–	0.18
Unspecified seed oils	0.17	–	–	**0.44**
Olive oil	0.12	**0.52**	0.11	0.11

[1] Correlations greater or equal to 0.25 (in absolute value) were shown in bold typeface; correlations smaller than 0.1 (in absolute value) were suppressed.

Table 4 shows the ORs and the corresponding 95% CIs for bladder cancer, according to quartile categories of the dietary patterns, and in continuum (per 1 standard deviation). Results refer to the composite models, including all the dietary patterns simultaneously. The "Vitamins and fiber" dietary pattern was inversely related to bladder cancer risk (OR = 0.70, 95% CI: 0.49–0.98, for the highest versus the lowest quartile category of consumption). Inverse relationships of borderline significance were also found for the "Animal products" dietary pattern (OR = 0.70, 95% CI: 0.48–1.01) and possibly for the "Animal unsaturated fatty acids" dietary pattern (OR = 0.81, 95% CI: 0.58–1.15). The remaining dietary pattern named "Starch-rich" was unrelated to bladder cancer risk (OR = 1.28, 95% CI: 0.90–1.81). Risk estimates in continuum were consistent with those in categories, indicating a mean risk reduction of 11% for the "Vitamins and fiber" pattern, and of 8% for the "Animal products" and "Animal unsaturated fatty acids" patterns. Risk estimates obtained from models fitted separately for each dietary pattern were comparable to those from the composite model (data not shown).

Table 4. Odds ratios (ORs) [1] of bladder cancer and corresponding 95% confidence intervals (CIs) on quartiles of factor scores from a principal component factor analysis.

Dietary Pattern	Quartile Category, OR (95% CI)				*p* Trend [3]	Per 1 SD [4]
	I [2]	II	III	IV		
Animal products	1	0.93 (0.66–1.30)	0.72 (0.50–1.03)	0.70 (0.48–1.01)	0.026	0.91 (0.80–1.04)
Vitamins and fiber	1	0.77 (0.55–1.09)	0.92 (0.66–1.30)	0.70 (0.49–0.98)	0.109	0.89 (0.79–1.01)
Starch-rich	1	1.25 (0.90–1.75)	1.50 (1.06–2.11)	1.28 (0.90–1.81)	0.107	1.02 (0.90–1.15)
Animal unsaturated fatty acids	1	0.82 (0.58–1.15)	0.58 (0.41–0.82)	0.81 (0.58–1.15)	0.084	0.91 (0.81–1.03)

[1] Estimates from an unconditional logistic regression model adjusted for age, sex, center of recruitment, education, tobacco smoking, alcohol drinking, occupational exposure, history of diabetes, history of cystitis, family history of bladder cancer, year of interview, body mass index. Results refer to the composite model, including all the four factors simultaneously. [2] Reference category. [3] *p*-value for linear trend. [4] SD: standard deviation.

4. Discussion

In this case-control study on bladder cancer, we identified four major dietary patterns that explained almost 80% of the nutritional variability in this population. Among these, the "Vitamins and fiber" and possibly the "Animal products" and "Animal unsaturated fatty acids" patterns were associated with a decreased risk of bladder cancer, after mutual adjustment for all the remaining patterns. These patterns allowed to recover the main characteristics of the Italian diet at the time of the present study, in agreement with the findings of a national food consumption survey conducted in Italy in the same period [19].

Correlation coefficients between the identified patterns and selected food groups confirmed the labeling of dietary patterns and provided further insight into their composition. The "Animal products" pattern was primarily characterized by consumption of cheese and milk—which had the highest correlation coefficients—while minor, dominant components were different types of meat, eggs, and desserts (mainly bakery products and ice cream). The dairy products correlating with this pattern, especially milk, may have driven its inverse association with bladder cancer [14]. Moreover, in Italy, in contrast to Northern Europe and North America, poorer diets are largely based on bread, pasta, and carbohydrate-rich foods, instead of meat products, and meat consumption tends to be less unfavorable than elsewhere [35]. The "Vitamins and fiber" pattern correlated highly with different types of fruit and vegetables, as well as with olive oil; these dietary components are major sources of fiber, carotenoids, vitamin C, and vitamin E, and flavonoids, which have antioxidant and anti-inflammatory properties against cancer development [36]. Moreover, this is in line with findings supporting a favorable role of fruit and vegetables on bladder cancer, as indicated in the World Cancer Research Fund updated report [7]. The "Animal unsaturated fatty acids" pattern was mainly characterized by consumption of fish, and, to a lesser extent, by seed oils and red meat.

Evidence on dietary habits and bladder cancer risk is mainly based on single nutrients, foods, or food groups [7]. In the Danish Diet, Cancer and Health Study prospective cohort, no association was found between vitamin C, E, or folate and urothelial carcinoma, and a protective effect of dietary, but not supplemental, total beta-carotene was found [37]. Likewise, the Melbourne Collaborative Cohort Study found no association between dietary intake of B-group vitamins and urothelial cell carcinoma [38]. In the European Prospective Investigation into Cancer and Nutrition, no association was found between dietary folate, vitamin B2, B6, and B12 and urothelial cell carcinoma [39]. A population-based case-control study reported no association for fruits or vegetables, but an inverse association for vitamin B12 [40]. In our study, vitamin C, E, B-group, folate, and beta-carotene equivalents are all part of the same pattern named "Vitamins and fiber", which was inversely related to bladder cancer risk.

The aforementioned case-control study also showed a positive association for processed meat intake [40]. A positive relationship between processed meat, and possibly red meat, and the risk of bladder cancer has also been reported in a meta-analysis, where, however, the association was evident in case-control but not prospective studies [41].

These findings confirm the difficulties of identifying a favorable or detrimental role of specific foods or nutrients on bladder cancer risk and the opportunity to consider diet as an overall exposure using the dietary pattern approach.

Only two papers from the same research group have reported on the role of food-based dietary patterns and bladder cancer risk [12,13]. In the first paper, the authors reported results of a case-control study, including 255 bladder cancer cases and 501 matched hospital controls from Uruguay [12]. They used factor analysis and identified 3 food-based dietary patterns, named "Sweet beverages", "Western", and "Prudent" patterns. The first one was characterized by high loadings on coffee, added sugar, and boiled eggs and was associated with an increased bladder cancer risk. In our study, we did not observe a pattern similar to that one. The "Western" pattern was characterized by positive loadings on red meat and wine and negative loadings on poultry and fish, and it was associated with an increased risk of bladder cancer. Given the common animal orientation, this pattern appears to have similarities with our "Animal products" and "Animal unsaturated fatty acids" patterns. However, unlike the Uruguayan pattern, ours were positively correlated with fish, and white and red meat; the presence of white meat and fish may explain the identified different results in terms of risk. The Uruguayan "Prudent" pattern was characterized by positive loadings on cooked and raw vegetables, citrus and other fruits, and desserts, and by negative loadings on French bread; this pattern was not related to bladder cancer risk. However, when the authors analyzed the separate effects on risk of the dominant food groups from the "Prudent" pattern, citrus fruit was inversely associated, whereas cooked vegetables were positively associated with bladder cancer; thus, the identified null association for that pattern appears to be the consequence of opposite effects on bladder cancer risk of its dominant food groups; an additional effect of condiments in cooked vegetables is likely to explain the positive association of this food group with bladder cancer risk. Another study was carried out by the same authors within a multi-site case-control design, including the aforementioned data on bladder cancer [13]. A factor analysis was carried out on two sets of male and female controls, and their association with cancer risk was evaluated separately by cancer site and sex. Four sex-specific and food-based dietary patterns called "Prudent", "Traditional", "Western", and "Drinker" were identified. While the "Western" and "Prudent" patterns showed associations in the same direction observed in the previous study, the "Traditional" and "Drinker" patterns were unrelated to bladder cancer risk; however, CIs were, in general, very wide. To our knowledge, no studies have been published considering *a posteriori* dietary patterns derived on nutrients.

A few other studies have investigated the role of *a priori* dietary patterns on bladder cancer. In a previous analysis within our case-control study, we found an inverse association between adherence to the Mediterranean diet, as assessed through the Mediterranean Diet Score (MDS) and bladder cancer risk [18]. In a pooled analysis of 13 prospective studies from the BLadder cancer Epidemiology and Nutritional Determinants (BLEND) consortium, a higher MDS was associated with a reduction in bladder cancer risk [11]. In the Melbourne Collaborative Cohort Study (MCCS), the MDS and the Healthy Eating Index (HEI) showed borderline inverse associations with invasive, but not superficial, bladder cancers [9]; no relationship was found in the same study with the Dietary Inflammatory Index, representing a pro-inflammatory diet [9]. Similarly, in the Nurses' Health Study (NHS) and in the Health Professionals Follow Up Study (HPFS), no association with bladder cancer was evident for the Empirical Dietary Inflammatory Pattern measuring the pro-inflammatory potential in diet [8]. A study within the Breast Cancer Detection Demonstration Project follow-up cohort reported an inverse association between the Recommended Food Score—a global measure of diet quality—and overall cancer risk, but no association was detected with bladder cancer [10].

Though these mixed results do not allow to draw firm conclusions, the role of diet in bladder carcinogenesis remains plausible; the biological argument targets those many compounds in foods and their metabolites that are excreted through the urinary tract [6].

Associations between dietary factors and disease may be influenced by selection and information bias, as well as confounding [42]. Hospital controls may not be fully representative of the general population. In our study, however, the low refusal rate and the comparable catchment areas of cases and controls avoided major selection biases. Furthermore, cases and controls were interviewed by the same trained personnel in the same hospital setting [43], and the FFQ was satisfactorily reproducible and valid [26,27], thus reducing the possibility of information bias. The FFQ aimed at assessing dietary habits two years before the enrolment in the study. While this time frame may be insufficient for the development of bladder cancer, it is implicitly assumed that most people do not appreciably change their dietary habits through their adult age; a bias may have occurred if this assumption did not hold. As for potential confounding, we were able to adjust for socioeconomic indicators, tobacco, and a number of other factors.

Further issues are related to the use of factor analysis to derive *a posteriori* dietary patterns [44–46]. This technique [47] requires subjective decisions at various levels of the analysis, including the type and number of dietary components to analyze, the number of factors to retain, the choice of applying a rotation method or not (and which method to use), and the interpretation of the identified factors [48]. For this reason, we performed several complementary analyses, which were reassuring and supported the (internal) reproducibility of the identified dietary patterns, as well as their interpretation.

5. Conclusions

The present work provides a comprehensive description of dietary habits in the Italian population considered, with the identification of four major dietary patterns explaining most of the variability in nutrient intakes. In line with the available evidence on dietary patterns and bladder cancer risk, the additional modeling effort required by dietary patterns has not ended up in stronger risk estimates for this cancer site, as compared to those observed for single dietary components. Our study confirms that scoring high on a fruit-and-vegetables pattern provides beneficial effects on bladder cancer risk.

Supplementary Materials: The following are available online at http://www.mdpi.com/2072-6643/12/6/1584/s1, Table S1: Distribution of 690 bladder cancer cases and 665 controls according to selected characteristics.

Italy 2003–2014, Table S2: Factor loading matrix and explained variances for the four major dietary patterns identified by principal component factor analysis.

Author Contributions: Conceptualization, V.E., C.L.V., M.F. and F.B.; methodology, V.E., M.F. and F.B.; data interpretation, V.E., C.L.V., M.D.M., M.P., M.F. and F.B.; formal analysis, V.E., M.D.M., M.F. and F.B.; investigation, C.L.V., A.C., J.P., M.L. and D.S.; data curation, C.L.V., A.C., J.P., M.L. and D.S.; writing—original draft preparation, V.E., M.F. and F.B.; writing—review and editing, V.E., C.L.V., M.D.M., A.C., J.P., M.L., M.P., D.S., M.F. and F.B.; funding acquisition, C.L.V., J.P. and M.F. All authors have read and agreed to the published version of the manuscript.

Funding: This research was funded by the Italian League Against Cancer and by Ministero della Salute Ricerca Corrente. Matteo Di Maso was partially supported by a grant from Fondazione Umberto Veronesi at the time the study was conducted. The APC was funded by funds from Ferraroni.

Acknowledgments: The authors wish to thank A. Tavani and R. Talamini for coordination of the data collection, and Vincenzo Tortora, who contributed to the analyses of the present work as part of his specialty thesis.

Conflicts of Interest: The authors declare no conflict of interest.

References

1. Ferlay, J.; Colombet, M.; Soerjomataram, I.; Mathers, C.; Parkin, D.M.; Pineros, M.; Znaor, A.; Bray, F. Estimating the global cancer incidence and mortality in 2018: Globocan sources and methods. *Int. J. Cancer* **2019**, *144*, 1941–1953. [CrossRef] [PubMed]
2. Van Osch, F.H.; Jochems, S.H.; van Schooten, F.J.; Bryan, R.T.; Zeegers, M.P. Quantified relations between exposure to tobacco smoking and bladder cancer risk: A meta-analysis of 89 observational studies. *Int. J. Epidemiol.* **2016**, *45*, 857–870. [CrossRef] [PubMed]
3. Al-Zalabani, A.H.; Stewart, K.F.; Wesselius, A.; Schols, A.M.; Zeegers, M.P. Modifiable risk factors for the prevention of bladder cancer: A systematic review of meta-analyses. *Eur. J. Epidemiol.* **2016**, *31*, 811–851.
4. Burger, M.; Catto, J.W.; Dalbagni, G.; Grossman, H.B.; Herr, H.; Karakiewicz, P.; Kassouf, W.; Kiemeney, L.A.; La Vecchia, C.; Shariat, S.; et al. Epidemiology and risk factors of urothelial bladder cancer. *Eur. Urol.* **2013**, *63*, 234–241.
5. Malats, N.; Real, F.X. Epidemiology of bladder cancer. *Hematol. Oncol. Clin. N. Am.* **2015**, *29*, 177–189.
6. Pelucchi, C.; Bosetti, C.; Negri, E.; Malvezzi, M.; La Vecchia, C. Mechanisms of disease: The epidemiology of bladder cancer. *Nat. Clin. Pract. Urol.* **2006**, *3*, 327–340. [CrossRef]
7. World Cancer Research Fund International/American Institute for Cancer Research. Continuous Update Project Report: Diet, Nutrition, Physical Activity and Bladder Cancer. 2015. Available online: Https://www.Wcrf.Org/dietandcancer/bladder-cancer (accessed on 21 February 2020).
8. Abufaraj, M.; Tabung, F.K.; Shariat, S.F.; Moschini, M.; Devore, E.; Papantoniou, K.; Yang, L.; Strohmaier, S.; Rohrer, F.; Markt, S.C.; et al. Association between inflammatory potential of diet and bladder cancer risk: Results of 3 united states prospective cohort studies. *J. Urol.* **2019**, *202*, 484–489.
9. Bravi, F.; Spei, M.E.; Polesel, J.; Di Maso, M.; Montella, M.; Ferraroni, M.; Serraino, D.; Libra, M.; Negri, E.; La Vecchia, C.; et al. Mediterranean diet and bladder cancer risk in italy. *Nutrients* **2018**, *10*, 1061. [CrossRef]
10. Dugue, P.A.; Hodge, A.M.; Brinkman, M.T.; Bassett, J.K.; Shivappa, N.; Hebert, J.R.; Hopper, J.L.; English, D.R.; Milne, R.L.; Giles, G.G. Association between selected dietary scores and the risk of urothelial cell carcinoma: A prospective cohort study. *Int. J. Cancer* **2016**, *139*, 1251–1260. [CrossRef]
11. Mai, V.; Kant, A.K.; Flood, A.; Lacey, J.V., Jr.; Schairer, C.; Schatzkin, A. Diet quality and subsequent cancer incidence and mortality in a prospective cohort of women. *Int. J. Epidemiol.* **2005**, *34*, 54–60. [CrossRef]
12. Witlox, W.J.A.; van Osch, F.H.M.; Brinkman, M.; Jochems, S.; Goossens, M.E.; Weiderpass, E.; White, E.; van den Brandt, P.A.; Giles, G.G.; Milne, R.L.; et al. An inverse association between the mediterranean diet and bladder cancer risk: A pooled analysis of 13 cohort studies. *Eur. J. Nutr.* **2020**, *59*, 287–296. [CrossRef] [PubMed]
13. De Stefani, E.; Boffetta, P.; Ronco, A.L.; Deneo-Pellegrini, H.; Acosta, G.; Mendilaharsu, M. Dietary patterns and risk of bladder cancer: A factor analysis in uruguay. *Cancer Causes Control.* **2008**, *19*, 1243–1249. [CrossRef] [PubMed]
14. De Stefani, E.; Deneo-Pellegrini, H.; Boffetta, P.; Ronco, A.L.; Aune, D.; Acosta, G.; Mendilaharsu, M.; Brennan, P.; Ferro, G. Dietary patterns and risk of cancer: A factor analysis in uruguay. *Int. J. Cancer* **2009**, *124*, 1391–1397. [CrossRef]

15. Di Maso, M.; Turati, F.; Bosetti, C.; Montella, M.; Libra, M.; Negri, E.; Ferraroni, M.; La Vecchia, C.; Serraino, D.; Polesel, J. Food consumption, meat cooking methods and diet diversity and the risk of bladder cancer. *Cancer Epidemiol.* **2019**, *63*, 101595. [CrossRef]
16. Augustin, L.S.A.; Taborelli, M.; Montella, M.; Libra, M.; La Vecchia, C.; Tavani, A.; Crispo, A.; Grimaldi, M.; Facchini, G.; Jenkins, D.J.A.; et al. Associations of dietary carbohydrates, glycaemic index and glycaemic load with risk of bladder cancer: A case-control study. *Br. J. Nutr.* **2017**, *118*, 722–729. [CrossRef] [PubMed]
17. Rossi, M.; Strikoudi, P.; Spei, M.E.; Parpinel, M.; Serraino, D.; Montella, M.; Libra, M.; La Vecchia, C.; Rosato, V. Flavonoids and bladder cancer risk. *Cancer Causes Control.* **2019**, *30*, 527–535. [CrossRef]
18. Shivappa, N.; Hebert, J.R.; Rosato, V.; Rossi, M.; Libra, M.; Montella, M.; Serraino, D.; La Vecchia, C. Dietary inflammatory index and risk of bladder cancer in a large italian case-control study. *Urology* **2017**, *100*, 84–89. [CrossRef]
19. Sette, S.; Le Donne, C.; Piccinelli, R.; Mistura, L.; Ferrari, M.; Leclercq, C.; Group, I.-S.S. The third national food consumption survey, inran-scai 2005-06: Major dietary sources of nutrients in Italy. *Int. J. Food Sci. Nutr.* **2013**, *64*, 1014–1021. [CrossRef]
20. Leclercq, C.; Arcella, D.; Piccinelli, R.; Sette, S.; Le Donne, C.; Turrini, A.; Group, I.-S.S. The italian national food consumption survey inran-scai 2005-06: Main results in terms of food consumption. *Public Health Nutr.* **2009**, *12*, 2504–2532. [CrossRef]
21. Polesel, J.; Bosetti, C.; Di Maso, M.; Montella, M.; Libra, M.; Garbeglio, A.; Zucchetto, A.; Turati, F.; Talamini, R.; La Vecchia, C.; et al. Duration and intensity of tobacco smoking and the risk of papillary and non-papillary transitional cell carcinoma of the bladder. *Cancer Causes Control.* **2014**, *25*, 1151–1158. [CrossRef]
22. Moch, H.; Humphrey, P.A.; Ulbright, T.M.; Reuter, V. *Who Classification of Tumours of the Urinary System and Male Genital Organs*; International Agency for Research on Cancer (IARC): Lyon, France, 2016.
23. Di Maso, M.; Bosetti, C.; Taborelli, M.; Montella, M.; Libra, M.; Zucchetto, A.; Turati, F.; Parpinel, M.; Negri, E.; Tavani, A.; et al. Dietary water intake and bladder cancer risk: An italian case-control study. *Cancer Epidemiol.* **2016**, *45*, 151–156. [CrossRef] [PubMed]
24. Gnagnarella, P.; Parpinel, M.; Salvini, S.; Franceschi, S.; Palli, D.; Boyle, P. The update of the italian food composition database. *J. Food Comp. Anal.* **2004**, *17*, 509–522. [CrossRef]
25. Salvini, S.; Parpinel, M.; Gnagnarella, P.; Maisonneuve, P.; Turrini, A. *Banca Di Composizione Degli Alimenti Per Studi Epidemiologici in Italia*; Istituto europeo di oncologia: Milano, Italia, 1998.
26. Decarli, A.; Franceschi, S.; Ferraroni, M.; Gnagnarella, P.; Parpinel, M.T.; La Vecchia, C.; Negri, E.; Salvini, S.; Falcini, F.; Giacosa, A. Validation of a food-frequency questionnaire to assess dietary intakes in cancer studies in italy. Results for specific nutrients. *Ann. Epidemiol.* **1996**, *6*, 110–118. [CrossRef]
27. Franceschi, S.; Negri, E.; Salvini, S.; Decarli, A.; Ferraroni, M.; Filiberti, R.; Giacosa, A.; Talamini, R.; Nanni, O.; Panarello, G.; et al. Reproducibility of an italian food frequency questionnaire for cancer studies: Results for specific food items. *Eur. J. Cancer* **1993**, *29A*, 2298–2305. [CrossRef]
28. Pett, M.A.; Lackey, N.R.; Sullivan, J.J. *Making Sense of Factor Analysis: The Use of Factor Analysis for Instrument Development in Health Care Research*; Sage: Sozende Oaks, CA, USA, 2003.
29. Johnson, R.A.; Wichern, D.W. *Applied Multivariate Statistical Analysis*, 5th ed.; Prentice Hall: Upper Saddler River, NJ, USA, 2002.
30. Comrey, A.; Lee, H.B. *A first Course in Factor Analysis*, 2nd ed.; Lawrence Erlbaum Associates: Hillsdale, NJ, USA, 1992.
31. Bravi, F.; Bertuccio, P.; Turati, F.; Serraino, D.; Edefonti, V.; Dal Maso, L.; Decarli, A.; Montella, M.; Zucchetto, A.; La Vecchia, C.; et al. Nutrient-based dietary patterns and endometrial cancer risk: An italian case-control study. *Cancer Epidemiol.* **2015**, *39*, 66–72. [CrossRef] [PubMed]
32. Cronbach, L.J. Coefficient alpha and the internal structure of tests. *Psychometrika* **1951**, *16*, 297–334. [CrossRef]
33. Revelle, W. Psych: Procedures for Psychological, Psychometric, and Personality Research. Northwestern University: Evanston, Illinois. Available online: http://CRAN.Rproject.org/package=psychR (accessed on 31 July 2019).
34. Bernaards, C.A.; Jennrich, R.I. Gradient projection algorithms and software for arbitrary rotation criteria in factor analysis. *Educ. Psychol. Meas.* **2005**, *65*, 676–696. [CrossRef]
35. Lucenteforte, E.; Scita, V.; Bosetti, C.; Bertuccio, P.; Negri, E.; La Vecchia, C. Food groups and alcoholic beverages and the risk of stomach cancer: A case-control study in italy. *Nutr. Cancer* **2008**, *60*, 577–584. [CrossRef]

36. Michaud, D.S.; Pietinen, P.; Taylor, P.R.; Virtanen, M.; Virtamo, J.; Albanes, D. Intakes of fruits and vegetables, carotenoids and vitamins a, e, c in relation to the risk of bladder cancer in the atbc cohort study. *Br. J. Cancer* **2002**, *87*, 960–965. [CrossRef]
37. Roswall, N.; Olsen, A.; Christensen, J.; Dragsted, L.O.; Overvad, K.; Tjonneland, A. Micronutrient intake and risk of urothelial carcinoma in a prospective danish cohort. *Eur. Urol.* **2009**, *56*, 764–770. [CrossRef]
38. Dugue, P.A.; Brinkman, M.T.; Hodge, A.M.; Bassett, J.K.; Bolton, D.; Longano, A.; Hopper, J.L.; Southey, M.C.; English, D.R.; Milne, R.L.; et al. Dietary intake of nutrients involved in one-carbon metabolism and risk of urothelial cell carcinoma: A prospective cohort study. *Int. J. Cancer* **2018**, *143*, 298–306. [CrossRef] [PubMed]
39. Vrieling, A.; Bueno-De-Mesquita, H.B.; Ros, M.M.; Kampman, E.; Aben, K.K.; Buchner, F.L.; Jansen, E.H.; Roswall, N.; Tjonneland, A.; Boutron-Ruault, M.C.; et al. One-carbon metabolism biomarkers and risk of urothelial cell carcinoma in the european prospective investigation into cancer and nutrition. *Int. J. Cancer* **2019**, *145*, 2349–2359. [PubMed]
40. Wu, J.W.; Cross, A.J.; Baris, D.; Ward, M.H.; Karagas, M.R.; Johnson, A.; Schwenn, M.; Cherala, S.; Colt, J.S.; Cantor, K.P.; et al. Dietary intake of meat, fruits, vegetables, and selective micronutrients and risk of bladder cancer in the new england region of the united states. *Br. J. Cancer* **2012**, *106*, 1891–1898.
41. Crippa, A.; Larsson, S.C.; Discacciati, A.; Wolk, A.; Orsini, N. Red and processed meat consumption and risk of bladder cancer: A dose-response meta-analysis of epidemiological studies. *Eur. J. Nutr.* **2018**, *57*, 689–701. [CrossRef] [PubMed]
42. Kogevinas, M.; Garcia-Closas, M.; Trichpoulos, D. Urinary bladder cancer. In *Cancer Epidemiology*, 2nd ed.; Adami, H.-O., Hunter, D., Trichpoulos, D., Eds.; Oxford Universiy Press: New York, NY, USA, 2008; pp. 573–596.
43. D'Avanzo, B.; La Vecchia, C.; Katsouyanni, K.; Negri, E.; Trichopoulos, D. An assessment, and reproducibility of food frequency data provided by hospital controls. *Eur. J. Cancer Prev.* **1997**, *6*, 288–293. [CrossRef]
44. Grosso, G.; Bella, F.; Godos, J.; Sciacca, S.; Del Rio, D.; Ray, S.; Galvano, F.; Giovannucci, E.L. Possible role of diet in cancer: Systematic review and multiple meta-analyses of dietary patterns, lifestyle factors, and cancer risk. *Nutr. Rev.* **2017**, *75*, 405–419.
45. Newby, P.K.; Tucker, K.L. Empirically derived eating patterns using factor or cluster analysis: A review. *Nutr. Rev.* **2004**, *62*, 177–203. [CrossRef]
46. Steck, S.E.; Murphy, E.A. Dietary patterns and cancer risk. *Nat. Rev. Cancer* **2020**, *20*, 125–138.
47. Hu, F.B. Dietary pattern analysis: A new direction in nutritional epidemiology. *Curr. Opin. Lipidol.* **2002**, *13*, 3–9. [CrossRef]
48. Edefonti, V.; De Vito, R.; Dalmartello, M.; Patel, L.; Salvatori, A.; Ferraroni, M. Reproducibility and validity of a posteriori dietary patterns: A systematic review. *Adv. Nutr.* **2020**, *11*, 293–326. [CrossRef]

nutrients

Article

Attenuating Effect of Peruvian Cocoa Populations on the Acute Asthmatic Response in Brown Norway Rats

Marta Périz [1,2], Francisco J. Pérez-Cano [1,2], Trinitat Cambras [1], Àngels Franch [1,2], Ivan Best [3,4], Santiago Pastor-Soplin [3], Margarida Castell [1,2] and Malén Massot-Cladera [1,2,*]

1 Secció de Fisiologia, Departament de Bioquímica i Fisiologia, Facultat de Farmàcia i Ciències de l'Alimentació, Universitat de Barcelona (UB), 08028 Barcelona, Spain; mperiz@ub.edu (M.P.); franciscoperez@ub.edu (F.J.P.-C.); cambras@ub.edu (T.C.); angelsfranch@ub.edu (À.F.); margaridacastell@ub.edu (M.C.)

2 Institut de Recerca en Nutrició i Seguretat Alimentària (INSA-UB), UB, 08921 Santa Coloma de Gramenet, Spain

3 Programa Cacao, Ingeniería Agroforestal, Facultad de Ciencias Ambientales, Universidad Científica del Sur, Lima 15842, Peru; ibest@usil.edu.pe (I.B.); spastor@ucientifica.edu.pe (S.P.-S.)

4 Unidad de Investigación en Nutrición, Salud, Alimentos Funcionales y Nutracéuticos, Universidad San Ignacio de Loyola, Lima 15024, Peru

* Correspondence: malen.massot@ub.edu

Received: 15 June 2020; Accepted: 24 July 2020; Published: 31 July 2020

Abstract: Cocoa contains bioactive components, which vary according to genetic and environmental factors. The present study aimed to ascertain the anti-allergic properties of native Peruvian cocoa populations ("Blanco de Piura" or BPC, "Amazonas Peru" or APC, "Criollo de Montaña" or CMC, "Chuncho" or CCC, and an ordinary cocoa or OC). To do so, after an initial in vitro approach, an in vivo study focused on the induction of an anaphylactic response associated with allergic asthma in Brown Norway rats was carried out. Based on their polyphenol content, antioxidant activity and in vitro effects, the APC and CMC were selected to be included in the in vivo study. Cocoa diets were tested in a model of allergic asthma in which anaphylactic response was assessed by changes in body temperature, motor activity and body weight. The concentration of specific immunoglobulin E (IgE), mast cell protease and leukotrienes was also quantified in serum and/or bronchoalveolar lavage fluid. CMC and OC populations exhibited a protective effect on the allergic asthma rat model as evidenced by means of a partial protection against anaphylactic response and, above all, in the synthesis of IgE and the release of mast cell protease.

Keywords: anaphylaxis; asthma; IgE; leukotriene; mast cell protease; methylxanthines; motor activity; polyphenols; temperature; *Theobroma cacao*

1. Introduction

Allergic asthma is a complex inflammatory disorder characterized by chronic airway inflammation and immune-mediated hypersensitivity reaction [1]. Asthmatic patients often present airflow limitation and suffer from variable respiratory symptoms such as wheezing, shortness of breath, chest tightness and cough [2]. Histologically, the airway of an asthmatic patient is characterized by eosinophils infiltration, degranulated mast cells together with alteration of epithelial cell tight junctions and hyperplasia of goblet cells [3,4]. Although allergic asthma often starts in childhood, its prevalence is also high in adults, and affects around 270 million people worldwide [5]. Peru is one of the countries with the highest prevalence of asthma in Latin America [6,7], where 19.6% of adolescents

(13–14 years old) suffer from asthma. However, this disease is underdiagnosed and the prevalence of physician-diagnosed asthma in Peru has been reported to be 33.1% [6].

Asthma therapy includes pharmacological interventions as well as treatment of associated comorbidities and modifiable lifestyle risk factors (e.g., avoidance of tobacco and weight loss). The pharmacological treatment consists of short-acting β_2-agonists and inhaled corticosteroids, although adherence is often poor. In addition, emerging biological therapies, such as monoclonal antibodies targeting several cytokines, show promising results [2]. Nevertheless, non-pharmacological strategies can reveal some benefits in reducing symptoms and corticosteroid use. In this context, it has been proposed that diet could represent a good complement to controlling allergic asthma disease.

Flavonoids, chemically belonging to the polyphenol class, are a large family of secondary products of plants that contribute to the blue, scarlet and orange colors of their leaves, flowers and fruits. They are found in seeds, nuts, grains and spices and in some derived beverages such as wine, tea, cocoa and beer [8]. Due to their chemical polyphenolic structure, flavonoids have demonstrated antioxidant properties. This is why, in the past few years, flavonoids have emerged as potential therapeutic/coadjuvant agents in several conditions, such as in cardiovascular diseases, in chronic inflammation, in cancer and also in allergies and asthma [8–11].

The cocoa bean and its derivates (e.g., cocoa powder, chocolate, etc.) have bioactive compounds and, among them, flavonoids [9]. In particular, cocoa contains (+)-catechin and (−)-epicatechin as monomers and procyanidins (from 2 to 10 monomeric units) as polymers [9]. These flavonoids together with other bioactive compounds, such as theobromine and fiber, confer on cocoa immunomodulatory properties both in vitro and in vivo [12–17]. In particular, cocoa consumption has previously shown anti-allergic properties in a model of systemic disease [18,19] and oral sensitization [20]. In addition, a protective effect on allergy has also been suggested when considering cocoa consumption in young people [21].

Effects of cocoa intake will vary depending on the amount of flavonoids and other bioactive compounds present in the product. Polyphenolic content of cocoa products varies greatly due to genetic factors, such as the variety or clone their beans come from [22,23]. However, genetics only partially determines the biochemical profile. The phenolic content of cocoa products also varies greatly due to environmental factors such as soil, water, cultivation latitude, management and post-harvest handling of the product (fermentation and drying). Thus, the result is a wide range of regional cocoa populations with singular qualities and quantities of these bioactive compounds [24–26].

Based on the above-mentioned, we hypothesized that the cocoa tree (*Theobroma cacao* L.) cultivated under a tropical climate, as in the North of Peru and South of Ecuador, which are considered as being the center of the origin and genetic diversity of cocoa [27,28], may include populations with different biological effects, such as those on the immune system reported so far. Thus, the present study aimed to ascertain the anti-allergic properties of Peruvian cocoa populations, firstly using an in vitro approach to select the most active populations and secondly using an in vivo study focused on the induction of an anaphylactic response associated with allergic asthma in Brown Norway rats.

2. Materials and Methods

2.1. Cocoa Population Characterization

Pastes made with beans from four Peruvian cocoa populations were used: "Blanco de Piura" (BPC) from the Piura region (latitude/longitude −5.270248, −79.964108), "Amazonas Peru" (APC) from the Amazonas region (−5.737422, −78.431114), "Criollo de Montaña" (CMC) from the Junín region (−11.335774, −74.533181), and "Chuncho" (CCC) from the Cusco region (−12.510664, −73.834577). As reference cocoa, CCN-51 ordinary cocoa paste from the same area as the CCC was included. With the exception of the CCN-51, these are the populations of Peruvian cocoa considered to be fine or flavor cocoa (Article 39, ICA, 2010) [29], and due to their morpho-agronomic and sensory properties, they are best known and characterized for their use in making high quality artisan chocolates [30,31]. The cocoa samples were obtained under prior informed consent (PIC), in agreement and signed with farmers, and in accordance with the Nagoya Protocol spirit of sharing the benefits arising from the utilization of genetic resources [32,33]. BPC comes from a coastal area, facing the Pacific Sea. It develops in a dry and warm environment, but under irrigation and neutral loamy soils. The other cocoa populations are found on the eastern side of the Andes, in the Amazon, and they develop in rain fed on acidic and clay loam soils. The cocoa pastes were made at the place of origin, based on a common protocol. Biochemical analysis, which was performed in triplicate, began with 100% pure cocoa paste.

2.1.1. Phenolic Compounds and Antioxidant Activity

The extraction of bioactive compounds from the different cocoa samples was carried out using the methodology proposed by Pedan et al. [34], with minor modifications. The cocoa paste was heated in a water bath until it reached a liquid state. To remove lipids, 20 mL of each sample was placed in a 250 mL flask and 80 mL of n-hexane was added (5 min at 20 °C) and then centrifuged (2880× *g*, 5 min). The defatting procedure was repeated five times, until the n-hexane extract remained colorless. After drying, 5 g of cocoa powder with an average particle size of less than 100 µm was extracted three times with 15 mL of acetone/water (50/50) (8 min at 50 °C) and then centrifuged (2880× *g*, 5 min). The supernatants obtained in each extraction step were mixed and used to measure the total levels of phenols and flavonoids using the Folin-Ciocalteu method [35] and the aluminum chloride colorimetric method [36], respectively.

In vitro antioxidant capacity was evaluated using the α,α-diphenyl-β-picrylhydrazyl (DPPH) radical scavenging assay [37] and the ferric-reducing/antioxidant power (FRAP) assay [38].

2.1.2. Methylxanthine Quantification

High-performance liquid chromatography (HPLC) determination of theobromine, theophylline, and caffeine was performed according to Srdjenovic et al. [39] with minor modifications. Firstly, an extract was prepared (2.5 g of cocoa powder with 10 mL of water) and then incubated in an ultrasonic bath (30 min at 60 °C). After centrifuging (4000× *g*, 10 min, 20 °C), 10 mL of the supernatants was purified using solid phase separation (SPE) with a Supelclean LC-18 SPE cartridge (Sigma-Aldrich, St. Louis, USA). Samples were run on a Chromaster 600 HPLC with a diode array detector (Hitachi, Tokyo, Japan), an autosampler and a C8 reverse-phase column (5 µm particle size, i.d. 4.6 × 150 mm). The mobile phase consisted of water-tetrahydrofuran (0.1% in water, pH 8)—acetonitrile (90:10, *v/v*); the run time was 8 min with a flow rate of 0.8 mL/min. Detection was performed at 273 nm using a photodiode array detector.

2.2. *Animals*

Four-week-old female Brown Norway rats were obtained from Envigo (Huntingdon, UK) and housed (3 rats per cage) in the animal facilities at the Faculty of Pharmacy and Food Science (University of Barcelona) in polycarbonate cages containing bedding of large fibrous particles (Souralit 1035, Bobadeb S.L., Santo Domingo de la Calzada, Spain) under controlled conditions of temperature and humidity in a 12:12 h light/dark cycle. The animals remained in quarantine for 1 week before experiments began.

All experimental procedures were conducted in accordance with the institutional guidelines for the Care and Use of Laboratory Animals and were approved by the Ethical Committee for Animal Experimentation of the University of Barcelona and the Catalonia Government (CEEA/UB ref. 414/16 and DAAM 9351, respectively), in full compliance with national legislation following the EU-Directive 2010/63/EU for the protection of animals used for scientific purposes.

2.3. In vitro Study

2.3.1. Peritoneal Macrophages and Lymphocytes Culture

Peritoneal macrophages and spleen mononuclear cells were obtained from six healthy rats under anesthesia with ketamine (90 mg/kg) (Merial Laboratories S.A, Barcelona, Spain) and xylazine (10 mg/kg) (Bayer A.G, Leverkusen, Germany).

Peritoneal macrophages were collected as previously described [40]. Briefly, after the injection of 40 mL of ice-cold sterile phosphate buffered saline (PBS, pH 7.2) into the peritoneal cavity, a 1 min massage was performed, and cell suspension was aspirated and centrifuged (538× *g*, 10 min, 4 °C). After removing possible erythrocytes by osmotic lysis (ammonium chloride), cells were resuspended with cold Roswell Park Memorial Institute (RPMI) medium without phenol red (Merck, Madrid, Spain), supplemented with 10% heat-inactivated fetal bovine serum (FBS), 100 IU/mL streptomycin-penicillin, 2 mM L-glutamine and 0.05 mM 2-mercaptoethanol (Sigma-Aldrich, Madrid, Spain). Macrophage counts were assessed using a Spincell hematology analyzer (MonLab Laboratories, Barcelona, Spain), properly calibrated for these cells, and were plated (10^6 cells/mL) at 37 °C overnight. After removing non-attached cells, macrophages were incubated with 10 µg/mL of each of the five cocoa extracts in dimethyl sulfoxide (DMSO) for 2 h. Afterwards, cells were stimulated with 100 ng/mL lipopolysaccharide (LPS) for 6 h. Stimulated macrophages with no cocoa addition were used as control. Cell viability was measured through determination of lactate dehydrogenase (LDH) enzyme released to the medium. Macrophages were also used to establish M1/M2 polarization.

Spleen mononuclear cells were isolated from rat spleens as previously described [41,42]. Firstly, spleen cell suspensions were obtained by passing the tissue through a cell strainer (40 µm, BD Biosciences, Heidelberg, Germany), and then erythrocytes were eliminated by osmotic lysis. A Countess™ Automated Cell Counter (Invitrogen™, Thermo Fisher Scientific, Waltham, MA, USA) was used for cell counting and the assessment of viability. Splenocytes (10^6 cells/well) were immediately incubated in the presence of 10 µg/mL of each of the five cocoa extracts in DMSO for 2 h. Afterwards, splenocytes were stimulated with 100 ng/mL LPS or remained nonstimulated for 24 h, and then, supernatants from both conditions were collected for TNF-α determination. In parallel, nonstimulated splenocytes were cultured for 96 h, whose supernatants were used to quantify the IgG. Both assays were performed in quadruplicate.

2.3.2. Radical Oxygen Species (ROS) Production

ROS production was quantified in isolated macrophages as previously described [42]. In brief, macrophages were plated (10^5 cells/well) and allowed to attach overnight. Then, they were washed with warm 1% FBS-supplemented RPMI medium without phenol red. Macrophages were incubated with 20 µM of reduced 2′,7′-dichlorofluorescein diacetate (H_2DCF-DA) probe (Invitrogen, Paisley, UK) for 30 min at 37 °C. Macrophage-derived ROS oxidized H_2DCF-DA to a fluorescent compound (20,70-dichlorofluorescein, DCF), which was quantified using the fluorimeter Modulus® Microplate Multimode Reader (excitation 538 nm, emission 485 nm, Turner BioSystems, CA, USA). ROS results are expressed as the time course from 0 to 130 min and also as the area under the curve (AUC) of this period of time.

2.3.3. M1 and M2 Characterization

After cocoa incubation and LPS stimulation, the M1 and M2 phenotype of macrophages were established by the exclusive expression of the molecules CD86 and CD206, respectively [43,44]. For M1 phenotype, an anti-rat CD86 conjugated to phycoerythrin (Biolegend, San Diego, CA, USA) was applied as in previous studies. For M2 phenotype, a primary rabbit polyclonal antibody (Ab) to mannose receptor (CD206) (Abcam plc, Cambridge, UK) was used, followed by blocking nonspecific signals, and a secondary Ab conjugated to Alexa-Fluor-647 (Abcam plc), as in previous studies [45]. A negative control using isotype-matched Ab was included for each sample. Data were acquired with a Gallios™ Cytometer (Beckman Coulter, Miami, FL, USA) in the Flow Cytometry Unit of the Scientific and Technological Centers of the UB (CCiTUB) and analyzed with FlowJo v.10 software (Tree Star, Inc., Ashland, OR, USA). Results are expressed as percentages of positive cells in the macrophage population previously selected according to their forward-scatter (*FSC*) and side-scatter (*SSC*) characteristics.

2.3.4. IgG and Tumor Necrosis Factor (TNF)-α Quantification by ELISA

TNF-α were quantified in the 24 h supernatants by stimulated splenocytes using Opt-EIA-set (BD Biosciences), as in previous studies [46]. IgG were quantified in 96 h supernatants by nonstimulated splenocytes with an enzyme-linked immunosorbent assay (ELISA) following the manufacturer's instructions (BD Biosciences), as previously described [46]. In both cases, absorbance was measured in a microplate photometer (LabSystems Multiskan) and analyzed using ASCENT version 2.6 software (Thermo Fisher Scientific, Waltham, MA, USA). TNF-α and IgG results are shown as percentage with respect to the control condition (without cocoa), which was considered as 100%.

2.4. *In vivo Study*

According to their polyphenol content, antioxidant activity and their in vitro effect on macrophages and splenocytes, two populations of Peruvian cocoa were selected: "Amazonas Peru" cocoa (APC) and "Criollo de Montaña" cocoa (CMC). The in vivo effects of these two populations were then established in a model of allergy in rats. The ordinary cocoa (OC) was also included to be considered as a reference cocoa.

2.4.1. Diets and Animal Groups

Four diets were elaborated: a standard diet based on the AIN-93M diet (Envigo) and three diets in which 90% of powdered AIN-93M was mixed with 10% of cocoa paste (OC, APC or CMC) previously pulverized. The mixture was pelletized and subsequently dried in a 40 °C oven for 36 h. Once dried, the pelleted diet was vacuum-packed to prevent oxidation and contamination and stored at 4 °C until used.

The animals were randomized into five experimental groups ($N = 9$ animals/group): the healthy reference group (REF) and asthmatic group (A) were both fed with the standard diet, and the three asthmatic groups received the OC, APC and CMC diets, respectively (CC, APC and CMC groups). The animals had free access to the experimental diet and water. The body weight and food and water intake were monitored every 2–3 days throughout the study. The Appraising Project Office's program from the Universidad Miguel Hernández de Elche (Alicante, Spain) was used to calculate the minimum number of animals providing statistically significant differences among groups, assuming that there is no dropout rate and type I error of 0.05 (two-sided). In addition, the sample size was adjusted following the University Ethical Committee guidelines and to apply the three Rs rule for experimenting in animals.

2.4.2. Sensitization and Induction of an Anaphylactic Response

At 1 week after the beginning of the experimental diet, asthma was induced using ovalbumin (OVA) as allergen, as previously described [47]. Briefly, on day 0, rats were firstly sensitized via intraperitoneal (i.p.) with 500 µL of a suspension containing 50 µg of OVA (grade V, Sigma-Aldrich, Madrid, Spain), 20 mg of alum (Imject®; Pierce, IL, USA) and 50 ng of *Bordetella pertussis* toxin (Sigma-Aldrich) and boosted a week later with 50 µg of OVA in 20 mg of alum (i.p.). A parallel group of non-sensitized rats (age and sex matched) was included.

At day 28, between 10 a.m. and 1 p.m., all rats received an intranasal (i.n) challenge with 300 µL of an OVA solution (50 mg/mL). Anaphylactic response was accurately assessed by changes in motor activity, body temperature, body weight and plasma protease concentration. Anti-OVA IgE was quantified in blood and bronchoalveolar lavage fluid (BALF) samples obtained 24 h later.

2.4.3. Body Temperature Monitoring

In order to monitor the body temperature, data loggers (Thermochron®, iButton type DS1921H-FS with a resolution of 0.125 °C) were used. For this, 1 week before the i.n. challenge, a logger was intraperitoneally implanted in each rat under isoflurane (Isoflo®, ECUPHAR, Barcelona, Spain) anesthesia (4–5% in the induction, 1–2% in the maintenance with an oxygen flow of 0.5–1.0 L/min). Meloxicam (1 mg/kg body weight, subcutaneous route) was administered subcutaneously immediately after the intervention and 24 h later. Animals were then housed in individual cages in an isolated room (see motor activity assessment section). Body temperature was recorded every minute from the night before the challenge (starting at 2 a.m.) until the day after the challenge, when the sensor was removed. The results of body temperature are expressed as the time course of the mean values during the registered period, the mean value every 2 h from 2 h before the challenge to 18 h after the i.n. challenge and as the AUC between 900 and 400 min after the challenge considering changes above 34 °C.

2.4.4. Motor Activity Assessment

The movement of animals housed in individual cages and placed in an isolated room were quantified using an activity meter, as previously performed [47,48]. The activity meters consisted of two infrared beams that crossed perpendicularly 7 cm above the floor of the cage. Every time the animal crossed one beam a count was detected. Number of movements was recorded every minute from 2 days before the i.n. challenge until 18 h after. To summarize the effects of anaphylactic response on motor activity, the total number of movements in the active period of the rats (darkness period from 8 p.m to 8 a.m) was considered, with the exception of the last 2 h in order to avoid the variations in motor activity in anticipation of light due to the normal circadian rhythm. The movements in the dark period before and after the i.n. challenge were also compared.

2.4.5. Sample Collection

One hour after the i.n. challenge, blood samples from the saphenous vein were obtained to quantify plasma mast cell protease.

Twenty-four hours after the i.n. challenge, the rats were anesthetized with ketamine (90 mg/kg) (Merial Laboratories S.A) and xylazine (10 mg/kg) (Bayer A.G). Urine samples obtained by direct puncture of the bladder were kept at −80 °C until quantification of the polyphenol concentration. Blood samples were collected by heart puncture and kept at −20 °C until anti-IgE determination.

2.4.6. Quantification of Plasma Rat Mast Cell Protease II

Plasma samples obtained 1 h after the i.n. challenge were used to quantify rat mast cell protease II (RMCPII) concentration using a commercial ELISA kit (Bionova, Madrid, Spain) following the manufacturer's instructions. Results are shown as absorbance units obtained from all samples analyzed in the same ELISA plate compared to that produced by asthmatic rats, which are considered as 100%.

2.4.7. Antibody Quantification

Anti-OVA specific IgE antibody isotype in serum and BALF samples were quantified using an antibody-capture ELISA, as previously performed [18,41]. A pool of positive sera was used as standard in each plate. Serum samples were diluted 1/10, whereas BALF samples were processed undiluted. Results are shown as mean percentage compared to the asthmatic group, which are considered as 100%.

IgE concentration in BALF samples was quantified using a sandwich ELISA, as previously described [47]. Results are shown as mean percentage compared to the asthmatic group, which are considered as 100%.

2.4.8. Quantification of Cysteinyl Leukotriene (CysLT)

The concentration of CysLT in BALF was quantified using an Cysteinyl Leukotriene ELISA kit (Enzo Life Sciences Inc., New York, NY, SUA) following manufacturer's instructions, with a prior extraction of leukotrienes as previously described [47]. Results are shown as the percentage from that produced by asthmatic rats, which are considered as 100%.

2.4.9. Urine Polyphenols

Total phenolic content in urine samples was determined according to Folin–Ciocalteu's method adapted to a microplate. Briefly, 250 μL of Folin–Ciocalteau's reagent (Sigma-Aldrich) and 1.25 mL of 20% Na_2CO_3 solution were added to 500 μL of diluted urine. After 2 h at room temperature, the absorbance was measured at 765 nm. A standard curve prepared with gallic acid (Sigma-Aldrich) was used.

2.5. Statistical Analysis

The Statistical Package for the Social Sciences (SPSS v22.0, IBM, Chicago, IL, USA) was used for statistical analysis. Data were tested for homogeneity of variance and normality distribution using the Levene's and Shapiro–Wilk tests, respectively. When data was homogeneous and had a normal behavior, a conventional two-way ANOVA test followed by the post hoc Bonferroni and paired t-test were used in order to assess significance for independent and related samples, respectively. Otherwise, the nonparametric Kruskal–Wallis test followed by the post hoc Mann–Whitney U test were performed. Significant differences were established when $p < 0.05$ for the paired t-test, whereas for multiple comparisons, the p value was adjusted following Bonferroni correction [49].

To explore the functional correlation between the antibody levels, CysLT concentration, RMCPII production, body temperature and motor activity changes, Spearman' correlation analyses were performed in all samples grouped together.

3. Results

3.1. Cocoa Peruvian Populations Characterization

The content of total phenolics, total flavonoids, theobromine and caffeine differed between the five cocoa samples considered (Table 1). The population with the highest content in phenolics and flavonoids was CMC, followed by APC, and CCC in the third place ($p < 0.005$ CMC, APC and CCC vs. OC; $p < 0.005$ APC vs. CMC; $p < 0.005$ CCC vs. BPC). The OC and BPC cocoa pastes contained the lowest levels of both phenolic and flavonoid content. With regard to methylxanthine content (Table 1), the CMC cocoa paste exhibited the highest amounts of theobromine and caffeine ($p < 0.00001$) followed by the CCC population.

With regard to the antioxidant capacity (Table 1), the APC cocoa paste was the one with the highest capacity, followed by the CMC and CCC populations ($p < 0.01$), whereas the BPC had the lowest antioxidant capacity.

Table 1. Content of total polyphenols and flavonoids, methylxanthines and antioxidant capacity in the cocoa populations considered in the study. APC: "Amazonas Peru" cocoa; BPC: "Blanco de Piura" cocoa; CCC: "Chuncho" cocoa; CMC: "Criollo de Montaña" cocoa; OC: ordinary cocoa. Results are represented as mean ± standard error of the mean from three independent experiments. Values not sharing letters denote significant differences between populations ($p < 0.01$) while values sharing the same letter did not differ.

	OC	BPC	APC	CMC	CCC
Total phenolics (mg gallic acid equivalents/g)	24.11 ± 1.15 ^{ab}	21.07 ± 0.98 ^{a}	28.69 ± 0.80 ^{c}	30.44 ± 0.56 ^{d}	25.30 ± 0.38 ^{b}
Total flavonoids (mg catechin equivalents/g)	34.82 ± 1.01 ^{a}	36.59 ± 0.98 ^{a}	50.35 ± 0.95 ^{b}	56.62 ± 1.12 ^{c}	45.60 ± 1.26 ^{d}
Theobromine (mg/100 g)	560.75 ± 0.45 ^{a}	491.39 ± 0.93 ^{b}	564.71 ± 0.37 ^{a}	604.19 ± 0.28 ^{d}	573.36 ± 0.30 ^{e}
Theophylline (mg/100 g)	1.41 ± 0.00 ^{a}	1.37 ± 0.00 ^{b}	1.40 ± 0.00 ^{c}	1.39 ± 0.00 ^{d}	1.54 ± 0.00 ^{e}
Caffeine (mg/100 g)	236.35 ± 0.06 ^{a}	280.74 ± 0.56 ^{b}	275.10 ± 0.11^{c}	360.53 ± 0.27 ^{d}	324.55 ± 0.15 ^{e}
FRAP activity (μmol Fe^{2+}/g)	275.29 ± 12.28 ^{ab}	261.39 ± 14.15 ^{a}	344.31 ± 11.80 ^{c}	309.55 ± 6.7 ^{b}	308.32 ± 4.79 ^{b}
DPPH activity (μg TEAC/g)	25.36 ± 0.24 ^{ab}	19.62 ± 0.38 ^{c}	29.77 ± 0.70 ^{a}	25.38 ± 0.94 ^{bd}	24.39 ± 0.07 ^{d}

3.2. *In Vitro Effects of Cocoa Peruvian Populations*

An approach to study the immunomodulatory effects of each cocoa population in vitro on spleen lymphocytes and peritoneal macrophages was carried out (Figure 1).

3.2.1. Effects on Spleen Cells

The viability of rat spleen mononuclear cells was not affected by the cocoa addition, which was around 90% in all cases. In these conditions, CMC and CCC cocoa populations were able to prevent the secretion of TNF-α after LPS stimulation ($p < 0.01$) (Figure 1a). Moreover, all cocoa populations lowered spontaneous IgG production compared to the control ($p < 0.01$) (Figure 1b).

3.2.2. Effects on Macrophages

Cell viability of rat macrophages did not decrease after cocoa addition but the ROS production was significantly reduced in cells incubated with both APC and CMC populations throughout the studied period, as observed in the time course (Figure 1c) as well as in the AUC ($p < 0.01$ vs. OC and CCC samples) (Figure 1d). In addition, when analyzing the proportion of M1 (pro-inflammatory) and M2 (anti-inflammatory) macrophages after LPS stimulation, it was observed that three of the samples tested (BPC, APC and CMC) decreased the proportions of M1 cells compared to the stimulated cells with no cocoa (Figure 1e). Moreover, the BPC population was able to significantly increase the proportion of M2 macrophages ($p < 0.01$) (Figure 1e). Overall, although all cocoa samples tended to lower the M1/M2 ratio, only the decrease caused by the BPC and CMC cocoas reached statistical significance ($p < 0.01$) (Figure 1f).

Based on their polyphenol content, antioxidant activity and their in vitro effects on macrophages and splenocytes, the APC and CMC Peruvian cocoa populations were selected to be included in the in vivo study. The OC cocoa was also included to be used as a reference cocoa.

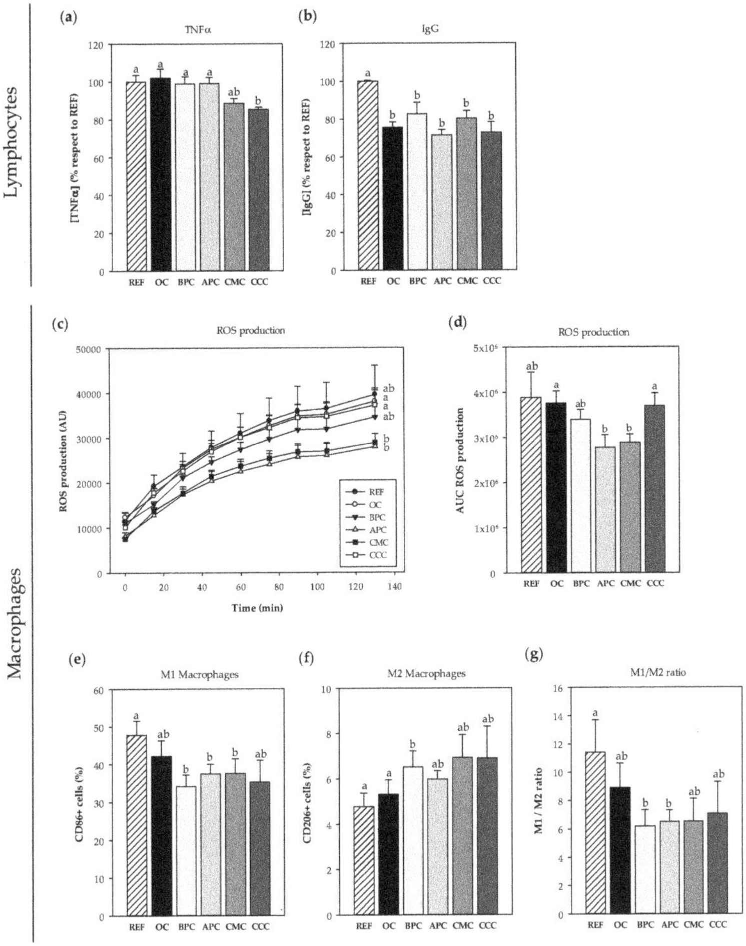

Figure 1. In vitro immunomodulatory effects of the five Peruvian cocoa samples. Effects of cocoa samples on TNF-α (**a**) and spontaneous IgG (**b**) production by splenocytes. Effects of cocoa samples on macrophages: oxygen reactive species (ROS) production (**c**) over time and as area under the curve (AUC) (**d**), and phenotype characterization: M1 (**e**), M2 (**f**) and M1/M2 ratio (**g**). REF: cells with no cocoa; OC: ordinary Peruvian cocoa; BPC: "Blanco de Piura"; APC: "Amazonas Peru" cocoa; CMC: "Criollo de Montaña" cocoa; CCC: "Chuncho" del Cusco. Results are represented as mean ± standard error of the mean ($N = 6$). Values not sharing letters denote significant differences between cocoa samples ($p < 0.01$), while values sharing the same letter did not differ.

3.3. *In Vivo Effects of Cocoa Peruvian Populations*

3.3.1. Body Weight and Food and Water Intake

At the beginning of the diets, 1 week before asthma induction, animals from all groups had a similar body weight (Table 2). Although the asthmatic animals' body weight was not significantly modified by either asthma induction or cocoa diets, it tended to be lower than that in the reference animals after the booster and it tended to be even lower with cocoa diets. These changes in body weight were not due to changes in either food or water consumption, which did not vary between diets and groups (Supplementary Tables S1 and S2).

Table 2. Body weight for all experimental groups throughout the study. REF: healthy reference group fed standard diet; A: asthmatic group fed standard diet; OC: asthmatic group fed 10% ordinary Peruvian cocoa; APC: asthmatic group fed 10% "Amazonas Peru" cocoa; CMC: asthmatic group fed 10% "Criollo de Montaña" cocoa. Results are represented as mean ± standard error of the mean ($N = 9$).

Time (days)	REF	A	OC	APC	CMC
−7	57.01 ± 5.04	57.52 ± 3.32	56.43 ± 3.59	56.46 ± 3.54	56.97 ± 2.63
−3	67.38 ± 5.37	68.60 ± 3.57	65.37 ± 3.86	64.84 ± 3.80	64.03 ± 2.90
0 [a]	75.96 ± 5.37	76.97 ± 3.55	72.44 ± 3.79	72.66 ± 4.03	71.91 ± 2.79
4	85.07 ± 4.80	84.76 ± 3.10	79.31 ± 3.65	79.21 ± 3.80	78.47 ± 2.59
7 [b]	91.57 ± 4.66	91.33 ± 3.57	82.73 ± 3.50	85.10 ± 3.94	84.48 ± 2.72
11	98.29 ± 5.52	93.70 ± 3.76	88.29 ± 3.34	89.31 ± 3.99	89.43 ± 2.67
14	104.03 ± 4.66	99.94 ± 3.74	94.59 ± 4.16	94.96 ± 4.28	94.07 ± 2.82
18	109.49 ± 4.88	105.90 ± 3.62	97.99 ± 4.68	97.98 ± 4.15	99.27 ± 2.92
21	115.80 ± 4.86	112.00 ± 3.72	104.26 ± 4.58	104.24 ± 4.49	105.77 ± 2.89
25	118.29 ± 5.58	115.68 ± 3.62	107.66 ± 4.93	108.10 ± 4.39	107.65 ± 2.52
28	123.73 ± 5.16	118.09 ± 2.46	109.46 ± 4.52	111.95 ± 4.03	112.97 ± 2.71

[a] day of sensitization; [b] day of booster.

3.3.2. Content of Polyphenols in Urine

Total polyphenol concentration was quantified in urine samples at the end of the study to verify the polyphenol absorption. As expected, urine samples from rats fed the cocoa diets showed higher values than those obtained from the reference and asthmatic rats fed the standard diet (Figure 2). Moreover, the highest polyphenol content was found in CMC- and APC-fed animals' urine samples, which was the population with the highest flavonoid content (Table 1).

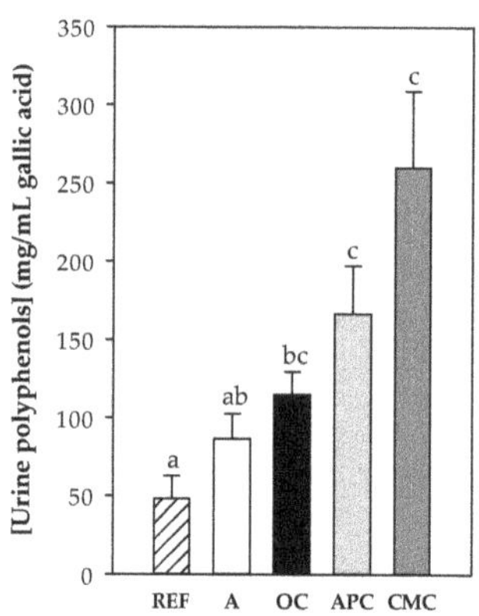

Figure 2. Polyphenol content (expressed as µg/mL gallic acid) in urine samples at the end of the study. REF: healthy reference group fed standard diet; A: asthmatic group fed standard diet OC: asthmatic animals fed with ordinary cocoa-enriched diet; APC: asthmatic animals fed with "Amazonas Peru" cocoa-enriched diet; CMC: asthmatic animals fed with "Criollo de Montaña" cocoa-enriched diet. Results are represented as mean ± standard error of the mean ($N = 9$). Values not sharing letters denote significant differences between groups ($p < 0.01$), while values sharing the same letter did not differ.

3.3.3. Changes in Body Temperature After i.n. Challenge

The body temperature (BT) was registered for each rat from 2 a.m. on the day of the i.n. challenge to 10 a.m. the day after (Figure 3). The mean value profile of BT registered every minute for this period with respect to the hour of the day (independently of the moment of challenge) is shown in Figure 3a, in which the darkness period is represented in gray. Also indicated is the period in which rats received the i.n. challenge and blood was collected. Table 3 summarizes the results of BT adjusted to the moment each rat was challenged and until 18 h after the i.n. challenge.

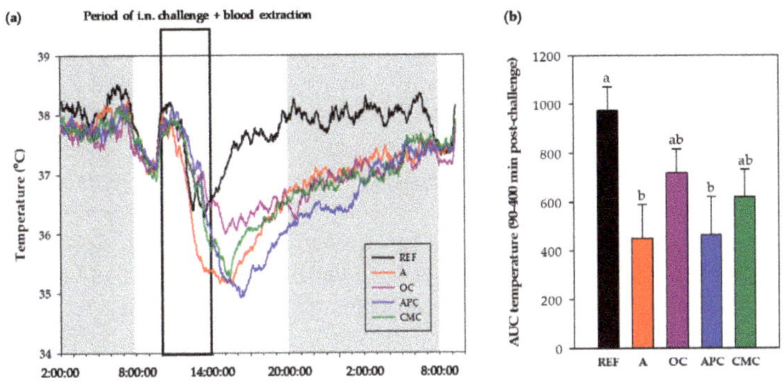

Figure 3. Changes in body temperature. Profile of body temperature from 2 a.m. before intranasal (i.n.) challenge and until 10 a.m the day after. Statistical differences not shown (**a**). Area under the curve (AUC) of body temperature (from 34 °C) in the period comprised between 90 and 400 min after the i.n. challenge (**b**). REF: healthy animals fed standard diet; A: asthmatic animals fed standard diet; OC: asthmatic animals fed with ordinary cocoa-enriched diet; APC: asthmatic animals fed with "Amazonas Peru" cocoa-enriched diet; CMC: asthmatic animals fed with "Criollo de Montaña" cocoa-enriched diet. Results are shown as mean (**a**) or as mean plus standard error of the mean (**b**) ($N = 9$). Values not sharing letters denote significant differences between groups in (**b**) ($p < 0.01$), while values sharing the same letter did not differ.

Table 3. Temperature means at 2 h intervals from 2 h before until 18 h after the intranasal challenge. REF: healthy animals fed standard diet; A: asthmatic animals fed standard diet; OC: asthmatic animals fed with ordinary cocoa-enriched diet; APC: asthmatic animals fed with "Amazonas Peru" cocoa-enriched diet; CMC: asthmatic animals fed with "Criollo de Montaña" cocoa-enriched diet. Results are represented as mean ± standard error of the mean ($N = 9$).

Time (h)	REF	A	OC	APC	CMC
−2 to 0	37.7 ± 0.11	37.6 ± 0.10	37.8 ± 0.18	37.9 ± 0.18	37.7 ± 0.13
0 to 2	37.1 ± 0.61	36.7 ± 0.29 *	37.1 ± 0.19	36.1 ± 0.44 *	36.2 ± 0.33 *
2 to 4	36.7 ± 0.51 *,a	35.1 ± 0.58 *,b	36.2 ± 0.24 *,a	35.3 ± 0.68 *,a	35.8 ± 0.32 *,a
4 to 6	37.5 ± 0.14 *,a	35.5 ± 0.41 *,b	36.2 ± 0.37 *,b	35.6 ± 0.44 *,b	36.1 ± 0.43 *,b
6 to 8	37.7 ± 0.11 a	36.2 ± 0.23 *,b	36.5 ± 0.44 *,b	36.1 ± 0.28 *,b	36.3 ± 0.39 *,b
8 to 10	37.9 ± 0.11 a	36.6 ± 0.17 *,b	36.4 ± 0.58 b	36.4 ± 0.36 *,b	36.7 ± 0.37 *,b
10 to 12	38.0 ± 0.09 a	36.9 ± 0.18 *,b	36.9 ± 0.50 a	36.5 ± 0.40 *,b	36.8 ± 0.39 b
12 to 14	37.9 ± 0.12 a	37.1 ± 0.18 *,b	37.0 ± 0.52 a	36.8 ± 0.30 *,b	36.9 ± 0.38 b
14 to 16	38.1 ± 0.09 a	37.2 ± 0.12 *,b	37.1 ± 0.60 a	37.2 ± 0.25 *,b	37.0 ± 0.40 b
16 to 18	38.0 ± 0.14 a	37.3 ± 0.18 *,b	37.2 ± 0.64 a	37.4 ± 0.24 a	37.4 ± 0.27 a

* represents statistical differences vs. their own basal values before challenge (−2 to 0) ($p < 0.05$). Values not sharing letters denote significant differences between groups during the same period of time ($p < 0.01$) while values sharing the same letter did not differ.

The i.n. challenge resulted in a reduced BT (Figure 3a and Table 3). However, the profile through the day shows that the REF group was able to recover the BT during the afternoon-evening and remained more or less constant the following night (Figure 3a). On the contrary, the asthmatic rats showed a slower BT recovery than the REF group and achieved the REF animals' BT at about 8 a.m. the day after the challenge.

Before challenging, the basal BT (mean value for 2 h before challenge) did not differ between groups (Table 3). After the i.n. challenge, all animals reduced their BT and reached the lowest values 2–4 h later. The REF group showed about 1 °C of BT reduction with respect to their own basal BT

($p < 0.05$), in the period between 2 and 6 h after the challenge. The asthmatic group fed a standard diet underwent a decrease of more than 2 °C of BT, which was already detected during the first 2 h after the i.n. challenge. Their BT remained significantly lower during all the period considered when comparing it either to their basal BT ($p < 0.05$) or to the REF animals' BT at the same time interval ($p < 0.01$).

The OC-fed asthmatic rats showed a BT reduction of about 1.5 °C during the interval of 2–8 h after the i.n. challenge. Their BT was significantly lower compared to their basal BT ($p < 0.05$) and to that in the REF group during the 4–10 h period of time after the challenge ($p < 0.05$). The APC-fed asthmatic rats also showed a reduction in BT (of about 2.5 °C) with respect to their basal values in the period comprised between 0 and 16 h after the challenge ($p < 0.01$) and with respect to the REF animals' BT in the period comprised between 4 and 16 h ($p < 0.01$). In the APC-fed group, the BT mean profile was the lowest (Figure 3a). Finally, the CMC-fed asthmatic rats also showed a decrease in BT (of about 2 °C) from immediately after the challenge until 10 h after, when it was compared to their basal values ($p < 0.05$), and during the period from 4 to 16 h after the challenge when compared to the REF animals' BT ($p < 0.05$).

The effects of diets on BT in the first hours after handling have been considered as AUC (Figure 3b). It can be observed that the BT was the lowest in the asthmatic rats fed either a standard diet or APC diet ($p < 0.01$)

3.3.4. Changes in Motor Activity After i.n. Challenge

The motor activity (MA) of each rat was registered from the day before the i.n. challenge to 18 h after. To summarize the results and avoid the period in which animals were handled, the MA in the dark period (activity period for rats) during the night before the i.n. challenge was compared to the MA during the night after the challenge (Figure 4). The challenge significantly decreased the MA in all the asthmatic groups compared to the REF group ($p < 0.05$). In particular, it resulted in a decrease of MA by about 60% in the asthmatic animals fed the standard diet with respect to their own basal MA ($p = 0.001$). However, the cocoa samples-fed asthmatic rats exhibited a lower reduction (of about 40–50%) in the MA in comparison to their MA before the challenge ($p < 0.05$). The MA of cocoa-fed animals did not differ from that of the asthmatic rats fed a standard diet.

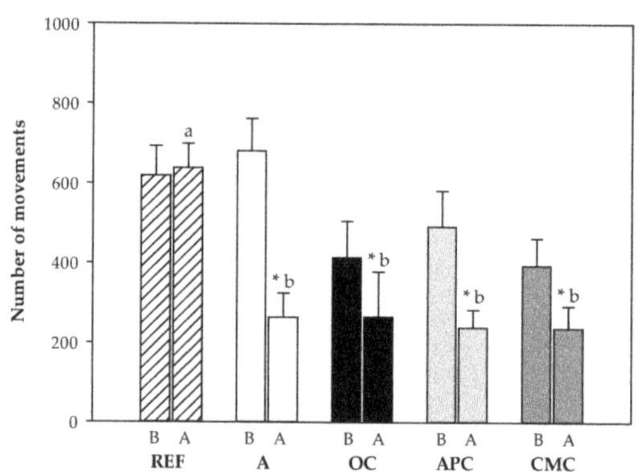

Figure 4. Number of movements during the active period (10 h of darkness) before (B) and after (A) intranasal challenge. REF: healthy animals fed standard diet; A: asthmatic animals fed standard diet; OC: asthmatic animals fed with ordinary cocoa-enriched diet; APC: asthmatic animals fed with "Amazonas Peru" cocoa-enriched diet; CMC: asthmatic animals fed with "Criollo de Montaña" cocoa-enriched diet. Results are shown as mean ± standard error ($N = 9$). * represents statistical differences from individual values before challenge (paired t-test) ($p < 0.05$). Values not sharing letters denote significant differences between groups ($p < 0.01$) while values sharing the same letter did not differ.

3.3.5. Changes in Body Weight After i.n. Challenge

One day after the i.n. challenge, all groups showed a significant decrease in body weight with respect to the body weight before the challenge (Figure 5). The asthmatic animals fed a standard diet decreased body weight by about 6%, whereas it decreased by up to 4% in cocoa-fed animals. Only the decrease in A group was significantly higher than that in the REF group ($p < 0.001$).

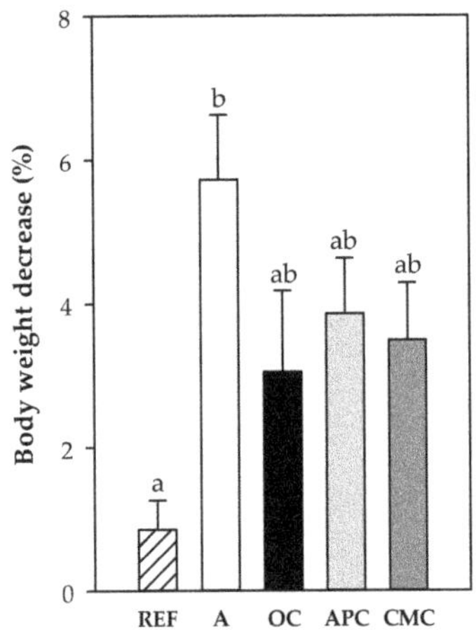

Figure 5. Body weight decrease (%) 24 h after the intranasal challenge with respect to the body weight before the challenge. REF: healthy reference group fed standard diet; A: asthmatic group fed standard diet; OC: asthmatic animals fed with ordinary cocoa-enriched diet; APC: asthmatic animals fed with "Amazonas Peru" cocoa-enriched diet; CMC: asthmatic animals fed with "Criollo de Montaña" cocoa-enriched diet. Results are represented as mean ± standard error of the mean ($N = 9$). Values not sharing letters denote significant differences between groups ($p < 0.01$), while values sharing the same letter did not differ.

3.3.6. Rat Mast Cell Protease II (RMCPII)

RMCPII concentration was determined in plasma samples collected 1 h after the i.n. challenge from all experimental groups. The asthmatic group (A group) showed higher RMCPII concentration than that in the REF animals (Figure 6) ($p < 0.01$). Both the APC and CMC diets tended to prevent such increase whereas the OC diet was able to maintain the RMCPII values similar to those found in the REF group.

3.3.7. IgE Antibodies

Specific anti-OVA IgE Ab concentration was quantified in serum and BALF samples obtained the day after the i.n. challenge. There were significant levels of specific IgE in the serum of asthmatic animals fed a standard diet ($p < 0.05$ vs. REF group) (Figure 7a). However, both OC and CMC diets were able to partially prevent such an increase, this reduction being significant only in the OC group, whose anti-OVA IgE levels were 50% lower than those in the A group ($p = 0.021$). In the BALF samples, both the OC and CMC groups showed a reduction of about 50% in the anti-OVA IgE levels in comparison to those observed in the asthmatic group ($p < 0.05$, Figure 7b). No significant changes were observed in the APC group in any of the sample types.

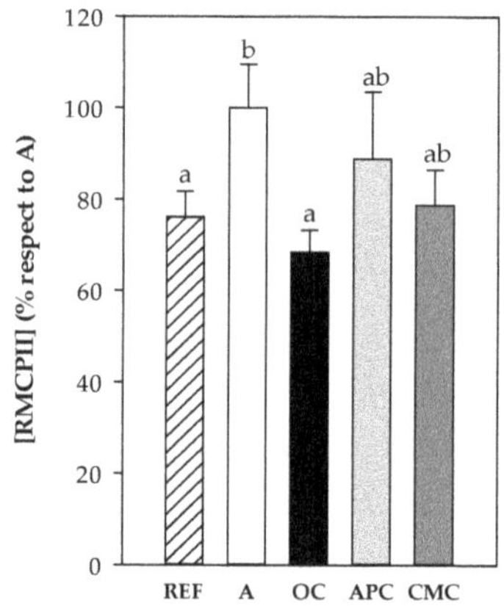

Figure 6. Rat mast cell protease II (RMCPII) in plasma obtained 1 h after the intranasal challenge. REF: healthy animals fed standard diet; A: asthmatic animals fed standard diet; OC: asthmatic animals fed with ordinary cocoa-enriched diet; APC: asthmatic animals fed with "Amazonas Peru" cocoa-enriched diet; CMC: asthmatic animals fed with "Criollo de Montaña" cocoa-enriched diet. Results are shown as mean ± standard error ($N = 9$). Results are expressed as mean ± standard error ($N = 9$) of absorbance units. Values not sharing letters denote significant differences between groups ($p < 0.01$), while values sharing the same letter did not differ.

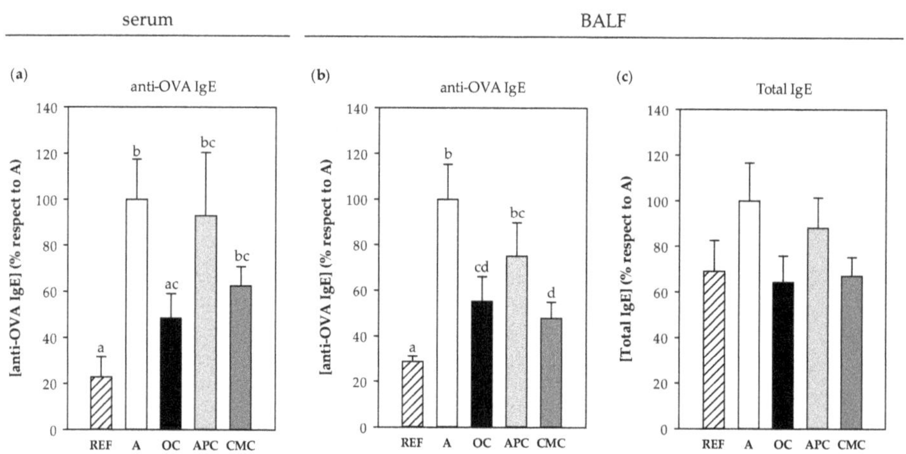

Figure 7. Anti-OVA IgE concentration in serum (**a**) and BALF (**b**) and total IgE content in BALF (**c**) obtained 24 h after the intranasal challenge. REF: healthy animals fed standard diet; A: asthmatic animals fed standard diet; OC: asthmatic animals fed with ordinary cocoa-enriched diet; APC: asthmatic animals fed with "Amazonas Peru" cocoa-enriched diet; CMC: asthmatic animals fed with "Criollo de Montaña" cocoa-enriched diet. Results are shown as mean ± standard error ($N = 9$). Values not sharing letters denote significant differences between groups ($p < 0.01$), while values sharing the same letter did not differ.

Total IgE content was also quantified in BALF, but in this case, although a similar profile to specific IgE was observed, no statistically significant differences were detected (Figure 7c).

There was a correlation between specific IgE levels in serum and BALF samples and the RMCPII content in plasma quantified in samples from 1 h after i.n. challenge ($r = 0.370$ $p = 0.029$ and $r = 0.440$

$p = 0.008$, respectively). However, no correlations were found between IgE levels and changes in MA and BT (data not shown).

3.3.8. Leukotrienes

Cys-LT concentration was determined in BALF samples collected 24 h after the i.n. challenge from all experimental groups. The asthmatic animals showed the highest values, but they did not reach statistical significance (Figure 8). This tendency to increase was partially reverted by all three diets containing cocoa samples.

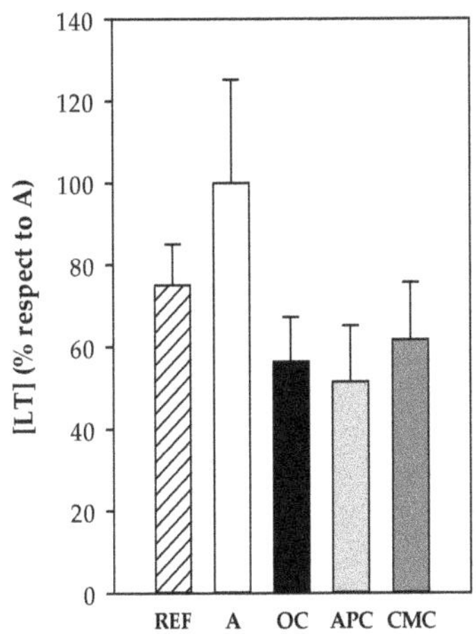

Figure 8. CysLT content in BALF samples obtained 24 h after the intranasal challenge. REF: healthy animals fed standard diet; A: asthmatic animals fed standard diet; OC: asthmatic animals fed with ordinary cocoa-enriched diet; APC: asthmatic animals fed with "Amazonas Peru" cocoa-enriched diet; CMC: asthmatic animals fed with "Criollo de Montaña" cocoa-enriched diet. Results are shown as mean ± standard error ($N = 9$).

No correlations were found between Cys-LT and RMCPII, IgE levels and changes in MA and BT.

4. Discussion

In the current study, the preventive potential of several Peruvian cocoa samples for allergic asthma has been approached. Cocoa has been considered beneficial for several chronic diseases [50–53], and it has been reported that such positive effects are mainly due to the composition of its bioactive compounds [54]. The content of such bioactive compounds in cocoa is largely dependent not only on the agricultural and postharvest practices and processing, but also on the cultivar and origins of cocoa [22,23]. In this sense, herein we have tested different Peruvian cocoa samples cultivated in various regions of the country, which could impact differently on human health due to their different bioactive compound contents. Therefore, the biochemical characterization along with the in vitro properties of four native Peruvian populations—all high quality cocoa used for making artisan chocolates—have been analyzed. At the same time, their properties were compared to an ordinary Peruvian cocoa but from an Ecuadorean clone [55]. It has been observed that the content of polyphenols, flavonoids and methylxanthines was distinct between the populations tested, thus confirming that environmental and genetic factors significantly influence their content, as previously reported [23]. The "Criollo de Montaña" (CMC) population was the richest in flavonoid (and polyphenol) concentration, as well as in methylxanthines content. This population grows in the Junín region (Satipo city) at 480 m.a.s.l. where the climate is predominantly hot and humid with a fluctuating temperature of between 18 and 35 °C. The second population with a higher content in flavonoids/polyphenols was the "Amazonas Peru"

(APC) cocoa, which also exhibited the highest antioxidant activity, followed by the CMC population. The antioxidant properties of both the APC and CMC populations were also observed in the in vitro ROS production by macrophages, which was significantly decreased by these populations. Therefore, the higher the content of polyphenols these populations have, the stronger the antioxidant effects they exert, as would be expected.

The CCC ("Chuncho") population exerted in vitro inhibitory effects on the inflammatory mediators' secretion, such as TNF-α, by LPS-stimulated splenocytes, in agreement with previous in vitro studies using a conventional cocoa and particular cocoa flavonoids [13]. The anti-inflammatory properties of cocoa samples were also evidenced when analyzing the phenotype of in vitro LPS-stimulated macrophages. In this sense, the results obtained show that the native populations from BPC ("Blanco de Piura") and APC had the potential to downregulate pro-inflammatory macrophage proportion while upregulating those from anti-inflammatory cells. Similar effects have been reported in the THP1 cell line cultured with a cocoa phenolic extract, which was able to induce a phenotypic switch in polarized macrophages in favor of the anti-inflammatory one [56].

On the other hand, all native populations tested in the present study, and also the ordinary cocoa, were able to downregulate the in vitro ability to produce immunoglobulins. This effect is in line with in vivo studies reporting a reduction of not only plasma IgG concentration, but also IgM and IgA in cocoa-fed animals [57,58]. This downregulation of immunoglobulin secretion seems to be due to an inhibitory B cell differentiation caused by the decrease in Th2 cytokines [59]. Overall, given that allergy is a Th2-associated response, these results encouraged us to evaluate the effects of cocoa in the present allergic asthma rat model.

As previously mentioned, based on their polyphenol content, their antioxidant activity and their in vitro effects, two cocoa populations ("Amazonas Peru" and "Criollo de Montaña") were selected to be used in vivo in an allergic asthma model. In fact, previous studies have demonstrated the antiallergic properties of cocoa in preclinical studies, in models of systemic disease [18,19] and oral sensitization [20], and in observational studies considering cocoa consumption habits in young people with allergy [21].

Herein, body temperature and motor activity variables have been used to assess the anaphylactic response after i.n. challenge as previously set up [47] and also as used earlier in a model of oral allergy [19,48]. In addition, the decrease in body weight was also considered as a variable to assess anaphylaxis. Our results evidenced that the anaphylactic response was accompanied by a reduction in motor activity, which was not modified by any cocoa diet, although this decrease in cocoa-fed animals was relatively lower. Previous studies had reported a reduction in motor activity in animals fed with cocoa and receiving an oral challenge [19]. Nevertheless, when changes in body temperature were considered, some differences appeared. The asthmatic group fed a standard diet showed a significant decrease in body temperature that appeared earlier than the other groups and was more long-lasting, while the OC- and CMC-fed asthmatic animals maintained their body temperature similar to that found in the REF group for a longer time. This partial prevention in body temperature decrease by OC and CMC cocoa samples does not match with the vasodilator properties reported for cocoa [52] and the effects observed on a food allergy model [19]. On the other hand, the protective effects of cocoa samples were also observed in the decrease in body weight after i.n. challenge. Furthermore, the increase in plasma RMCPII released by activated mucosal mast cells [60] was partially prevented by both OC and CMC cocoas. In summary, the CMC population appears to be the native Peruvian population with the most potential to prevent anaphylactic response after i.n. challenge, and this ability is shared and even higher with the OC clone.

Diets containing Peruvian cocoa samples were able to differently influence the synthesis of specific IgE antibodies. Again, the CMC cocoa paste was the most effective native population at decreasing both the serum and bronchoalveolar lavage fluid IgE concentration, but its effects did not differ from the non-native OC cocoa. The protective effect of cocoa on IgE synthesis found here has already been described in other allergy models [8,18], and it confirms the immunomodulatory effects of a 10% cocoa diet in preclinical studies. Similarly, the anti-asthmatic effect of some extracts rich in flavonoids has also been demonstrated in an asthmatic animal model [61,62] and also in a limited number of clinical trials [63]. In agreement with these effects, the consumption of polyphenol-containing apple extracts was associated with an alleviation of some allergic symptoms, such as runny nose and nasal congestion in subjects suffering from allergic rhinitis [64].

Leukotrienes are other inflammatory molecules released when allergen binds to the mast cell-coupled IgE in allergy, which are primarily responsible for the bronchoconstriction during asthma attacks [60]. In our model, its concentration tended to increase in the asthmatic animals fed a standard diet, but all cocoa diets tended to prevent it and showed similar content as the reference group. All three cocoa samples tested in the present study contain theophylline, a methylxanthine naturally present in small amounts (1.40 mg/100 g) in cocoa beans that came out as a clinical treatment for asthma and other respiratory diseases once its bronchodilator effects had been identified [65]. In fact, it has been found that theophylline can act as suppressant of leukotriene production [66].

From the results obtained in in vivo experiments, it can be hypothesized that the main mechanism by which cocoa diets can exert a protective effect against allergic asthma response is by attenuating the synthesis of Th2-related antibodies. In particular, the cocoa diets have shown their anti-allergic potential mainly reducing the anti-OVA IgE levels. Therefore, the lower the IgE production is, the lower the amount of this antibody which can bind to mast cells in the airway. Consequently, when a new allergen contact is produced, the allergen may bind to only a little mast cell-bound IgE. For this reason, the release of mediators such as proteases and leukotrienes in the bronchoalveolar compartment is low and a weak anaphylactic response can be observed. The mechanisms produced by the effective cocoa populations must be addressed in further studies.

Polyphenol content in urine was also determined for approaching flavonoid absorption. The animals fed the CMC diet, which was elaborated with the native Peruvian population that exerted the most protective effects, had the highest polyphenol concentration in urine. Nevertheless, the OC diet also showed a protective response against anaphylactic response and showed the lowest urine polyphenol concentration. Therefore, it is not just the flavonoid content that seems to be important in playing a protective role in this model, but it may also be the type of flavonoids (for example monomers or polymeric forms) present in each population. Anyway, as CMC has the most immunomodulatory properties of the Peruvian "fine aroma" cocoa populations evaluated in this study, further experiments focused on particular cocoa flavonoids or other cocoa bioactive compounds in this population must shed light on this issue.

The cocoa used as raw material for chocolate is a source of differences in terms of its sensorial quality, as can be observed between ordinary cocoa and fine aroma cocoa. Ordinary cocoa has the most widespread use in the industry, being used for common chocolates, while fine aroma cocoa is used for fine or artisan chocolates. The result of this study shows that the origin of the cocoa populations from which the chocolate pastes are obtained determines differences that are expressed at the biochemical level and in their bioactivity. Chocolate pastes, made with cocoa taken from different populations in the same country (Peru), are biochemically differentiable and have different bioactivity, as demonstrated in the present study.

5. Conclusions

Overall, it has been shown that particular populations of Peruvian cocoa exhibit a protective effect on a rat model of acute allergic asthma response. This effect can be observed by means of a partial protection against anaphylactic response and, above all, in the synthesis of IgE and the release of mast cell protease. These results show that the origin of cocoa is relevant and should be taken into account and declared in these types of studies, and probably also when it comes to be used in the making of dark chocolate or as nutraceuticals.

Supplementary Materials: The following are available online at http://www.mdpi.com/2072-6643/12/8/2301/s1, Table S1: Food intake for all experimental groups throughout the study. Table S2: Water consumption for all experimental groups throughout the study.

Author Contributions: Conceptualization, F.J.P.-C., I.B., S.P.-S., M.C. and M.M.-C.; methodology, M.P., T.C., À.F. and M.M.-C.; formal analysis, M.P., F.J.P.-C., T.C., M.C. and M.M.-C.; writing—original draft preparation, F.J.P.-C., M.C. and M.M.-C.; writing—review and editing, M.P, T.C., À.F., I.B., S.P.-S.; supervision, M.C. and M.M.-C.; funding acquisition, F.J.P.-C., I.B, S.P.-S. and M.C. All authors have read and agreed to the published version of the manuscript.

Funding: This research was funded by the National Fund for Scientific, Technology and Technological Innovation Development (FONDECYT) of the National Council of Science, Technology and Technological Innovation (CONCYTEC), Contract 137-2017-FONDECYT, and Universidad Científica del Sur. Open access fees were supported by the Universidad Cientifica del Sur (Lima, Peru).

Acknowledgments: The authors would like to thank Mariona Camps and Mariano Nicola Llorente for their technical assistance in the in vitro studies and Nerea Moreno, Andrea Barranco and Anna Cutrina for their technical assistance in the in vivo studies.

Conflicts of Interest: The authors declare no conflict of interest.

References

1. World Health Organization. *Prevention of Allergy and Allergic Asthma*; World Health Organization: Geneva, Switzerland, 2003.
2. Papi, A.; Brightling, C.; Pedersen, S.E.; Reddel, H.K. Asthma. *Lancet* **2018**, *391*, 783–800. [CrossRef]
3. Mishra, V.; Banga, J.; Silveyra, P. Oxidative stress and cellular pathways of asthma and inflammation: Therapeutic strategies and pharmacological targets HHS Public Access. *Pharmacol. Ther.* **2018**, *181*, 169–182. [CrossRef] [PubMed]
4. Curtis, J.L. Cell-mediated Adaptive Immune Defense of the Lungs. *Proc. Am. Thorac. Soc.* **2005**, *2*, 412–416. [CrossRef] [PubMed]
5. James, S.L.; Abate, D.; Abate, K.H.; Abay, S.M.; Abbafati, C.; Abbasi, N.; Abbastabar, H.; Abd-Allah, F.; Abdela, J.; Abdelalim, A.; et al. Global, regional, and national incidence, prevalence, and years lived with disability for 354 diseases and injuries for 195 countries and territories, 1990–2017: A systematic analysis for the Global Burden of Disease Study 2017. *Lancet* **2018**, *392*, 1789–1858. [CrossRef]
6. Solé, D.; Aranda, C.S.; Wandalsen, G.F. Asthma: Epidemiology of disease control in Latin America—Short review. *Asthma Res. Pract.* **2017**, *3*, 4–9. [CrossRef]
7. Mallol, J.; Solé, D.; Baeza-Bacab, M.; Aguirre-Camposano, V.; Soto-Quiros, M.; Baena-Cagnani, C. Regional variation in asthma symptom prevalence in Latin American children. *J. Asthma* **2010**, *47*, 644–650. [CrossRef]
8. Castell, M.; Pérez-Cano, F.J.; Abril-Gil, M.; Franch, À. Flavonoids on Allergy. *Curr. Pharm. Des.* **2014**, *20*, 973–987. [CrossRef]
9. Shahidi, F.; Ambigaipalan, P. Phenolics and polyphenolics in foods, beverages and spices: Antioxidant activity and health effects. *J. Funct. Foods* **2015**, 1–78. [CrossRef]
10. Joskova, M.; Sadlonova, V.; Nosalova, G.; Novakova, E.; Franova, S. Polyphenols and their components in experimental allergic asthma. In *Respiratory Regulation—The Molecular Approach*; Pokorski, M., Ed.; Springer: Dordrecht, The Netherlands, 2013; pp. 91–98.
11. Ayala-Mata, F.; Barrera-Mendoza, C.C.; Cortés-Rojo, C.; Montoya-Pérez, R.d.C.; García-Pérez, M.E.; Rodríguez-Orozco, A.R. Antioxidants in asthma: Polyphenols. *Med. Interna Mex.* **2019**, *35*, 223–234.
12. Ramiro, E.; Franch, À.; Castellote, C.; Andrés-Lacueva, C.; Izquierdo-Pulido, M.; Castell, M. Effect of Theobroma cacao flavonoids on immune activation of a lymphoid cell line. *Br. J. Nutr.* **2005**, *93*, 859–866. [CrossRef]
13. Ramiro, E.; Franch, A.; Castellote, C.; Pérez-Cano, F.; Permanyer, J.; Izquierdo-Pulido, M.; Castell, M. Flavonoids from Theobroma cacao down-regulate inflammatory mediators. *J. Agric. Food Chem.* **2005**, *53*, 8506–8511. [CrossRef]
14. Ramiro-Puig, E.; Castell, M. Cocoa: Antioxidant and immunomodulator. *Br. J. Nutr.* **2009**, *101*, 931–940. [CrossRef] [PubMed]

15. Castell, M.; Franch, A.; Castellote, C. Effect of a diet rich in cocoa flavonoids on experimental acute inflammation. In *Flavonoids: Biosynthesis, Biological Effects and Dietary Sources*; Keller, R.B., Ed.; Nova Science Publishers Inc.: New York, NY, USA, 2009; Volume 6, pp. 213–229. ISBN 9781607416227.
16. Massot-Cladera, M.; Franch, À.; Pérez-Cano, F.J.; Castell, M. Cocoa and cocoa fibre differentially modulate IgA and IgM production at mucosal sites. *Br. J. Nutr.* **2016**, *115*, 1539–1546. [CrossRef] [PubMed]
17. Camps-Bossacoma, M.; Massot-Cladera, M.; Abril-Gil, M.; Franch, A.; Pérez-Cano, F.J.; Castell, M. Cocoa Diet and Antibody Immune Response in Preclinical Studies. *Front. Nutr.* **2017**, *4*, 1–14. [CrossRef] [PubMed]
18. Abril-Gil, M.; Massot-Cladera, M.; Pérez-Cano, F.J.; Castellote, C.; Franch, A.; Castell, M. A diet enriched with cocoa prevents IgE synthesis in a rat allergy model. *Pharmacol. Res.* **2012**, *65*, 603–608. [CrossRef] [PubMed]
19. Abril-Gil, M.; Pérez-Cano, F.J.; Franch, À.; Castell, M. Effect of a cocoa-enriched diet on immune response and anaphylaxis in a food allergy model in Brown Norway rats. *J. Nutr. Biochem.* **2016**, *27*, 317–326. [CrossRef] [PubMed]
20. Camps-Bossacoma, M.; Abril-Gil, M.; Saldaña-Ruiz, S.; Franch, À.; Pérez-Cano, F.J.; Castell, M. Cocoa diet prevents antibody synthesis and modifies lymph node composition and functionality in a rat oral sensitization model. *Nutrients* **2016**, *8*, 242. [CrossRef] [PubMed]
21. Rodríguez-Lagunas, M.J.; Vicente, F.; Pereira, P.; Castell, M.; Pérez-Cano, F.J. Relationship between cocoa intake and healthy status: A pilot study in university students. *Molecules* **2019**, *24*, 812. [CrossRef]
22. Febrianto, N.A.; Zhu, F. Diversity in Composition of Bioactive Compounds among 26 Cocoa Genotypes. *J. Agric. Food Chem.* **2019**, *67*, 9501–9509. [CrossRef]
23. Oracz, J.; Zyzelewicz, D.; Nebesny, E. The Content of Polyphenolic Compounds in Cocoa Beans (Theobroma cacao L.), Depending on Variety, Growing Region, and Processing Operations: A Review. *Crit. Rev. Food Sci. Nutr.* **2015**, *55*, 1176–1192. [CrossRef]
24. Bertoldi, D.; Barbero, A.; Camin, F.; Caligiani, A.; Larcher, R. Multielemental fingerprinting and geographic traceability of Theobroma cacao beans and cocoa products. *Food Control* **2016**, *65*, 46–53. [CrossRef]
25. Caligiani, A.; Marseglia, A.; Prandi, B.; Palla, G.; Sforza, S. Influence of fermentation level and geographical origin on cocoa bean oligopeptide pattern. *Food Chem.* **2016**, *211*, 431–439. [CrossRef] [PubMed]
26. D'Souza, R.N.; Grimbs, S.; Behrends, B.; Bernaert, H.; Ullrich, M.S.; Kuhnert, N. Origin-based polyphenolic fingerprinting of Theobroma cacao in unfermented and fermented beans. *Food Res. Int.* **2017**, *99*, 550–559. [CrossRef] [PubMed]
27. Thomas, E.; van Zonneveld, M.; Loo, J.; Hodgkin, T.; Galluzzi, G.; van Etten, J. Present Spatial Diversity Patterns of Theobroma cacao L. in the Neotropics Reflect Genetic Differentiation in Pleistocene Refugia Followed by Human-Influenced Dispersal. *PLoS ONE* **2012**, *7*, e47676. [CrossRef] [PubMed]
28. Zarrillo, S.; Gaikwad, N.; Lanaud, C.; Powis, T.; Viot, C.; Lesur, I.; Fouet, O.; Argout, X.; Guichoux, E.; Salin, F.; et al. The use and domestication of Theobroma cacao during the mid-Holocene in the upper Amazon. *Nat. Ecol. Evol.* **2018**, *2*, 1879–1888. [CrossRef]
29. United Nations Conference on Trade and Development. *International Cocoa Agreement*; United Nations: Geneva, Switzerland, 2010; ISBN 9780857091253.
30. Eskes, A.B.; Rodriguez, C.A.C.; Cruz Condori, D.; Seguine, E.; Garcia Carrion, L.F.; Lachenaud, P. Large genetic diversity for fine-flavor traits unveiled in cacao (Theobroma cacao L.) with special attention to the native Chunco variety in Cusco, Peru. *Agrotrópica (Itabuna)* **2018**, *30*, 157–174. [CrossRef]
31. García Carrión, L.F. *Catálogo de Cultivares de cacao del Perú*; Ministerio de Agricultura y Riego, Dirección General de Competitividad Agraria: Lima, Peru, 2010.
32. Secretaria del Convenio sobre la Diversidad Biológica. *Protocolo de Nagoya Sobre Acceso a los Recursos Genéticos y Participación Justa y Equitativa en los Beneficios que se Deriven de su Utilización al Convenio Sobre la Diversidad Biológica*; Secretaría del Convenio sobre la Diversidad Biológica: Montreal, QC, Canada, 2011; ISBN 92-9225-310-7.
33. Laird, S.; Wynberg, R.; Rourke, M.; Humphries, F.; Muller, M.R.; Lawson, C. Rethink the expansion of access and benefit sharing. *Science* **2020**, *367*, 1200–1202. [CrossRef]
34. Pedan, V.; Fischer, N.; Rohn, S. An online NP-HPLC-DPPH method for the determination of the antioxidant activity of condensed polyphenols in cocoa. *Food Res. Int.* **2016**, *89*, 890–900. [CrossRef]
35. Shetty, K.; Curtis, O.F.; Levin, R.E.; Witkowsky, R.; Ang, W. Prevention of Vitrification Aßociated with in vitro Shoot Culture of Oregano. (Origanum vulgare) by Pseudomonas spp. *J. Plant Physiol.* **1995**, *147*, 447–451. [CrossRef]

36. Xu, G.; Ye, X.; Chen, J.; Liu, D. Effect of heat treatment on the phenolic compounds and antioxidant capacity of citrus peel extract. *J. Agric. Food Chem.* **2007**, *55*, 330–335. [CrossRef]
37. Brand-Williams, W.; Cuvelier, M.E.; Berset, C. Use of a free radical method to evaluate antioxidant activity. *LWT Food Sci. Technol.* **1995**, *28*, 25–30. [CrossRef]
38. Marinova, D.; Ribarova, F.; Atanassova, M. Total Phenolics and Total Flavonoids in Bulgarian Fruits and Vegetables. *J. Univ. Chem. Technol. Metall.* **2005**, *40*, 255–260.
39. Srdjenovic, B.; Djordjevic-Milic, V.; Grujic, N.; Injac, R.; Lepojevic, Z. Simultaneous HPLC determination of caffeine, theobromine, and theophylline in food, drinks, and herbal products. *J. Chromatogr. Sci.* **2008**, *46*, 144–149. [CrossRef] [PubMed]
40. Ramos-Romero, S.; Pérez-Cano, F.J.; Ramiro-Puig, E.; Franch, A.; Castell, M. Cocoa intake attenuates oxidative stress associated with rat adjuvant arthritis. *Pharmacol. Res.* **2012**, *66*, 207–212. [CrossRef]
41. Abril-Gil, M.; Garcia-Just, A.; Pérez-Cano, F.J.; Franch, À.; Castell, M. Development and characterization of an effective food allergy model in Brown Norway rats. *PLoS ONE* **2015**, *10*, e0125314. [CrossRef]
42. Estruel-Amades, S.; Massot-Cladera, M.; Garcia-Cerdà, P.; Pérez-Cano, F.J.; Franch, À.; Castell, M.; Camps-Bossacoma, M. Protective effect of hesperidin on the oxidative stress induced by an exhausting exercise in intensively trained rats. *Nutrients* **2019**, *11*, 783. [CrossRef]
43. Wu, L.; Magaz, A.; Maughan, E.; Oliver, N.; Darbyshire, A.; Loizidou, M.; Emberton, M.; Birchall, M.; Song, W. Cellular responses to thermoresponsive stiffness memory elastomer nanohybrid scaffolds by 3D-TIPS. *Acta Biomater.* **2019**, *85*, 157–171. [CrossRef]
44. Yu, T.; Gao, M.; Yang, P.; Liu, D.; Wang, D.; Song, F.; Zhang, X.; Liu, Y. Insulin promotes macrophage phenotype transition through PI3K/Akt and PPAR-γ signaling during diabetic wound healing. *J. Cell. Physiol.* **2019**, *234*, 4217–4231. [CrossRef]
45. Rodríguez-Palmero, M.; Franch, À.; Castell, M.; Pelegrí, C.; Pérez-Cano, F.J.; Kleinschnitz, C.; Stoll, G.; Hünig, T.; Castellote, C. Effective treatment of adjuvant arthritis with a stimulatory CD28-specific monoclonal antibody. *J. Rheumatol.* **2006**, *33*, 110–118.
46. Ramiro-Puig, E.; Pérez-Cano, F.J.; Ramírez-Santana, C.; Castellote, C.; Izquierdo-Pulido, M.; Permanyer, J.; Franch, A.; Castell, M. Spleen lymphocyte function modulated by a cocoa-enriched diet. *Clin. Exp. Immunol.* **2007**, *149*, 535–542. [CrossRef]
47. Périz, M.; Pérez-cano, F.J.; Rodríguez-lagunas, M.J.; Cambras, T. Development and characterization of an allergic asthma rat model for interventional studies. *Int. J. Mol. Sci.* **2020**, *21*, 3841. [CrossRef]
48. Abril-Gil, M.; Garcia-Just, A.; Cambras, T.; Pérez-Cano, F.J.; Castellote, C.; Franch, À.; Castell, M. Motor activity as an unbiased variable to assess anaphylaxis in allergic rats. *Exp. Biol. Med.* **2015**, *240*, 1373–1377. [CrossRef] [PubMed]
49. Jafari, M.; Ansari-Pour, N. Why, when and how to adjust your P values? *Cell J.* **2019**, *20*, 604–607.
50. Ellam, S.; Williamson, G. Cocoa and Human Health. *Annu. Rev. Nutr.* **2013**, *33*, 105–128. [CrossRef] [PubMed]
51. Ferri, C.; Desideri, G.; Ferri, L.; Proietti, I.; Di Agostino, S.; Martella, L.; Mai, F.; Di Giosia, P.; Grassi, D. Cocoa, Blood Pressure, and Cardiovascular Health. *J. Agric. Food Chem.* **2015**, *63*, 9901–9909. [CrossRef]
52. Desch, S.; Schmidt, J.; Kobler, D.; Sonnabend, M.; Eitel, I.; Sareban, M.; Rahimi, K.; Schuler, G.; Thiele, H. Effect of Cocoa Products on blood pressure: Systematic review and meta-analysis. *Am. J. Hypertens.* **2010**, *23*, 97–103. [CrossRef]
53. Seem, S.A.; Yuan, Y.V.; Tou, J.C. Chocolate and chocolate constituents influence bone health and osteoporosis risk. *Nutrition* **2019**, *65*, 74–84. [CrossRef] [PubMed]
54. Andújar, I.; Recio, M.C.; Giner, R.M.; Ríos, J.L. Cocoa polyphenols and their potential benefits for human health. *Oxid. Med. Cell. Longev.* **2012**, *2012*, 1–23. [CrossRef] [PubMed]
55. Pallares Pallares, A.; Estupiñán A, M.R.; Perea Villamil, J.A.; López Giraldo, L.J. Impacto de la fermentación y secado sobre el contenido de polifenoles y capacidad antioxidante del clon de cacao CCN-51. *Rev. ION Bucaramanga* **2016**, *29*, 7–21.
56. Dugo, L.; Belluomo, M.G.; Fanali, C.; Russo, M.; Cacciola, F.; Maccarrone, M.; Sardanelli, A.M. Effect of Cocoa Polyphenolic Extract on Macrophage Polarization from Proinflammatory M1 to Anti-Inflammatory M2 State. *Oxidative Med. Cell. Longev.* **2017**, *2017*, 1–11. [CrossRef]

57. Ramiro-Puig, E.; Pérez-Cano, F.J.; Ramos-Romero, S.; Pérez-Berezo, T.; Castellote, C.; Permanyer, J.; Franch, À.; Izquierdo-Pulido, M.; Castell, M. Intestinal immune system of young rats influenced by cocoa-enriched diet. *J. Nutr. Biochem.* **2008**, *19*, 555–565. [CrossRef] [PubMed]
58. Massot-Cladera, M.; Abril-Gil, M.; Torres, S.; Franch, A.; Castell, M.; Pérez-Cano, F.J. Impact of cocoa polyphenol extracts on the immune system and microbiota in two strains of young rats. *Br. J. Nutr.* **2014**, *112*, 1944–1954. [CrossRef] [PubMed]
59. Pérez-Berezo, T.; Ramiro-Puig, E.; Pérez-Cano, F.J.; Castellote, C.; Permanyer, J.; Franch, A.; Castell, M. Influence of a cocoa-enriched diet on specific immune response in ovalbumin-sensitized rats. *Mol. Nutr. Food Res.* **2009**, *53*, 389–397. [CrossRef]
60. Gibson, S.; Mackeller, A.; Newlands, G.F.J.; Miller, H.R.P. Phenotypic expression of mast cell granule proteinases. Distribution of mast cell proteinases I and II in the rat digestive system. *Immunology* **1987**, *62*, 621–627. [PubMed]
61. Das, M.; Ram, A.; Ghosh, B. Luteolin alleviates bronchoconstriction and airway hyperreactivity in ovalbumin sensitized mice. *Inflamm. Res.* **2003**, *52*, 101–106.
62. Choi, J.R.; Lee, C.M.; Jung, I.D.; Lee, J.S.; Jeong, Y.I.; Chang, J.H.; Park, H.J.; Choi, I.W.; Kim, J.S.; Shin, Y.K.; et al. Apigenin protects ovalbumin-induced asthma through the regulation of GATA-3 gene. *Int. Immunopharmacol.* **2009**, *9*, 918–924. [CrossRef]
63. Tanaka, T.; Takahashi, R. Flavonoids and asthma. *Nutrients* **2013**, *5*, 2128–2143. [CrossRef]
64. Enomoto, T.; Nagasako-Akazome, Y.; Kanda, T.; Ikeda, M.; Dake, Y. Clinical effects of apple polyphenols on persistent allergic rhinitis: A randomized double-blind placebo-controlled parallel arm study. *J. Investig. Allergol. Clin. Immunol.* **2006**, *16*, 283–289.
65. Cooling, D.S. Theophylline toxicity. *J. Emerg. Med.* **1993**, *11*, 415–425. [CrossRef]
66. Kraft, M.; Torvik, J.A.; Trudeau, J.B.; Wenzel, S.E.; Martin, R.J. Theophylline: Potential antiinflammatory effects in nocturnal asthma. *J. Allergy Clin. Immunol.* **1996**, *97*, 1242–1246. [CrossRef]

nutrients

MDPI

Article

The Association of Dietary Macronutrients with Lung Function in Healthy Adults Using the Ansan-Ansung Cohort Study

Sang-Ah Lee [1,2,*], Pankaj Joshi [1], Yeonjin Kim [1], Daehee Kang [3] and Woo Jin Kim [4,5,*]

1 Department of Preventive Medicine, Kangwon National University School of Medicine, 1, Kangwondaehak-gil, Chuncheon-si, Gangwon-do 24341, Korea; daccpankaj@gmail.com (P.J.); bbkeye1229@naver.com (Y.K.)

2 Division of Epidemiology, Vanderbilt University Medical Center, 1211 Medical Center Dr, Nashuville, TN 37232, USA

3 Department of Preventive Medicine, Seoul National University College of Medicine, 103, Daehak-ro, Jongno-gu, Seoul 03080, Korea; dhkang@snu.ac.kr

4 Department of Internal Medicine, School of Medicine, Kangwon National University, 1, Kangwondaehak-gil, Chuncheon-si, Gangwon-do 24341, Korea

5 Environmental Health Center, 156, Baengnyeong-ro, Chuncheon-si, Gangwon-do 24289, Korea

* Correspondence: sangahlee@kangwon.ac.kr (S.-A.L.); plumo2@kangwon.ac.kr (W.J.K.); Tel.: +84-33-250-8871 (S.-A.L.)

Received: 24 July 2020; Accepted: 31 August 2020; Published: 3 September 2020

Abstract: This study is aimed to examine the association between macronutrient intake and lung function in healthy adults (n = 5880) using the Ansan-Ansung cohort study. To identify the index of lung function, we used the percentage difference of predicted Forced Expiratory Volume (%FEV_{1_diff}) between baseline and follow-up. Based on the median %FEV_{1_diff}, subjects were classified by two groups as "decreased vs. unchanged/improved". The dietary macronutrients were estimated and validated using the food-frequency questionnaire. Multiple logistic regression models were used to evaluate the association after adjusting for confounders. Advanced analysis examined the association after stratifying by age and obesity. The average of %FEV_1 is 114.1 and 112.5 at baseline and follow-up, respectively. The positive association of protein and fiber intake with lung function was observed in men. Low fat and high carbohydrate intake decreased the lung function in women only. After stratification by age, the association of protein, fat, and carbohydrate intake with lung function was observed in young men and old women only. Otherwise, the association of protein and fiber with lung function was influenced by abdominal obesity. In conclusion, the lung function was positively associated with high protein and fat intake, but was negatively associated with high carbohydrate intake, which could be influenced by age and obesity.

Keywords: lung function of healthy population; difference of FEV_1; macronutrient; longitudinal study; obese

1. Introduction

Respiratory dysfunction is a life-threatening but treatable chronic disorder of the lung. Nutrition has been suggested as an important aspect in the care of respiratory disease. Malnutrition adversely affects lung function by diminishing respiratory muscle strength, altering ventilator capacity, and impairing immune function [1]. Even among men without COPD (Chronic Obstructive Pulmonary Disease), the lung function is associated with blood markers of nutritional status [2]. A prospective study of middle-aged men revealed a significant negative association between total energy intake and lung function. Regression coefficient suggested lung function (FEV_1) was 48.8 mL (95% CI 21.4 to 76.3) lower for total energy intakes one standard deviation (597 kcal/day) apart [3]. Root et al. [4] summarized that macronutrient intake is correlated with lung function, and highlighted the positive association of animal protein with forced expiratory volume/forced vital capacity (FEV_1/FVC (Forced Vital Capacity)). A report with COPD patents suggested that accurate evaluation of protein and energy requirements should be included in the goals of medical treatment of COPD patients [5].

Most Asians, including Koreans, consume relatively large amounts of carbohydrates (e.g., refined rice) as a staple food compared to Western countries but ingest low amounts of animal protein sources. Adults consume 62.2% and 14.9% of total energy intake from dietary carbohydrate and protein, respectively (https://knhanes.cdc.go.kr/knhanes/sub01/sub01_05.do#s5_03) as reported in the Korean Health Statistics 2016. Lee et al. reported that a low intake of protein is associated with airway obstruction in patients [6]. Otherwise, dietary fat is associated with impaired lung function in older men, which was attributed to induce innate immune activation and IL-6 (InterLeukin-6) release [7]. Cai et al., however, demonstrated experimentally that oral supplementation of high-fat and low-carbohydrate diets significantly increased lung function [8]. A study of Korean women reported that a high consumption of refined diets (high intake of natural carbohydrates) and low intake of fiber were associated with a decrease in lung function [9]. Otherwise, a high intake of dietary fiber was significantly associated with a higher percentage of individuals with normal lung function [10].

Although the effects of nutritional status on the respiratory system have been mainly focused on the prevention and prognosis of lung dysfunction, it is also very important to examine the association of macronutrients with pulmonary function in healthy adults. This is because an effect of nutrition on the respiratory disease incidence is accumulation over a long period of time. Therefore, this study is aimed to examine the association between macronutrient intake and lung function in healthy adults and to evaluate the association after stratification by age and both general and abdominal obesity.

2. Methods

2.1. Study Population

The study used the data from one of the community-based Korean Genome Epidemiology Study (KoGES) cohorts, namely the Ansan-Ansung cohort, and the detailed study design and procedures are reported elsewhere [11]. Spirometry was measured by three well-trained pulmonary technologists in concordance with the 1994 American Thoracic Society recommendation, using a spirometer (Vmax-229,Sensor-Medics, Yorba Linda, CA, USA) for all subjects. The predicted forced expiratory volume in one second (FEV1) were measured using a standardized method. To evaluate the lung function, we used the difference of FEV_1 ($_{diff}$ FEV_1) from baseline (from May 2001 to February 2003) to follow-up (from February 2005 to November 2007) data. Prior informed consent was obtained from each participant during the recruitment and subsequent visits, which were approved by the Human Subjects Committee of Korea. The Korean National Institute of Health Institutional Review Committee approved the study procedures. Study procedures were in accordance with institutional guidelines and were approved by the Kangwon National University Hospital institutional review committee (IRB No. 2013-07-006) for the ethical approval. From the baseline (N = 10,030), subject with lost to follow-up (F/U) (N = 2515) couldn't be included in the study. The distribution of ecological factors between eligible population and follow-up loss was presented in Supplementary Table S1. We excluded the participant who had a history of COPD at baseline (N = 66), without lung function information (FEV_1) at baseline or F/U (N = 515), with unacceptable energy intake (less than 800 kcal or more than 4000 kcal for men and less than 500 kcal or more than 3500 kcal for women, N = 213), and without dietary information at baseline (N = 236). In addition, participants with COPD diagnosed at recruitment (n = 562) were excluded to eliminate a latent period bias. Finally, 5880 healthy subjects were included in this study (Figure 1).

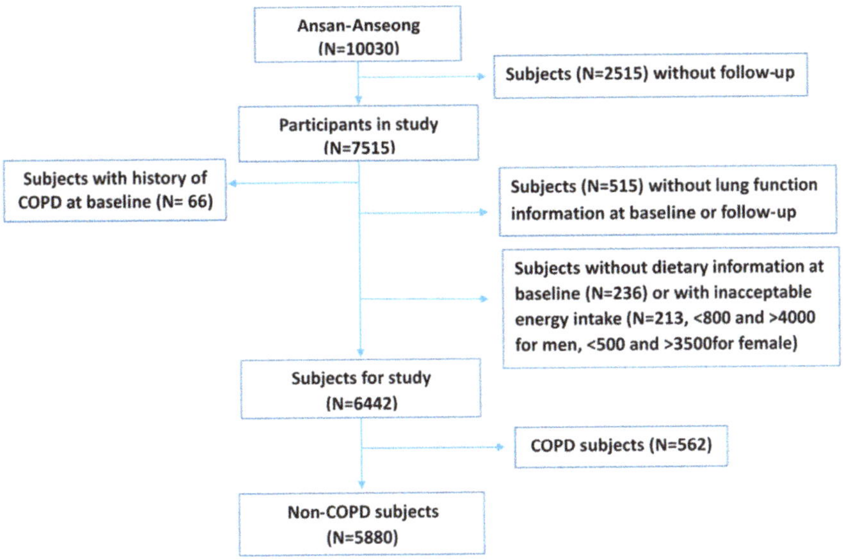

Figure 1. Flow diagram of analytical sample in current study using KoGES_Ansan and Ansung cohort. COPD: Chronic Obstructive Pulmonary Disease.

2.2. Covariates

Age was grouped by 5-year interval and education was categorized as two group, namely educated below than 12 years and more than 12 years. Household income per month was divided in two as below $2000 and more than $2000. Job was classified four groups (sedentary worker and labor, farmer, and housekeeper/other) from original eight job classification; (1) sedentary worker included officer, self-employment, and professional worker; (2) labor included sale officer and blue-colored worker; (3) farmer; and (4) housekeeper/other. Moreover, we separately made a farmer category because many subjects in the Ansung province among the cohort community engaged in agriculture (26.8%). BMI (Body Mass Index) was calculated by height (m) and weight (kg) (kg/m^2) and categorized based on WHO (World Health Organization) criteria as less than <25, 25–29.9, and ≥30. Waist circumference was taken the Asian guidelines by National Cholesterol Education Program Adult Treatment Panel III (NCEP-ATP III) (men > 90, women > 80). We calculated the waist-to-hip ratio (WHR) using waist and hip circumference. For the lifestyle factors, cigarette smoking and alcohol drinking included three groups as never, former, and current. Participants were classified in terms of regular exercise based on the "yes" and "no" answer to the following question, "Do you currently engage in regular exercise strenuous enough to cause you to break into a sweat at least once per week?".

2.3. Lung Function Measurement

To test a pulmonary function, Spirometry (VMAX2130 from Sensormedics Corporation, Yorba, CA, USA) was used to measure Forced expiratory volume (FEV_1, the amount of air that can force from lungs in one second) at baseline and follow-up visits. Spirometry was performed by trained technicians according to American Thoracic Society/European Respiratory Society guidelines [12]. Lung function was presented as percentage of predicted FEV_1 (%FEV_1), which calculated using measured and predicted FEV1 for each person (% FEV_1 = base FEV_1*100/predicted FEV_1). To evaluate the lung function, we calculated the difference of %FEV1 (%FEV_{1_diff}) between at baseline and follow-up. Based on the distribution of %FEV_{1_diff} (median = −1), we divided two groups for the lung function (decreased vs. unchanged/improved). The median value was involved in the unchanged/improved lung function group due to the maintenance of the $\%FEV_1$ despite the fact that aging means the lung function is relatively healthy.

2.4. Dietary Assessment

Dietary information was collected at baseline using validated and reproducible [13] semi-quantitative Food-Frequency Questionnaire (FFQ). Detailed information pertaining to dietary assessment is available in our previous report [13,14]. Briefly, FFQ consisted 103 food items and contained 9 serving categories for each item (never or seldom, once a month, 2–3 times a month, 1–2 times a week, 3–4 times a week, 5–6 times a week, once a day, twice a day, and 3 times or more every day) and the serving portion size (small, medium, and large). For the food items with different seasonal availability, we requested the participants to mark one on how long they ate among four categories: 3, 6, 9, and 12 months. The portion size was determined depending on the median value of each food determined from the 24-h recall data obtained from the Korean Health and Nutrition Examination Survey (KHANES). The portion size of each food item was classified as follows: small (0.5), medium (1.0), or large (1.5). For easy understanding of portion size, we provided pictures on serving size for food items on their own pages. Nutrient intake of each food item was converted based on the weight derived from the consumption frequency and the portion size in each food item. Daily nutrient intakes of everyone were the sum of the nutrient intake of each food item, which were calculated using DS24 (Human Nutrition Lab, Seoul National University & AI/DB Lab., Sookmyung Women's University, 1996). The food composition table used in the two calculations was the 7th edition Food Composition Table of Korea (The Korean Nutrition Society, 2000).

2.5. Statistical Analysis

Ecological and lifestyle factors were presented the distribution using descriptive statistics and evaluated the association with lung function using logistic regression model after adjustment for age, education, income, marriage status, height, job, history of asthma, and history of tuberculosis as covariates. To examine the association between macronutrients and the lung function, a multiple logistic regression model was undertaken for the determination of odds ratios (OR) and the 95% confidence interval as increased interquartile range (IQR) after adjustment for age (continuous), education (<12 vs. ≥12 years), job, BMI (continuous), WHR (% of more than 0.9 and 0.85 for men and women respectively), smoking status (none, former, and current smoker), and energy intake (continuous), which were observed the association with the lung function in this population. If alpha = 0.05, total subject number = 5880, and estimated odds ratio = 1.1, the power for the association between decreased and unchanged/improved is 0.928 based on the logistic regression analysis. To evaluate the association of macronutrients as continuous variable, we had examined odds ratio as increased interquartile range of each macronutrient (OR_{IQR}). We examined the advanced analyses after stratification by age, BMI and WHR (Waist-Hip Ratio). The age was classified two groups based on the median age (50 years old) of study population and was added the analysis for the elder (≥65 years old). For the general obese, we divided two groups as normal (BMI < 25) and overweight/obese (BMI ≥ 25). For the WHR, we divided two groups based on the cut point by NCEP-ATP III (cut-point = 0.90 and 0.85 for men and women, respectively) A p-value of <0.05 was considered as the level of significance. Two-sided probability tests were employed using SAS statistical software (version 9.2, SAS Institute Inc., Cary, NC, USA).

3. Results

Comparing the distribution of selected characteristics between the eligible population and those lost to follow-up, it was not statistically different to age, sex, BMI, and %FEV1 at baseline between eligible population and follow-up loss. Otherwise, those lost to follow-up were likely to less educated, lower income, not married, occupied as labor or housekeeper, and had lower abdominal obesity, compare to the eligible population. The average of $\%FEV_1$ is 114.1 ± 16.3 and 112.5 ± 15.6 (109.3 ± 14.5

and 107.6 ± 13.8 for men, 118.2 ± 16.6 and 116.6 ± 15.9 for women) at baseline and follow-up, respectively. The distribution of $\%FEV_1$ by lung function according to gender presented in Table 1. The lung function in men was negatively associated with age (elder), job (farmer), and abdominal obesity, and current smoking, but positively associated with low education (for women only) and general obesity (high BMI) (Table 1).

The percentage of energy from protein, fat, and carbohydrate is 13.8%, 15.7%, and 70.4% for men and 13.4%, 13.5%, and 73.1% for women, respectively. The distribution of energy percentage from each macronutrient was presented in Supplementary Table S2 The association between macronutrients intake and lung function was presented in Table 2 (ORs for IQR) and Supplementary Table S3 (ORs for quintile categories). The inverse association of protein (OR_{IQR} = 0.78) and fiber (OR_{IQR} = 0.85) intake with lung function was found in men. Fat (OR_{IQR} = 0.83) and carbohydrate (OR_{IQR} = 1.38) intake was associated with women's lung function, but the total energy intake was not related to the lung function (Table 2).

To evaluate the modified effect of age (Table 3) and obese status (Table 4) on the association between macronutrients and lung function, we analyzed the advanced analysis after stratified by age, BMI, and WHR. Although lung function was not associated with energy intake in both men and women, an inverse association of protein and fat but positive association of carbohydrate with the decreased lung function was observed in young men (<50 years old, OR_{IQR} = 0.72, 0.80 and 1.47 for protein, fat and carbohydrate, in order) and elderly women (both ≥50 and ≥65 years old, OR_{IQR} = 0.74, 0.56 and 3.21 for protein, fat and carbohydrate, in order) (Table 3). Otherwise, fiber intake was shown in the opposite direction as age group. High fiber intake was inversely associated with unchanged/improved lung function in younger women (OR_{IQR} = 1.44) but was positively associated with the lung function in older men (OR_{IQR} = 0.78).

In advanced analysis after stratified by BMI, no association of each macronutrients with lung function with exception of fat intake (OR_{IQR} = 0.78 in women with normal weight). Otherwise, after stratified by waist-to-hip ratio, we observed the inverse association of protein (OR_{IQR} = 0.67) and fiber (OR_{IQR} = 0.76) intake with decreased lung function in men with normal waist-hip ratio. Among women with normal waist-hip ratio, the inverse association of total energy intake (OR_{IQR} = 0.85) with decreased lung function. Furthermore, the decreased lung function was associated inversely with fat (OR_{IQR} = 0.81), but associated positively with carbohydrate (OR_{IQR} = 1.37) intake among women with abdominal obesity.

Table 1. The distribution of difference of %FEV_1 (lung function) and the association with general and lifestyle factors according to gender.

	Men				Women			
	Unchanged/Improved	Decreased	Crude OR	Adj OR	Unchanged/Improved	Decreased	Crude OR	Adj OR
	(n = 1432)	(n = 1270)	95% CI	95% CI	(n = 1583)	(n = 1595)	95% CI	95% CI
% FEV_1 at baseline	105.7 ± 13.0	113.4 ± 15.0			113.9 ± 15.9	122.6 ± 16.3		
% FEV_1 at follow-up	110.5 ± 12.9	104.3 ± 13.9			120.3 ± 15.6	113.0 ± 15.3		
Ecological Factors								
Age (age, mean ± SD)	49.1 ± 7.67	51.4 ± 8.33	1.04 (1.03–1.05)	1.02 (1.00–1.03)	51.6 ± 8.81	52.3 ± 8.64	1.01 (1.00–1.02)	1.00 (0.98–1.01)
Education (>12 years, %)	69.7	59.5	0.63 (0.54–0.74)	0.94 (0.76–1.16)	37.9	29.7	0.70 (0.60–0.81)	0.81 (0.66–0.99)
Income (≥2000$/M, %)	54.0	41.9	0.62 (0.53–0.72)	0.98 (0.80–1.20)	34.0	28.2	0.76 (0.66–0.89)	0.98 (0.81–1.19)
Marriage status (No)	3.3	4.4	1.36 (0.92–2.02)	1.14 (0.71–1.81)	13.8	13.7	0.99 (0.81–1.21)	1.04 (0.81–1.34)
Job								
Sedentary worker	47.2	35.7	Reference	Reference	11.9	8.8	Reference	Reference
Labor	9.7	6.8	0.97 (0.73–1.22)	1.06 (0.78–1.44)	6.3	6.9	0.59 (0.29–1.21)	0.72 (0.33–1.58)
Farmer	18.6	37.6	2.54 (2.10–3.09)	2.16 (1.66–2.80)	17.7	26.5	1.64 (1.38–1.95)	1.49 (1.18–1.89)
Housekeeper/Others	24.5	20.0	0.91 (0.75–1.11)	0.92 (0.74–1.15)	64.1	57.8	0.86 (0.66–1.12)	0.78 (0.57–1.05)
BMI (kg/m^2)	24.7 ± 2.72	24.2 ± 2.74	0.93 (0.91–0.96)	0.94 (0.91–0.97)	25.0 ± 3.24	24.9 ± 3.22	0.99 (0.97–1.02)	0.98 (0.95–1.00)
Abdominal obesity * (%)	37.8	50.1	1.65 (1.41–1.92)	1.32 (1.07–1.63)	53.2	63.5	1.53 (1.33–1.76)	1.33 (1.09–1.61)
His of Asthma (yes)	1.12	0.79	0.70 (0.32–1.55)	1.05 (1.07–1.63)	1.83	2.26	1.24 (0.75–2.03)	1.18 (0.65–1.22)
His of Tuberculosis (yes)	4.96	6.30	1.29 (0.93–1.79)	1.23 (0.85–1.79)	3.85	3.20	0.83 (0.57–1.20)	0.77 (0.50–1.19)
Lifestyle Factors								
Cigarette smoking								
Never	23.0	20.3	Reference	Reference	96.0	95.7	Reference	Reference
Former	35.8	30.6	0.91 (0.75–1.11)	0.91 (0.73–1.14)	1.5	1.0	0.65 (0.34–1.25)	0.54 (0.25–1.17)
Current	41.3	49.1	1.28 (1.06–1.55)	1.28 (1.03–1.59)	2.5	3.4	1.35 (0.80–2.27)	1.23 (0.68–2.20)
Alcohol drinking								
Never	17.2	18.6	Reference	Reference	70.9	71.7	Reference	Reference
Former	9.0	9.4	0.99 (0.79–1.24)	1.11 (0.91–1.35)	2.7	2.6	1.05 (0.89–1.22)	0.95 (0.79–1.15)
Current	73.9	72.0	1.00 (0.73–1.37)	1.08 (0.83–1.40)	26.5	25.8	1.00 (0.64–1.53)	0.91 (0.54–1.53)
Exercise (yes)	33.3	31.5	0.92 (0.78–1.08)	1.03 (0.86–1.25)	28.1	25.0	0.85 (0.73–1.00)	0.96 (0.80–1.16)

Unchanged and improved: unchanged or inclined lung function, more than median (−1) of the difference between %FEV_1 at baseline and follow-up. Decreased: under the median (−1) of the difference between %FEV1 at baseline and follow-up. % FEV1: Percentage of the annual change in lung function due to aging predicted FEV_1. Abdominal obesity *: waist-to-hip ratio ≥0.9 or ≥0.85 for men and women, respectively. Adj OR: OR after adjusted for age, education, income, marriage status, BMI, waist-to hip ratio, job, history of asthma, and history of tuberculosis.

Table 2. The association between macronutrient intake and the lung function among healthy population.

	Men			Women		
	Unchanged/Improved (*n* = 1432)	Decreased (*n* = 1270)	OR_{IQR} (95% CI)	Unchanged/Improved (N = 1583)	Decreased (N = 1595)	OR_{IQR} (95% CI)
Energy (kcal/day)	1934 *	1901	0.98 (0.88–1.09)	1776	1739	0.96 (0.88–1.05)
Protein (g/day)	67.3	63.8	0.78 (0.64–0.96)	59.7	56.2	0.87 (0.72–1.06)
Fat (g/day)	34.2	31.3	0.97 (0.83–1.13)	26.0	23.5	0.83 (0.71–0.96)
Carbohydrate (g/day)	331.2	330.3	1.14 (0.89–1.46)	318.2	313.1	1.38 (1.09–1.75)
Fiber (g/day)	6.44	6.19	0.85 (0.72–0.99)	6.43	6.41	1.15 (0.99–1.34)

*: median of each macronutrients. OR_{IQR}: OR as increased IQR (Q3–Q1) after adjusted for age, education, BMI, waist-to-hip ratio, job, smoking status and total energy intake; Unchanged and improved: more than median (−1) of the difference between %FEV1 at baseline and follow-up. Decreased: under the median (−1) of the difference between %FEV1 at baseline and follow-up.

Table 3. The association between macronutrient intake and the lung function by median age (50 years old) and among elderly participants.

	Men			Women		
	Unchanged/Improved	Decreased	OR_{IQR} (95% CI)	Unchanged/Improved	Decreased	OR_{IQR} (95% CI)
Below Median (Unchanged or improved = 1688, Decreased = 1376)						
Energy (kcal/day)	1945 *	1908	0.92 (0.79–1.06)	1821	1783	0.96 (0.84–1.09)
Protein (g/day)	68.8	65.2	0.72 (0.54–0.95)	62.5	59.9	1.01 (0.77–1.34)
Fat (g/day)	35.9	32.7	0.80 (0.64–1.00)	30.0	28.2	0.83 (0.68–1.03)
Carbohydrate (g/day)	331.1	329.0	1.47 (1.04–2.08)	322.9	315.4	1.33 (0.94–1.88)
Fiber (g/day)	6.36	6.07	0.93 (0.74–1.16)	6.26	6.45	1.44 (1.15–1.80)
Above Median (Unchanged or improved = 1063, Decreased = 1185)						
Energy (kcal/day)	1915	1888	1.14 (0.90–1.21)	1721	1683	0.96 (0.85–1.08)
Protein (g/day)	64.9	62.7	0.86 (0.63–1.17)	56.3	52.9	0.74 (0.56–0.97)
Fat (g/day)	30.3	29.4	1.17 (0.93–1.47)	22.1	20.3	0.81 (0.65–0.99)
Carbohydrate (g/day)	331.7	332.0	0.87 (0.60–1.25)	314.7	310.2	1.47 (1.05–2.05)
Fiber (g/day)	6.60	6.40	0.78 (0.62–0.98)	6.64	6.36	0.98 (0.80–1.19)
Elder (≥65 years old, Unchanged or improved = 264, Decreased = 304)						
Energy (kcal/day)	1790	1901	1.03 (0.74–1.45)	1699	1584	0.86 (0.67–1.10)
Protein (g/day)	59.2	57.7	0.90 (0.43–1.90)	53.8	47.2	0.48 (0.24–0.96)
Fat (g/day)	25.4	26.4	1.69 (0.97–2.92)	19.0	16.6	0.56 (0.31–0.98)
Carbohydrate (g/day)	327.6	341.9	0.62 (0.26–1.50)	316.2	302.5	3.12 (1.21–8.02)
Fiber (g/day)	6.78	6.14	0.88 (0.51–1.53)	6.46	6.08	1.05 (0.70–1.57)

*: median of each macronutrients. OR_{IQR}: OR as increased IQR (Q3–Q1) after adjusted for age, education, BMI, waist-to-hip ratio, job, smoking status and total energy intake; Unchanged or improved: more than median (−1) of the difference between %FEV1 at baseline and follow-up. Decreased: under the median (−1) of the difference between %FEV1 at baseline and follow-up.

Table 4. The association between macronutrient intake and the lung function by general (BMI) and abdominal (WHR) obesity.

	Men			Women		
	Unchanged/Improved	Decreased	OR_{IQR} (95% CI)	Unchanged/Improved	Decreased	OR_{IQR} (95% CI)
General Obesity						
BMI < 25 (kg/m^2)						
Energy (kcal/day)	1914 *	1875	0.98 (0.86–1.11)	1775	1736	0.91 (0.82–1.01)
Protein (g/day)	65.9	62.4	0.80 (0.61–1.03)	59.7	55.9	0.84 (0.67–1.07)
Fat (g/day)	33.1	30.6	1.02 (0.84–1.23)	26.0	23.4	0.79 (0.66–0.95)
Carbohydrate (g/day)	330.2	326.9	1.05 (0.77–1.42)	319.2	311.2	1.44 (1.08–1.93)
Fiber (g/day)	6.42	6.12	0.82 (0.67–1.01)	6.51	6.29	1.12 (0.93–1.34)
BMI ≥ 25 (kg/m^2)						
Energy (kcal/day)	1955	1970	0.98 (0.82–1.18)	1777	1759	1.07 (0.92–1.25)
Protein (g/day)	69.8	67.2	0.74 (0.53–1.03)	59.6	57.1	0.91 (0.66–1.27)
Fat (g/day)	35.1	32.9	0.86 (0.65–1.14)	25.8	24.0	0.89 (0.68–1.16)
Carbohydrate (g/day)	332.9	336.9	1.35 (0.88–2.09)	314.8	318.4	1.28 (0.84–1.96)
Fiber (g/day)	6.51	6.49	0.89 (0.68–1.17)	4.82	6.69	1.24 (0.96–1.60)
Abdominal Obesity						
WHR (<0.9 and <0.8 for men and women, respectively)						
Energy (kcal/day)	1937	1877	0.89 (0.77–1.03)	1805	1774	0.85 (0.73–0.98)
Protein (g/day)	68.3	63.6	0.67 (0.49–0.90)	62.5	59.7	0.86 (0.61–1.21)
Fat (g/day)	35.4	31.5	0.94 (0.75–1.18)	30.0	27.8	0.89 (0.70–1.14)
Carbohydrate (g/day)	327.7	325.6	1.23 (0.85–1.77)	318.3	311.9	1.28 (0.86–1.92)
Fiber (g/day)	6.44	6.12	0.76 (0.59–0.98)	6.41	6.21	1.23 (0.95–1.60)
WHR (≥0.9 and ≥0.8 for men and women, respectively)						
Energy (kcal/day)	1927	1933	1.07 (0.92–1.25)	1742	1719	1.02 (0.92–1.14)
Protein (g/day)	65.4	64.2	0.90 (0.68–1.19)	55.5	54.1	0.90 (0.71–1.14)
Fat (g/day)	32.4	31.2	1.00 (0.81–1.25)	22.3	21.5	0.81 (0.68–0.98)
Carbohydrate (g/day)	335.9	337.4	1.04 (0.73–1.47)	317.8	313.6	1.37 (1.02–1.85)
Fiber (g/day)	6.46	6.32	0.90 (0.73–1.12)	6.42	6.54	1.12 (0.93–1.33)

*: median of each macronutrients. Bold: statistically significant association ($p < 0.05$). OR_{IQR}: OR as increased IQR (Q3–Q1) after adjusted for age, education, BMI, waist-to-hip ratio, job, smoking status and total energy intake; The adjustment for BMI and WHR in advanced analysis after stratified by BMI and WHR, we had taken each confounder as crossing each other. Unchanged/improved: more than median (−1) of the difference between %FEV1 at baseline and follow-up. Deceased: under the median (−1) of the difference between %FEV1 at baseline and follow-up.

4. Discussion

This study suggested that protein and fat intake was inversely associated with lung function decline, but carbohydrate intake was positively associated. The association was influenced by age and both general and abdominal obese. The inverse association of protein and fat was shown in men with below median age, but in women with above median age. Additionally, low fat intake was decreased the lung function in only women with abdominal obesity. Lung function was associated with high carbohydrate intake in women, predominantly older women (both >50 and >65 years old) and was declined in women with normal BMI or abdominal obese as carbohydrate intake increased. The association of fiber intake with lung function decline is inversely associated in men with above median age or with normal WHR, but positively associated with in the women with below median age.

Inverse correlation with unchanged/improved lung function (%FEV_{1_diff}) with age was observed only in men. The working in farm was associated with the decreased lung function which could be explained to exposure to inorganic and organic dust, supported by other studies [15,16]. We found the negative correlation of BMI with %FEV_{1_diff} consistent with a recent report [17]; although the CARDIA study (Coronary Artery Risk Development In young Adults study) suggested that the obesity epidemic threatens the lung health of the general population [18]. Several studies have reported that abdominal obesity is related to decreased pulmonary function, consistent to ours result [19]. A recent clinical study using directly measured visceral adipose tissue and subcutaneous adipose tissue through magnetic resonance imaging, dual-energy radiography absorptiometry, and computed tomography (CT) reported that decreasing abdominal visceral obesity could increase lung function despite ageing [20]. The epidemiologic and clinical evidence over the past 50 years suggested the biological plausibility of a link between cigarette smoking and adverse respiratory outcomes [21,22]. A recent report using the South Carolina Behavioral Risk Factor Surveillance System (BRFSS) discusses the adverse effects of smoking duration and number of pack-years on lung function [23]. In this study population, the association of smoking with lung function was found in men only.

Although Butland et al. found an inverse association between lung function and energy intake [3], the total energy intake was not significantly associated with increase in lung function in our study, with the exception of the protective effect in females with normal BMI. Therefore, the role of energy intake on the lung function is controversial and needs further investigation. In the present study, we observed a significant association between higher protein intake and better lung function in both healthy men and women supported by Beasley's results [24], although gender differences play a vital role in relation to lung function [25]. After stratified by age group, the association was predominantly observed in women with above median age, even in elderly women, similar to the result from a previous report [26]. This finding could be supported by the fact that the protein needs of elderly females could be higher [27] and the high protein intake leads to decrease health problems in elderly women [28].

The effect of fat intake on the decreased lung function was controversial and inconsistent because of the bi-directionally mechanical role on the pulmonary system. The harmful effect of fat was explained by bronchial hyper-responsiveness [29], incidence of asthma [30], a higher risk for COPD [31] and innate immune activation [7], which can lead to a systemic inflammatory response. The systemic inflammatory response includes the high level of circulating pro-inflammatory mediators. Otherwise, the beneficial effect of fat on the lung function was reported with many possible mechanisms; (1) omega-3 fatty acids protect the lung function by inhibiting the production of prostaglandin E2, thereby preventing allergic sensitization [32], (2) the intake of omega-6 fatty acids associated with the decreased risk of chronic nonspecific lung disease [32], (3) fat helps in digestion, absorption, and transportation of fat-soluble antioxidant vitamins; vitamin A, D, E and K [33], which could improve lung function [14,34]. and (4) fat metabolism generates less CO_2 which has a lower respiratory quotient [35]. Our study suggested the protective effect of fat intake on lung function of relatively young men, elderly women, and abdominal obese women, contrary to one of the published results [7] which suggested a reduced lung function in older men as increased proportion of fat in the diet. Although our results suggest an inverse association of fat intake with lung function decline, it is necessary to advanced analysis considering the several limitations, such as the lack information of fat intake including animal vs. plant fat, the composition of trans-fat, saturated vs. polyunsaturated fatty acid, *w*-3 vs. *w*-6, etc. Therefore, advanced analyses considering these factors are needed to explain the protective effect on the lung function.

In the healthy population, carbohydrate metabolism increases CO_2 production and respiratory quotient [35]. Although individuals with normal lungs eliminate excess loads of CO_2 easily [36], the long-term effects suggest deterioration in lung function, especially in the elderly. Therefore, a lower carbohydrate diet might prevent or increase respiratory health [37]. The possibility that altered respiratory variables after ingestion of carbohydrate were nonspecific and unrelated remote for the following reasons. The metabolic and respiratory responses to ingestion of carbohydrate confirm qualitatively and quantitatively to previous observations and predictions [38–42]. Consistent with previous reports [31,43–45], a carbohydrate-rich diet is negatively associated with predicted FEV_1, consistent with our result as well as another report with Korean women [46].

Several studies had reported an inverse association of dietary intake of fiber and lung function decline [10,47]; we found the association in older men only (more than 50 years old). For younger women, otherwise, high fiber intake was positively associated with the decreased lung function in this study population. To illustrate the positive association between high fiber intake and lung function decline in women, it is valuable to consider the source of dietary fiber in Korean diet. The food groups that contributed most to dietary fiber intake were (in descending order) cereals, vegetables, seasoning, and fruits in Korean and the fiber-containing food items consumed most were cabbage- kimchi, cooked rice, instant noodles, and cabbage [48]. Unfortunately, the difference between sex and age in Korean dietary fiber sources has not been studied yet. Therefore, further studies are needed to explain the relationship of dietary fiber to lung function.

Our study has several limitations. First, although it is acceptable to use FVE_1 for the assessment of the pulmonary function in the epidemiology study, the association between macronutrients intake and the lung function could be underestimated due to the non-differential misclassification of the lung function. In addition, the association should be illustrated with consideration of the different follow-up time among the participants and annual change in lung function as aging. Therefore it is difficult to accurately adjust the effect of pulmonary function decline as increasing age. Second, we used FFQ that depended on the subjects' memories of dietary intake and possible difficulty in accurate recall of frequency and food portion size. To overcome this limitation, we used a closed format and the ability to exercise different options based on food illustrations to facilitate easier recall. Besides, dietary information was collected only at baseline and changes during follow up period were not taken into account. Third, the findings cannot be generalized to populations other than Korean, especially in the Western countries with high-protein diets, because most of our participants were Korean. Possibly, our result could apply to Asia countries where the primary food are carbohydrate-rich (especially, starch-based refined rice). Forth, we had considered only macronutrient intake to evaluate the effect on the lung function. However, an advanced analysis is required considering the quality of the diet, such as plant vs. animal protein, sugar vs. complex carbohydrate, the very low fiber in the diet. Fifth, smokers at baseline may have quit smoking between the lung function measurements, which may have had impact on lung function. In addition, we couldn't consider the change of the general and abdominal obesity status during follow-up. Finally, the high proportion of follow-up loss could be affected by selection bias, although we tried to adjust for education, married status, job abdominal obesity, which were observed a different distribution between the eligible population and follow-up loss.

Aside from these limitations, our study has several strengths. First, the use of data from a large prospective cohort allows the possibility of a temporal relationship, although causality cannot be assessed unless some strong assumptions are made (e.g., missing at random conditional on observed variables). Second, we included well-known confounding factors in the analysis such as age, sex, marital status, BMI, WHR, history of asthma and tuberculosis, and cigarette smoking, because of the large sample size. Third, FFQ was developed based on nationwide dietary data, and hence, the use of a validated FFQ strengthened the reproducibility of our results. Fourth, to avoid the latent period bias and to preclude the bias related with altered lifestyle factors, because of inconvenience lung function without a diagnosis of respiratory disease, we excluded subjects who were diagnosed with COPD at recruitment. Fifth, this report examined advanced analytics to evaluate the advanced analysis after stratification by age and both general and abdominal obese status on the association between macronutrient intake and the lung function decline.

5. Conclusions

In conclusion, this study suggested a positive association of protein and fat intake but a negative association of carbohydrate intake with the lung function in the healthy population. Furthermore, the association between macronutrients and lung function could be attenuated by age and obese status.

Supplementary Materials: The following are available online at http://www.mdpi.com/2072-6643/12/9/2688/s1, Table S1: Distribution of ecological factors between eligible study population and follow-up loss, Table S2: Distribution of macronutrients intake and percentage of total energy intake by age group, BMI, and WHR, Table S3: The association between macronutrient intake and lung function among healthy population using five categorical analysis.

Author Contributions: The authors' responsibilities were as follows—S.-A.L., W.J.K., and D.K.: designed and conducted the research, S.-A.L., P.J. and Y.K.: analyzed the data and performed the statistical analyses; S.-A.L.: wrote the manuscript and had primary responsibility for the final content of the manuscript; and all authors: read and approved the final manuscript. None of the authors reported a conflict of interest related to the study. The funding source was not involved in the design, implementation, analyses, and interpretation of the data. All authors have read and agreed to the published version of the manuscript.

Funding: This research received no external funding.

Acknowledgments: This study has been worked with the support by a research grant of Kangwon National University in 2017–2018 and by Ministry of Environment. This research received no specific grant from any funding agency, commercial or not-for-profit sectors.

Conflicts of Interest: The authors declare no conflict of interest.

References

1. Pingleton, S.K. Enteral nutrition in patients with respiratory disease. *Eur. Respir. J.* **1996**, *9*, 364–370. [CrossRef]
2. Shiozawa, N.; Hayashimoto, K.; Suzuki, E.; Kikuchi, H.; Takata, S.; Ashida, K.; Watanabe, M.; Hosaki, Y.; Mitsunobu, F. Lung function and blood markers of nutritional status in non-COPD aging men with smoking history: A cross-sectional study. *Int. J. Chronic Obstr. Pulm. Dis.* **2010**, *5*, 233–240.
3. Butland, B.K.; Fehily, A.M.; Elwood, P.C. Diet, lung function, and lung function decline in a cohort of 2512 middle aged men. *Thorax* **2000**, *55*, 102–108. [CrossRef]
4. Root, M.M.; Houser, S.M.; Anderson, J.J.B.; Dawson, H.R. Healthy Eating Index 2005 and selected macronutrients are correlated with improved lung function in humans. *Nutr. Res.* **2014**, *34*, 277–284. [CrossRef]
5. Yazdanpanah, L.; Shidfar, F.; Moosavi, A.J.; Heidarnazhad, H.; Haghani, H. Energy and protein intake and its relationship with pulmonary function in chronic obstructive pulmonary disease (COPD) patients. *Acta Med. Iran.* **2010**, *48*, 374–379.
6. Lee, J.H.; Sim, Y.S.; Suh, G.Y.; Ryu, J.S.; Shin, D.H.; Koh, K.H.; Kim, Y.J.; Park, W.; Yoon, H.K.; Lee, M.J.; et al. Diet and airway obstruction: A cross sectional study from the second Korean National Health and Nutrition Examination Survey. *Korean J. Intern. Med.* **2010**, *25*, 132–139. [CrossRef]
7. Wood, L.G.; Attia, J.; McElduff, P.; McEvoy, M.; Gibson, P.G. Assessment of dietary fat intake and innate immune activation as risk factors for impaired lung function. *Eur. J. Clin. Nutr.* **2010**, *64*, 818–825. [CrossRef]
8. Cai, B.; Zhu, Y.; Ma, Y.; Xu, Z.; Zao, Y.; Wang, J.; Lin, Y.; Comer, G.M. Effect of supplementing a high-fat, low-carbohydrate enteral formula in COPD patients. *Nutrition* **2003**, *19*, 229–232. [CrossRef]
9. Jang, J.-Y.; Kim, T.-W.; Park, H.; Park, S.-H.; Lee, J.; Choi, H.-J.; Han, E.S.; Kang, M.; Kim, H.J. Safety Evaluation of Heavy Metal in Salted Vegetable Foods from Diverse Origin in Korea. *J. Food Hyg. Saf.* **2014**, *29*, 146–151. [CrossRef]
10. Hanson, C.; Lyden, E.; Rennard, S.; Mannino, D.M.; Rutten, E.P.; Hopkins, R.; Young, R. The Relationship between Dietary Fiber Intake and Lung Function in the National Health and Nutrition Examination Surveys. *Ann. Am. Thorac. Soc.* **2016**, *13*, 643–650. [CrossRef]
11. Shin, C.; Abbott, R.D.; Lee, H.; Kim, J.; Kimm, K. Prevalence and correlates of orthostatic hypotension in middle-aged men and women in Korea: The Korean Health and Genome Study. *J. Hum. Hypertens.* **2004**, *18*, 717–723. [CrossRef]
12. Miller, M.R.; Hankinson, J.; Brusasco, V.; Burgos, F.; Casaburi, R.; Coates, A.; Crapo, R.; Enright, P.; van der Grinten, C.P.; Gustafsson, P.; et al. Standardisation of spirometry. *Eur. Respir. J.* **2005**, *26*, 319–338. [CrossRef]
13. Ahn, Y.; Kwon, E.; Shim, J.E.; Park, M.K.; Joo, Y.; Kimm, K.; Park, C.; Kim, D.H. Validation and reproducibility of food frequency questionnaire for Korean genome epidemiologic study. *Eur. J. Clin. Nutr.* **2007**, *61*, 1435–1441. [CrossRef]
14. Joshi, P.; Kim, W.J.; Lee, S.A. The effect of dietary antioxidant on the COPD risk: The community-based KoGES (Ansan-Anseong) cohort. *Int. J. Chronic Obstr. Pulm. Dis.* **2015**, *10*, 2159–2168. [CrossRef]
15. Eduard, W.; Pearce, N.; Douwes, J. Chronic bronchitis, copd, and lung function in farmers: The role of biological agents. *Chest* **2009**, *136*, 716–725. [CrossRef]
16. Oh, Y.M.; Bhome, A.B.; Boonsawat, W.; Gunasekera, K.D.; Madegedara, D.; Idolor, L.; Roa, C.; Kim, W.J.; Kuo, H.P.; Wang, C.H.; et al. Characteristics of stable chronic obstructive pulmonary disease patients in the pulmonology clinics of seven Asian cities. *Int. J. Chronic Obstr. Pulm. Dis.* **2013**, *8*, 31–39.
17. Banerjee, J.; Roy, A.; Singhamahapatra, A.; Dey, P.K.; Ghosal, A.; Das, A. Association of Body Mass Index (BMI) with Lung Function Parameters in Non-asthmatics Identified by Spirometric Protocols. *J. Clin. Diagn. Res. JCDR* **2014**, *8*, 12–14. [CrossRef]

18. Thyagarajan, B.; Jacobs, D.R.; Apostol, G.G.; Smith, L.J.; Jensen, R.L.; Crapo, R.O.; Barr, R.G.; Lewis, C.E.; Williams, O.D. Longitudinal association of body mass index with lung function: The CARDIA Study. *Respir. Res.* **2008**, *9*, 31. [CrossRef]
19. Ochs-Balcom, H.M.; Grant, B.J.; Muti, P.; Sempos, C.T.; Freudenheim, J.L.; Trevisan, M.; Cassano, P.A.; Iacoviello, L.; Schunemann, H.J. Pulmonary function and abdominal adiposity in the general population. *Chest* **2006**, *129*, 853–862. [CrossRef]
20. Choe, E.K.; Kang, H.Y.; Lee, Y.; Choi, S.H.; Kim, H.J.; Kim, J.S. The longitudinal association between changes in lung function and changes in abdominal visceral obesity in Korean non-smokers. *PLoS ONE* **2018**, *13*, e0193516. [CrossRef]
21. Kim, W.J.; Lee, C.Y. Environmental exposures and chronic obstructive pulmonary disease. *Mol. Cell. Toxicol.* **2017**, *13*, 251–255. [CrossRef]
22. Alberg, A.J.; Shopland, D.R.; Cummings, K.M. The 2014 Surgeon General's report: Commemorating the 50th Anniversary of the 1964 Report of the Advisory Committee to the US Surgeon General and updating the evidence on the health consequences of cigarette smoking. *Am. J. Epidemiol.* **2014**, *179*, 403–412. [CrossRef] [PubMed]
23. Liu, Y.; Pleasants, R.A.; Croft, J.B.; Wheaton, A.G.; Heidari, K.; Malarcher, A.M.; Ohar, J.A.; Kraft, M.; Mannino, D.M.; Strange, C. Smoking duration, respiratory symptoms, and COPD in adults aged ≥45 years with a smoking history. *Int. J. Chronic Obstr. Pulm. Dis.* **2015**, *10*, 1409–1416. [CrossRef] [PubMed]
24. Beasley, J.M.; Wertheim, B.C.; LaCroix, A.Z.; Prentice, R.L.; Neuhouser, M.L.; Tinker, L.F.; Kritchevsky, S.; Shikany, J.M.; Eaton, C.; Chen, Z.; et al. Biomarker-Calibrated Protein Intake and Physical Function in the Women's Health Initiative. *J. Am. Geriatr. Soc.* **2013**, *61*, 1863–1871. [CrossRef] [PubMed]
25. Olafsdottir, I.S.; Gislason, T.; Thjodleifsson, B.; Olafsson, I.; Gislason, D.; Jogi, R.; Janson, C. Gender differences in the association between C-reactive protein, lung function impairment, and COPD. *Int. J. Chronic Obstr. Pulm. Dis.* **2007**, *2*, 635–642.
26. Walrand, S.; Boirie, Y. Optimizing protein intake in aging. *Curr. Opin. Clin. Nutr. Metab. Care* **2005**, *8*, 89–94. [CrossRef]
27. Morse, M.H.; Haub, M.D.; Evans, W.J.; Campbell, W.W. Protein requirement of elderly women: Nitrogen balance responses to three levels of protein intake. *J. Gerontol. Ser. A Biol. Sci. Med Sci.* **2001**, *56*, M724–M730. [CrossRef]
28. Vellas, B.J.; Hunt, W.C.; Romero, L.J.; Koehler, K.M.; Baumgartner, R.N.; Garry, P.J. Changes in nutritional status and patterns of morbidity among free-living elderly persons: A 10-year longitudinal study. *Nutrition* **1997**, *13*, 515–519. [CrossRef]
29. Soutar, A.; Seaton, A.; Brown, K. Bronchial reactivity and dietary antioxidants. *Thorax* **1997**, *52*, 166–170. [CrossRef]
30. Strom, K.; Janzon, L.; Mattisson, I.; Rosberg, H.E.; Arborelius, M. Asthma but not smoking-related airflow limitation is associated with a high fat diet in men: Results from the population study "Men born in 1914", Malmo, Sweden. *Monaldi Arch. Chest Dis. Arch. Monaldi per le Malattie Del Torace* **1996**, *51*, 16–21.
31. Varraso, R.; Fung, T.T.; Hu, F.B.; Willett, W.; Camargo, C.A. Prospective study of dietary patterns and chronic obstructive pulmonary disease among US men. *Thorax* **2007**, *62*, 786–791. [CrossRef] [PubMed]
32. Black, P.N.; Sharpe, S. Dietary fat and asthma: Is there a connection? *Eur. Respir. J.* **1997**, *10*, 6–12. [CrossRef] [PubMed]
33. Thompson, G.R. Absorption of fat-soluble vitamins and sterols. *J. Clin. Pathol.* **1971**, *5*, 85–89. [CrossRef]
34. Gilbert, C.R.; Arum, S.M.; Smith, C.M. Vitamin D deficiency and chronic lung disease. *Can. Respir. J.* **2009**, *16*, 75–80. [CrossRef]
35. Frankfort, J.D.; Fischer, C.E.; Stansbury, D.W.; McArthur, D.L.; Brown, S.E.; Light, R.W. Effects of high- and low-carbohydrate meals on maximum exercise performance in chronic airflow obstruction. *Chest* **1991**, *100*, 792–795. [CrossRef]
36. Saltzman, H.A.; Salzano, J.V. Effects of carbohydrate metabolism upon respiratory gas exchange in normal men. *J. Appl. Physiol.* **1971**, *30*, 228–231. [CrossRef]
37. Obase, Y.; Mouri, K.; Shimizu, H.; Ohue, Y.; Kobashi, Y.; Kawahara, K.; Oka, M. Nutritional deficits in elderly smokers with respiratory symptoms that do not fulfill the criteria for COPD. *Int. J. Chronic Obstr. Pulm. Dis.* **2011**, *6*, 679–683. [CrossRef]

38. Bachmann, G.; Haldi, J. A comparative study of the respiratory quotient following the ingestion of glucose and of fructose as affected by the lactic acid and carbon dioxide changes in the blood. *J. Nutr.* **1937**, *13*, 157–178. [CrossRef]
39. Carpenter, T.M.; Lee, R.C. The effect of glucose and of fructose on the human respiratory quotient and alveolar air. *J. Nutr.* **1933**, *6*, 55–82. [CrossRef]
40. Edwards, H.T.; Bensley, E.H.; Dill, D.B.; Carpenter, T.M. Human respiratory quotients in relation to alveolar carbon dioxide and blood lactic acid after ingestion of glucose, fructose, or galactose. *J. Nutr.* **1944**, *27*, 241–251. [CrossRef]
41. Rahn, H.; Fenn, W.O. *A Graphical Analysis of the Respiratory Gas Exchange: The Ob2-COb2 Diagram*; American Physiological Society: Washington, DC, USA, 1955.
42. Sachs, B.; Sternfeld, L.; Kraus, G. Essential fructosuria: Its pathophysiology. *Am. J. Dis. Child.* **1942**, *63*, 252–269. [CrossRef]
43. McKeever, T.M.; Lewis, S.A.; Cassano, P.A.; Ocke, M.; Burney, P.; Britton, J.; Smit, H.A. Patterns of dietary intake and relation to respiratory disease, forced expiratory volume in 1 s, and decline in 5-y forced expiratory volume. *Am. J. Clin. Nutr.* **2010**, *92*, 408–415. [CrossRef] [PubMed]
44. Shaheen, S.O.; Jameson, K.A.; Syddall, H.E.; Sayer, A.A.; Dennison, E.M.; Cooper, C.; Robinson, S.M.; Hertfordshire Cohort Study Group. The relationship of dietary patterns with adult lung function and COPD. *Eur. Respir. J.* **2010**, *36*, 277–284. [CrossRef] [PubMed]
45. Varraso, R.; Fung, T.T.; Barr, R.G.; Hu, F.B.; Willett, W.; Camargo, C.A., Jr. Prospective study of dietary patterns and chronic obstructive pulmonary disease among US women. *Am. J. Clin. Nutr.* **2007**, *86*, 488–495. [CrossRef]
46. Cho, Y.; Chung, H.-K.; Kim, S.-S.; Shin, M.-J. Dietary patterns and pulmonary function in Korean women: Findings from the Korea National Health and Nutrition Examination Survey 2007–2011. *Food Chem. Toxicol.* **2014**, *74*, 177–183. [CrossRef]
47. Varraso, R.; Willett, W.C.; Camargo, C.A., Jr. Prospective study of dietary fiber and risk of chronic obstructive pulmonary disease among US women and men. *Am. J. Epidemiol.* **2010**, *171*, 776–784. [CrossRef]
48. Park, S.; Na, W.; Kim, M.; Kim, E.; Sohn, C. Correlation between Intake of Dietary Fiber and Adherence to the Korean National Dietary Guidelines in Adolescents from Jeonju. *Prev. Nutr. Food Sci.* **2012**, *17*, 254–260. [CrossRef]

nutrients

MDPI

Article

Trends and Disparities of Energy Intake and Macronutrient Composition in China: A Series of National Surveys, 1982–2012

Zhenni Zhu [1,2], Xiaoguang Yang [1], Yuehui Fang [1], Jian Zhang [1], Zhenyu Yang [1], Zhu Wang [1], Ailing Liu [1], Li He [1], Jing Sun [1], Yiyao Lian [1], Gangqiang Ding [1] and Yuna He [1,*]

1 National Institute for Nutrition and Health, Chinese Center for Disease Control and Prevention, 27 Nanwei Road, Xicheng District, Beijing 100050, China; zhuzhenni@scdc.sh.cn (Z.Z.); yangxg@ninh.chinacdc.cn (X.Y.); fangyh@ninh.chinacdc.cn (Y.F.); zhangjian@ninh.chinacdc.cn (J.Z.); yangzy@ninh.chinacdc.cn (Z.Y.); wangzhu@ninh.chinacdc.cn (Z.W.); liual@ninh.chinacdc.cn (A.L.); heli@ninh.chinacdc.cn (L.H.); sunjing@ninh.chinacdc.cn (J.S.); lianyy@ninh.chinacdc.cn (Y.L.); dinggq@chinacdc.cn (G.D.)

2 Division of Health Risk Factors Monitoring and Control, Shanghai Municipal Center for Disease Control and Prevention, 1380 West Zhongshan Road, Shanghai 20036, China

* Correspondence: heyn@ninh.chinacdc.cn; Tel.: +86-10-6623-7233

Received: 22 June 2020; Accepted: 20 July 2020; Published: 22 July 2020

Abstract: Background: China's diet transition might offer guidance to undeveloped countries on the way to prosperity. This report describes the trends and disparities in energy and macronutrient composition among Chinese adults, and between subpopulations. Methods: Data for the current study were obtained from the 1982, 1992, 2002, and 2010–2012 China National Nutrition Survey (CNNS) rounds, which were nationally representative cross-sectional surveys. We applied 24-h dietary recall and food weighing to assess dietary intake. Results: There were 204,877 participants aged 20 years or older included in the current analysis. From 1982 to 2012, the estimated energy intake declined from 2614.7 kcal to 2063.9 kcal. The trend in the estimated percentage of energy intake from fat showed a spike. It increased from 16.3% to 33.1% (1992 vs. 1982 difference, 7.6%; 95% CI 7.4% to 7.7%; 2002 vs. 1992 difference, 7.7%; 95% CI 7.6% to 7.9%; 2012 vs. 2002 difference, 1.6%; 95% CI 1.4% to 1.7%; $p < 0.01$ for trend). The trends coincided in all the subgroups (all $p < 0.01$ for trend) except for the subgroup of those educated over 15 years, whose percentage of energy intake from fat declined from 37.4% to 36.6% (2012 vs. 2002 difference, −0.8%; 95% CI −1.6% to 0.0%). The estimated percentage of energy intake from carbohydrates declined from 74.0% to 55.0%. The ranges of the estimated percentage of energy intake from fat, within population subgroups stratified by education level, area and Gross national product (GNP) level, were narrowed. Conclusions: Quick improvements in society and the economy effectively curbed undernutrition, but easily triggered overnutrition. Disparities persistently existed between different subpopulations, while the gaps would narrow if comprehensive efforts were made. Education might be a promising way to prevent overnutrition during prosperous progress. The low-social profile populations require specific interventions so as to avoid further disease burdens.

Keywords: dietary energy; macronutrient composition; trend; disparity; subpopulation

1. Introduction

China has been one of the fastest-growing countries over the past three decades. It implemented major social and economic reforms in 1979, and achieved tremendous economic and agricultural productivity improvement [1]. Changes in the economy, food supply and nutrition-related policies can affect diet quality at the population level. The mass Chinese population consume diets that have developed from scarcity to prosperity within only one decade or two, but this has cost a lot with regards to health outcomes, in that the burden of diet-related non-communicable disease has increased [2]. Malnutrition covers two broad groups of conditions: undernutrition and overnutrition [3]. Many developing countries work on the problem of undernutrition, while overnutrition soon emerges [4]. China's diet transition might give guidance for the developing countries on the way to prosperity. We try to take a close look at this dietary transition during this extraordinary time in China. This report describes data from four rounds of the China National Nutrition Survey (CNNS), from 1982 to 2012. We examined the trends in energy and macronutrient composition among the Chinese population, and we also determined the disparities in dietary quality between subpopulations in terms of area, education level and economic background.

2. Materials and Methods

2.1. Study Population and Sampling

Data for the current study were obtained from the 1982, 1992, 2002 and 2010–2012 CNNS rounds, which were nationally representative cross-sectional surveys conducted by the Chinese Center for Disease Control and Prevention in order to assess the health and nutrition of the Chinese population [5]. The design, sampling and dietary data collection methods of each round were homogeneous. The survey design and methods have been presented in detail previously [5]. A stratified and multistage cluster randomized sampling method was applied. There were initially 238,124, 100,201, 247,464 and 188,622 participants recruited in the surveys from 1982 to 2012, respectively. The response rate was 87.9% in 2002 and 76.5% in 2010–2012. Response rates were not recorded in the 1982 and 1992 surveys. All CNNS rounds collected identical data from household and dietary interviews, body measurements and laboratory tests. Some participants were selected to participate in certain survey items, while others participated in another. For this analysis, we restricted the study sample to adults aged 20 years or older with dietary intake data.

Education level was defined as years the participant had received education. Area was defined as urban or rural because China is a two-class society with rural–urban distinctions in many aspects. The urban sector has gained more benefits from social and economic reforms than the rural sector has. Life style and dietary pattern were distinguishing between the two sectors. Gross national product (GNP) level was classified by provincial level according to the GNP quartiles across provinces. In 1982, the first to fourth classes were classified as ≥284, (244, 284), (194, 244) and <194, respectively, in USD; in 1992, the first to fourth classes were classified as USD ≥482, (354, 482), (268, 354) and <268; in 2002, the first to fourth classes were classified as USD ≥1569, (958, 1569), (743, 958) and <743; and in 2012, the first to fourth classes were classified as USD ≥8510, (5761, 8510), (4670, 5761) and <4670.

The series of national surveys was approved by the ethics committee of the National Institute for Nutrition and Health at the Chinese Center for Disease Control and Prevention.

2.2. Dietary Assessment

The field work of each round was launched in autumn, considering the comparability between survey rounds. Dietary information was collected for 5 days in 1982 by trained investigators who weighed all available foods in the participants' homes at the beginning of the first day, recorded (and weighed if necessary) all new foods brought into the homes during the 5 days and weighed all leftovers at the end of the survey to calculate the total amount of food consumed by participants during those 5 days. In the 1992, 2002 and 2010–2012 surveys, diet was assessed via 3 consecutive days (including two weekdays and one weekend) of 24-h dietary recall, in addition to weighing household cooking oil and condiments. For each dietary recall day, investigators went to participants' homes and helped to record food intake during the past 24 h. Investigators also weighed the household cooking oil and condiments at the beginning and end of each 24 h dietary survey. Nutrient intakes were calculated with the China Food Composition tables (FCTs) [6–8], which are continuously updated with commonly consumed foods and changes in nutrient composition. FCT-1981 [8] was used for dietary data from the 1982 round, FCT-2002 [6] for those from the 1992 and 2002 rounds, and FCT-2009 [7] for those from the 2010–2012 round.

2.3. Statistical Analyses

The post-stratification population sampling weights were applied to the estimated nationally representative population levels for intakes of energy and macronutrients. In order to compare dietary intake across years, the weights were derived from the sampling probability of the 2010 Chinese population aged 20 years or older (based on census data) and applied to estimate the representative dietary intake in each survey round. Means and 95% confidence intervals (CIs) of energy, and the percentages of macronutrients contributing to energy, were determined by adjustment for the sample weights. General linear regression models were used to determine the dietary trends across the survey rounds and the dietary differences between and within years. Regarding the difference between years, the year of each survey was treated as an ordinal variable and as the dependent variable. Regarding the difference within years, the subgroup of the two ends within each group (classified by education level, area, GNP level, sex and age group) was treated as an ordinal variable and as the dependent variable. Energy and macronutrient composition were treated as continuous variables and as the independent variable respectively in each model. A two-sided $p < 0.05$ was considered to indicate statistical significance. Statistical analyses were conducted using SAS statistical software (v. 9.4; SAS Institute, Cary, NC, USA).

3. Results

3.1. Participant Characteristics

There were 204,877 participants aged 20 years or older included in the current analysis. In the survey rounds of 1982, 1992, 2002 and 2010–2012, dietary intake data were available for 39,084, 58,316, 52,426 and 55,051 participants, respectively. The age structure of the participants was assorted across the survey rounds in accordance with the structure of actual change among the Chinese population. The sex ratios were balanced in the samples. Participants in each round had higher education level than the former round. Urban participants gradually accounted for greater percentages of the samples in the survey rounds, due to urbanization progress in China (Table 1).

3.2. Trends of Energy and Macronutrient Composition

From 1982 to 2012, the estimated energy intake declined from 2614.7 kcal to 2063.9 kcal (1992 vs. 1982 difference, −82.6; 95% CI −92.5 to −72.7; 2002 vs. 1992 difference, −335.4; 95% CI −344.2 to −326.7; 2012 vs. 2002 difference, −132.7; 95% CI −141.5 to −123.9; $p < 0.01$ for trend). These were the trends in the population subgroups (Table 2).

The trend of estimated percentage of energy intake from fat showed a spike. It increased from 16.3% to 33.1% (1992 vs. 1982 difference, 7.6%; 95% CI 7.4% to 7.7%; 2002 vs. 1992 difference, 7.7%; 95% CI 7.6% to 7.9%; 2012 vs. 2002 difference, 1.6%; 95% CI 1.4% to 1.7%; $p < 0.01$ for trend). The trends coincided in the subgroups (all $p < 0.01$ for trend) except for the subgroup of those educated for over 15 years. In the most recent two survey rounds, the estimated percentage of energy intake from fat among the well-educated population declined from 37.4% to 36.6% (2012 vs. 2002 difference, −0.8%; 95% CI −1.6% to 0.0%) (Table 3).

The estimated percentage of energy intake from carbohydrates declined from 74.0% to 55.0% (1992 vs. 1982 difference, −10.5%; 95% CI −10.7% to −10.4%; 2002 vs. 1992 difference, −7.4%; 95% CI −7.5% to −7.2%; 2012 vs. 2002 difference, −1.0%; 95% CI −1.2% to −0.9%; $p < 0.01$ for trend). The trends in the subgroups were the same (Table 4).

The estimated percentage of energy intake from protein increased between the first and second rounds, from 10.9% to 12.8%, and slightly declined to 12.3% in the successive two rounds (1992 vs. 1982 difference, 1.9%; 95% CI 1.9% to 1.9%; 2002 vs. 1992 difference, −0.3%; 95% CI −0.4% to −0.3%; 2012 vs. 2002 difference, −0.1%; 95% CI −0.1% to 0.0%) (Table 5).

3.3. Disparities of Macronutrient Composition in Population Subgroups

The estimated percentages of energy intake from fat within population subgroups, stratified by education level, were 10.6% (95% CI 10.1–11.1%) in 1992, 9.5% (95% CI 8.8–10.1%) in 2002 and 6.3% (95% CI 5.8–6.7%) in 2010–2012. Those stratified by area were 6.8% (95% CI 6.7–7.0%) in 1982, 10.8% (95% CI 10.6–10.9%) in 1992, 8.9% (95% CI 8.7–9.1%) in 2002 and 7.4% (95% CI 7.2–7.6%) in 2010–2012. Those stratified by GNP level were 2.0% (95% CI 1.8–2.3%) in 1982, 7.3% (95% CI 7.1–7.6%) in 1992, 2.9% (95% CI 2.6–3.1%) in 2002 and 3.7% (95% CI 3.5–4.0%) in 2010–2012. The ranges of the estimated percentage of energy intake from carbohydrates within population subgroups stratified by education level were 13.1% (95% CI 12.6–13.6%) in 1992, 12.4% (95% CI 11.8–13.0%) in 2002 and 9.5% (95% CI 9.0–10.0%) in 2010–2012. Those stratified by area were 8.2% (95% CI 8.0–8.3%) in 1982, 12.9% (95% CI 12.7–13.0%) in 1992, 10.8% (95% CI 10.7–11.0%) in 2002 and 9.6% (95% CI 9.4–9.8%) in 2010–2012. Those stratified by GNP level were 2.9% (95% CI 2.6–3.2%) in 1982, 9.2% (95% CI 9.0–9.5%) in 1992, 4.8% (95% CI 4.5–5.1%) in 2002 and 5.8% (95% CI 5.5–6.1%) in 2010–2012 (Tables 2–5 and Figure 1). The trends and disparities stratified by age and sex are given in Tables A1–A4.

Table 1. Sociodemographic Characteristics of Participants by China National Nutrition Survey (CNNS) Rounds, 1982–2012.

		1982	1992	2002	2012
n		39,084	58,316	52,426	55,051
Age Group, year					
	20–29	12,642 (32.4)	16,116 (27.6)	7531 (14.4)	5310 (9.7)
	30–39	8729 (22.3)	13,840 (23.7)	12,959 (24.7)	7894 (14.3)
	40–49	6540 (16.7)	11,440 (19.6)	11,745 (22.4)	12,420 (22.6)
	50–59	5533 (14.2)	8429 (14.5)	10,201 (19.5)	12,828 (23.3)
	60–69	3662 (9.4)	5598 (9.6)	6630 (12.7)	10,308 (18.7)
	≥70	1978 (5.1)	2893 (5.0)	3360 (6.4)	6291 (11.4)
Sex					
	Male	19,432 (49.7)	28,010 (48.0)	24,709 (47.1)	25,278 (45.9)
	Female	19,652 (50.3)	30,306 (52.0)	27,717 (52.9)	29,773 (54.1)
Education Level					
	under 6 years		23,479 (40.3)	6567 (12.5)	6901 (12.5)
	6 years		8477 (14.5)	15,686 (29.9)	15,866 (28.8)
	9 years		15,620 (26.8)	18,075 (34.5)	19,064 (34.6)
	12 years		7766 (13.3)	8433 (16.1)	8454 (15.4)
	15 years		1514 (2.6)	2452 (4.7)	2740 (5.0)
	over 15 years		1148 (2.0)	1115 (2.1)	2026 (3.7)
	No answer		312 (0.5)	98 (0.2)	0 (0.0)
Area					
	Urban	13,766 (35.2)	17,633 (30.2)	17,530 (33.4)	27,471 (49.9)
	Rural	25,318 (64.8)	40,683 (69.8)	34,896 (66.6)	27,580 (50.1)
GNP Level [1]					
	First Class	7195 (18.4)	16,635 (28.5)	15,374 (29.3)	15,384 (27.9)
	Second Class	11,758 (30.1)	13,802 (23.7)	10,916 (20.8)	14,244 (25.9)
	Third Class	8508 (21.8)	14,544 (24.9)	14,693 (28.0)	13,818 (25.1)
	Fourth Class	11,623 (29.7)	13,335 (22.9)	11,443 (21.8)	11,605 (21.1)

CNNS, China National Nutrition Survey; GNP, gross national product. Data are numbers of participants (%), unless otherwise indicated. [1] GNP level was classified at provincial level according to the GNP quartiles across provinces. In 1982, the first to fourth classes were classified as USD ≥284, (244, 284), (194, 244) and <194, respectively; in 1992, the first to fourth class were classified as USD ≥482, (354, 482), (268, 354) and <268; in 2002, the first to fourth classes were classified as USD ≥1569, (958, 1569), (743, 958) and <743; and in 2012, the first to fourth classes were classified as USD ≥8510, (5761, 8510), (4670, 5761) and <4670.

Table 2. Trends and Disparities in the Daily Energy Intake of Adults Aged 18 Years or Older by CNNS Round, 1982–2012 [1].

	Daily Energy Intake-Survey-Weighted Mean, kcal (95% CI)				*p* Value for Trend	Difference between Rounds, kcal (95% CI)		
	1982	1992	2002	2012		1992 vs. 1982	2002 vs. 1992	2012 vs. 2002
All	2614.7 (2606.5–2622.8)	2532.1 (2525.9–2538.2)	2196.6 (2190.4–2202.8)	2063.9 (2057.7–2070.2)	<0.01	−82.6 (−92.5 to −72.7)	−335.4 (−344.2 to −326.7)	−132.7 (−141.5 to −123.9)
Education level								
under 6 years		2488.7 (2478.3–2499.1)	2055.7 (2038.5–2072.8)	1882.0 (1865.7–1898.4)	<0.01		−433.0 (−454.2 to −411.8)	−173.6 (−197.4 to −149.9)
6 years		2618.8 (2602.5–2635.1)	2288.5 (2277.2–2299.8)	2126.4 (2114.4–2138.3)	<0.01		−330.3 (−349.9 to −310.8)	−162.1 (−178.5 to −145.6)
9 years		2589.0 (2577.4–2600.6)	2269.2 (2258.6–2279.9)	2155.5 (2144.5–2166.5)	<0.01		−319.8 (−335.6 to −303.9)	−113.8 (−129.1 to −98.4)
12 years		2494.3 (2479.0–2509.6)	2123.3 (2108.2–2138.4)	2015.2 (2000.3–2030.2)	<0.01		−371.0 (−392.6 to −349.5)	−108.1 (−129.4 to −86.7)
15 years		2500.7 (2467.0–2534.4)	2052.3 (2025.4–2079.2)	1911.1 (1886.9–1935.3)	<0.01		−448.4 (−491.6 to −405.1)	−141.2 (−177.3 to −105.2)
over 15 years		2472.2 (2436.3–2508.0)	2035.2 (1996.9–2073.5)	1883.3 (1855.6–1911.1)	<0.01		−437.0 (−489.5 to −384.5)	−151.9 (−198.6 to −105.2)
Range within subgroups		146.6 (109.8–183.4)	253.3 (220.9–285.6)	273.4 (250.7–296.2)				
Area								
Urban	2531.0 (2518.1–2543.9)	2423.6 (2413.5–2433.7)	1983.9 (1974.2–1993.7)	1897.2 (1889.4–1905.0)	<0.01	−107.4 (−123.5 to −91.2)	−439.7 (−453.8 to −425.7)	−86.7 (−98.9 to −74.6)
Rural	2703.9 (2693.3–2714.5)	2647.7 (2639.8–2655.6)	2423.5 (2416.0–2431.0)	2241.8 (2232.5–2251.1)	<0.01	−56.2 (−68.9 to −43.4)	−224.2 (−235.1 to −213.3)	−181.7 (−193.5 to −169.9)
Range within subgroups	172.9 (156.7–189.1)	224.1 (211.9–236.2)	439.6 (427.8–451.4)	344.6 (332.5–356.7)				
GNP level								
First class	2624.2 (2605.9–2642.5)	2443.4 (2432.9–2453.8)	2173.6 (2162.9–2184.3)	1963.9 (1953.2–1974.6)	<0.01	−180.8 (−200.0 to −161.6)	−269.7 (−284.7 to −254.8)	−209.7 (−224.8 to −194.5)
Second class	2555.3 (2540.6–2570.0)	2514.8 (2502.3–2527.4)	2140.9 (2127.7–2154.2)	1936.6 (1925.8–1947.5)	<0.01	−40.5 (−59.8 to −21.1)	−373.9 (−392.1 to −355.7)	−204.3 (−221.2 to −187.4)
Third class	2629.1 (2611.2–2647.1)	2538.4 (2526.3–2550.5)	2191.5 (2179.7–2203.4)	2141.6 (2128.8–2154.4)	<0.01	−90.7 (−111.3 to −70.1)	−346.9 (−363.9 to −329.9)	−49.9 (−67.3 to −32.5)
Fourth class	2654.6 (2639.7–2669.6)	2681.6 (2667.2–2696.0)	2299.3 (2285.0–2313.6)	2260.3 (2245.1–2275.6)	<0.01	27.0 (6.1 to 48.0)	−382.3 (−402.7 to −361.9)	−39.0 (−59.9 to −18.1)
Range within subgroups	99.3 (78.4–120.3)	238.3 (220.7–255.8)	158.4 (138.9–177.9)	323.7 (305.5–342.0)				

CNNS, China National Nutrition Survey; GNP, gross national product. [1] Data were adjusted for CNNS weights to be nationally representative. Values may not equal the difference between two years', or the highest and lowest subgroups', estimates because of rounding.

Table 3. Trends and Disparities in the Estimated Percentage of Energy Intake from Fat of Adults Aged 18 Years or Older by CNNS Round, 1982–2012 [1].

		Estimated Percentage of Energy Intake from Fat, Survey-Weighted % (95% CI)				*p* Value for Trend	Difference between Rounds, % (95% CI)		
		1982	1992	2002	2012		1992 vs. 1982	2002 vs. 1992	2012 vs. 2002
All		16.3 (16.2–16.4)	23.8 (23.7–23.9)	31.6 (31.5–31.7)	33.1 (33.0–33.2)	<0.01	7.6 (7.4 to 7.7)	7.7 (7.6 to 7.9)	1.6 (1.4 to 1.7)
Education level									
	under 6 years		19.9 (19.8–20.0)	28.0 (27.7–28.3)	30.4 (30.1–30.7)	<0.01		8.1 (7.8 to 8.4)	2.5 (2.0 to 2.9)
	6 years		22.2 (22.0–22.5)	28.9 (28.7–29.1)	31.1 (31.0–31.3)	<0.01		6.6 (6.3 to 6.9)	2.3 (2.0 to 2.5)
	9 years		25.2 (25.0–25.3)	31.0 (30.8–31.2)	32.9 (32.7–33.1)	<0.01		5.8 (5.6 to 6.1)	1.9 (1.7 to 2.1)
	12 years		28.2 (28.0–28.4)	34.8 (34.6–35.1)	35.1 (34.8–35.3)	<0.01		6.6 (6.2 to 6.9)	0.3 (−0.1 to 0.6)
	15 years		30.3 (29.8–30.8)	36.5 (36.1–37.0)	36.7 (36.3–37.1)	<0.01		6.2 (5.5 to 6.9)	0.1 (−0.5 to 0.7)
	over 15 years		30.5 (30.0–31.0)	37.4 (36.8–38.0)	36.6 (36.2–37.1)	<0.01		6.9 (6.1 to 7.7)	−0.8 (−1.6 to 0.0)
	Range within subgroups		10.6 (10.1–11.1)	9.5 (8.8–10.1)	6.3 (5.8–6.7)				
Area									
	Urban	19.6 (19.4–19.7)	29.0 (28.9–29.2)	35.8 (35.7–36.0)	36.7 (36.6–36.8)	<0.01	9.5 (9.3 to 9.7)	6.8 (6.6 to 7.0)	0.9 (0.7 to 1.1)
	Rural	12.7 (12.6–12.8)	18.3 (18.2–18.4)	27.0 (26.9–27.1)	29.3 (29.2–29.4)	<0.01	5.5 (5.4 to 5.7)	8.7 (8.6 to 8.9)	2.3 (2.1 to 2.5)
	Range within subgroups	6.8 (6.7–7.0)	10.8 (10.6–10.9)	8.9 (8.7–9.1)	7.4 (7.2–7.6)				
GNP level									
	First class	17.0 (16.8–17.2)	27.5 (27.4–27.7)	33.0 (32.8–33.1)	35.0 (34.9–35.2)	<0.01	10.5 (10.3 to 10.8)	5.4 (5.2 to 5.7)	2.1 (1.8 to 2.3)
	Second class	15.8 (15.6–15.9)	23.7 (23.5–23.9)	31.2 (31.0–31.4)	33.3 (33.1–33.5)	<0.01	8.0 (7.7 to 8.2)	7.5 (7.2 to 7.7)	2.1 (1.8 to 2.4)
	Third class	15.1 (14.9–15.3)	20.2 (20.0–20.4)	31.3 (31.1–31.5)	31.3 (31.1–31.5)	<0.01	5.1 (4.8 to 5.3)	11.1 (10.8 to 11.4)	0.0 (−0.3 to 0.3)
	Fourth class	17.1 (17.0–17.3)	21.9 (21.7–22.1)	30.1 (29.9–30.3)	32.5 (32.3–32.7)	<0.01	4.8 (4.5 to 5.1)	8.2 (7.9 to 8.5)	2.4 (2.1 to 2.7)
	Range within subgroups	2.0 (1.8–2.3)	7.3 (7.1–7.6)	2.9 (2.6–3.1)	3.7 (3.5–4.0)				

CNNS, China National Nutrition Survey; GNP, gross national product. [1] Data were adjusted for CNNS weights to be nationally representative. Values may not equal the difference between two years', or the highest and lowest subgroups', estimates because of rounding.

Table 4. Trends and Disparities in the Estimated Percentage of Energy Intake from Carbohydrates of Adults Aged 18 Years or Older by CNNS Round, 1982–2012 [1].

		Estimated Percentage of Energy Intake from Carbohydrates, Survey-Weighted % (95% CI)				*p* Value for Trend	Difference between Rounds, % (95% CI)		
		1982	1992	2002	2012		1992 vs. 1982	2002 vs. 1992	2012 vs. 2002
All		74.0 (73.8–74.1)	63.4 (63.3–63.5)	56.0 (55.9–56.1)	55.0 (54.9–55.1)	<0.01	−10.5 (−10.7 to −10.4)	−7.4 (−7.5 to −7.2)	−1.0 (−1.2 to −0.9)
Education level									
	under 6 years		68.1 (67.9–68.2)	60.5 (60.2–60.8)	59.1 (58.8–59.4)	<0.01		−7.6 (−7.9 to −7.3)	−1.4 (−1.8 to −1.0)
	6 years		65.4 (65.2–65.6)	59.5 (59.3–59.7)	57.7 (57.5–57.9)	<0.01		−5.9 (−6.2 to −5.6)	−1.8 (−2.0 to −1.5)
	9 years		61.9 (61.8–62.1)	56.8 (56.6–57.0)	55.4 (55.2–55.6)	<0.01		−5.1 (−5.4 to −4.9)	−1.4 (−1.7 to −1.2)
	12 years		58.1 (57.9–58.4)	51.9 (51.7–52.2)	52.4 (52.1–52.6)	<0.01		−6.2 (−6.5 to −5.8)	0.4 (0.1 to 0.8)
	15 years		55.4 (54.9–55.9)	49.6 (49.2–50.0)	50.0 (49.6–50.4)	<0.01		−5.8 (−6.5 to −5.1)	0.4 (−0.2 to 1.0)
	over 15 years		55.0 (54.4–55.5)	48.1 (47.5–48.7)	49.6 (49.2–50.1)	<0.01		−6.9 (−7.7 to −6.1)	1.5 (0.8 to 2.3)
	Range within subgroups		13.1 (12.6–13.6)	12.4 (11.8–13.0)	9.5 (9.0–10.0)				
Area									
	Urban	70.0 (69.8–70.2)	57.2 (57.0–57.3)	50.8 (50.6–51.0)	50.4 (50.2–50.5)	<0.01	−12.8 (−13.0 to −12.6)	−6.4 (−6.6 to −6.2)	−0.4 (−0.6 to −0.2)
	Rural	78.2 (78.0–78.3)	70.1 (70.0–70.2)	61.6 (61.5–61.7)	60.0 (59.8–60.1)	<0.01	−8.1 (−8.2 to −7.9)	−8.4 (−8.6 to −8.3)	−1.7 (−1.8 to −1.5)
	Range within subgroups	8.2 (8.0–8.3)	12.9 (12.7–13.0)	10.8 (10.7–11.0)	9.6 (9.4–9.8)				
GNP level									
	First class	72.7 (72.5–73.0)	58.7 (58.5–58.8)	53.6 (53.4–53.7)	51.6 (51.4–51.8)	<0.01	−14.1 (−14.3 to −13.8)	−5.1 (−5.3 to −4.9)	−1.9 (−2.2 to −1.7)
	Second class	74.5 (74.3–74.7)	63.7 (63.5–63.9)	56.5 (56.3–56.8)	55.4 (55.2–55.6)	<0.01	−10.8 (−11.0 to −10.5)	−7.2 (−7.5 to −6.9)	−1.1 (−1.4 to −0.8)
	Third class	75.6 (75.4–75.8)	67.9 (67.7–68.0)	56.8 (56.6–57.0)	57.4 (57.2–57.6)	<0.01	−7.8 (−8.0 to −7.5)	−11.1 (−11.3 to −10.8)	0.6 (0.3 to 0.9)
	Fourth class	72.9 (72.7–73.1)	65.9 (65.7–66.1)	58.4 (58.1–58.6)	56.3 (56.0–56.5)	<0.01	−7.0 (−7.3 to −6.7)	−7.6 (−7.9 to −7.2)	−2.1 (−2.4 to −1.8)
	Range within subgroups	2.9 (2.6–3.2)	9.2 (9.0–9.5)	4.8 (4.5–5.1)	5.8 (5.5–6.1)				

CNNS, China National Nutrition Survey; GNP, gross national product. [1] Data were adjusted for CNNS weights to be nationally representative. Values may not equal the difference between two years', or the highest and lowest subgroups', estimates because of rounding.

Table 5. Trends and Disparities in the Estimated Percentage of Energy Intake from Protein of Adults Aged 18 Years or Older by CNNS Round, 1982–2012 [1].

		Estimated Percentage of Energy Intake from Protein, Survey-Weighted % (95% CI)				p Value for Trend	Difference between Rounds, % (95% CI)		
		1982	1992	2002	2012		1992 vs. 1982	2002 vs. 1992	2012 vs. 2002
All		10.9 (10.8–10.9)	12.8 (12.7–12.8)	12.4 (12.4–12.4)	12.3 (12.3–12.4)	<0.01	1.9 (1.9 to 1.9)	−0.3 (−0.4 to −0.3)	−0.1 (−0.1 to 0.0)
Education level									
	under 6 years		12.0 (12.0–12.0)	11.5 (11.5–11.6)	11.2 (11.2–11.3)	<0.01		−0.5 (−0.6 to −0.4)	−0.3 (−0.4 to −0.2)
	6 years		12.4 (12.3–12.4)	11.7 (11.6–11.7)	11.5 (11.4–11.5)	<0.01		−0.7 (−0.8 to −0.6)	−0.2 (−0.2 to −0.1)
	9 years		12.9 (12.9–13.0)	12.2 (12.2–12.2)	12.1 (12.0–12.1)	<0.01		−0.7 (−0.8 to −0.6)	−0.1 (−0.2 to −0.1)
	12 years		13.7 (13.6–13.7)	13.2 (13.2–13.3)	13.2 (13.1–13.2)	<0.01		−0.4 (−0.5 to −0.3)	−0.1 (−0.2 to 0.0)
	15 years		14.2 (14.0–14.4)	13.8 (13.7–14.0)	14.1 (14.0–14.3)	0.83		−0.4 (−0.6 to −0.2)	0.3 (0.1 to 0.5)
	over 15 years		14.5 (14.3–14.7)	14.5 (14.3–14.7)	14.6 (14.4–14.8)	0.40		0.0 (−0.3 to 0.2)	0.1 (−0.1 to 0.4)
	Range within subgroups		2.5 (2.4–2.6)	3.0 (2.8–3.1)	3.4 (3.3–3.5)				
Area									
	Urban	11.2 (11.2–11.3)	13.8 (13.7–13.8)	13.4 (13.3–13.4)	13.5 (13.4–13.5)	<0.01	2.6 (2.5 to 2.6)	−0.4 (−0.5 to −0.3)	0.1 (0.0 to 0.2)
	Rural	10.5 (10.5–10.5)	11.7 (11.6–11.7)	11.4 (11.4–11.4)	11.2 (11.1–11.2)	<0.01	1.2 (1.1 to 1.2)	−0.3 (−0.3 to −0.2)	−0.2 (−0.3 to −0.2)
	Range within subgroups	0.7 (0.7–0.8)	2.1 (2.1–2.1)	2.0 (1.9–2.0)	2.3 (2.3–2.4)				
GNP level									
	First class	11.0 (10.9–11.0)	13.8 (13.8–13.9)	13.5 (13.4–13.5)	13.6 (13.5–13.7)	<0.01	2.8 (2.8 to 2.9)	−0.3 (−0.4 to −0.3)	0.1 (0.0 to 0.2)
	Second class	11.2 (11.2–11.2)	12.5 (12.5–12.6)	12.3 (12.2–12.3)	12.0 (12.0–12.1)	<0.01	1.3 (1.3 to 1.4)	−0.3 (−0.3 to −0.2)	−0.3 (−0.3 to −0.2)
	Third class	10.6 (10.5–10.6)	12.0 (11.9–12.0)	11.9 (11.9–12.0)	11.8 (11.8–11.9)	<0.01	1.4 (1.3 to 1.4)	0.0 (−0.1 to 0.0)	−0.1 (−0.2 to 0.0)
	Fourth class	10.7 (10.7–10.7)	12.1 (12.1–12.2)	11.5 (11.5–11.6)	11.7 (11.6–11.8)	<0.01	1.4 (1.4 to 1.5)	−0.6 (−0.7 to −0.5)	0.2 (0.1 to 0.2)
	Range within subgroups	0.6 (0.5–0.7)	1.9 (1.8–1.9)	1.9 (1.9–2.0)	1.9 (1.8–2.0)				

CNNS, China National Nutrition Survey; GNP, gross national product. [1] Data were adjusted for CNNS weights to be nationally representative. Values may not equal the difference between two years', or the highest and lowest subgroups', estimates because of rounding.

Figure 1. Trends and Disparities between Two Ends of Subgroups with regards to Estimated Energy Intake from Macronutrients of Adults Aged 18 Years or Older by CNNS Round, 1982–2012, Stratified by Area, Education Level and GNP Level. The two polygonal lines in each graph represent the two ends of subgroups. (**A–C**) show percentages of energy intake from fat across survey rounds. (**D–F**) show percentages of energy intake from carbohydrates across survey rounds. (**G–I**) show percentages of energy intake from protein across survey rounds.

4. Discussion

China has made substantial progress in improving nutrition. Diet quality improved remarkably from 1982 to 2012 in China. The trends of energy intake constantly decreased in the survey rounds due to the fast pace of modernization and urbanization. The percentage of fat's contribution to energy spiked, that of carbohydrates fell all the way, and that of protein stabilized within a small range. The macronutrient composition went from poor, to ideal, and then to far from ideal again. Though the composition was not satisfying at the beginning round of CNNS, in 1982, which featured excessive carbohydrates and a lack of fat, it became more ideal in the 1992 survey round. The macronutrient composition was within the national recommendations among most subpopulations around that period [9]. However, in the most recent two surveys, the macronutrient composition dropped out of the ideal range, which led to health conditions diametrically opposed to malnutrition, i.e., overnutrition, potentially contributing to the prevalence of nutrition-related non-communicable chronic diseases (NCDs) nation-wide [10]. We considered that different fat compositions at the same level of energy intake could have diverse impacts on the development of obesity. It seemed a paradox in China that overweight and obesity dramatically increased since 1980s, despite energy intake constantly decreasing [1,11]. Reduced physical activity could explain the increasing prevalence of obesity, but most developed countries, like America and Korea, also experience both obesity prevalence and raising energy intakes [12–15]. Indeed, few countries, like Japan, had a similar situation to ours, whereby the obesity rate went up as the energy intake decreased [16]. New studies suggested that the percentage of fat contributing to energy could be the cause of adiposity, but not carbohydrates or protein [17]. In fact, the proportion of fat in the diet kept going up worldwide, as did the prevalence of obesity [13,16,18,19]. The current findings were based on massive samples and observations over the long-term, which might provide new thoughts as to the cause of obesity.

The great achievements following the social and economic shifts after 1979 had a tremendous impact on the diet of the Chinese population [20]. It took no more than one decade for the Chinese people to go from lacking various foods, to having plenty of every food. There was a big leap in nutrition improvement, and diet patterns changed most in the 1980s and 1990s. The macronutrient composition rapidly reached the ideal range at that time. The pace of the change of macronutrient composition slowed down, and it has been unsatisfying in recent years. The promoter of the diet has shifted. Economy and food supply were still continuing to improve, but it was contributing only a little to diet improvement in China. Other things, like nutrition policy retargeting or the availability of nutrition education and knowledge, might be the key to promoting diet quality in China.

The disparities persistently existed in different subpopulations across China, but the gaps narrowed in recent years. The Chinese government has put huge effort into poverty reduction, transportation system construction and raising the agricultural yield, which all potentially increased the equity of access to various foods by people with different background. Especially in the most recent survey round, the percentages of fats' and carbohydrates' contributions to energy were getting closer between the two ends of the subpopulations as regards area, education level and economic background. It was obvious that the subpopulations with better social profiles (living in urban areas, well-educated and wealthy economic background) were leading the diet trends, and the rest followed in the next decade or two. Nevertheless, the macronutrient compositions of those with better social profiles had been moving toward the overnutrition pattern since around 2002, which was probably a major cause of nutrition-related NCDs prevailing in China [3,10,21]. If people with low social profiles continue to follow the diet trend, there might be another surge of nutrition-related NCDs in China. Moreover, inequalities in health resource access have existed for some time in China [22]. It would deepen the social contradictions if those who suffered from diseases could not be able to access necessary health resources. More governmental interventions should be launched into the subpopulation with low social profiles in order to slow down or even curb their movement into overnutrition.

One promising trend was discovered in the well-educated subpopulation. In the survey round of 2012, the macronutrient composition distinctively retuned to the recommended ranges among these people. "Eat well" was linked to "live well" in Chinese culture, but people always confused "eat well" with "eat whatever one wants". Actually, "eat well" means "eat properly" in the modern nutritional theory, and it leads to "live well". It is clear that some risks of nutrition-related NCDs can be modified though education improvement [23]. Well-educated people have greater volition and ability to acquire health information which might help them regulate dietary behaviors, rather than following their instinctive appetite or preference. Health education would probably be a useful tool to help China get through the possible dilemma of a further potential surge of NCDs in the subpopulations with low social profiles.

This study has several limitations. First, 3-day 24-h dietary recalls were used to obtain food consumption information, and so the accuracy of dietary intake was mostly dependent on the participants' recall and estimation. Second, for the individual income information variabilities in different survey rounds, the classification of GNP level was applied. It was based on each province's GNP in the survey year, which might not classify each participant meticulously. Third, a recent study mentioned that the quality or food sources of macronutrients might lead to different health outcomes [24]. Diet quality in the current study was determined based on the macronutrient composition, which might cause bias without taking food composition into consideration. Fourth, the inference of macronutrient composition and consequential health outcome in the discussion was only derived from reports on the national level in an ecological way, rather than the relationships among CNNS participants.

5. Conclusions

Quick improvements in society and the economy effectively curbed undernutrition, but easily triggered larger-scaled overnutrition soon after in China. Disparities have persistently existed in different subpopulations, while these gaps would narrow if major efforts were made. Populations with low social profiles might lag behind the trend of diet transition, but would be more vulnerable to the side-effects of the trend. Education might be a promising way of preventing overnutrition during the prosperous progress of developing countries. Low social profile populations require specific interventions so as to avoid the further burdens of diet-related non-communicable diseases, in order to maintain social stability.

Author Contributions: Conceptualization, Y.H. and Z.Z.; Methodology, X.Y. and Y.H.; Investigation, Y.F., J.Z., Z.Y., Z.W., A.L., L.H., J.S. and Y.L.; Data Analysis and Writing: Original Draft Preparation, Z.Z.; Writing: Review and Editing, G.D. and Y.H.; Supervision and manuscript review, Y.H. All authors have read and agreed to the published version of the manuscript.

Funding: This study was supported by the National Health Commission of China, the National Key Research and Development Program of China (2018YFC1315303) and Ministry of Science and Technology of China, and Foundation of Shanghai Health Commission (201740073).

Acknowledgments: We thank all the team members and participants involved in the China National Nutrition Surveys. The 1982 CNNS was supported by the Major Program of National Medical and Health Research of China's Ministry of Health. The Ministry of Health in China provided special funding support to the 1992 CNNS, which was organized by the Ministry of Health, Ministry of Agriculture, Ministry of Public Security, and National Bureau of Statistics. The 2002 CNNS was supported by the Ministry of Health and the Ministry of Science and Technology in China (2001-DEA30035, 2003-DIA6N008), UNICEF, WHO, Unilever China and Danone Nutrition Institute China. The 2010–2012 CNNS was supported by the Special Fund for Health-Scientific Research in the Public Interest (20120212) from the National Health and Family Planning Commission of the People's Republic of China.

Conflicts of Interest: The authors declare no conflict of interest.

Appendix A

Table A1. Trends and Disparities in the Daily Energy Intake of Adults Aged 18 Years or Older by CNNS Round, 1982–2012 [1] (kcal).

		Daily Energy Intake, Survey-Weighted Mean, kcal (95% CI)				p Value for Trend	Difference between Rounds, kcal (95% CI)		
		1982	1992	2002	2012		1992 vs. 1982	2002 vs. 1992	2012 vs. 2002
All		2614.7 (2606.5, 2622.8)	2532.1 (2525.9, 2538.2)	2196.6 (2190.4, 2202.8)	2063.9 (2057.7, 2070.2)	<0.01	−82.6 (−92.5 to −72.7)	−335.4 (−344.2 to −326.7)	−132.7 (−141.5 to −123.9)
Age group-y									
	20–29	2843.2 (2829.0, 2857.5)	2639.9 (2627.8, 2651.9)	2209.3 (2193.0, 2225.6)	2054.3 (2034.8, 2073.9)	<0.01	−203.4 (−221.8 to −184.9)	−430.6 (−449.7 to −411.4)	−154.9 (−180.0 to −129.9)
	30–39	2798.9 (2782.0, 2815.8)	2625.7 (2613.5, 2638.0)	2262.8 (2250.3, 2275.4)	2128.8 (2111.7, 2145.9)	<0.01	−173.2 (−193.4 to −153.1)	−362.9 (−380.5 to −345.3)	−134.0 (−154.5 to −113.6)
	40–49	2700.0 (2681.3, 2718.8)	2594.4 (2581.1, 2607.7)	2254.8 (2241.6, 2267.9)	2164.5 (2150.9, 2178.1)	<0.01	−105.6 (−127.6 to −83.7)	−339.7 (−358.3 to −321.0)	−90.3 (−109.2 to −71.4)
	50–59	2504.3 (2483.3, 2525.3)	2526.9 (2511.3, 2542.5)	2238.8 (2225.0, 2252.7)	2092.7 (2080.0, 2105.4)	<0.01	22.6 (−2.8 to 48.0)	−288.1 (−308.8 to −267.3)	−146.1 (−164.9 to −127.4)
	60–69	2181.6 (2160.0, 2203.3)	2360.2 (2341.4, 2378.9)	2103.9 (2087.4, 2120.5)	1955.1 (1942.5, 1967.7)	<0.01	178.5 (150.4 to 206.7)	−256.3 (−281.2 to −231.4)	−148.8 (−169.0 to −128.6)
	≥70	1963.5 (1937.0, 1990.1)	2002.9 (1978.1, 2027.7)	1836.3 (1814.8, 1857.8)	1695.5 (1680.6, 1710.4)	<0.01	39.4 (3.3 to 75.5)	−166.6 (−199.2 to −134.0)	−140.8 (−165.5 to −116.1)
	Range within subgroups	879.7 (851.1, 908.3)	637.0 (612.2, 661.7)	426.5 (402.0, 451.1)	469.0 (444.8, 493.2)				
Sex									
	Male	2857.7 (2845.9, 2869.4)	2738.7 (2729.7, 2747.7)	2382.6 (2373.4, 2391.8)	2241.6 (2232.0, 2251.2)	<0.01	−119.0 (−133.4 to −104.5)	−356.1 (−369.0 to −343.2)	−141.0 (−154.3 to −127.7)
	Female	2366.9 (2356.9, 2377.0)	2321.4 (2313.7, 2329.2)	2007.1 (1999.4, 2014.7)	1882.8 (1875.3, 1890.4)	<0.01	−45.5 (−57.9 to −33.2)	−314.4 (−325.2 to −303.5)	−124.2 (−134.9 to −113.5)
	Range within subgroups	490.7 (475.2, 506.2)	417.3 (405.5, 429.1)	375.5 (363.6, 387.5)	358.8 (346.7, 370.8)				

CNNS, China National Nutrition Survey. [1] Data were adjusted for CNNS weights to be nationally representative. Values may not equal the difference between two years', or the highest and lowest subgroups', estimates because of rounding.

Table A2. Trends and Disparities in the Estimated Percentage of Energy Intake from Fat of Adults Aged 18 Years or Older by CNNS Round, 1982–2012 [1].

		Estimated Percentage of Energy Intake from Fat, Survey-Weighted % (95% CI)				*p* Value for Trend	Difference between Rounds, % (95% CI)		
		1982	1992	2002	2012		1992 vs. 1982	2002 vs. 1992	2012 vs. 2002
All		16.3 (16.2, 16.4)	23.8 (23.7, 23.9)	31.6 (31.5, 31.7)	33.1 (33.0, 33.2)	<0.01	7.6 (7.4 to 7.7)	7.7 (7.6 to 7.9)	1.6 (1.4 to 1.7)
Age group-y									
	20–29	16.9 (16.8, 17.1)	24.4 (24.2, 24.6)	32.0 (31.7, 32.3)	33.8 (33.5, 34.2)	<0.01	7.5 (7.2 to 7.7)	7.6 (7.3 to 7.9)	1.8 (1.4 to 2.2)
	30–39	16.6 (16.4, 16.8)	24.7 (24.5, 24.9)	31.5 (31.3, 31.7)	33.4 (33.2, 33.7)	<0.01	8.1 (7.8 to 8.4)	6.8 (6.5 to 7.0)	2.0 (1.6 to 2.3)
	40–49	15.7 (15.5, 15.9)	23.3 (23.1, 23.5)	31.5 (31.3, 31.8)	33.3 (33.1, 33.5)	<0.01	7.6 (7.3 to 7.9)	8.3 (8.0 to 8.6)	1.8 (1.5 to 2.1)
	50–59	16.1 (15.8, 16.3)	23.6 (23.4, 23.8)	31.4 (31.2, 31.7)	32.8 (32.6, 33.0)	<0.01	7.5 (7.2 to 7.9)	7.8 (7.5 to 8.2)	1.4 (1.1 to 1.7)
	60–69	16.2 (15.9, 16.5)	23.1 (22.8, 23.4)	31.3 (31.0, 31.6)	31.9 (31.7, 32.1)	<0.01	6.9 (6.5 to 7.3)	8.2 (7.8 to 8.6)	0.6 (0.3 to 1.0)
	≥70	15.6 (15.2, 16.0)	22.7 (22.3, 23.2)	31.1 (30.7, 31.5)	31.7 (31.4, 32.0)	<0.01	7.1 (6.5 to 7.7)	8.3 (7.8 to 8.9)	0.7 (0.2 to 1.2)
	Range within subgroups	1.3 (1.0, 1.7)	2.0 (1.6, 2.4)	0.9 (0.4, 1.5)	2.1 (1.6, 2.6)				
Sex									
	Male	16.2 (16.1, 16.4)	23.6 (23.5, 23.7)	31.4 (31.3, 31.6)	32.9 (32.8, 33.1)	<0.01	7.4 (7.2 to 7.6)	7.8 (7.6 to 8.0)	1.5 (1.3 to 1.7)
	Female	16.3 (16.2, 16.4)	24.1 (23.9, 24.2)	31.7 (31.5, 31.8)	33.3 (33.2, 33.5)	<0.01	7.8 (7.6 to 7.9)	7.6 (7.4 to 7.8)	1.6 (1.4 to 1.8)
	Range within subgroups	0.1 (−0.1, 0.3)	0.5 (0.3, 0.6)	0.3 (0.0, 0.5)	0.4 (0.2, 0.6)				

CNNS, China National Nutrition Survey. [1] Data were adjusted for CNNS weights to be nationally representative. Values may not equal the difference between two years', or the highest and lowest subgroups', estimates because of rounding.

Table A3. Trends and Disparities in the Estimated Percentage of Energy Intake from Carbohydrates of Adults Aged 18 Years or Older by CNNS Round, 1982–2012 [1].

		Estimated Percentage of Energy Intake from Carbohydrates, Survey-Weighted % (95% CI)				*p* Value for Trend	Difference between Rounds, % (95% CI)		
		1982	1992	2002	2012		1992 vs. 1982	2002 vs. 1992	2012 vs. 2002
All		74.0 (73.8, 74.1)	63.4 (63.3, 63.5)	56.0 (55.9, 56.1)	55.0 (54.9, 55.1)	<0.01	−10.5 (−10.7 to −10.4)	−7.4 (−7.5 to −7.2)	−1.0 (−1.2 to −0.9)
Age group-y									
	20–29	73.2 (73.0, 73.4)	62.9 (62.7, 63.1)	55.4 (55.2, 55.7)	54.2 (53.9, 54.6)	<0.01	−10.3 (−10.5 to −10.0)	−7.5 (−7.8 to −7.2)	−1.2 (−1.7 to −0.8)
	30–39	73.6 (73.4, 73.8)	62.3 (62.1, 62.5)	56.1 (55.9, 56.3)	54.6 (54.3, 54.8)	<0.01	−11.3 (−11.6 to −11.0)	−6.2 (−6.5 to −5.9)	−1.5 (−1.9 to −1.2)
	40–49	74.6 (74.3, 74.8)	64.0 (63.8, 64.3)	56.1 (55.8, 56.3)	54.8 (54.6, 55.0)	<0.01	−10.5 (−10.9 to −10.2)	−8.0 (−8.3 to −7.7)	−1.2 (−1.6 to −0.9)
	50–59	74.2 (73.9, 74.5)	63.7 (63.4, 64.0)	56.3 (56.1, 56.5)	55.3 (55.1, 55.5)	<0.01	−10.5 (−10.9 to −10.1)	−7.4 (−7.7 to −7.0)	−1.0 (−1.3 to −0.7)
	60–69	74.1 (73.7, 74.4)	64.2 (63.9, 64.5)	56.4 (56.1, 56.7)	56.5 (56.3, 56.8)	<0.01	−9.9 (−10.3 to −9.4)	−7.8 (−8.3 to −7.4)	0.2 (−0.2 to 0.5)
	≥70	74.6 (74.2, 75.1)	64.5 (64.1, 65.0)	56.5 (56.1, 56.9)	56.7 (56.4, 57.0)	<0.01	−10.1 (−10.8 to −9.5)	−8.0 (−8.6 to −7.4)	0.3 (−0.2 to 0.8)
	Range within subgroups	1.5 (1.1, 1.8)	2.2 (1.8, 2.6)	1.0 (0.5, 1.6)	2.5 (2.0, 3.0)				
Sex									
	Male	73.9 (73.8, 74.1)	63.7 (63.5, 63.8)	56.2 (56.0, 56.4)	54.7 (54.6, 54.9)	<0.01	−10.3 (−10.5 to −10.1)	−7.5 (−7.7 to −7.2)	−1.5 (−1.7 to −1.3)
	Female	74.0 (73.8, 74.1)	63.2 (63.0, 63.3)	55.9 (55.7, 56.0)	55.3 (55.2, 55.5)	<0.01	−10.8 (−11.0 to −10.6)	−7.3 (−7.5 to −7.1)	−0.5 (−0.7 to −0.3)
	Range within subgroups	0.0 (−0.2, 0.2)	0.5 (0.3, 0.7)	0.3 (0.1, 0.6)	0.6 (0.4, 0.8)				

CNNS, China National Nutrition Survey. [1] Data were adjusted for CNNS weights to be nationally representative. Values may not equal the difference between two years', or the highest and lowest subgroups', estimates because of rounding.

Table A4. Trends and Disparities in the Estimated Percentage of Energy Intake from Protein of Adults Aged 18 Years or Older by CNNS Round, 1982–2012 [1].

		Estimated Percentage of Energy Intake from Protein, Survey-Weighted % (95% CI)				p Value for Trend	Difference between Rounds, % (95% CI)		
		1982	1992	2002	2012		1992 vs. 1982	2002 vs. 1992	2012 vs. 2002
All		10.9 (10.8, 10.9)	12.8 (12.7, 12.8)	12.4 (12.4, 12.4)	12.3 (12.3, 12.4)	<0.01	1.9 (1.9 to 1.9)	−0.3 (−0.4 to −0.3)	−0.1 (−0.1 to 0.0)
Age group-y									
	20–29	10.9 (10.9, 10.9)	12.7 (12.6, 12.7)	12.5 (12.5, 12.6)	12.8 (12.7, 12.9)	<0.01	1.8 (1.7 to 1.8)	−0.1 (−0.2 to −0.1)	0.2 (0.1 to 0.4)
	30–39	10.9 (10.9, 11.0)	13.0 (12.9, 13.0)	12.4 (12.4, 12.5)	12.5 (12.4, 12.6)	<0.01	2.1 (2.0 to 2.1)	−0.6 (−0.6 to −0.5)	0.1 (0.0 to 0.2)
	40–49	10.9 (10.8, 10.9)	12.7 (12.6, 12.7)	12.4 (12.3, 12.5)	12.2 (12.1, 12.2)	<0.01	1.8 (1.7 to 1.9)	−0.3 (−0.4 to −0.2)	−0.2 (−0.3 to −0.1)
	50–59	10.8 (10.7, 10.8)	12.7 (12.6, 12.8)	12.3 (12.2, 12.3)	12.1 (12.0, 12.1)	<0.01	1.9 (1.8 to 2.0)	−0.4 (−0.5 to −0.3)	−0.2 (−0.3 to −0.1)
	60–69	10.8 (10.7, 10.9)	12.7 (12.6, 12.8)	12.3 (12.3, 12.4)	12.0 (11.9, 12.0)	<0.01	1.9 (1.8 to 2.0)	−0.4 (−0.5 to −0.2)	−0.3 (−0.4 to −0.3)
	≥70	10.8 (10.7, 10.9)	12.8 (12.6, 12.9)	12.4 (12.3, 12.6)	12.1 (12.1, 12.2)	<0.01	1.9 (1.8 to 2.1)	−0.3 (−0.5 to −0.1)	−0.3 (−0.4 to −0.2)
	Range within subgroups	0.1 (0.1, 0.2)	0.3 (0.2, 0.4)	0.3 (0.2, 0.4)	0.8 (0.7, 0.9)				
Sex									
	Male	10.9 (10.9, 10.9)	12.7 (12.7, 12.8)	12.4 (12.3, 12.4)	12.3 (12.3, 12.4)	<0.01	1.9 (1.8 to 1.9)	−0.4 (−0.4 to −0.3)	0.0 (−0.1 to 0.0)
	Female	10.9 (10.8, 10.9)	12.8 (12.7, 12.8)	12.5 (12.4, 12.5)	12.4 (12.3, 12.4)	<0.01	1.9 (1.9 to 2.0)	−0.3 (−0.4 to −0.3)	−0.1 (−0.1 to 0.0)
	Range within subgroups	0.0 (0.0, 0.1)	0.0 (0.0, 0.1)	0.1 (0.0, 0.1)	0.0 (0.0, 0.1)				

CNNS, China National Nutrition Survey. [1] Data were adjusted for CNNS weights to be nationally representative. Values may not equal the difference between two years', or the highest and lowest subgroups', estimates because of rounding.

References

1. Du, S.F.; Wang, H.J.; Zhang, B.; Zhai, F.Y.; Popkin, B.M. China in the period of transition from scarcity and extensive undernutrition to emerging nutrition-related non-communicable diseases, 1949–1992. *Obes Rev.* **2014**, *15*, 8–15. [CrossRef] [PubMed]
2. Zhou, M.; Wang, H.; Zeng, X.; Yin, P.; Zhu, J.; Chen, W.; Li, X.; Wang, L.; Wang, L.; Liu, Y.; et al. Mortality, morbidity, and risk factors in China and its provinces, 1990–2017: A systematic analysis for the Global Burden of Disease Study 2017. *Lancet* **2019**. [CrossRef]
3. He, Y.; Li, Y.; Yang, X.; Hemler, E.C.; Fang, Y.; Zhao, L.; Zhang, J.; Yang, Z.; Wang, Z.; He, L.; et al. The dietary transition and its association with cardiometabolic mortality among Chinese adults, 1982–2012: A cross-sectional population-based study. *Lancet Diabetes Endocrinol.* **2019**. [CrossRef]
4. Piernas, C.; Wang, D.; Du, S.; Zhang, B.; Wang, Z.; Su, C.; Popkin, B.M. The double burden of under- and overnutrition and nutrient adequacy among Chinese preschool and school-aged children in 2009–2011. *Eur. J. Clin. Nutr.* **2015**, *69*, 1323–1329. [CrossRef] [PubMed]
5. He, Y.; Zhao, W.; Zhang, J.; Zhao, L.; Yang, Z.; Huo, J.; Yang, L.; Wang, J.; He, L.; Sun, J.; et al. Data resource profile: China national nutrition surveys. *Int. J. Epidemiol.* **2019**, *48*, 368. [CrossRef] [PubMed]
6. Yang, Y.; Wang, G.; Pan, X. *Chinese Food Composition Book 1*, 2nd ed.; Peking University Medical Press: Beijing, China, 2009; pp. 45–192. ISBN 9787811167276.
7. Yang, Y.; Wang, G.; Pan, X. *Chinese Food Composition Book 2*, 1st ed.; Peking University Medical Press: Beijing, China, 2005; pp. 75–216. ISBN 9787810716789.
8. Institute of Health, Chinese Academy of Medical Science. *China Food Composition Table 1981*; China's Medical Publishing House: Beijing, China, 1981.
9. Chinese Nutrition Society. *Chinese Dietary Reference Intakes*; Science Press: Beijing, China, 2013.
10. Yang, G.; Wang, Y.; Zeng, Y.; Gao, G.F.; Liang, X.; Zhou, M.; Wan, X.; Yu, S.; Jiang, Y.; Naghavi, M.; et al. Rapid health transition in China, 1990-2010: Findings from the Global Burden of Disease Study 2010. *Lancet* **2013**, *381*, 1987–2015. [CrossRef]
11. Tian, Y.; Jiang, C.; Wang, M.; Cai, R.; Zhang, Y.; He, Z.; Wang, H.; Wu, D.; Wang, F.; Liu, X.; et al. BMI, leisure-time physical activity, and physical fitness in adults in China: Results from a series of national surveys, 2000–2014. *Lancet Diabetes Endocrinol.* **2016**, *4*, 487–497. [CrossRef]
12. Ford, E.S.; Dietz, W.H. Trends in energy intake among adults in the United States: Findings from NHANES. *Am. J. Clin. Nutr.* **2013**, *97*, 848–853. [CrossRef] [PubMed]
13. Afshin, A.; Forouzanfar, M.H.; Reitsma, M.B.; Sur, P.; Estep, K.; Lee, A.; Marczak, L.; Mokdad, A.H.; Moradi-Lakeh, M.; Naghavi, M.; et al. Health Effects of Overweight and Obesity in 195 Countries over 25 Years. *N. Engl. J. Med.* **2017**, *377*, 13–27. [CrossRef] [PubMed]
14. Yun, S.; Kim, H.J.; Oh, K. Trends in energy intake among Korean adults, 1998-2015: Results from the Korea National Health and Nutrition Examination Survey. *Nutr. Res. Pract.* **2017**, *11*, 147–154. [CrossRef] [PubMed]
15. Flegal, K.M.; Kruszon-Moran, D.; Carroll, M.D.; Fryar, C.D.; Ogden, C.L. Trends in Obesity Among Adults in the United States, 2005 to 2014. *JAMA* **2016**, *315*, 2284. [CrossRef] [PubMed]
16. Saito, A.; Imai, S.; Htun, N.C.; Okada, E.; Yoshita, K.; Yoshiike, N.; Takimoto, H. The trends in total energy, macronutrients and sodium intake among Japanese: Findings from the 1995–2016 National Health and Nutrition Survey. *Brit. J. Nutr.* **2018**, *120*, 424–434. [CrossRef] [PubMed]
17. Hu, S.; Wang, L.; Yang, D.; Li, L.; Togo, J.; Wu, Y.; Liu, Q.; Li, B.; Li, M.; Wang, G.; et al. Dietary fat, but not protein or carbohydrate, regulates energy intake and causes adiposity in mice. *Cell Metab.* **2018**, *28*, 415–431. [CrossRef] [PubMed]
18. Shan, Z.; Rehm, C.D.; Rogers, G.; Ruan, M.; Wang, D.D.; Hu, F.B.; Mozaffarian, D.; Zhang, F.F.; Bhupathiraju, S.N. Trends in dietary carbohydrate, protein, and fat intake and diet quality among US adults, 1999–2016. *JAMA* **2019**, *322*, 1178. [CrossRef]
19. Micha, R.; Khatibzadeh, S.; Shi, P.; Fahimi, S.; Lim, S.; Andrews, K.G.; Engell, R.E.; Powles, J.; Ezzati, M.; Mozaffarian, D. Global, regional, and national consumption levels of dietary fats and oils in 1990 and 2010: A systematic analysis including 266 country-specific nutrition surveys. *BMJ* **2014**, *348*, g2272. [CrossRef] [PubMed]
20. Kroeber, A.R. *China's Economy: What Everyone Needs to Know*; Oxford University Press: New York, NY, USA, 2016; ISBN 0190239034.

21. Yan, R.; Li, W.; Yin, L.; Wang, Y.; Bo, J.; Liu, L.; Liu, B.; Hu, B.; Chen, C.; Guo, J.; et al. Cardiovascular diseases and risk-factor burden in urban and rural communities in high-, middle-, and low-income regions of china: A large community-based epidemiological study. *J. Am. Heart Assoc.* **2017**, *6*. [CrossRef] [PubMed]
22. Song, S.; Yuan, B.; Zhang, L.; Cheng, G.; Zhu, W.; Hou, Z.; He, L.; Ma, X.; Meng, Q. Increased inequalities in health resource and access to health care in rural china. *Int. J. Environ. Res. Public Health* **2019**, *16*, 49. [CrossRef] [PubMed]
23. Yusuf, S.; Joseph, P.; Rangarajan, S.; Islam, S.; Mente, A.; Hystad, P.; Brauer, M.; Kutty, V.R.; Gupta, R.; Wielgosz, A.; et al. Modifiable risk factors, cardiovascular disease, and mortality in 155 722 individuals from 21 high-income, middle-income, and low-income countries (PURE): A prospective cohort study. *Lancet* **2020**, *395*, 795–808. [CrossRef]
24. Shan, Z.; Guo, Y.; Hu, F.B.; Liu, L.; Qi, Q. Association of Low-Carbohydrate and Low-Fat diets with mortality among US adults. *J. AMA Int. Med.* **2020**. [CrossRef] [PubMed]

nutrients

MDPI

Article

Association of Home Food Availability with Prediabetes and Diabetes among Adults in the United States

Jennifer R. McAtee [1], Meng-Hua Tao [2], Christian King [3] and Weiwen Chai [1,*]

1 Department of Nutrition and Health Sciences, University of Nebraska-Lincoln, 1700 N 35th Street, Lincoln, NE 68583, USA; jennifer.mcatee@huskers.unl.edu
2 Department of Biostatistics and Epidemiology, University of North Texas Health Science Center, 3500 Camp Bowie Boulevard., Fort Worth, TX 76107, USA; Menghua.Tao@unthsc.edu
3 Department of Health Management and Informatics, University of Central Florida, 500 W Livingston Street, Suite 402G, Orlando, FL 32801, USA; Christian.King@ucf.edu
* Correspondence: wchai2@unl.edu; Tel.: +1-402-472-7822

Received: 26 March 2020; Accepted: 20 April 2020; Published: 25 April 2020

Abstract: This study examined associations of home food availabilities with prediabetes and diabetes among 8929 adults (20–70 years) participating in 2007–2010 National Health and Nutrition Examination Surveys. Odds ratios (ORs) and 95% confidence intervals (95% CIs) were estimated by logistic regression. Relative to non-diabetic participants (individuals without diabetes or prediabetes), prediabetes participants were associated with lower availabilities of green vegetables (OR = 0.82; 95% CI = 0.73–0.91; $p = 0.0006$) and fat-free/low-fat milk (OR = 0.80, 95% CI = 0.65–0.89; $p = 0.001$) and higher sugary drink availability (OR = 1.24, 95% CI = 1.04–1.48; $p = 0.02$), adjusting for age, sex, and ethnicity (Model 1). The associations remained significant for vegetables ($p = 0.005$) and fat-free/low-fat milk ($p = 0.02$) adjusting for additional confounders (body mass index, education, Model 2). Adjusting for dietary components did not change the above results (in model 2) significantly. Participants with high healthy food availability scores had approximately 31% reduction ($p = 0.003$) in odds of prediabetes compared to those with low scores in Model 1. No associations were detected for diabetes except for fat-free/low-fat milk availability, for which an inverse association was observed in Model 1 (OR = 0.80, 95% CI = 0.65–0.99; $p = 0.04$). The results show prediabetes participants had lower availability of healthy foods and higher availability of unhealthy foods, suggesting the need to improve healthy food availability at home for this population.

Keywords: home food availability; diabetes; prediabetes; healthy food; unhealthy food; healthy food availability

1. Introduction

Diabetes is the seventh leading cause of death in the United States. In 2015, approximately 30.3 million (9.4% of the population) Americans had diabetes [1]. In addition, approximately one third of the U.S. adults have prediabetes [1], which is characterized by blood glucose levels that are elevated but not high enough to be diagnosed as diabetes [1]. However, only eleven percent of these individuals reported being aware of their prediabetes status [1]. Without proper treatment and prevention efforts, a majority (~70%) of persons with prediabetes will eventually become diabetic [2].

To date, the prevention of type 2 diabetes has mainly focused on behavioral modification and weight management, as causes of the disease are attributed largely to obesity, lack of physical activity, and poor dietary patterns [3–6]. Home food availability defined as the presence or absence of healthy and unhealthy food items at home is an important avenue to investigate since it may reflect people's food choices and their dietary intake [7–9]. It has been reported the availability of unhealthy foods in the home was significantly associated with energy intake in both adults and their adolescent children [7]. Among older adults (50 years old), home "obesogenic foods'" scores were predictive of intakes of nutrients such as saturated fat, sugar, and fiber [9]. In addition, previous work found that obese individuals had fewer healthy foods available in their homes than that of non-obese participants [10]. Considering the established relationship between diet, obesity, and metabolic diseases such as type 2 diabetes [3–6,11,12], it is likely that the home food environment may, in turn, contribute to the prevalence and development of prediabetes and/or type 2 diabetes. Research has shown the availability of healthy foods at home was associated with lower rates of prediabetes and diabetes among adolescents [13]. To our knowledge, no studies have examined these associations in adults using a nationally representative sample. Further, the home food environment and dietary habits could differ between persons with type 2 diabetes and those with prediabetes given the consideration that many prediabetes individuals are not aware of their prediabetes status [1]. Thus, when assessing the relationship between home food availability and metabolic disease conditions, it is important to distinguish individuals with prediabetes from those with diabetes. Therefore, the primary objective of this study was to examine associations of the availabilities of individual healthy (fruits, dark green vegetables, and fat-free/low-fat milk) and unhealthy foods (salty snacks and sugary drinks) at home as well as overall home healthy food availability with the presence of prediabetes or diabetes among adults aged 20–70 years using a nationally representative sample from the 2007–2010 National Health and Nutrition Examination Surveys (NHANES). In addition, we examined whether the above associations, if present were independent of individuals' dietary intake.

2. Materials and Methods

2.1. Study Population

NHANES is an ongoing program of studies intended to assess the health and nutritional status of approximately 5000 adults and children in the United States each year. NHANES uses a complex, multistage, probability sampling design to select participants who are geographically dispersed and representative of the civilian, noninstitutionalized US population [14].

The 2007–2010 data were used since questions regarding home food availability were only assessed in 2007–2010 NHANES surveys. There were 10,044 age-eligible respondents (20–70 years). Sequential exclusions included missing data for food availability, hemoglobin A1C (HbA1C) values, and response to diabetes diagnosis questions, with overlap from participants who were missing data in more than one category (N = 919). Additionally, individuals who were categorized as non-diabetic (participants without the presence of diabetes or prediabetes) but reported taking insulin or diabetes medications (N = 34), as well as participants who reported being pregnant (N = 114), were excluded from analysis. Furthermore, we also excluded participants who had missing data on key covariates (age, sex, race/ethnicity, body mass index [BMI], education) (N = 48). There were no statistically significant differences in sociodemographic characteristics between participants with and without these missing values. The final analytic sample included 8929 adults including 4448 men and 4481 women.

2.2. Relevant Measures and Independent/Dependent Variables

2.2.1. Home Food Availability of Individual Food Items

The home food availability questions from NHANES Consumer Behavior questionnaire measured the frequency of availability of certain healthy and unhealthy food and beverage items in the home [14]. Although the questions did not provide a specific reference evaluation period, it can be assumed that participants were asked "in general" or on a "typical day/week/month". Items considered "healthy" were fruits (fresh, dried, canned, and frozen fruits), dark green vegetables (fresh, dried, canned, and frozen vegetables), and fat-free/low-fat milk (1%, skim or fat-free; excluding 2%). Items considered "unhealthy" were salty snacks (such as chips and crackers; excluding nuts) and sugary drinks (soft drinks, fruit-flavored drinks, or fruit punch; excluding diet drinks, 100% juice, or sport drinks). The classification of "healthy" and "unhealthy" foods and beverages were derived from previous home food inventory tools and literature [7,15].

A five-point scale describing how often the food items were available in the home (always, most of the time, sometimes, rarely or never available) was used for survey responses and was coded on a scale of 1–5 with "1" referring "always" and "5" referring "never available" [14]. We further classified the above scale into two categories stressing the importance of contrasting the two values reflecting the frequency of home food availabilities: (1) "high availability" group (recoded as "1") for participants who responded with "always" or "most of the time available"; and (2) "low availability" group (recoded as "2") for participants who responded with "sometimes", "rarely", or "never available".

2.2.2. Overall Healthy Food Availability at Home

Using the responses to the questions of availabilities of the aforementioned five food items from NHANES Consumer Behavior questionnaire [14], a scoring system for the overall home healthy food availability [HFA] was created by summing the number of positive responses from 0 to 5. A participant scored 5 (maximum points) if he/she had all the positive responses (i.e., high availability for fruits, vegetables, and fat-free/low-fat milk; low availability for salty snacks and sugary drinks). Scores 0–4 were given corresponding to 0–4 (out of 5) positive responses. We further combined the scores into high, medium, and low categories. A high score (overall healthy) was defined as having an HFA score of 5 or 4; a medium score (borderline) was defined as having an HFA score of 3 or 2; and a low score (overall unhealthy) was defined as having an HFA score of 1 or 0. Unhealthy items (salty snacks, sugary drink) were reversely scored. High, medium, and low categories were recoded as "3", "2" and "1", respectively.

2.2.3. Prediabetes and Diabetes Status

The presence of diabetes was defined as having any of the following: (1) Hemoglobin A1c (HbA1c) concentration ≥6.5% [16]; or (2) self-report of diabetes diagnosis (yes on the question "did your doctor tell you that you have diabetes?"). Prediabetes was defined as: (1) HbA1c between 5.7% to 6.4% [16]; and (2) participants reporting "no" on the diabetes diagnosis question. Participants were categorized as non-diabetic (without the presence of diabetes or prediabetes) if they reported "no" on their diabetes status, were not taking any diabetes medications, and had a HbA1c laboratory measure lower than 5.7%.

2.2.4. Nutrient Intake Assessments

The NHANES dietary interview component gathers detailed dietary intake from participants. On two separate occasions, participants reported their food and beverage intake over the past 24 h using the USDA's Automated Multiple-Pass Method [17–19]. The two 24-h recalls were conducted in NHANES 2007 to 2010. The first dietary recall was collected in person by trained interviewers in NHANES Mobile Exam Center (MEC) and the second dietary recall was completed by trained interviewers via telephone 3–10 days after the MEC interview [18]. The data collected from each participant's two 24-h recall interviews were coded and linked to a database of foods and beverages and their nutrient compositions. The database was used to estimate the types and amounts of food and beverages (including water) consumed, as well as to estimate energy, nutrients, and other components from those food and beverage items. Dietary recall data were collected on both weekdays and weekend days. Dietary intakes can vary by day of the week. Thus, special dietary weights were computed for each survey cycle that adjust for the differences in the proportion of recalls on weekdays compared with weekend days [17,18]. The current study used both dietary recall interviews to establish a mean estimate of daily dietary intake for total energy, carbohydrate, protein, total fat, saturated fat, dietary fiber, and total sugar.

2.3. Statistical Analysis

The "Survey" procedure in SAS 9.4 software (SAS Institute, Cary, NC, USA) was used to estimate variance after incorporating the weights for the sample population in NHANES [20]. Participants' characteristics and dietary intake of total energy and major nutrients (carbohydrate, protein, total fat, saturated fat, dietary fiber, and sugar) were compared between diabetes, prediabetes, and non-diabetic participants using t test for continuous variables and chi-square test for categorical variables. Odds ratios (ORs), 95% confidence intervals (95% CIs) and p_{trend} for associations of home food availabilities with the presence of diabetes or prediabetes were estimated using logistic regression accounting for survey design and weights. The models were adjusted for age (continuous), sex, and race/ethnicity (whites, blacks, Hispanics, or other race/ethnicity) (Model 1), and additionally adjusted for BMI (continuous) and education (non-college graduates or college graduates) (Model 2). We also considered smoking status and physical activities as covariates but did not include it in the final models because it did not alter the estimates substantially. Previous research found indicators of socioeconomic status including both monetary factors such as income/poverty [21,22] and food security [23] and non-monetary factors such as education [22] can be important predictors of home food availability. However, comparing to economic resources such as income to poverty ratio, an individual's education background appeared to be a more significant predictor for the availability of both healthy and unhealthy foods at home from NHANES [22]. Thus, to avoid over-adjustment in this cross-sectional analysis, we only included the most relevant socioeconomic indicator (college education) as a covariate in the full models (Model 2).

To test whether associations of home food availabilities with diabetes or prediabetes were independent of individuals' dietary intake, our models were further included dietary variables as covariates (Model 3). Thus, Model 3 was adjusted for all the covariates in Model 2 as well as dietary variables including total energy, carbohydrate, sugar, fat (total) and saturated fat. Dietary two-day sample weights were applied in the analyses. All of the reported p-values were two-tailed, and statistical significance was set at 0.05. Since the NHANES database is publicly available, no Institutional Review Board approval was required.

3. Results

The current study included 8929 participants (4448 men and 4481 women) aged 20–70 years, of which 1197 participants (13.4%) had diabetes and 2075 (23.2%) had prediabetes. Study participants with prediabetes or diabetes were more likely to be older and blacks, have higher BMI and lower education achievement compared to participants without prediabetes or diabetes (non-diabetic participants). Moreover, participants with diabetes were less likely to be current smokers than those with prediabetes or non-diabetic individuals (Table 1). With respect to dietary intake, diabetes group consumed less total energy, carbohydrate, and sugar compared to non-diabetic or prediabetes group, while no significant differences in dietary intakes of major nutrients and total energy were observed between non-diabetic and prediabetes groups (Table 1).

Results of the availabilities of individual food items at home with the presence of prediabetes or diabetes were shown in Table 2. Compared to non-diabetic individuals, prediabetes participants were more likely to have low availabilities of dark green vegetables (OR = 0.82; 95% CI =0.73–0.91; $p = 0.0006$) and fat-free/low-fat milk (OR = 0.80; 95% CI = 0.65–0.89; $p = 0.001$), and high availability of sugary drinks (OR = 1.24; 95% CI = 1.04–1.48; $p = 0.02$) in model 1 adjusting for age, sex, and ethnicity. The associations remained statistically significant for green vegetables (OR = 0.86; 95% CI = 0.78–0.95; $p = 0.005$) and fat-free/low-fat milk availabilities (OR = 0.82; 95% CI = 0.69–0.97, $p = 0.02$) in Model 2 adjusting for additional confounders such as BMI and education. Home food availability was not associated with the presence of diabetes with the exception of fat-free/low-fat milk as an inverse association was observed for the availability of this food item in Model 1 (OR = 0.80, 95% CI = 0.65–0.99; $p = 0.04$).

Table 1. Characteristics and dietary nutrient intake of study participants.

	Non-Diabetic *	Diabetes	Prediabetes
Characteristics			
N	5657	1197	2075
Age (year)	40.1 ± 0.3 [a]	54.5 ± 0.5 [b]	50.4 ± 0.4 [c]
Sex, %			
Male	48.9	54.4	49.6
Female	51.1	45.6	50.4
Race/ethnicity, %			
Black	8.6 [a]	17.7 [b]	16.2 [b]
Hispanic	13.7	16.1	14.5
White	71.3 [a]	58.3 [b]	62.0 [b]
Other	6.5	8.0	7.3
Body mass index (kg/m^2)	27.5 ± 0.1 [a]	34.0 ± 0.3 [b]	30.4 ± 0.2 [c]
College graduate, %	30.8 [a]	17.5 [b]	22.3 [b]
Current smoker, %	52.8 [a]	40.7 [b]	53.2 [a]
Dietary intake			
Total energy (kcal/day)	2188 ± 19 [a]	1947 ± 29 [b]	2123 ± 30 [a]
Carbohydrate (g/day)	266 ± 2 [a]	226 ± 4 [b]	259 ± 3 [a]
Protein (g/day)	85.1 ± 0.8	81.8 ± 1.2	83.2 ± 1.2
Total fat (g/day)	81.7 ± 1.0	78.2 ± 1.6	81.3 ± 1.5
Saturated fat (g/day)	26.9 ± 0.4	25.3 ± 0.6	26.8 ± 0.6
Dietary fiber (g/day)	17.0 ± 0.3	16.2 ± 0.4	16.9 ± 0.3
Total sugar (g/day)	125.1 ± 2.0 [a]	97.9 ± 4.0 [b]	119.3± 2.0 [a]

Note: Values are presented as weighted mean ± SE and weighted percentage (%); Values within a row with different superscript letters (a, b, c) are significantly different ($p < 0.05$). * Non-Diabetic: Individuals without the presence of diabetes or prediabetes.

We also assessed whether the associations of home food availabilities with prediabetes or diabetes were influenced by an individual's dietary intake with further adjustment for dietary variables (Model 3) in addition to the covariates included in Model 2. The associations with prediabetes remained significant for dark green vegetables (OR = 0.87; 95% CI = 0.77–0.99; $p = 0.03$) and fat-free/low-fat milk (OR = 0.79; 95% CI = 0.67–0.93; $p = 0.007$) availabilities after adjusting for dietary intakes of total energy, carbohydrate, sugar, fat (total) and saturated fat. There was a borderline association between sugary drink availability and prediabetes in Model 2 (OR = 1.17; 95% CI = 0.99–1.37; $p = 0.06$) and this borderline association appeared to be attenuated after accounting for above dietary factors ($p = 0.26$) (Table 2).

Likewise, overall home healthy food availability (HFA) scores were inversely associated with the presence of prediabetes in Model 1 adjusting for age, sex, and ethnicity ($P_{trend} = 0.004$). Participants with high HFA scores (overall healthy) had approximately 31% reduction in odds of prediabetes compared to those with low scores (overall unhealthy) (OR = 0.69; 95% CI = 0.55–0.87; $p = 0.003$). The association was attenuated in Model 2 adjusting for additional confounders including BMI and education ($P_{trend} = 0.07$) as well as in Model 3 ($P_{trend} = 0.08$) adjusting for confounders in Model 2 plus dietary components (total energy, carbohydrate, sugar, fat [total], saturated fat). However, there was still a significant reduction (22%) in odds of prediabetes for those with high HFA scores versus those with low HFA scores (OR = 0.78; 95% CI = 0.61–0.99; $p = 0.04$) when further adjusting for dietary components in Model 3 (Table 3). In addition, among participants with diabetes (N = 1197), 27.2% were characterized as having high HFA scores (overall healthy), 63.0% were characterized as having medium HFA scores (borderline) and 9.8% were characterized as having low HFA scores (overall unhealthy). The corresponding distributions were 28.3%, 60.5% and 11.2% for participants with prediabetes (N = 2075) and were 29.3%, 59.2% and 11.5% for non-diabetic individuals (N = 5657).

Table 2. Associations of the availabilities of individual food items with diabetes and prediabetes.

High Availability *	All (N)	Cases (N)	Model 1 † OR (95% CI) ¶	p ¶	Model 2 ‡ OR (95% CI) ¶	p ¶	Model 3 § OR (95% CI) ¶	p ¶
Diabetes (vs. non-diabetic ‖)								
Fruits	6854	1197	0.85 (0.65–1.11)	0.21	0.89 (0.68–1.17)	0.39	0.80 (0.57–1.11)	0.17
Dark green vegetables	6854	1197	0.82 (0.65–1.04)	0.11	0.92 (0.72–1.17)	0.47	0.91 (0.69–1.21)	0.51
Fat-free/low-fat milk	6854	1197	0.80 (0.65–0.99)	0.04	0.86 (0.71–1.06)	0.16	0.85 (0.67–1.07)	0.17
Salty snacks	6854	1197	1.04 (0.85–1.26)	0.70	1.03 (0.85–1.25)	0.79	0.99 (0.80–1.22)	0.90
Sugary drinks	6854	1197	0.98 (0.79–1.21)	0.83	0.89 (0.72–1.11)	0.29	0.95 (0.76–1.20)	0.68
Prediabetes (vs. non-diabetic ‖)								
Fruits	7732	2075	0.90 (0.74–1.10)	0.30	0.96 (0.80–1.16)	0.68	0.88 (0.71–1.09)	0.24
Dark green vegetables	7732	2075	0.82 (0.73–0.91)	0.0006	0.86 (0.78–0.95)	0.005	0.87 (0.77–0.99)	0.03
Fat-free/low-fat milk	7732	2075	0.80 (0.65–0.89)	0.001	0.82 (0.69–0.97)	0.02	0.79 (0.67–0.93)	0.007
Salty snacks	7732	2075	0.97 (0.85–1.11)	0.62	0.98 (0.82–1.13)	0.59	0.96 (0.80–1.16)	0.68
Sugary drinks	7732	2075	1.24 (1.04–1.48)	0.02	1.17 (0.99–1.37)	0.06	1.11 (0.92–1.32)	0.26

* High availability = always/most of time available; reference group = low availability (not always/most of time available). † Model 1 was adjusted for age, sex, and race/ethnicity. ‡ Model 2 was adjusted for covariates in Model 1 and additionally adjusted for body mass index and education. § Model 3 was adjusted for covariates in Model 2 and additionally adjusted for dietary intakes of total energy, total carbohydrate, total sugar, total fat and total saturated fat. ¶ Odds ratios (ORs), 95% confidence intervals (95% CIs) and p values were estimated using logistic regression modeling accounting for complex survey design and sample weighting. ‖ Non-Diabetic: Individual without the presence of diabetes or prediabetes.

Table 3. Associations of overall home healthy food availability sores with diabetes and prediabetes.

Healthy Food Availability Score * (Overall)	Participants N	Model 1 † OR (95% CI) ¶	p ¶	Model 2 ‡ OR (95% CI) ¶	p ¶	Model 3 § OR (95% CI) ¶	p ¶
Diabetes (vs. non-diabetic ‖)							
Overall healthy (high score)	1981	0.76 (0.54–1.07)	0.11	0.92 (0.67–1.26)	0.58	0.86 (0.60–1.25)	0.42
Borderline (medium score)	4100	0.89 (0.63–1.24)	0.46	0.98 (0.73–1.31)	0.87	0.95 (0.70–1.29)	0.71
Overall unhealthy (low score)	773	1.00		1.00		1.00	
		$p_{trend} = 0.07$		$p_{trend} = 0.71$		$p_{trend} = 0.45$	
Prediabetes (vs. non-diabetic ‖)							
Overall healthy (high score)	2241	0.69 (0.55–0.87)	0.003	0.80 (0.63–1.00)	0.05	0.78 (0.61–0.99)	0.04
Borderline (medium score)	4603	0.80 (0.64–1.01)	0.06	0.87 (0.71–1.06)	0.16	0.83 (0.67–1.03)	0.09
Overall unhealthy (low score)	888	1.00		1.00		1.00	
		$p_{rend} = 0.004$		$p_{trend} = 0.07$		$p_{trend} = 0.08$	

* The overall healthy food availability score was defined based on the overall availability score of healthy and unhealthy food items at home. † Model 1 was adjusted for age, sex, and race/ethnicity. ‡ Model 2 was adjusted for covariates in Model 1 and additionally adjusted for body mass index and education. § Model 3 was adjusted for covariates in Model 2 and additionally adjusted for dietary intakes of total energy, total carbohydrate, total sugar, total fat and total saturated fat. ¶ Odds ratios (ORs), 95% confidence intervals (95% CIs) and p values were estimated using logistic regression modeling accounting for complex survey design and sample weighting. ‖ Non-Diabetic: Individual without the presence of diabetes or prediabetes.

4. Discussion

To our knowledge, the current study was the first that examined the relationships between home food availability and prediabetes and diabetes in adults using a nationally representative sample. The current results suggest that the availability of healthy foods such as green vegetables or fat-free/low-fat milk was inversely associated with the presence of prediabetes, whereas, a positive association between the availability of unhealthy foods such as sugary drink and prediabetes was detected adjusting for age, sex, and ethnicity. The results remained significant for green vegetables and fat-free/low-fat milk when adjusting for additional covariates including BMI and education. Furthermore, participants with high scores of overall home healthy food availability had approximately 31% decreases in odds of prediabetes compared to those with low scores, adjusting for age, sex, and ethnicity although the association was attenuated when including BMI and education as additional covariates. In contrast, no associations were detected of home food availabilities with diabetes except for fat-free/low-fat milk for which an inverse association was observed in the age, sex, and ethnicity-adjusted model.

Since the temporal relations between home food availabilities, diabetes, and prediabetes cannot be determined due to the nature of cross-sectional study design, we were not able to examine whether the associations between food availabilities and metabolic conditions (diabetes or prediabetes) were mediated by dietary intake. That being said, in the current study, we found that the significant associations of green vegetable and fat-free/low-fat milk availabilities and overall healthy food availability scores with the presence of prediabetes were independent of dietary intakes as the associations did not change substantially after further adjusting for dietary variables such as intakes of total energy, carbohydrate, sugar, fat (total) and saturated fat. The only exception was sugary drink availability with prediabetes, as adjusting for dietary components attenuated the borderline association observed in the models that was not adjusted for dietary variables ($p = 0.06$ in Model 2 vs. $p = 0.26$ in Model 3). Additionally, adjusting for dietary factors did not affect the null associations between home food availabilities and the presence of diabetes. Thus, with the estimated high rates of undiagnosed prediabetes [1], our results may suggest it is possible the individuals with prediabetes have yet to be diagnosed and were not aware of their health conditions, and therefore have not begun to make changes to their diets and home food environments. It also could be that individuals who were aware of their prediabetes status might not be fully informed about the consequences of this condition. It would be interesting and clinically meaningful to perform stratified analyses according to participants with or without a previous diagnosis of prediabetes (e.g., ever told having prediabetes) to determine whether there were differences in outcomes between those who were aware of the condition and those who were not. Unfortunately, the sample size was not sufficient (i.e., large missing data on the question "ever told you have prediabetes" in NHANES) to perform such analyses.

Our study examined whether dietary intakes of energy and major nutrients would moderate the associations between home food availabilities and the presence of metabolic diseases such as diabetes or prediabetes. We did not further examine how eating behaviors may influence the observed associations. For instance, would eating out/food prepared away from home have any influence on the current results since research has suggested having meals prepared at home more often was associated with lower risk of type 2 diabetes [24]? However, the 24-h dietary recall method is considered as the gold standard measure in nutritional epidemiologic studies and its use in NHANES to assess dietary intake is supported by expert panel discussions in "Strategies to Optimize the Impact of Nutrition Surveys and Epidemiological Studies" symposium [25]. Thus, we would assume our dietary data from multiple 24-h recalls (two recalls with an interval of 3 to 10 days) were likely to have captured the total energy and major nutrients consumed from foods/meals that were prepared away from home. Future research is warranted to use prospective cohorts to examine: (1) whether the availabilities of healthy and unhealthy food items at home are prospectively associated with the risk of prediabetes and type 2 diabetes; and (2) whether these prospective associations, if present are mediated by dietary intake or influenced by diet/nutrition-related behaviors such as frequently having meals away from home.

Previous work has extensively investigated how dietary intake of individual foods such as vegetables [26,27], low-fat dairy products [28–30], and sugar-sweetened beverages [31,32] as well as dietary patterns [11,12] may influence the risk of type 2 diabetes through prospective cohort studies. However, very few population-based studies were conducted to determine whether a diagnosis of type 2 diabetes or prediabetes would affect individual's dietary behaviors and to date no studies examined how disease diagnosis would further influence the home food environment, for example, the availabilities of healthy and unhealthy foods at home. One study reported that participants who had received a diagnosis of type 2 diabetes were less likely to consume sugary drinks particularly regular soda but more likely to drink bottled waters and artificially-sweetened beverages suggesting receiving a diagnosis of diabetes might have motivated people to make healthier food choices such as choosing non-sugary beverages instead [33]. The basic concepts of several health behavior theories support the assumption that individuals are likely to change health behaviors and lifestyles after a diagnosis of a chronic health condition. For instance, according to the stage of change model [34], the onset of a serious illness could at least lead to the initial stage of change which is to acknowledge the problem. The health belief model hypothesizes that health behaviors depend on people's beliefs about health problems [35,36]. Therefore, lifestyles are expected to improve if one recognizes the severity of the illness and clear benefits of actions. The theory of reasoned action suggests individuals' pre-existing attitudes and behavioral intentions predict how they behave. One of the key determinants is subjective norms that refers to perceived social pressure to perform or not perform the behavior [37]. Subjective norms supporting change of behaviors after a diagnosis of a chronic health condition may play an important role for individuals to make healthy choices. In our study when comparing dietary nutrients intakes among diabetes, prediabetes and non-diabetic groups, diabetes group consumed less total energy, carbohydrate, and sugar compared to non-diabetic or prediabetes groups, while no significant differences in dietary intakes of major nutrients and total energy were observed between non-diabetic and prediabetes groups. Since this was a cross-sectional study design and based on the above behavior change theories [34–37], the current dietary results were more likely to suggest that participants with diabetes may be more aware and have made changes of their dietary intakes particularly dietary components that are more relevant to glycemic control such as dietary total energy, carbohydrate, and sugar compared to prediabetes and non-diabetic participants. These results may also to some extent support our primary findings that the associations of home food availabilities with the presence of prediabetes or diabetes were not mainly influenced by dietary factors such as total energy and major nutrient intake.

The present study has a number of strengths, including the use of a large population-based study with a nationally representative sample. Blood samples were collected to measure HbA1c levels by trained research staff to provide objective indicators for prediabetes and diabetes status. However, there were limitations to our study. One limitation was the nature of cross-sectional design of NHANES study, which prevented us from drawing causal relationships. As all prevalent case-control studies, the temporal sequence was unclear. Second, due to the self-reported measures utilized to assess food availability and dietary intakes, recall bias may occur. Third, we were not able to distinguish between type 1 and type 2 diabetes; however, the bias of misclassification would be minimum, since more than 95% of adult diabetes cases are type 2 diabetes [1]. Fourth, we collapsed the original five food availability scores into two or three categories and used odds ratios as outcome estimates in the current cross-sectional study. Thus, our outcomes could be overestimated in this regard. In addition, since the home food availability measures were only included in the 2007–2008 and 2009–2010 waves of NHANES, our results may not completely reflect the most updated situation in terms of how the availabilities of healthy and unhealthy foods at home were associated with prediabetes and diabetes. Finally, the home food availability questions in NHANES measured only five food and beverage items, which may provide a limited picture of the overall home food environment related to healthy food availability.

5. Conclusions

Findings from the current study suggest the availabilities of healthy foods such as green vegetables and fat-free or low-fat milk at home were inversely associated with the presence of prediabetes. Furthermore, participants with low scores of overall healthy food availability were more likely to have prediabetes. These findings were independent of dietary intakes as the results remained significant after adjusting for dietary components including total energy, carbohydrate, sugar, fat (total) and saturated fat. Thus, in addition to continuing the educational efforts made for diabetes patients, it is equally important to devote our preventative efforts to individuals who are at risk of developing type 2 diabetes, for example, those with prediabetes since the current results suggest there is a potential need to improve the home food environment related to healthy food availability for this population. Providing these individuals with useful and practical tools to monitor their household foods, food purchase habits and food choices may help to prevent and reduce the onset of type 2 diabetes. Longitudinal studies including prospective cohort studies and clinical trials are necessary to further elucidate the temporal associations between home food availability and metabolic conditions such as type 2 diabetes and prediabetes, as well as the underlying mechanisms for these associations.

Author Contributions: All of the authors made substantial contributions to the study concept and design or analysis and interpretation of the data. Specifically, J.R.M. designed the study, analyzed data, and was the primary author of every section of the text. W.C. and M.-H.T. were instrumental in the design of the study and helped to draft the manuscript. C.K. helped to design the study's analytic strategy and commented on the manuscript. All authors have read and agreed to the published version of the manuscript.

Funding: This research received no external funding.

Conflicts of Interest: The authors declare no conflict of interest.

References

1. Centers for Disease Control and Prevention. National Diabetes Statistics Report. Atlanta, GA: Centers for Disease Control and Prevention, U.S. Department of Health and. Human Services 2017. Available online: https://www.cdc.gov/diabetes/pdfs/data/statistics/national-diabetestatistics-report.pdf (accessed on 30 June 2019).
2. Nathan, D.M.; Davidson, M.B.; DeFronzo, R.A.; Heine, R.J.; Henry, R.R.; Pratley, R.; Zinman, B. American Diabetes A: Impaired fasting glucose and impaired glucose tolerance: Implications for care. *Diabetes Care* **2007**, *30*, 753–759. [CrossRef] [PubMed]
3. Hu, F.B. Globalization of diabetes: The role of diet, lifestyle, and genes. *Diabetes Care* **2011**, *34*, 1249–1257. [CrossRef] [PubMed]
4. Mokdad, A.H.; Bowman, B.A.; Ford, E.S.; Vinicor, F.; Marks, J.S.; Koplan, J.P. The continuing epidemics of obesity and diabetes in the United States. *JAMA* **2001**, *286*, 1195–1200. [CrossRef] [PubMed]
5. Mozaffarian, D.; Kamineni, A.; Carnethon, M.; Djousse, L.; Mukamal, K.J.; Siscovick, D. Lifestyle risk factors and new-onset diabetes mellitus in older adults: The cardiovascular health study. *Arch. Intern. Med.* **2009**, *169*, 798–807. [CrossRef]
6. Schulze, M.B.; Hu, F.B. Primary prevention of diabetes: What can be done and how much can be prevented? *Ann. Rev. Public Health* **2005**, *26*, 445–467. [CrossRef]
7. Fulkerson, J.A.; Nelson, M.C.; Lytle, L.; Moe, S.; Heitzler, C.; Pasch, K.E. The validation of a home food inventory. *Int. J. Behav. Nutr. Phys. Act.* **2008**, *5*, 55. [CrossRef]
8. Kratt, P.; Reynolds, K.; Shewchuk, R. The role of availability as a moderator of family fruit and vegetable consumption. *Health Educ. Behav.* **2000**, *27*, 471–482. [CrossRef]
9. Grant, E.; Gearry, R.B.; Wilson, R.; Pearson, J.; Skidmore, P.M.L. Home availability of fruit and vegetables and obesogenic foods as an indicator of nutrient intake in 50 year olds from Canterbury, New Zealand. *Asia Pac. J. Clin. Nutr.* **2017**, *26*, 524–530.
10. Emery, C.F.; Olson, K.L.; Lee, V.S.; Habash, D.L.; Nasar, J.L.; Bodine, A. Home environment and psychosocial predictors of obesity status among community-residing men and women. *Int. J. OBES* **2015**, *39*, 1401–1407. [CrossRef]

11. Maghsoudi, Z.; Ghiasvand, R.; Salehi-Abargouei, A. Empirically derived dietary patterns and incident type 2 diabetes mellitus: A systematic review and meta-analysis on prospective observational studies. *Public Health Nutr.* **2016**, *19*, 230–241. [CrossRef]
12. Ericson, U.; Brunkwall, L.; Alves Dias, J.; Drake, I.; Hellstrand, S.; Gullberg, B.; Sonestedt, E.; Nilsson, P.M.; Wirfalt, E.; Orho-Melander, M. Food patterns in relation to weight change and incidence of type 2 diabetes, coronary events and stroke in the Malmo Diet and Cancer cohort. *Eur. J. Nutr.* **2019**, *58*, 1801–1814. [CrossRef] [PubMed]
13. Nagarajan, S.; Khokhar, A.; Holmes, D.S.; Chandwani, S. Family consumer behaviors, adolescent prediabetes and diabetes in the National Health and Nutrition Examination Survey (2007–2010). *J. Am. Coll. Nutr.* **2017**, *36*, 520–527. [CrossRef] [PubMed]
14. National Center for Health Statistics. National Health and Nutrition Examination Survey. Available online: http://www.cdc.gov/nchs/nhanes/ (accessed on 30 June 2019).
15. Boles, R.E.; Scharf, C.; Filigno, S.S.; Saelens, B.E.; Stark, L.J. Differences in home food and activity environments between obese and healthy weight families of preschool children. *J. Nutr. Educ. Behav.* **2013**, *45*, 222–231. [CrossRef] [PubMed]
16. American Diabetes Association. American Diabetes Association Position Statement: Standards of medical care in diabetes—2016. *Diabetes Care* **2016**, *39* (Suppl. 1), S1–S112.
17. National Health and Nutrition Examination Survey. MEC In-Person Dietary Interviewers Procedures Manual. 2007. Available online: https://wwwn.cdc.gov/nchs/data/nhanes/2007-2008/manuals/manual_dietarymec.pdf (accessed on 30 June 2019).
18. Ahluwalia, N.; Dwyer, J.; Terry, A.; Moshfegh, A.; Johnson, C. Update on NHANES dietary data: Focus on collection, release, analytical considerations, and uses to inform public policy. *Adv. Nutr.* **2016**, *7*, 121–134. [CrossRef]
19. Moshfegh, A.J.; Rhodes, D.G.; Baer, D.J.; Murayi, T.; Clemens, J.C.; Rumpler, W.V.; Paul, D.R.; Sebastian, R.S.; Kuczynski, K.J.; Ingwersen, L.A.; et al. The US Department of Agriculture Automated Multiple-Pass Method reduces bias in the collection of energy intakes. *Am. J. Clin. Nutr.* **2008**, *88*, 324–332. [CrossRef]
20. Johnson, C.L.; Paulose-Ram, R.; Ogden, C.L.; Carroll, M.D.; Kruszon-Moran, D.; Dohrmann, S.M.; Curtin, L.R. National health and nutrition examination survey: Analytic guidelines, 1999–2010. *Vital Health Stat. 2* **2013**, *161*, 1–24.
21. Masters, M.A.; Stanek Krogstrand, K.L.; Eskridge, K.M.; Albrecht, J.A. Race/ethnicity and income in relation to the home food environment in US youth aged 6 to 19 years. *J. Acad. Nutr. Diet* **2014**, *114*, 1533–1543. [CrossRef]
22. Chai, W.; Fan, J.X.; Wen, M. Association of individual and neighborhood factors with home food availability: Evidence from the National Health and Nutrition Examination Survey. *J. Acad. Nutr. Diet* **2018**, *118*, 815–823. [CrossRef]
23. McAtee, J.; King, C.; Chai, W. Food Insecurity is inversely associated with healthy food availability among adults in the United States. *Diabesity* **2019**, *5*, 17–22. [CrossRef]
24. Zong, G.; Eisenberg, D.M.; Hu, F.B.; Sun, Q. Consumption of meals prepared at home and risk of type 2 diabetes: An analysis of two prospective cohort studies. *PLoS Med.* **2016**, *13*, e1002052. [CrossRef] [PubMed]
25. Webb, D.; Leahy, M.M.; Milner, J.A.; Allison, D.B.; Dodd, K.W.; Gaine, P.C.; Matthews, R.A.; Schneeman, B.O.; Tucker, K.L.; Young, S.S. Strategies to optimize the impact of nutritional surveys and epidemiological studies. *Adv. Nutr.* **2013**, *4*, 545–547. [CrossRef] [PubMed]
26. Carter, P.; Gray, L.J.; Troughton, J.; Khunti, K.; Davies, M.J. Fruit and vegetable intake and incidence of type 2 diabetes mellitus: Systematic review and meta-analysis. *BMJ* **2010**, *341*, c4229. [CrossRef] [PubMed]
27. Cooper, A.J.; Forouhi, N.G.; Ye, Z.; Buijsse, B.; Arriola, L.; Balkau, B.; Barricarte, A.; Beulens, J.W.; Boeing, H.; Buchner, F.L.; et al. Fruit and vegetable intake and type 2 diabetes: EPIC-InterAct prospective study and meta-analysis. *Eur. J. Clin. Nutr.* **2012**, *66*, 1082–1092. [CrossRef] [PubMed]
28. Choi, H.K.; Willett, W.C.; Stampfer, M.J.; Rimm, E.; Hu, F.B. Dairy consumption and risk of type 2 diabetes mellitus in men: A prospective study. *Arch. Intern. Med.* **2005**, *165*, 997–1003. [CrossRef]
29. Grantham, N.M.; Magliano, D.J.; Hodge, A.; Jowett, J.; Meikle, P.; Shaw, J.E. The association between dairy food intake and the incidence of diabetes in Australia: The Australian Diabetes Obesity and Lifestyle Study (AusDiab). *Public Health Nutr.* **2013**, *16*, 339–345. [CrossRef]

30. Liu, S.; Choi, H.K.; Ford, E.; Song, Y.; Klevak, A.; Buring, J.E.; Manson, J.E. A prospective study of dairy intake and the risk of type 2 diabetes in women. *Diabetes Care* **2006**, *29*, 1579–1584. [CrossRef]
31. Malik, V.S.; Popkin, B.M.; Bray, G.A.; Despres, J.P.; Willett, W.C.; Hu, F.B. Sugar-sweetened beverages and risk of metabolic syndrome and type 2 diabetes: A meta-analysis. *Diabetes Care* **2010**, *33*, 2477–2483. [CrossRef]
32. Ma, J.; Jacques, P.F.; Meigs, J.B.; Fox, C.S.; Rogers, G.T.; Smith, C.E.; Hruby, A.; Saltzman, E.; McKeown, N.M. Sugar-sweetened beverage but not diet soda Consumption is positively associated with progression of insulin resistance and rrediabetes. *J. Nutr.* **2016**, *146*, 2544–2550. [CrossRef]
33. Miller, C.; Ettridge, K.; Wakefield, M.; Pettigrew, S.; Coveney, J.; Roder, D.; Durkin, S.; Wittert, G.; Martin, J.; Dono, J. Consumption of sugar-sweetened beverages, juice, artificially-sweetened soda and bottled water: An Australian population study. *Nutrients* **2020**, *12*, 817. [CrossRef]
34. Prochaska, J.O.; Prochaska, J.M. An update on maximum impact practices from a transtheoretica approach. In *Best Practices in the Behavioral Management of Chronic Disease*; Trafton, J.A., Gordon, W.P., Eds.; Institute for Disease Management: Los Altos, CA, USA, 2005; Volume 1, pp. 1–6.
35. Becker, M.H. The Health Belief Model and personal health behavior. *Health Educ. Monographs.* **1974**, *2*, 324–508. [CrossRef]
36. Rosenstock, I.M. Historical origins of the health belief model. *Health Educ. Monographs.* **1974**, *2*, 328–335. [CrossRef]
37. Ajzen, I.; Albarracín, D. Predicting and changing behavior: A reasoned action approach. In *Prediction and Change of Health Behavior: Applying the Reasoned Action Approach*; Ajzen, I., Albarracín, D., Hornik, R., Eds.; Erlbaum: Mahwah, NJ, USA, 2007; pp. 3–21.

nutrients

MDPI

Article

Recommended Intake of Key Food Groups and Cardiovascular Risk Factors in Australian Older, Rural-Dwelling Adults

Alice J. Owen [1,*], Michael J. Abramson [1], Jill F. Ikin [1], Tracy A. McCaffrey [2], Sylvia Pomeroy [1], Brigitte M. Borg [1,3], Caroline X. Gao [1], David Brown [1] and Danny Liew [1]

1 School of Public Health and Preventive Medicine, Monash University, 553 St Kilda Rd, Melbourne, VIC 3004, Australia; michael.abramson@monash.edu (M.J.A.); jill.blackman@monash.edu (J.F.I.); sylvia.pomeroy@monash.edu (S.P.); brigitte.borg@monash.edu (B.M.B.); caroline.gao@monash.edu (C.X.G.); david.brown@monash.edu (D.B.); danny.liew@monash.edu (D.L.)

2 Department of Nutrition, Dietetics and Food, Monash University, Melbourne, VIC 3168, Australia; tracy.mccaffrey@monash.edu

3 Respiratory Medicine, Alfred Hospital, Melbourne, VIC 3004, Australia

* Correspondence: alice.owen@monash.edu; Tel.: +61-3-9903-0045

Received: 25 February 2020; Accepted: 21 March 2020; Published: 23 March 2020

Abstract: This study examined the relationship between diet quality scores and cardiometabolic risk factors in regionally-dwelling older Australian adults with increased cardiovascular risk. This study was a cross-sectional analysis of demographic, anthropometric, and cardiometabolic risk factor data from 458 participants of the Cardiovascular Stream of the Hazelwood Health Study. Participants completed a 120 item semi-quantitative food frequency questionnaire. Multivariable linear regression adjusting for age, sex, smoking, physical activity, education, diabetes, and body mass index was used to examine the relationship between diet and cardiometabolic risk factors. Mean (SD) age of participants was 71 (8) years, and 55% were male. More than half of men and women did not meet recommended intakes of fibre, while 60% of men and 42% of women exceeded recommended dietary sodium intakes. Higher diet quality in terms of intake of vegetables, grains, and non-processed meat, as well as intake of non-fried fish, was associated with more favourable cardiometabolic risk profiles, while sugar-sweetened soft drink intake was strongly associated with adverse cardiometabolic risk factor levels. In older, regionally-dwelling adults, dietary public health strategies that address whole grain products, vegetable and fish consumption, and sugar-sweetened soft-drink intake may be of benefit in reducing cardiometabolic risk.

Keywords: diet quality; cardiometabolic risk; sugar-sweetened beverages; food groups

1. Introduction

Cardiometabolic diseases remain a major cause of mortality and morbidity across the globe, with high blood pressure, smoking, elevated plasma glucose, and high body mass index (BMI, kg/m^2), the top four risk factors for attributable disability-adjusted life years [1]. Dietary intake is a well-known risk factor for non-communicable diseases, and the Global Burden of Disease Study recently estimated that 11 million deaths were attributable to dietary risk factors in 2017 [2]. In examining the mortality attributable to poor quality diet, it was estimated that the top five risks were diets high in sodium, low in whole grains, low in fruits, low in nuts and seeds, and low in vegetables [2]. However, limited availability of geographically representative data remains a barrier to a clearer understanding of dietary risks, and to the development of effective local interventions to reduce the cardiometabolic disease risk conferred by inadequate dietary intake.

Age-and sex-specific dietary intake guidelines for maintaining health have been developed in many countries, including Australia, largely on the basis of findings from observational and prospective cohort data [3]. However, for many foods and nutrients, adherence to these guidelines is low [4]. In the 2011/12 Australian Health Survey, less than 1 in 25 adults met recommended guideline intakes of vegetables and legumes [4]. While among those aged 51–70 years, only 5% of men and less than 1% of women met guideline-recommended intakes of dairy foods/alternatives [4]. In addition, there was evidence that dietary risk factors may be influenced by socioeconomic and geographic factors, with rural residential status, education, and other socioeconomic status markers previously reported to be related to dietary quality and fibre intake [5–7]. There is also some evidence of gender differences in adherence to dietary quality and guideline adherence in non-metropolitan areas [6].

This study aimed to characterise diet quality in two rural Australian towns with a high burden of cardiovascular disease [8] and to examine associations with cardiometabolic risk factors using a recently developed dietary quality score [9], which allowed key food group diet quality to be explored.

2. Materials and Methods

2.1. Participants

The sample for this study comprised study participants from the Cardiovascular Stream of the Hazelwood Health Study, who additionally agreed to complete a dietary survey. Recruited between October 2017 and May 2018, Cardiovascular Stream participants were drawn from a weighted random sample of 1133 people who had previously completed the Hazelwood Health Study Adult Survey [10], lived in the rural Victorian towns of Morwell or Sale, and were males aged 55–89 years or females 60–89 years (Figure 1). Those who identified any underlying cardiovascular condition on the Adult Survey were oversampled. Years of education was captured as the highest educational qualification and classified as up to year 10, upper secondary (to year 11–12), trade certifications, or university/tertiary education. Residential area-related socioeconomic status was determined through the linkage of participant residential postcode with the Index of Relative Socioeconomic Advantage and Disadvantage (IRSAD) [11].

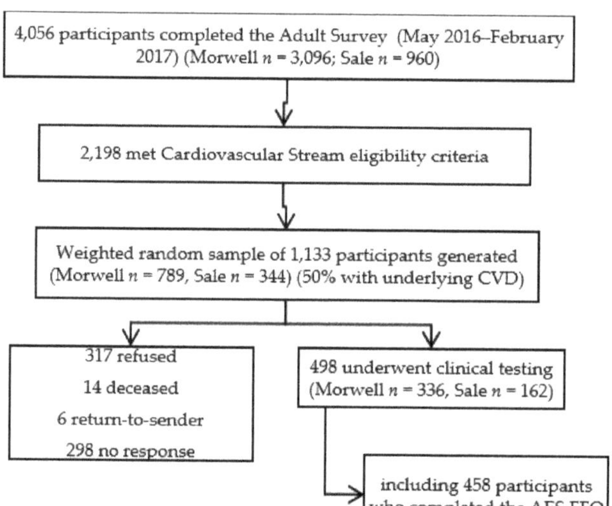

Figure 1. Flow diagram of participant recruitment. CVD = cardiovascular disease, AES = Australian Eating Survey, FFQ = food frequency questionnaire

2.2. Measures

2.2.1. Cardiometabolic Risk Factors

Participants attended a clinic during which a number of health assessments were made. Anthropometric measures included height and weight, from which body mass index (BMI) was calculated. Hip and waist circumference were also measured. Height (to nearest 0.1 cm) was measured using a wall-mounted stadiometer, waist circumference (to nearest 0.1 cm) was measured at the midpoint of the last palpable rib and top of the hip bone, and hip circumference (to nearest 0.1 cm) was taken as the point of maximum circumference around the buttocks. Weight was measured on calibrated standing scales to the nearest 0.1 kg. World Health Organization criteria were used to categorise BMI into overweight (BMI from ≥25 to <30 kg/m^2) and obesity (BMI ≥30 kg/m^2) [12]. Other cardiometabolic risk factors assessed were blood pressure, heart rate, plasma cholesterol, haemoglobinA1c, and creatinine, from which the estimated glomerular filtration rate (eGFR) was calculated as a marker of renal function. Blood pressure (mmHg) was measured in a seated position three times using a digital automatic blood pressure monitor (Omron, Matsusaka, Japan) with a one-minute rest between readings. The average of the last two measurements was used in the analysis. A non-fasting blood sample was taken for measurement of plasma cholesterol and haemoglobinA1c (HbA1c). eGFR was calculated using the Chronic Kidney Disease Epidemiology Collaboration (CKD-EPI) formula, which used age, gender, and blood creatinine to estimate renal function (expressed in mL/min/1.73 m^2) [13]. The self-reported behavioural risk factors collected were smoking status, history of diabetes and cardiovascular disease and physical activity. The presence of diabetes was determined by self-reported doctor diagnosis, use of diabetes medications or HbA1c ≥ 6.5%. History of cardiovascular disease was self-reported doctor diagnosis. Smoking status was determined by self-reported smoking of at least 100 cigarettes, or a similar amount of tobacco, in a participant's lifetime, and reporting of current or former smoking. Self-reported physical activity was assessed using the validated Active Australia survey, an eight-item questionnaire that captured information on time spent undertaking walking, household physical activity, vigorous physical activity, and moderate physical activity in the past week [14]. Participants were considered to have engaged in 'adequate physical activity' if they reported having undertaken any vigorous physical activity or at least 150 min of moderate physical activity over the previous seven days.

2.2.2. Dietary Intake Assessment

The Australian Eating Survey Food Frequency Questionnaire (AES FFQ), a 120 item semi-quantitative FFQ previously validated in community-dwelling adults aged 30–70 years [9], was used to assess dietary intake. The dietary assessment was a voluntary component of clinic visits, and in some cases, completion was not undertaken due time constraints or participant preference. Nutrient intakes were computed against the AusNut Database [15]. The Australian Recommended Food Score (ARFS), a diet quality index that captures the dietary quality of key food groups, was calculated from the AES FFQ, as previously described [9]. The ARFS is computed as a total score, as well as subscales relating to intakes of vegetables, fruit, meat, non-meat protein, grains (breads and cereals), dairy, water, and spreads/sauces. Reported food items within the food sub-groups were awarded points for frequency of consumption based upon Australian national dietary guidelines, and the ARFS scores were calculated by summing the points for each item, as previously described [9]. To determine proportions with inadequate intakes of key macro- and micronutrients, estimated intakes were compared to age- and sex-specific estimated average requirements (EAR) or, in the case of sodium and potassium, suggested dietary target (SDT) and adequate intake (AI), respectively [16].

2.3. Statistical Analysis

Data analysis was undertaken using the Statistical Program for Social Sciences (IBM SPSS; Armonk, NY, USA), version 25. Standard descriptive statistics were used to examine cohort characteristics following assessment of normality. Associations between variables were explored using Pearson and Spearman's correlations, as appropriate. Multivariable linear regression was used to examine the associations between diet quality indices and cardiovascular risk factors. Interactions between gender and dietary predictor variables were assessed by the inclusion of gender–diet interaction terms in models. Associations were initially examined in a minimally-adjusted model (Model 1, adjusted for age and sex), followed by multivariable-adjusted modelling adjusted for age, sex, smoking, physical activity, education, and diabetes for models examining waist:hip, and the addition of BMI as a covariate for all other risk factors (Model 2). History of cardiovascular disease, antihypertensive use, and lipid lowering therapy use were tested for inclusion in models. Lipid lowering therapy use was significant in models examining total cholesterol and was included in final models for this variable. For linear regression analyses examining fish and beverage intakes, Model 3 additionally included total dietary quality (ARFS total). To adjust for potential misreporting, regression analyses were repeated, excluding those with total dietary energy below 2000 kJ or above 15,000 kJ [17].

2.4. Ethics

The Hazelwood Health Study Cardiovascular Stream protocol was reviewed and approved by the Monash University Human Research Ethics Committee (project#1078). All participants provided informed written consent to participate, and this research was conducted in accordance with the Declaration of Helsinki.

3. Results

A flow diagram showing recruitment from the Adult Survey through to the Cardiovascular Stream is shown in Figure 1. From 498 Cardiovascular Stream participants who attended the clinic, 458 completed the AES FFQ and were included in the analysis.

3.1. Participant Characteristics

The mean (SD) age of participants was 71 (8) years, and over 55% were male (Table 1). There was a high prevalence of overweight and obesity in the cohort, with 83% of men and 77% of women having a BMI ≥25 kg/m^2, and 46% of men and 48% of women having a BMI ≥30 kg/m^2. Based on residential area, two-thirds of participants were categorised in the first (most disadvantaged) quintile of the IRSAD [11] (Table 1). Eighteen per cent of participants reported a history of diabetes, and 49.3% reported a history of cardiovascular disease (Table 1).

Table 1. Demographic and cardiometabolic risk factor characteristics.

	Men (n = 256)	Women (n = 202)	p
Age (years)	70 ± 9	73 ± 7	<0.001
Area level disadvantage: n (%) first IRSAD quintile	164 (64.1)	141 (69.8)	0.482
School education to year 10 or below	73 (28.6)	99 (49.3)	<0.001
Meeting physical activity guidelines (self-reported)	123 (48.4)	91 (45.3)	0.510
Body Mass Index (kg/m^2)	29.9 ± 5.0	30.3 ± 6.5	0.36
Waist Circumference (cm)	108.9 ± 13.4	100.3 ± 15.1	<0.001
Waist to Hip ratio	1.04 ± 0.06	0.90 ± 0.07	<0.001
Diabetes	53 (20.8)	29 (14.6)	0.110
History of CVD	131 (49.6)	95 (45.6)	0.408
Current smoker	19 (7.5)	14 (7.0)	0.096
Systolic/Diastolic Blood Pressure (mmHg)	133/72 ±17/11	135/72 ± 19/11	0.878
Heart rate (bpm)	63 ± 11	67 ± 11	0.003
Total cholesterol (mmol/L)	4.36 ± 0.99	4.85 ± 0.98	<0.001
HDL cholesterol (mmol/L)	1.21 ± 0.33	1.47 ± 0.37	<0.001
LDL cholesterol (mmol/L)	2.33 ± 0.88	2.58 ± 0.89	0.004
HbA1c (%)	6.1 ± 1.1	5.9 ± 0.8	0.128
Estimated glomerular filtration rate (eGFR) (mL/min/1.73 m^2)	71.2 ± 15.6 [a]	71.3 ± 15.7 [b]	0.955

Values given as Mean ± SD or n (% of reporting population) unless otherwise specified. p: differences between men and women. Smaller sample size for estimated glomerular filtration rate [a] n = 173, [b] n = 141.

3.2. Dietary Intake

Mean diet quality (assessed by the ARFS) was 29.3/73 for men and 32.8/73 for women. Compared with men, women reported higher age- and education-adjusted diet quality scores for ARFS total ($p < 0.001$) and for some specific food groups: ARFS vegetables ($p < 0.001$), ARFS fruit ($p = 0.004$), and ARFS dairy ($p = 0.01$) (Table 2). When compared to Nutrient Reference Values for Australians [16], intakes of protein, iron, and vitamins were mostly adequate, while more than half of participants reported inadequate intake of fibre and more than 40% reported inadequate folate and calcium intakes. A greater proportion of men than women reported above recommended intakes of sodium and below recommended intakes of potassium (Table 3).

Table 2. Dietary characteristics for men and women.

	Men (n = 256)	Women (n = 202)	p
Energy (kJ/day)	9727 (9326–10128)	7989 (7537–8441)	<0.001
ARFS-total (score/73)	29.0 (27.8–30.3)	33.1 (31.6–34.5)	<0.001
ARFS-Vegetable (score/21)	11.4 (10.8–12.0)	13.5 (12.8–14.2)	<0.001
ARFS-Fruit (score/12)	4.7 (4.4–5.1)	5.5 (5.1–5.9)	0.004
ARFS-Grain (score/13)	4.0 (3.8–4.2)	4.2 (4.0–4.5)	0.198
ARFS-Meat (score/7)	2.7 (2.5–2.9)	2.9 (2.7–3.0)	0.230
ARFS-Alternate Protein (score/6)	1.7 (1.5–1.8)	1.9 (1.7–2.0)	0.071
ARFS-Dairy (score/11)	3.6 (3.4–3.8)	4.0 (3.7–4.2)	0.011
ARFS-Spreads-Sauces (score/2)	0.8 (0.7–0.9)	0.8 (0.7–0.9)	0.736
ARFS-Water (score/1)	0.3 (0.3–0.4)	0.5 (0.4–0.6)	0.001
Carbohydrate (% of total energy)	45.4 (44.5–46.3)	42.7 (41.7–43.7)	<0.001
Protein (% of total energy)	18.7 (18.3–19.1)	20.5 (20.0–21.0)	<0.001
Fat (% of total energy)	32.5 (31.8–33.2)	34.3 (33.5–35.0)	0.001
Saturated fat (% of total energy)	14.1 (13.7–14.5)	15.0 (14.5–15.4)	0.008
Polyunsaturated fat (% of total energy)	3.8 (3.7–3.9)	4.0 (3.8–4.1)	0.032
Confectionery (% total energy)	5.6 (4.9–6.2)	4.6 (3.8–5.3)	0.047
Baked sweet products (% total energy)	6.1 (5.5–6.7)	5.9 (5.2–6.5)	0.294
Takeaway foods (% total energy)	7.2 (6.7–7.8)	5.9 (5.3–6.5)	0.002
Alcoholic beverage (% of total energy)	5.0 (4.0–5.9)	4.0 (3.0–4.0)	0.177
Sugar-sweetened drinks (% total energy)	3.6 (3.0–4.2)	1.7 (1.1–2.4)	<0.001
Soft drinks (% consuming ≥ 1/week)	48 (18.8)	22 (10.9)	0.252
Sugar-sweetened drinks (%consuming≥ 1/week)	76 (29.9)	38 (18.8)	0.008
Fresh fish (% consuming ≥ 1/week)	55 (21.5)	71 (35.2)	0.01
Canned fish (% consuming ≥ 1/week)	78 (30.5)	75 (37.2)	0.379
Crumbed/battered fish (% consuming ≥ 1/week)	50 (19.6)	50 (24.7)	0.036
Sodium (mg/day)	2329 (2218–2440)	1984 (1857–2109)	<0.001
Potassium (mg/day)	3927 (3757–4097)	3526 (3334–3718)	0.003

Values given as adjusted means (95% CI) or n (%) in those with complete dietary data. Means are adjusted for age and education. The denominator of ARFs scores vary as indicated (score/denominator) [9]. *p*: Significance of sex-differences.

Table 3. Prevalence of inadequate nutrient intakes #.

	Men (n = 256)	Women (n = 202)
Protein	21 (8.2)	8 (4.0)
Fibre	148 (57.8)	101 (50.2)
Folate	104 (40.6)	100 (49.8)
Vitamin A	31 (12.1)	17 (8.5)
Vitamin C	5 (2)	1 (0.5)
Thiamine	29 (11.3)	25 (12.4)
Riboflavin	16 (6.3)	12 (6)
Sodium	155 (60.5)	85 (42.3)
Potassium	128 (50)	57 (28.4)
Magnesium	88 (34.4)	40 (19.9)
Calcium	103 (40.2)	106 (52.7)
Iron	10 (3.9)	5 (2.5)
Zinc	92 (35.9)	12 (6)

Values given as n (%). # Proportions NOT meeting age- and gender-specific estimated average requirements (EARs) [16], except for sodium (% above suggested dietary target) and potassium (% below adequate intake).

Consumption of sugar-sweetened beverages (soft drinks and cordials) at least once per week was reported by 32.0% of men and 19.3% of women (Table 2), with men having a substantially higher proportion of total dietary energy derived from sweetened drinks compared to women 3.5% vs. 1.9%, respectively ($p < 0.001$). There was also a sex-difference in fresh fish intake ($p = 0.01$), with women more likely to consume fresh fish at least once a week (Table 2).

3.3. *Associations between Diet Quality and Cardiometabolic Risk Factors*

Key cardiometabolic risk factors associated with diet quality were examined (Table 4). No associations were observed between AFRS total or component scores and HbA1c, (low density lipoprotein) LDL-cholesterol, blood pressure, or heart rate. There were no significant interactions between gender and dietary predictors evident in the models, except in the cases of the eGFR and ARFS score for alternate sources of protein ($p = 0.006$), and soft drinks and HbA1c ($p = 0.012$). In both age- and gender-adjusted models, and models adjusting for comorbidities, there was strong evidence that ARFS total, fruit, and grain scores were negatively associated with central adiposity (waist:hip) (Table 4). Higher quality of vegetable and meat intakes were positively associated with higher (high density lipoprotein) HDL-cholesterol. The dairy score was negatively associated with total cholesterol in minimally and fully adjusted models (Table 4), but there was no association between total cholesterol and percentage of total energy derived from all dairy products (results not shown). The association between ARFs alternate protein sources was significant in women (unstandardised beta coefficient B = 2.024, 95% CI: 0.677, 3.371, $p = 0.003$) but not men.

Table 4. Australian Recommended Food Scores and modifiable cardiometabolic risk factors.

	Age- and Gender-adjusted (Model 1) [a]	Multivariable-Adjusted [b] (Model 2)	Model 2 excl. Potential Misreporting [c]
Waist:hip			
ARFS total	−0.001 (−0.001, 0.000) **	−0.001 (−0.001, 0.000) *	−0.001 (−0.001, 0.000) *
ARFS vegetable	−0.001 (−0.003, 0.000) *	n.s.	n.s.
ARFS fruit	−0.003 (−0.005, −0.001) **	−0.002 (−0.004, 0.000) *	−0.003(−0.005, −0.001) *
ARFS grain	−0.006 (−0.009, −0.003) ***	−0.004 (−0.007, −0.001) *	−0.004(−0.007,−0.001) **
ARFS alt. protein	−0.005 (−0.009, 0.000)*	n.s.	n.s.
HDL cholesterol			
ARFS vegetable	0.008 (0.002, 0.015) **	n.s.	0.007 (0.000, 0.013) *
ARFS meat	0.045 (0.023, 0.068) ***	0.037 (0.015, 0.059) **	0.044 (0.021, 0.067) ***
eGFR			
ARFS meat	1.210 (0.002, 2.418)	1.365 (0.164, 2.566) *	1.305 (0.022, 2.587) *
ARFS alt. protein [d]	1.339 (0.001, 2.677)	1.342 (0.021, 2.664) *	n.s.
Total cholesterol			
ARFS dairy	−0.080 (−0.129, −0.030) **	−0.067 (−0.109, −0.024) **	−0.073 (−0.118,−0.028) ***

Values are Unstandardised Beta Coefficients (95% CI), n.s. = not significant. alt. protein= non-meat protein sources. [a] Model 1 adjusted for age and gender (n = 458). [b] In Model 2, all variables are adjusted for age, gender, smoking, physical activity, education, and BMI, except waist:hip, which is adjusted for age, gender, smoking, physical activity, education and diabetes, and total cholesterol, which is additionally adjusted for lipid-lowering therapy use. [c] Model 2, excluding those with total dietary energy <2000 kJ and >15000 kJ (total n = 432). [d] Significant interaction between dietary scores and gender, $p < 0.01$. * $p < 0.05$, ** $p < 0.01$, *** $p \leq 0.001$.

Intake of fresh fish was associated with higher eGFR, while intake of crumbed/battered fish was associated with a higher waist:hip ratio (Table 5). Canned fish intake was associated with higher HDL-cholesterol and lower HbA1c, although this finding was not significant when potential dietary misreporters were excluded (Table 5).

Table 5. Fish intake and modifiable cardiometabolic risk factors.

	Age- and Gender-Adjusted (Model 1) [a]	Multivariable Adjusted [b] (Model 2)	Model 2 excl. Potential Misreporting [c]
Waist:hip			
Crumbed/battered fish	0.003 (0.000, 0.006)	0.003 (0.000, 0.006) *	0.004 (0.001, 0.007) **
HDL cholesterol			
Canned fish	0.022 (0.009, 0.035) ***	0.023 (0.010, 0.036) ***	0.026 (0.013, 0.040) ***
eGFR			
Fresh fish	0.966 (0.239, 1.692) **	1.001 (0.273, 1.730) **	1.128 (0.373, 1.184) **
HbA1c			
Canned fish	−0.038 (−0.076,−0.001) *	−0.032 (−0.064,−0.001) *	−0.030 (−0.63, 0.002)

Values are Unstandardised Beta Coefficients (95%CI). [a] Model 1 adjusted for age and gender (n = 458). [b] In Model 2 all variables are adjusted for age, gender, smoking, physical activity, education, and BMI, except waist:hip, which is adjusted for age, gender, smoking, physical activity, education, and diabetes. [c] Model 2 excluding those with total dietary energy <2000 kJ and >15000 kJ. * $p < 0.05$, ** $p < 0.01$, *** $p \leq 0.001$.

In both minimally and fully adjusted models, the proportion of dietary energy derived from sugar-sweetened drinks was positively associated with central adiposity and HbA1c but negatively associated with HDL-cholesterol and eGFR (Table 6). No associations between sugar-sweetened beverage intake and LDL or total cholesterol were observed. After adjusting for behavioural risk factors, demographic factors, and overall diet quality, each additional daily consumption of soft drink was associated with a 0.06 mmol/L decrease in HDL-cholesterol, and a 0.18 unit increase in HbA1c (Table 6). However, there was a soft drink–gender interaction seen for HbA1c, with the association significant in men (B = 0.166, 95% CI: 0.085, 0.245, $p < 0.001$) but not women. No associations were observed between cardiometabolic risk markers and other discretionary sweet food consumption, such as confectionery or sweet baked goods.

Table 6. Beverage intakes and modifiable cardiometabolic risk factors.

	Age- and Sex-Adjusted (Model 1) [a]	Multivariable Adjusted [b] (Model 2)	Multivariable Adjusted [c] (Model 3)	Model 3 Excluding Potential Misreporting [d]
Waist:hip				
Sugar-sweetened beverages (%E)	0.002 (0.001, 0.003) **	0.002 (0.001, 0.003) **	0.002 (0.001, 0.003) **	0.002 (0.001, 0.003) **
Soft drink (number consumed/day)	0.010 (0.002, 0.017) *	n.s.	n.s.	n.s.
HbA1c				
Sugar-sweetened beverage (%E)	0.032 (0.014–0.051) ***	0.035 (0.019–0.050) ***	0.035 (0.019–0.050) ***	0.029 (0.013–0.045) ***
Soft drink (number consumed/day)[e]	0.262 (0.143, 0.380) ***	0.174 (0.076, 0.272) ***	0.178 (0.078, 0.277) ***	0.159 (0.054, 0.264) **
Diet soft drink (times consumed)	0.058 (0.033, 0.083) ***	0.029 (0.008, 0.050) **	0.029 (0.009, 0.050) **	0.030 (0.009, 0.051) **
HDL cholesterol				
Sugar-sweetened beverage (%E)	−0.011 (−0.017,−0.004) ***	−0.009 (−0.015,−0.002) ***	−0.008 (−0.015,−0.001) *	−0.008 (−0.015, −0.001) *
Soft drink (number consumed/day)	−0.093 (−0.135,−0.051) ***	−0.062 (−0.104,−0.021) **	−0.061(−0.103,−0.019) **	−0.071 (−0.115, −0.027) **
Diet soft drink intake	−0.011 (−0.020, −0.002) *	n.s.	n.s.	n.s.
eGFR				
Sugar-sweetened beverage (%E)	−0.419 (−0.748, −0.091) *	−0.400 (−0.744, −0.055) *	−0.363 (−0.709, −0.017) *	−0.396 (−0.759, −0.032) *
Soft drink (number consumed/day)	−3.802 (−6.099,−1.504) ***	−3.110 (−5.464, −0.756) *	−2.794 (−5.184, −0.404) *	−3.351 (−5.909, −0.792) *
Diet soft drink intake	−0.484 (−0.941, −0.028) *	n.s.	n.s.	n.s.

Values are Unstandardised Beta Coefficients (95% CI), n.s. = not significant. Beverage intake as % total energy intake (%E), or number consumed. [a] Model 1 adjusted for age and sex ($n = 458$). [b] In Model 2, all variables are adjusted for age, sex, smoking, physical activity, education, and BMI, except waist:hip, which is adjusted for age, sex, smoking, physical activity, education, and diabetes. [c] In Model 3, all variables are adjusted for age, sex, smoking, physical activity, education, diabetes, BMI, and ARFS (total), except waist:hip, which is adjusted for age, sex, smoking, physical activity, education, diabetes, and ARF total score. [d] Model 3 excluding those with total dietary energy <2000 kJ and >15000 kJ ($n = 432$). [e] Significant interaction between dietary scores and gender, $p = 0.012$

4. Discussion

In this cohort of older adults living in a regional area of south-eastern Australia, diet quality was on average lower than that previously reported for another Australian regionally-located cohort [9]. The Global Burden of Disease study has identified key dietary risk factors for non-communicable disease mortality as diets high in sodium, low in whole grains, and low in fruits and vegetables [2], all dietary risk patterns evident in this cohort. Intake of sugar-sweetened beverages was adversely associated with cardiometabolic risk factors, while intake of fresh and canned fish was beneficially associated with cardiometabolic risk factors. Prevalence of overweight and obesity in our cohort was higher than previously reported for those aged 65–74 years in Australia, which in 2015 was 80% in men and 69% in women [18]. The proportion of our cohort with diabetes was similar to that previously reported for Australians for aged 65 years and above (18.1% in this cohort compared to 17.4% in the 2014/15 Australian National Health Survey) [19]. Intake of sodium by this cohort was comparable to that reported in the Australian Health Survey (AHS) for women (1972 mg/d for 51–70-year-old women in AHS versus 1984 mg/d herein) but slightly lower in this cohort than population data previously reported for 51–70-year-old men in Australia (2510 mg/d in AHS vs. 2329 mg/d herein) [20]. However, dietary survey methods for assessing sodium intake are well-recognised to under-report sodium intake when compared to 24 h urinary sodium excretion studies [21].

Historically, dietary epidemiology has had a strong focus on the intake of individual nutrients and their relationship to health outcomes. More recently, methods to assess overall diet quality have been employed as an attempt to capture not only the quantity of nutrient intake but also dietary diversity and how well an individual's dietary pattern adheres to dietary guidelines [22]. However, while validation of these scores is often undertaken against micronutrient intake, the association between diet quality scores and chronic disease biomarkers is less consistent [9,22]. In the present study, associations were noted between dietary quality scores and 'metabolic' health markers (abdominal obesity and HDL-cholesterol), but not other 'cardiovascular' health markers (blood pressure and heart rate).

Consistent with a recent finding from the CHARGE consortium [23], we noted an inverse association between quality of dairy intake and total cholesterol, but interestingly, this was not observed when total dairy intake was examined as a percentage of total energy. In line with Australian dietary guidelines, the ARFS dairy score calculation allocates a higher score for low-fat milk [9], as dairy fats are a source of saturated fat and there has been concern about adverse effects of this saturated fat intake on cholesterol levels and subsequent cardiovascular risk. However, there remains a lack of clear evidence that the consumption of low-fat dairy products is associated with lower cardiovascular risk when compared to high-fat dairy [24,25].

In middle-aged women, greater dietary quality of vegetable intake (ARFS vegetable score, which encompasses both variety and quantity of vegetable intake) was associated with fewer Medicare (health service) claims [26]. However, dietary quality using this measure was not found to be related to the subsequent development of obesity in a previous study [27]. In the present cohort, total diet quality, as well as the quality of dietary intake of fruits and grains, but not vegetables, was associated with a marker of abdominal obesity (waist to hip ratio).

Sugar-sweetened beverage consumption among those aged 65 years and over in the most recent Australian National Health Survey was 16% for women and 22.4% for men [28]. Thus, consumption of sugar-sweetened beverages in this cohort was higher than national average intakes, consistent with greater consumption by those living outside major metropolitan areas and in areas of greater socioeconomic disadvantage [19]. The cluster of cardiometabolic risk factors associated with intake of sugar-sweetened beverages in the present study were those that form the criteria for metabolic syndrome. This is consistent with findings from cross-sectional studies that have suggested an association between sugar-sweetened beverage intake and metabolic syndrome, although this has not consistently been observed in prospective studies [29].

Strengths and Limitations

The diet quality scores used in this study (ARFS) are relatively newly developed, and this is one of the first studies to have examined ARFS and cardiometabolic disease risk markers. However, this was a cross-sectional analysis; thus, causality cannot be inferred. Nutrition or dietary epidemiology has some well-known limitations in terms of sources of error: (1) with diet being time-varying (e.g., due to seasonal, health, or economic factors), and (2) omission of foods (e.g., because dietary instruments rely on memory, epidemiological scale instruments may not capture all foods, or bias conferred by tendencies to misreport foods perceived as either 'unhealthy' or 'healthy') [30]. Of the dietary assessment tools available to researchers and clinicians, FFQs are less expensive and have a low participant burden, thus validated FFQs are often the most practical option for large-scale studies. While FFQs tend to give higher values relative to food diaries or 24 h recalls, FFQs are better able to capture seasonably consumed foods and capture usual or habitual intake. Comparison to Australian national data is limited by the differences in dietary assessment methodology, with an FFQ used in this study compared to a 24-h dietary recall in the Australian National Health Survey [19]. Furthermore, the participants were not a truly random sample of the source population, as the sample was over-represented by people with a history of cardiovascular disease. There were other potential sources of bias relating to dietary intake and cardiometabolic risk that were not accounted for in these models, including non-cardiovascular medication use, cultural factors, living alone, income, work status, and other comorbidities.

5. Conclusions

Among older, regionally-dwelling adults, potentially modifiable dietary risk factors for cardiometabolic disease are common, namely inadequate intakes of fibre and folate, and excessive sodium intake. Women have higher dietary quality scores for total diet, vegetable, fruit, and dairy intake compared to men. Public health strategies aiming to reduce intake of sugar-sweetened beverages may be of particular benefit in this population.

Author Contributions: Conceptualization, A.J.O., M.J.A. and D.L.; Data curation, J.F.I., D.B.; Formal analysis, A.J.O. and C.X.G. ; funding acquisition, M.J.A.; investigation, M.J.A., J.F.I., S.P., B.M.B., and D.L.; methodology, A.J.O., M.J.A., T.A.M. and B.M.B.; project administration, J.F.I., S.P. and D.B.; supervision, D.L.; Validation, T.A.M.; writing—original draft, A.J.O.; writing—review and editing, M.J.A., J.F.I, T.A.M., S.P., B.M.B., C.X.G., D.B. and D.L. All authors have read and agreed to the published version of the manuscript.

Funding: The Hazelwood Study was funded by the Victorian Department of Health and Human Services, Australia. The paper presents the views of the authors and does not represent the views of the Department of Health and Human Services.

Acknowledgments: We would like to thank participants in the Hazelwood Health Study, as well as Shantelle Allgood, Susan Denny, Melanie Reeves, Kylie Sawyer, Andrea Taggert and Kristina Thomas for their assistance with recruitment and conducting the testing in the Hazelwood study. We would like to thank Emma Herron for data entry and Clare Collins, Megan Rollo, Tracy Schumacher and Rebecca Haslam for assistance in the preparation of diet quality scoring.

Conflicts of Interest: M.J.A. holds investigator initiated grants from Pfizer and Boehringer-Ingelheim for unrelated research. He has undertaken an unrelated consultancy for and received assistance with conference attendance from Sanofi. He has also received a speaker's fee from GSK. The authors declare no conflict of interest.

References

1. GBD 2017 Risk Factor Collaborators. Global, regional, and national comparative risk assessment of 84 behavioural, environmental and occupational, and metabolic risks or clusters of risks for 195 countries and territories, 1990–2017: A systematic analysis for the Global Burden of Disease Study 2017. *Lancet* **2018**, *392*, 1923–1994.
2. GBD 2017 Diet Collaborators. Health effects of dietary risks in 195 countries, 1990–2017: A systematic analysis for the Global Burden of Disease Study 2017. *Lancet* **2019**, *393*, 1958–1972. [CrossRef]
3. National Health & Medical Research Council. *Eat for Health: Australian Dietary Guidelines*; National Health and Medical Research Council: Canberra, Australia, 2013.
4. Australian Bureau of Statistics. Australian Health Survey: Consumption of food groups from the Australian Dietary Guidelines. Australian Bureau of Statistics: Canberra, Australia, 2016. Available online: https://www.abs.gov.au/AUSSTATS/abs@.nsf/DetailsPage/4364.0.55.0122011-12 (accessed on 3 December 2019).
5. Fayet-Moore, F.; Cassettari, T.; Tuck, K.; McConnell, A.; Petocz, P. Dietary fibre intake in Australia. Paper I: Associations with demographic, socio-economic, and anthropometric factors. *Nutrients* **2018**, *10*, 599. [CrossRef] [PubMed]
6. Thorpe, M.G.; Milte, C.M.; Crawford, D.; McNaughton, S.A. A revised Australian Dietary Guideline Index and its association with key sociodemographic factors, health behaviors and body mass index in peri-retirement aged adults. *Nutrients* **2016**, *8*, 160. [CrossRef] [PubMed]
7. Bennett, E.; Peters, S.A.E.; Woodward, M. Sex differences in macronutrient intake and adherence to dietary recommendations: Findings from the UK Biobank. *BMJ Open* **2018**, *8*, e020017. [CrossRef] [PubMed]
8. Heart Foundation of Australia. Heart Health Map for Latrobe. 2019. Available online: https://www.heartfoundation.org.au/for-professionals/heart-maps/australian-heart-maps (accessed on 19 December 2019).
9. Collins, C.E.; Burrows, T.L.; Rollo, M.E.; Boggess, M.M.; Watson, J.F.; Guest, M.; Duncanson, K.; Pezdirc, K.; Hutchesson, M.J. The comparative validity and reproducibility of a diet quality index for adults: The Australian Recommended Food Score. *Nutrients* **2015**, *7*, 785–798. [CrossRef] [PubMed]
10. Hazelwood Health Study. Adult Survey Volume 1 Comparison of Morwell and Sale. 2017. Available online: https://hazelwoodhealthstudy.org.au/study-findings/study-reports (accessed on 3 December 2019).

11. Australian Bureau of Statistics. 033.0.55.001-Census of Population and Housing: Socio-Economic Indexes for Areas (SEIFA), Australia. 2016. Available online: https://www.abs.gov.au/ausstats/abs@.nsf/Lookup/by%20Subject/2033.0.55.001~{}2016~{}Main%20Features~{}IRSAD~{}20 (accessed on 3 December 2019).
12. World Health Organization. Body Mass Index. Available online: http://www.euro.who.int/en/health-topics/disease-prevention/nutrition/a-healthy-lifestyle/body-mass-index-bmi (accessed on 13 November 2019).
13. Levey, A.S.; Stevens, L.A.; Schmid, C.H.; Zhang, Y.L.; Castro III, A.F.; Feldman, H.I.; Kusek, J.W.; Eggers, P.; Van Lente, F.; Greene, T.; et al. A new equation to estimate glomerular filtration rate. *Ann. Intern. Med.* **2009**, *150*, 604–612. [CrossRef] [PubMed]
14. Australian Institute of Health and Welfare. The Active Australia Survey: A guide and manual for implementation, analysis and reporting. 2003. Available online: https://www.aihw.gov.au/reports/physical-activity/active-australia-survey/contents/table-of-contents (accessed on 3 December 2019).
15. Australian Government Publishing Service. *AUSNUT Australian Food and Nutrient Database*; Australian New Zealand Food Authority; Australian Government Publishing Service: Canberra, Australia, 1999.
16. Australian Government National Health & Medical Research Council. Nutrient Reference Values for Australia and New Zealand. 2017. Available online: www.nrv.gov.au (accessed on 3 December 2019).
17. Banna, J.C.; McCrory, M.A.; Fialkowski, M.K.; Boushey, C. Examining plausibility of self-reported energy intake data: Considerations for method selection. *Front. Nutr.* **2017**, *4*, 45. [CrossRef] [PubMed]
18. Australian Institute of Health and Welfare. A picture of overweight and obesity in Australia 2017. Canberra. 2017. Available online: www.aihw.gov.au (accessed on 3 December 2019).
19. Australian Bureau of Statistics. National Health Survey: First results, 2014–2015. Canberra; 2015. Catalogue number: 4364.0. Available online: https://www.abs.gov.au/AUSSTATS/abs@.nsf/DetailsPage/4364.0.55.0012014-15 (accessed on 3 December 2019).
20. Australian Bureau of Statistics. Australian Health Survey: Nutrition First Results-Foods and Nutrients, 2011–2012. Canberra; 2014. Catalogue Number 4364.0.55.007. Available online: https://www.abs.gov.au/ausstats/abs@.nsf/lookup/4364.0.55.007main+features12011-12 (accessed on 23 March 2020).
21. McLean, R.M.; Farmer, V.L.; Nettleton, A.; Cameron, C.M.; Cook, N.R.; Campbell, N.R.C.; TRUE Consortium (International Consortium for Quality Research on Dietary Sodium/Salt). Assessment of dietary sodium intake using a food frequency questionnaire and 24-hour urinary sodium excretion: A systematic literature review. *J. Clin. Hypertens.* **2017**, *19*, 1214–1230. [CrossRef] [PubMed]
22. Waijers, P.M.; Feskens, E.J.; Ocke, M.C. A critical review of predefined diet quality scores. *Br. J. Nutr.* **2007**, *97*, 219–231. [CrossRef] [PubMed]
23. Mendelian Randomization of Dairy Consumption Working Group; CHARGE Consortium. Dairy Intake and Body Composition and Cardiometabolic Traits among Adults: Mendelian Randomization Analysis of 182041 Individuals from 18 Studies. *Clin. Chem.* **2019**, *65*, 751–760. [CrossRef] [PubMed]
24. Dehghan, M.; Mente, A.; Rangarajan, S.; Sheridan, P.; Mohan, V.; Iqbal, R.; Gupta, R.; Lear, S.; Wentzel-Viljoen, E.; Avezum, A.; et al. Association of dairy intake with cardiovascular disease and mortality in 21 countries from five continents (PURE): A prospective cohort study. *Lancet* **2018**, *392*, 2288–2297. [CrossRef]
25. Alexander, D.D.; Bylsma, L.C.; Vargas, A.J.; Cohen, S.S.; Doucette, A.; Mohamed, M.; Irvin, S.R.; Miller, P.E.; Watson, H.; Fryzek, J.P. Dairy consumption and CVD: A systematic review and meta-analysis. *Br. J. Nutr.* **2016**, *115*, 737–750. [CrossRef] [PubMed]
26. Patterson, A.; Hure, A.; Burrows, T.; Jackson, J.; Collins, C. Diet quality and 10-year healthcare costs by BMI categories in the mid-age cohort of the Australian Longitudinal Study on Women's Health. *J. Hum. Nutr. Diet.* **2018**, *31*, 463–472. [CrossRef] [PubMed]
27. Aljadani, H.M.; Sibbritt, D.; Patterson, A.; Collins, C. The Australian Recommended Food Score did not predict weight gain in middle-aged Australian women during six years of follow-up. *Aust. N. Z. J. Public Health* **2013**, *37*, 322–328. [CrossRef] [PubMed]
28. Australian Bureau of Statistics. National Health Survey: First results 2017–18. Sugar sweetened drinks and diet drinks. 2019. Available online: https://www.abs.gov.au/ausstats/abs@.nsf/Lookup/by%20Subject/4364.0.55.001~{}2017-18~{}Main%20Features~{}Sugar%20sweetened%20and%20diet%20drink%20consumption~{}110 (accessed on 23 March 2020).
29. Narain, A.; Kwok, C.S.; Mamas, M.A. Soft drink intake and the risk of metabolic syndrome: A systematic review and meta-analysis. *Int. J. Clin. Pract.* **2017**, *71*, e12927. [CrossRef] [PubMed]

30. Satija, A.; Yu, E.; Willett, W.C.; Hu, F.B. Understanding nutritional epidemiology and its role in policy. *Adv. Nutr.* **2015**, *6*, 5–18. [CrossRef] [PubMed]

MDPI
St. Alban-Anlage 66
4052 Basel
Switzerland
Tel. +41 61 683 77 34
Fax +41 61 302 89 18
www.mdpi.com

Nutrients Editorial Office
E-mail: nutrients@mdpi.com
www.mdpi.com/journal/nutrients